E.R. Barnea   J. Hustin   E. Jauniaux   (Eds.)

The First Twelve Weeks of Gestation

With Contributions by
E.R. Barnea, D. Bider, Z. Blumenfeld, B. Brambati, J.M. Brandes,
M. Bronshtein, J.N. Bulmer, G.J. Burton, S. Campbell, M. Camus,
T. Chard, C.B. Coulam, P. Devroey, T.K.A.B. Eskes, R.F. Feinberg,
R.L. Fischer, C. Fourneau, H. Fox, P. Franchimont, Y. Gillerot,
J.G. Grudzinskas, G.M. Hartshorne, J. Hustin, E. Jauniaux, C.J.P. Jones,
D. Jurkovic, H.J. Kliman, N. Laufer, M.C. Macnamee, T. Maruo,
S. Mashiach, M. Mochizuki, J.G. Moscoso, G. Oelsner, F. Pierre,
E.A. Reece, J.-P. Schaaps, J.G. Schenker, R. Schurtz-Svirsky,
J.H. Soutoul, R.P.M. Steegers-Theunissen, M.H. Valkenburg,
L. van Maldergem, A.C. van Steirteghem, R.J. Wapner, M. Wouters,
J.S. Younis

With 230 Figures, Some in Color

Springer-Verlag Berlin Heidelberg GmbH

E.R. Barnea, MD, FACOG
Department of Obstetrics and Gynecology
Robert Wood Johnson Medical School at Camden
University of Medicine and Dentistry of New Jersey
Three Cooper Plaza, Camden, NJ 08103, USA

J. Hustin, MD, PhD
Histopathologic Institute
Institut de Morphologie Pathologique Loverval (I.M.P.L.)
Allée des Templiers 41, 6280 Gerpinnes (Loverval), Belgium

E. Jauniaux, MD
Department of Obstetrics and Gynecology
University Hospital Erasme
Free University of Brussels (ULB)
Route de Lennik 808, 1070 Brussels, Belgium

ISBN 978-3-662-08519-6   ISBN 978-3-662-08518-9 (eBook)
DOI 10.1007/978-3-662-08518-9

Library of Congress Cataloging-in-Publication Data. The First twelve weeks of gestation/E.R. Barnea,
J. Hustin, E. Jauniaux (eds.); with contributions by E.R. Barnea . . . [et al.]. p. cm. Includes biblio-
graphical references and index. ISBN-13:978-3-642-84387-7 (alk. paper).
1. Embryology, Human. 2. Abnormalities, Human. I. Barnea, E.R. II. Hustin, J. III. Jauniaux, E.
[DNLM: 1. Pregnancy Complications. 2. Pregnancy Trimester, First – physiology. WQ 200 F527]
QM601.F59 1992 612.6'4—dc20 DNLM/DLC for Library of Congress 92-2152 CIP

Typesetting: Best-set Typesetter Ltd., Hong Kong
21/3130-5 4 3 2 1 0 – Printed on acid-free paper

Foreword

The Challenge of the Nineties

There has been an immense acceleration in the development of new concepts and technology in science and medicine. We would be so much the poorer intellectually today without the many specific and accurate molecular techniques, without monoclonal antibodies, lectins, methods of cell sorting, new assay methods and so on, which have revolutionized our means of analysing tissues and cells. Medicine would be so much worse off without the introduction of ultrasound, lasers, magnetic resonance, new hormone preparations and countless drugs of astonishing specificity and accuracy. Indeed, some of the major advances in human reproduction could not have occurred without all those new kits or packages, including statistical packages, which enhance the output of so many investigators.

The impact of these advances is apparent in the pages of this book. Contributions are included from scientists and doctors, and the new methodologies are used to assess physiological processes, disease and factors causing fetal death and malformation. The move to the molecular level is clear, even though the need for medical care makes it extremely difficult to plan these studies. Disorders which were once diagnosed by the sheer skill of physicians are now observed routinely using ultrasound or the most delicate forms of endoscopy. The new technology is also used in comprehensive accounts of the use of new substrates to analyse trophoblast invasiveness, the formation of the placenta, and the prenatal diagnosis of genetic disease. Many syndromes are analysed at the macroscopic, microscopic and molecular levels, providing fresh opportunities to understand the true nature of the problems and improve patient care.

Nor could these new technologies have come at a more timely moment. We now have the opportunity, methods and much of the knowledge to begin to understand a most fundamental aspect of biology—the factors involved in the differentiation and organogenesis of a mammalian and even of a human fetus, and to try to apply this knowledge in obstetrical practice. The formation and fate of literally every cell in the body can be assessed in some invertebrates and amphibians, though it will be a long time before this is done with a mouse fetus or any other mammal. Nevertheless, much has been done to understand organogenesis and its errors in laboratory mammals, and some chapters in the present book illustrate the advances in diagnosis of human fetal growth and its anomalies. Although such work is described, quite correctly, in the matter-of-fact language of science and medicine, it is not hard to imagine the feelings of the mother or father, and perhaps the doctor too, as conception goes well or badly, or a fetus is found to have anomalous growth during pregnancy.

This book is primarily a medical story, of establishing and monitoring pregnancies, of spontaneous and repeated abortions, of how fetuses develop incorrectly, implant in the wrong place or result in the birth of a child with inherited disease. Averting disasters such as these can be achieved only by gaining knowledge about ourselves, through painstaking studies on human pregnancy, applying to humans concepts gained from the study of animals and maintaining the highest standards of medical care. The topic, the first 12 weeks of gestation, encompasses some of the most fundamental parts of human life, a period of life at one time almost inaccessible to medicine, yet one that must and will be brought under medical care. Many techniques are described which reflect new forms of investigation, detection and better methods of diagnosis. The book fulfills a purpose, being largely concerned with clinical issues of these early stages of life and supplementing the many books devoted to mammalian or experimental embryology. It is all the more welcome for that.

Pride of place in the book is given to the normal differentiation and properties of the trophoblast, essential to sustain embryonic growth from the onset of pregnancy and a most fascinating story in itself. Various trophoblastic tissues each have specific roles in implantation and pregnancy, their own distinct morphology and possibly their own specific substrates as they interact with endometrium and embryos during implantation. The placenta is established and regulated by the coordinated actions of numerous growth factors, binding proteins and enzymes produced opportunely to open pathways for migration into the endometrial stroma or the movement of embryonic tissues. All this is of great significance in controlling uterine responses to the implanting embryo, and perhaps in limiting the invasiveness of the trophoblast.

The anatomy and physiology of placental and embryonic development is, justifiably, the largest section in the book. This is perhaps where medicine exerts its greatest impact on early pregnancy at the present time, and it opens fascinating aspects of embryonic differentiation. The formation of chorionic villi and the villous tree includes longstanding matters of contention, e.g. on the formation of endothelial cells and capillaries in the villi, and it now seems that the former are derived from extraembryonic mesoderm. Clinical assessments of these stages of growth rely heavily on ultrasound, one of the most widely used forms of non-invasive diagnosis today; and embryological phenomena, once matters for textbooks, are now routinely recorded with outstanding clarity. Some advances deserve the adjective "breathtaking", including the observation and recording of the fetal heart and its rapid beat from the moment it forms as a primitive tube at 22–23 days post-fertilization in embryos only 2–3 mm in length. Blood flow along ovarian and uterine arteries is measured by Doppler ultrasound to assess the vascularization of the mother's reproductive system, and the umbilical vessels of the fetus are assessed at 7 weeks to measure fetal well-being. An unexpected turn of events here has shown how the intervillous space in the developing placenta is not transfused with maternal blood, for the maternal vessels do not gain access to it during early stages of pregnancy.

The placenta is essential to many aspects of pregnancy, and it has two quite distinct systems, the yolk sac and the chorioallantoic placenta. Metabolites, hormones and ions are transferred across the placenta in different ways: oxygen, carbon dioxide, urea and steroids by passive diffusion; glucose by facilitated diffusion; and amino acids and ions by active transport. The placenta metabolizes glucose and lipids, etc., defends the fetus by destroying toxins and environmental pollutants, has a major endocrine role in pregnancy and perhaps fends off the mother's immune response. It produces several protein hormones, and steroids after 6–7 weeks, and has its own regulatory systems. Its properties are yielding to investigative analyses, e.g. the gene for alpha-human chorionic gonadotrophin is transcribed in cytotrophoblast before the gene for beta-human chorionic gonadotrophin, luteinizing hormone-releasing hormone is produced locally among other factors to regulate placental endocrinology, and a myriad of enzyme systems have been identified and clarified. The placenta exists in a close relationship to the decidua with its array of leucocytes, macrophages and lymphocytes of various types, each expressing different combinations of differentiation antigens. These cell types may be essential for the maternal recognition of the embryo, response, to protect the embryo from a hostile maternal immune response, as a defence against inflammation and to regulate placental growth via cytokines and other factors.

There is little that can be done in these early stages of growth to study or repair a deformed embryo. One chapter casts a glance at knowledge on the regulation of embryogenesis in other phyla, at the homeo box genes, the molecular basis of cell adhesion and cell junctions, and the role of substrates in early embryonic growth. Such work has not yet impinged on the care of the early human fetus, yet there are some enormous advances in clinical studies not matched elsewhere. Some of these are primarily clinical, e.g. the diverse roles and properties of the yolk sac: it is of great importance in the diagnosis of normal growth by ultrasound, a placental organ with distinct nutritive properties in early human growth, exhibiting biochemical similarities to liver, and a passing source of haemopoietic cells and germinal cells in the early embryo. The origin of the hydatidiform mole provides another example of the contribution of clinical science to fundamental embryology, for it was the first example of genomic imprinting, and its androgenetic origin was described in detail long before similar syndromes were established experimentally in laboratory animals.

A book such as this would be incomplete without a discussion on assisted human conception which has clarified so many aspects of human implantation and pregnancy. It is essential for the care of the infertile couple and it has revealed some astonishing phenomena in early pregnancy. Artificial cycles can be established in agonadal and anovulatory patients by giving steroids or gonadotrophin-releasing hormone and its analogues to dampen the pituitary gland, and then various steroids to establish a secretory endometrium. Very high rates of pregnancy are established by donating an oocyte to these patients. Who would have forecast that women with gonadal dysgenesis or other causes of agonadism would be delivering children today? New methods of regulating the endocrine system, replacing embryos cryopreserved over many years, and the use of methods with endless acronyms such as GIFT or ZIFT, or the non-invasive intrafallopian deposition of embryos via the vagina have outstripped equivalent work in laboratory or domestic animals.

What opportunities can medicine and surgery offer to improve the chances of normal pregnancy, to cope with the tragedies of pregnancy, abortion, repeated abortion, fetal malformation? Some of these familiar problems have been exacerbated in modern society as women conceive at older ages, live more stressful lives and seek artificial conception when they cannot conceive naturally. The well-known syndromes include endocrine disorders, endometriosis, systemic disorders and genetic disorders. Chromosome imbalance is the most common embryonic anomaly, often defined by villous pathologies, except in trisomy 21, monosomy X and tetraploidies, although many pathologists believe that villous diagnoses are largely meaningless without karyotypic analyses. Luteal deficiencies, lupus anticoagulant and anticardiolipin antibodies are associated with abortion or repeated abortion, and disorders in the basal plate impairing implantation can cause early abortion.

Ectopic pregnancy is relatively common when salpingitis occurs as a result of infection, and one astonishing facet of this condition is the overall similarity of implantation and decidualization in organs such as the ovary and oviduct with those in the uterus. Molar pregnancies can be a threat to the mother, and trophoblastic overgrowth might be detected by the production of very high levels of human chorionic gonadotrophin. Approximately 3% of children are born with minor or major malformations, 15% of them incompatible with life, and causes are classified into malformations arising early in life, disruptions arising through secondary changes in developing systems, syndromes, dysplasias and others, but much more knowledge is needed about their ontogeny.

What hope is there for detecting or alleviating such conditions? Diagnostic methods are well advanced. Assays on plasma for steroids, protein hormones, and pregnancy proteins help to diagnose a threatened or spontaneous miscarriage, an anembryonic pregnancy, trophoblastic disease or an ectopic pregnancy. Ultrasound is again invaluable to measure the gestational sac, to assess the status of the yolk sac, and to identify many severe and potentially lethal deformities in the fetus before 13 weeks. Amniocentesis and chorionic villous sampling enable fetal tissue or cells to be analysed for genetic disease, from before 8 weeks until the 2nd trimester, and risks to the fetus or the ongoing pregnancy are low. Perhaps the routine diagnosis of embryos before implantation will be introduced before too long.

Therapies are few and far between. Preconception treatments could ensure everything is as normal as possible before the pregnancy begins, by assessing disease, the use of drugs including those taken medically, smoking, diabetes, and anomalies in the reproductive system. The few available treatments after conception include the use of steroids or human chorionic gonadotrophin for threatened abortion, a few surgical techniques such as the reduction of high-order multiple pregnancies to twins or singletons or the induction of abortion. Some doctors recommend immunotherapy in cases of threatened abortion in the belief that the maternal recognition of trophoblast is essential for normal pregnancy, and that immunopathological situations involving the placenta arise during normal pregnancies. Recurrent abortion is held to be a disorder arising from antigenic similarity between husband and wife, to be overcome by immunotherapy using tissue from the husband, seminal plasma, trophoblast or a donor, but this treatment remains highly controversial.

Where will the new advances and opportunities arise in care during the first weeks of human gestation? Some authors write on topics such as the value of animal models in understanding human pregnancy and its problems. So many myriad forms of embryonic growth and placentation exist in animals that only our most closely related species are likely to provide data of fundamental clinical value, but these species are too precious for such a use today; fortunately, laboratory rodents seem to resemble humans in many aspects of fertilization, blastulation and implantation. Improved diagnoses might help to devise new therapies, but diagnoses are often virtually impossible because a mixture of placental and decidual tissue is all that is available after spontaneous or recurrent abortion. These diagnoses might be useful only to confirm that a pregnancy existed and did not involve a hydatidiform mole. Many clinicians recommend that dysfunctions in early pregnancy should be diagnosed by all means available and preventative measures applied for the next pregnancy, but the disorder should not be treated because therapy could save many abnormal fetuses and lead to greater troubles later in gestation. And, of course, among these complex ethical issues, the world outside watches over medicine and science. How interesting to discover there is no legal definition of an embryo—it might even be a "chattel" in the possession of its mother. Nor does clarity of thought arise from the enormous differences in values between major religions, so that surrogacy and gamete donation are accepted by one and rejected by another.

Many, and perhaps most, new therapies for the next decade will arise from the persistent enquiry of doctors into the health of their patients, using new hormone assays, magnetic resonance imaging, and better laser technology to excise or ablate tissue or to trap the least motile spermatozoa and so improve a semen sample. Perhaps new methods of separating X and Y spermatozoa in rabbits and pigs will be applied to humans and extended to identify genes in the sperm head. What a wonderful approach this would be to avert the conception of embryos with genetic disease. High rates of implantation might be possible after in vitro fertilization, based on studies on agonadal women or high responders to ovarian stimulation who seem to be unusually fertile. And we have hardly begun to utilize the regenerative power of cell lines grown as outgrowths from embryos in vitro or taken from aborted fetuses to repair defects in fetuses, newborns or even adults. The insertion or excision of genes from targeted sites in these cells has already led to the synthesis of new cell types and to animal models of human disease, which will one day be highly relevant to our field of study.

Three things are certain about this brave new world. It will depend on close collaboration between scientists and doctors to apply the exquisite sensitivity of molecular and other techniques in medicine. It will almost certainly be expensive. But, perhaps more important, new advances will reinforce those earlier demands of ordinary men and women to have access to new treatments as witnessed by their insistence on availing themselves of the opportunities raised by contraception, assisted conception, fetal screening and abortion. One day, it might be mandatory to use new technology to avert disease and deformity in children, just as the vaccination of children against serious disease can be enforced in some societies and under some circumstances today. There will doubtless be opposition to new therapies, and it will be fascinating to see who will claim the right to prevent their use. Perhaps the world will be a happier place for these new advances, perhaps not. The best answer is to wait and see.

Churchill College, Cambridge   R.G. EDWARDS

Contents

Assisted Conception

Failures of Placentation and Embryonic Development

Diagnostic Methods

Treatments

Future Guidelines

List of Contributors

Barnea, E.R., Department of Obstetrics and Gynecology,
Robert Wood Johnson Medical School at Camden,
University of Medicine and Dentistry of New Jersey,
Three Cooper Plaza, Camden, NJ 08103, USA

Bider, D., Department of Obstetrics and Gynecology, The Chaim Sheba
Medical Center, Tel Hashomer 52621, Israel

Blumenfeld, Z., Department of Obstetrics and Gynecology,
Rambam Medical Center, Technion – Faculty of Medicine,
Haifa 31096, Israel

Brambati, B., Prenatal Unit, Second Institute of Obstetrics and
Gynecology, University of Milan, Via Commenda 12, 20122 Milan, Italy

Brandes, J.M., Department of Obstetrics and Gynecology,
Rambam Medical Center, Technion – Faculty of Medicine,
Haifa 31096, Israel

Bronshtein, M., Department of Obstetrics and Gynecology,
Rambam Medical Center, Technion – Faculty of Medicine,
Haifa 31096, Israel

Bulmer, J.N., Department of Pathology, University of Leeds,
Leeds LS2 9JT, UK

Burton, G.J., Subdepartment of Veterinary Anatomy, University
of Cambridge, Downing Street, Cambridge CB12 1QS, UK

Campbell, S., King's College Hospital School of Medicine and Dentistry,
Department of Obstetrics and Gynecology, University of London,
Denmark Hill, London SE5 8RX, UK

Camus, M., Centre for Reproductive Medicine, Free University of Brussels
(ULB), Laarbeeklaan 101, 1090 Brussels, Belgium

Chard, T., Departments of Obstetrics, Gynecology and Reproductive
Physiology, St. Bartholomew's Hospital Medical College and the London
Hospital Medical College, West Smithfield, London EC1A 7BE, UK

Coulam, C.B., Center for Reproduction and Transplantation Immunology,
   Methodist Hospital of Indiana, 1633 North Capitol Avenue, Indianapolis,
   IN 46202, USA

Devroey, P., Centre for Reproductive Medicine, Free University of Brussels
   (ULB), Laarbeeklaan 101, 1090 Brussels, Belgium

Eskes, T.K.A.B., Department of Obstetrics and Gynecology, Sint Radboud
   Hospital, Perinatal Research Group, Division of Prevention
   of Malformations, Catholic University Nijmegen, Geert Grooteplein
   Zuid 14, 6525 GA Nijmegen, The Netherlands

Feinberg, R.F., University of Pennsylvania School of Medicine,
   106 Dilles Building, 3400 Spruce Street, Philadelphia, PA 19104, USA

Fischer, R.L., Antepartum Diagnostic Center, Cooper Hospital/University
   Medical Center, One Cooper Plaza, Camden, NJ 08103, USA

Fourneau, C., Institut de Morphologie Pathologique de Loverval
   (I.M.P.L.), Allée des Templiers 41, 6280 Gerpinnes (Loverval), Belgium

Fox, H., Department of Pathological Sciences, University of Manchester,
   Stopford Building, Oxford Road, Manchester M13 9PT, UK

Franchimont, P., Department of Radioimmunology, University of Liège –
   CHU, Sart Tilman, 4000 Liège, Belgium

Gillerot, Y., Institut de Morphologie Pathologique de Loverval (I.M.P.L.),
   Allée des Templiers 41, 6280 Gerpinnes (Loverval), Belgium

Grudzinskas, J.G., Departments of Obstetrics, Gynecology and
   Reproductive Physiology, St. Bartholomew's Hospital Medical College
   and the London Hospital Medical College, West Smithfield, London
   EC1A 7BE, UK

Hartshorne, G.M., Bourn Hall Clinic, Bourn, Cambridge CB3 7TR, UK

Hustin, J., Institut de Morphologie Pathologique Loverval (I.M.P.L.),
   Allée des Templiers, 41, 6280 Gerpinnes (Loverval), Belgium

Jauniaux, E., Department of Obstetrics and Gynecology,
   University Hospital Erasme, Free University of Brussels (ULB),
   1070 Brussels, Belgium

Jones, C.J.P., Department of Pathological Sciences, University of
   Manchester, Stopford Building, Oxford Road, Manchester M13 9PT, UK

Jurkovic, D., The Ultrasonic Institute, University of Zagreb,
   P. Miskine, 64, 41000 Zagreb, Yugoslavia

Kliman, H.J., Department of Pathology, Yale University School
   of Medicine, 130 Brady Hall, 310 Cedar Street, New Haven,
   CT 06510-8023, USA

Laufer, N., IVF Unit, Department of Obstetrics and Gynecology, Hadassah University Hospital, Ein-Karem, Jerusalem 91120, Israel

Macnamee, M.C., Bourn Hall Clinic, Bourn, Cambridge CB3 7TR, UK

Maruo, T., Department of Obstetrics and Gynecology, Kobe University School of Medicine, 5-1 Kusunoki, Cho 7 Chome, Chuo-ku, Kobe 650, Japan

Mashiach, S., Department of Obstetrics and Gynecology, The Chaim Sheba Medical Center, Tel Hashomer 52621, Israel

Mochizuki, M., Department of Obstetrics and Gynecology, Kobe University School of Medicine, 5-1 Kusunoki, Cho 7 Chome, Chuo-ku, Kobe 650, Japan

Moscoso, J.G., King's College Hospital, Fetal and Perinatal Pathology Unit, Department of Morbid Anatomy, University of London, Denmark Hill, London SE5 8RX, UK

Oelsner, G., Department of Obstetrics and Gynecology, The Chaim Sheba Medical Center, Tel Hashomer 52621, Israel

Pierre, F., Department of Gynecology and Obstetrics, Maternité du Beffroi, Avenue de Roubaix 23, 37100 Tours, France

Reece, E.A., Department of Obstetrics, Gynecology and Reproductive Sciences, Temple University – Health Sciences Center School of Medicine, 3401 N. Broad Street, Philadelphia, PA 19140, USA

Schaaps, J.-P., Department of Obstetrics and Gynecology, University of Liège – CHU, C.H.R. Citadelle, Bd, du 12ème de Ligne 1, 4000 Liège, Belgium

Schenker, J.G., IVF Unit, Department of Obstetrics and Gynecology, Hadassah University Hospital, Ein-Kerem, Jerusalem 91120, Israel

Schurtz-Svirsky, R., Feto-Placental Endocrine Unit, Rappaport Institute, Department of Gynecology and Obstetrics, Rambam Medical Center, Technion – Faculty of Medicine, Haifa 31096, Israel

Soutoul, J.H., Department of Gynecology and Obstetrics, Maternité du Beffroi, Avenue de Roubaix 23, 37100 Tours, France

Steegers-Theunissen, R.P.M., Department of Obstetrics and Gynecology, Sint Radboud Hospital, Perinatal Research Group, Division of Prevention of Malformations, Catholic University Nijmegen, Geert Grooteplein Zuid 14, 6525 GA Nijhmegen, The Netherlands

Valkenburg, M.H., Centre for Reproductive Medicine, Free University of Brussels (ULB), Laarbeeklaan 101, 1090 Brussels, Belgium

van Maldergem, L., Institut de Morphologie Pathologique de Loverval
(I.M.P.L.), Allée des Templiers 41, 6280 Gerpinnes (Loverval), Belgium

van Steirteghem, A.C., Centre for Reproductive Medicine, Free University
of Brussels (ULB), Laarbeeklaan 101, 1090 Brussels, Belgium

Wapner, R.J., Department of Obstetrics and Gynecology, Division
of Maternal-Fetal Medicine, Jefferson Medical College, 1025 Walnut
Street, Philadelphia, PA 19107, USA

Wouters, M., Department of Obstetrics and Gynecology, Sint Radboud
Hospital, Perinatal Research Group, Division of Prevention
of Malformations, Catholic University Nijmegen, Geert Grooteplein
Zuid 14, 6525 GA Nijmegen, The Netherlands

Younis, J.S., IVF Unit, Department of Obstetrics and Gynecology,
Hadassah University Hospital, Ein-Karem, Jerusalem 91120, Israel

Peri-implantation Events

1 Differentiation of the Trophoblast

H.J. Kliman and R.F. Feinberg

Human Trophoblasts In Vivo: Three Differentiation Pathways

Trophoblasts are unique cells, derived from the outer cell layer of the blastocyst, which mediate implantation and placentation [27]. Depending on their subsequent function in vivo, undifferentiated cytotrophoblasts can develop into (a) hormonally active villous syncytiotrophoblasts, (b) extra-villous anchoring trophoblastic cell columns, or (c) invasive intermediate trophoblasts (Fig. 1). Interestingly, within the villi of the human placenta – at all gestational ages – there exists a population of cytotrophoblasts which remain undifferentiated. The purpose of this chapter is to review our knowledge about the various differentiated functions of human trophoblasts and to suggest control mechanisms for these differentiation pathways. Furthermore, by studying the differentiation behavior of villous cytotrophoblasts in culture, it is now possible to develop in vitro correlates for elucidating the biology of the human trophoblast.

Villous Syncytiotrophoblast. The hormones secreted by the villous syncytiotrophoblast are critical for maintaining pregnancy [10, 65]. Early in gestation, human chorionic gonadotropin (hCG) is essential to maintain corpus luteum progesterone production. Near the end of the first trimester, the mass of villous syncytiotrophoblast is large enough to make sufficient progesterone and estrogen to maintain the pregnancy. During the third trimester, large quantities of placental lactogen are produced, a hormone purported to have a role as a regulator of lipid and carbohydrate metabolism in the mother. Other syncytiotrophoblast products include: pregnancy specific β_1-glycoprotein [35], plasminogen activator inhibitor type 2 [17], growth hormone [28], collagenases [54], thrombomodulin [52, 60], and growth factor receptors [31, 69, 87]. The factors responsible for the regulated synthesis of these compounds has been the subject of a great deal of investigations, some of which will be reviewed below.

Anchoring Trophoblasts. The premature loss of attachment of the developing conceptus or placenta to the uterus can terminate the gestation. Therefore, the anchoring trophoblast cell columns and the extracellular

Cytotrophoblast

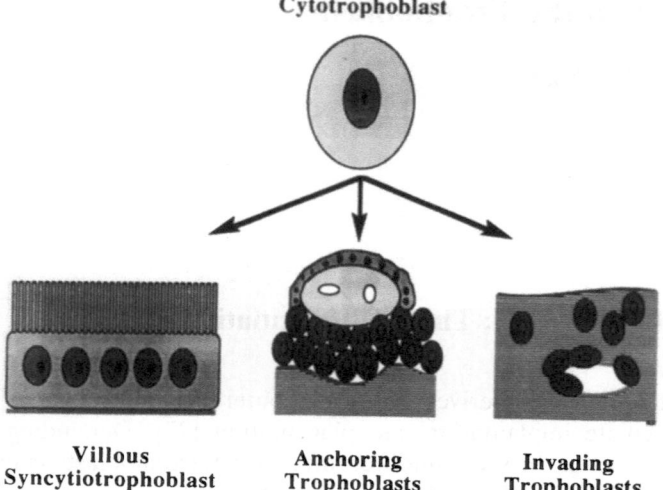

| Villous | Anchoring | Invading |
| Syncytiotrophoblast | Trophoblasts | Trophoblasts |

Fig. 1. Model of pathways of cytotrophoblast differentiation. All placental and extraplacental trophoblasts derive from undifferentiated cytotrophoblasts. The signals that direct cytotrophoblasts to the three major classes of differentiated trophoblast are not known. Autocrine, paracrine, extracellular matrix-mediated, and release of inhibition mechanisms have been suggested

matrix proteins which mediate this attachment are critical to the developing pregnancy. It has been generally accepted that some form of cell-extracellular matrix interaction takes place at the attachment interface between these trophoblasts and the uterus. Some have considered Nitabuch's layer as being related to this function. In addition to the anchoring cell columns of the placenta, the extravillous trophoblasts of the external membranes (chorion laeve) play a critical role in maintaining attachment of the external membranes to the endometrial surface. Recently, a specific type of fibronectin – trophouteronectin – has been implicated as the protein responsible for the attachment of anchoring, extravillous trophoblasts to the uterus throughout gestation [20].

Invading Trophoblasts. As human gestation progresses, invasive populations of extravillous trophoblasts attach to and interdigitate through the extracellular spaces of the endo- and myometrium. The endpoint for this invasive behavior is penetration of maternal spiral arteries within the uterus [68]. Histologically, trophoblast invasion of maternal blood vessels results in disruption of extracellular matrix components and development of dilated capacitance vessels within the uteroplacental vasculature. Biologically, trophoblast-mediated vascular remodeling within the placental bed allows for marked distensibility of the uteroplacental vessels, thus accommodating the increased blood flow needed during gestation. Abnormalities in this invasive process have been correlated with early and mid-trimester pregnancy loss, preeclampsia and eclampsia, and intrauterine growth retardation [68, 68a, 75a].

In Vitro Model Systems to Study Trophoblast Differentiation

The most commonly utilized approaches for examining the regulation of hormone production by trophoblasts has come from in vitro studies. Model systems developed to study placental and trophoblast function have included placental organ and explant culture, trophoblast culture, chorion laeve culture, choriocarcinoma cell line culture, and placental perfusion studies. Recently, most investigators have turned to trophoblast cell culture since it eliminates the complications of more heterogeneous cell systems. Since the cytotrophoblast is the precursor of all other trophoblasts, a variety of methods have been proposed to purify this cell type from the human placenta [4, 5, 8, 13, 23, 35, 47, 48, 82a, 86, 90].

Where does one find cytotrophoblasts? Cytotrophoblasts are present in the human placenta throughout pregnancy as undifferentiated mononuclear cells [7a]. Cytotrophoblasts can be found as early as the blastocyst, through all stages of placental development, and within the chorion laeve of the extraplacental membranes. Few investigators have utilized human blastocyst trophoblasts because of practical and ethical limitations, but all other starting material has been utilized to purify the trophoblast stem cell, the cytotrophoblast. For example, Fisher et al. [22] have utilized first-trimester placentae to yield 1×10^6 cells per gram of starting material. Yagel et al. [90] have also started with first-trimester placentae to purportedly produce passable trophoblast cell lines. Shorter et al. [82a] started with external term membranes and reportedly separated out trophoblasts using flow cytometry and cell sorting. As Loke [46] has pointed out, a critical part of trophoblast purification is often overlooked – adequate markers to prove what cell has been purified. This continues to be a valid caution and must be considered when evaluating the efficacy of any particular purification protocol and the claims made by the investigators about the cells they are examining.

Beginning with third-trimester placentae, we have been able to obtain almost pure cultures of human cytotrophoblasts [35] (Fig. 2). Processing 60 g of villous tissue with yields of $2-5 \times 10^6$ cells per gram generates up to 300 million cells. Using similar methods, cytotrophoblasts can also be purified from first- and second-trimester placentae. Heeding Loke's [46] caution, we asked: are these cells trophoblasts? We utilized a variety of immuno-chemical markers, electron-microscopic examination, and biochemical assay of secreted products to substantiate our claim that these cells represent cytotrophoblasts which are capable of differentiating along multiple tropho-blast pathways [17, 21, 29, 34–36, 39]. We have demonstrated by time-lapse

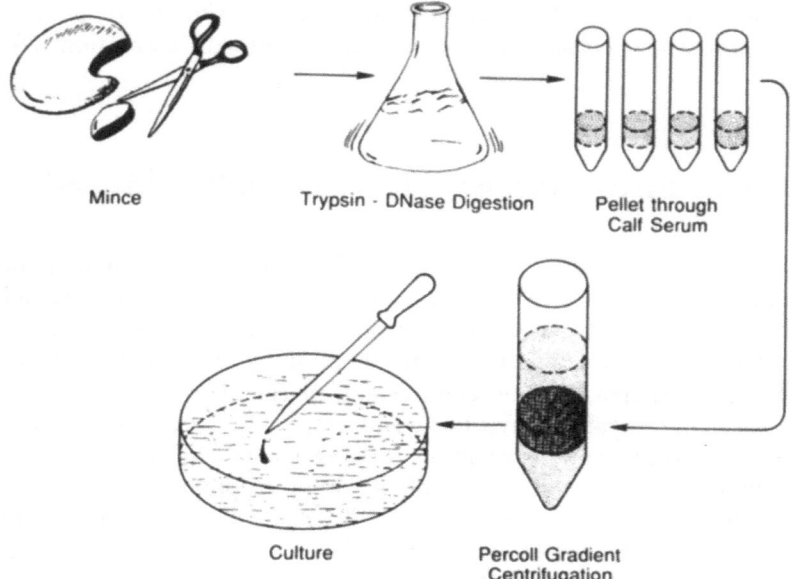

Mince   Trypsin · DNase Digestion   Pellet through
Calf Serum

Culture   Percoll Gradient
Centrifugation

Fig. 2. Purification of human trophoblasts as described by Kliman et al. [35]. Minced villous tissue is subjected to trypsin-deoxyribonuclease (*DNase*) digestion. Trypsin is inactivated by spinning the digest supernatants through a calf serum layer. The digest-supernatants are centrifuged on a 5%–70% Percoll gradient, yielding a band of pure cytotrophoblasts. (From [74])

cinematography that when these mononuclear cytotrophoblasts are placed in Dulbecco's modified Eagles' medium (DMEM) containing 20% (v/v) heat-inactivated fetal calf serum (FCS), they flatten onto the culture surface within 3–12 h, migrate toward each other to form aggregates within the first 24 h, and form syncytiotrophoblasts over the next 24 h of culture (Fig. 3). Concomitant with these morphologic changes, the trophoblasts synthesize and secrete a number of cell products, including (a) protein hormones: chorionic gonadotropin, placental lactogen, and pregnancy-specific β_1-glycoprotein [21, 35]; and (b) steroid hormones: estrogen and progesterone [35]. We and others have utilized these cells to elucidate the products of trophoblast differentiation and to explore the mechanisms by which their synthesis and secretion are regulated.

Trophoblasts as Endocrine Cells

Trophoblasts synthesize and secrete a vast array of endocrine products (for reviews see [7, 10, 65, 73, 83]).

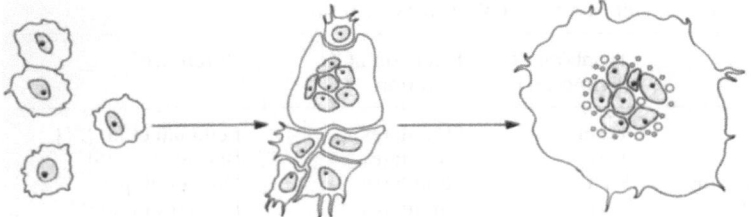

Fig. 3. Formation of syncytiotrophoblasts in vitro. Initially, the isolated cytotrophoblasts are mononuclear single cells. At 24 h, the dominant form is the multicellular aggregate. After 24 h, increasing numbers of syncytia are seen until eventually they become the dominant form. (From [35])

Production and Regulation of Protein Hormones

Chorionic Gonadotropin. The most widely studied trophoblast product is human chorionic gonadotropin (hCG). This glycoprotein is critical to pregnancy since it rescues the corpus luteum from involution, thus maintaining progesterone secretion by the ovarian granulosa cells. Its usefulness as a diagnostic marker of pregnancy stems from the fact that it may be one of the earliest secreted products of the conceptus. Ohlsson et al. [59] have demonstrated by in situ hybridization that β-hCG transcripts are present in human blastocyst trophoblasts prior to implantation. Placental production of hCG peaks during the 10th–12th weeks of gestation and tends to plateau at a lower level for the remainder of pregnancy. This difference in the rate of hCG secretion may be mimicked to some extent by trophoblasts cultured from first- versus third-trimester placentae. Kato and Braunstein [30] have demonstrated that trophoblasts from first trimester placentae secrete greater amounts of hCG than trophoblasts purified from term placentae, suggesting that cultured trophoblasts may retain the regulatory effects of their in situ milieu even after several days of culture. What regulates hCG synthesis and secretion in the trophoblast? Workers have attempted to answer this question by examining those factors that can regulate hCG synthesis and secretion in vitro. Table 1 summarizes our current knowledge of the regulatory factors that appear to modulate hCG secretion in cultured trophoblasts.

Human Placental Lactogen. This potent glycoprotein is made throughout gestation, increasing progressively until the 36th week, when it can be found in the maternal serum at a concentration of 5–15 μg/ml, the highest concentration of any known protein hormone. The major source of human placental lactogen (hPL) appears to be the villous syncytiotrophoblasts, where it is made at a constant level throughout gestation [80]. In addition to the villous syncytiotrophoblast, hPL has been identified in invasive inter-

Table 1. Regulation of cultured trophoblast hCG secretion

Factor	Trophoblasts (Trimester)	Effect on hCG secretion	Reference
cAMP	Term	Stimulates	Feinman et al. [21]
GnRH	Term	Stimulates	Belisle et al. [6]
β-Adrenergic agonists	First	Stimulates	Oike et al. [61]
Dexamethasone	Term	Stimulates	Ringler et al. [75]
Inhibin	Term	Inhibits	Petraglia et al. [62]
Activin	Term	Potentiates GnRH simulation of hCG secretion	Petraglia et al. [63]
EGF	First, term	Stimulates	Maruo et al. [51]
Interleukin 1	First	Stimulates	Yagel et al. [91]
Interleukin 6	First	Stimulates	Nishino et al. [56]
Basement membrane	First	Stimulates	Truman and Ford [85]
Decidual protein	Term	Inhibits	Ren and Braunstein [72]
Prolactin	Term	Inhibits	Yuen et al. [94]

mediate trophoblasts during the first trimester [27a, 27b, 41a]. In addition to identifying hPL within trophoblasts in situ, cultured first-trimester trophoblasts have been shown to secrete hPL in vitro [13]. Sakbun et al. [80] have also identified hPL mRNAs in cultured trophoblasts. Hoshina et al. [27b], working with choriocarcinoma cell lines, have proposed that hPL gene expression occurs after α-hCG and β-hCG gene expression, suggesting that hPL is a product of a more differentiated trophoblast. We have also shown that intracytoplasmic α-hCG appears prior to intracytoplasmic hPL in cultured term trophoblasts [36]. The factors that regulate hPL synthesis and secretion are not as well studied as for hCG. Kato and Braunstein [30a] have demonstrated that the secretion of hCG and hPL are discordant during the first 5 days of term trophoblast culture, suggesting different regulatory pathways for these hormones. Dodeur et al. [13] demonstrated that dibutyryl cyclic adenosine monophosphate (cAMP) stimulated hPL secretion from cultured first-trimester trophoblasts. Maruo et al. [51] have shown that epidermal growth factor (EGF), in addition to increasing hCG secretion by cultured human trophoblasts, also augments hPL secretion by these cells. Handwerger et al. [26] showed that high-density lipoproteins stimulate the release of hPL from human placental explants. Finally, Petit et al. [61a] have demonstrated that angiotensin II stimulates hPL release by cultured trophoblasts.

Prolactin. Al and Fox [1] and Sakbun et al. [79] have demonstrated by immunohistochemical studies that villous syncytiotrophoblasts contain prolactin. The significance of this finding is unclear at this time.

Relaxin. Sakbun et al. [79], using anti-peptide antibodies, demonstrated immunoreactivity for the C peptide and/or prorelaxin in villous cytotrophoblasts. More recently, Sakbun et al. [81] have demonstrated relaxin secretion by cultured trophoblasts. Trophoblast-derived relaxin may, therefore, play an important role in maternal extracellular matrix (ECM) modification as parturition approaches.

Chorionic Adrenocorticotropin. An adrenocorticotropic hormone (ACTH)-like protein, lipotropin, and β-endorphin have all been identified in placental extracts [58], presumably all derived from the common precursor pro-opiomelanocortin [40]. Liotta et al. [43] demonstrated that ACTH is synthesized by cultured placental cells, and Al and Fox [1] have demonstrated ACTH within villous syncytiotrophoblasts by immunohistochemistry. Mulder et al. [55] demonstrated that isoproterenol stimulated ACTH secretion by placental explant cultures. The physiologic role of placental ACTH is unclear. As with other placental hormones, it may represent a shift from maternal to placental control of this metabolic pathway.

Production and Regulation of Steroid Hormones

Progesterone. The significance of placental elaboration of progesterone was revealed by Diczfalusy and Troen [12], who showed that bilateral oophorectomy between 7 and 10 weeks of gestation had little impact on the conceptus or urinary pregnanediol levels. More recently, we have been able to demonstrate progesterone secretion by cultured term trophoblasts [35]. In addition, we have identified various components of the steroidogenic machinery necessary for progesterone biosynthesis within cultured trophoblasts [57]. Like hCG, progesterone synthesis and secretion seem to be upregulated by cAMP agonists [21, 39, 57].

Estrogen. The placenta does not have all the necessary enzymes to make estrogens from cholesterol, or even progesterone. Human trophoblasts lack 17α-hydroxylase and therefore can not convert C_{21} steroids to C_{19} steroids, the immediate precursors of estrogen. To bypass this deficit, dehydroisandrosterone sulfate from fetal adrenal is converted to estradiol-17β by trophoblasts [82b]. Not surprisingly, trophoblasts contain the necessary enzymes to make this conversion [10], namely sulfatase, 3β-hydroxysteroid dehydrogenase/$\Delta^{5\rightarrow4}$-isomerase (3βHSD), and aromatase. Lobo and Bellino [45] have demonstrated that cultured trophoblasts synthesize aromatase and that cAMP appears to stimulate aromatase production by these cells. Recently, Nestler [55a] demonstrated that insulin-like growth factor II inhibits aromatase in cultured human trophoblasts.

Production and Regulation of Hypothalamic-Pituitary Hormones

The placenta appears to produce a number of hypothalamic-pituitary hormones, including gonadotropin-releasing hormone (GnRH) and corticotropin-releasing hormone (CRH) (for a recent review see [83]).

GnRH was first identified within villous cytotrophoblasts by immuno-chemical staining of intact placentae [32]. More recently, Petraglia et al. [66] have demonstrated GnRH secretion by cultured trophoblasts and have shown that estrogen augments cAMP induction of trophoblast GnRH secretion. CRH is also made and secreted by cultured trophoblasts [78]. Robinson et al. [76] have demonstrated that glucocorticoids stimulate CRH release by cultured trophoblasts. Adding a further level of complexity to the regulatory signals impinging on the placenta, Petraglia et al. [64] have shown that neurotransmitters and peptides modulate the release of immuno-reactive CRH, and interleukin 1-β increases both CRH and ACTH release from cultured human trophoblasts [67]. In addition to the hypothalamic factors GnRH and CRH, pituitary growth hormone is synthesized and secreted by first- and third-trimester cultured trophoblasts [15]. It appears from these studies that the placenta, in addition to replacing much of the woman's pituitary function during pregnancy, also replaces critical hypothalamic functions so as to maintain control and feedback loop mech-anisms close to the conceptus.

Trophoblasts as Attachment Cells

Examination of the junction between the human placenta and uterus reveals a population of trophoblasts grouped together in dense masses between the chorionic villi and the uterine stroma (Fig. 4). These cells appear to grow out of the cytotrophoblast layer of the nearby chorionic villi. In addition to being extravillous trophoblasts, they appear to function specifically as the junctional apparatus of the conceptus. In some places they form elongated structures, often referred to as "cell columns." Are these cells simply pro-liferating villous cytotrophoblasts, or possibly invasive intermediate tropho-blasts in transit to penetrate the uterus? Recent work suggests that they are truly a unique form of trophoblast, with their own differentiation markers [20].

Previous studies have identified fibronectin staining in a variety of locations in pregnancy tissue, including ECM of the placental-uterine junction, uterine stroma, connective tissue core of placental villi, fetal membranes, and in walls of fetal blood vessels [14, 88, 92]. These immuno-localization studies, while specific for fibronectin, were performed with antibodies which may have reacted with more ubiquitous, less well-characterized fibronectin epitopes. The specificity of staining that we have identified with monoclonal anti-human oncofetal fibronectin (FDC-6) (Fig. 4), unlike other anti-fibronectin antibodies, implicates the oncofetal domain or closely adjacent portions of the type III repeat-connecting segment as critical moieties associated with implantation and trophoblast attachment [20].

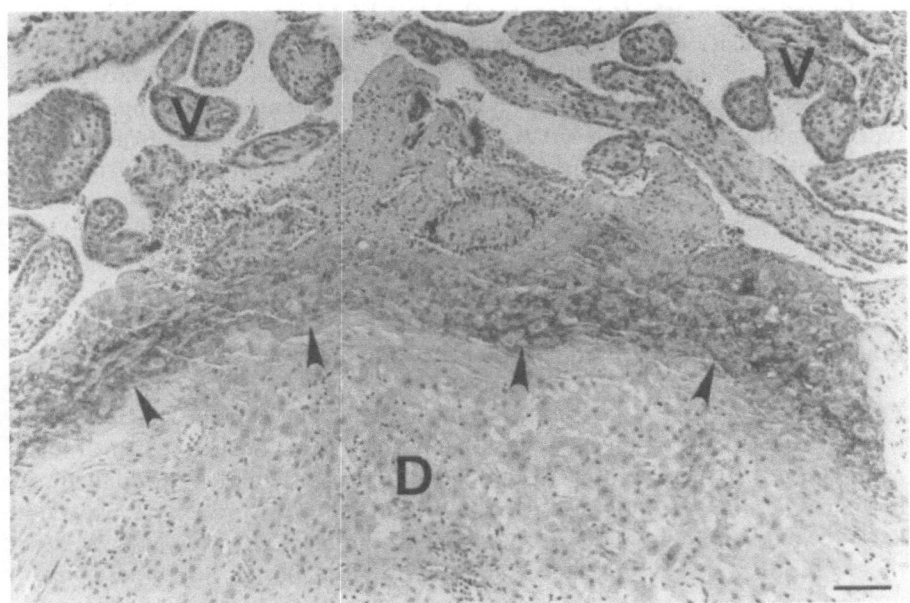

Fig. 4. Trophouteronectin (TUN) immunohistochemistry. A Bouin's-fixed, paraffin-embedded tissue section from a 6-week gestation was immunohistochemically stained with monoclonal anti-human oncofetal fibronectin (FDC-6). The uteroplacental junction exhibits a distinct band of TUN staining (*arrow heads*) at the zone of contact between the extravillous trophoblasts and the decidualized endometrium (*D*). Note the negatively stained chorionic villi (*V*). *Scale bar*, 100 μm

During the differentiation process in vitro, cytotrophoblasts have the ability to attach to types I and IV collagen, fibronectin, and laminin. We have shown that these associations can be specifically reversed with Ary-Gly-Asp containing peptides and antibodies to fibronectin, suggesting that human cytotrophoblasts have specific receptors to ECM proteins [29].

Trophoblasts as Invasive Cells

When purified human trophoblasts were co-incubated with endometrial explants in a suspension-culture system [38], the trophoblasts attached to the endometrial fragments, showed evidence of invasion into the tissue, and induced zones of necrosis at the points of contact with the exposed stromal surfaces. These results suggested to us that these trophoblasts possessed the machinery necessary for invasion. The ability to attach to ECM proteins is the first step in a series of steps necessary for cellular invasion [44].

In order to better focus on trophoblast-ECM interactions, without the added complexity of human endometria, we cultured trophoblasts with the ECM material Matrigel (Collaborative Research Bedford, MA). Matrigel is a mixture of solubilized basement membrane components containing laminin, type IV collagen, heparan sulfate proteoglycan, and entactin from mouse Engelbreth-Holm-Swarm tumor [33]. In our initial system [34], Matrigel was layered into a Millicell (Millipore, Bedford, MA) filter assembly, trophoblasts were added to the top chamber and incubated for up to 7 days. After 120 h, there was evidence of significant trophoblast penetration into the Matrigel layer.

Matribeach: A Biologic Assay of ECM Degradation and Invasion

The Millicell-Matrigel invasion system revealed that after 120 h human trophoblasts could invade into ECM material, based on detailed histologic examination of the invading trophoblasts in cross section. Although these findings gave us the impetus to continue our invasion studies, we were not able to readily manipulate or quantify trophoblast invasive behavior by the Millicell-Matrigel invasion system. As Albini et al. [2] have shown, undiluted Matrigel can form a significant barrier to invasion, even by aggressive tumor cells. Since we were also interested in examining the morphologic and degradative features of trophoblasts during the early stages of the invasive process, we developed a novel slope of undiluted Matrigel in order to examine the characteristics of ECM invasion as a function of time and Matrigel thickness [34]. This system had the advantage of allowing us to quantify trophoblast-ECM interactions, morphology, and proteolytic activity during the invasion process in vitro at the light- and electron-microscopic level. Using this thickness gradient of Matrigel with an 8° slope (Matribeach), we discovered that ECM thickness itself affected trophoblast morphology as well as the ability of these cells to degrade the ECM. In zone 2 of the Matribeach (see below), we observed cellular processes not observed in the Millicell-Matrigel invasion system.

Cytotrophoblasts from first- or third-trimester placentae, cultured on Matribeach, yielded a typical pattern of trophoblast morphology (Fig. 5) [34]. After 48 h, the trophoblasts on the glass surface had flattened, aggregated, and begun to form syncytia, as has been described previously [35]. On the thin part of the beach – zone 1 (defined as the Matrigel extending from the visible edge to 4 μm in thickness) – the cells flattened out in a fashion very similar to the flattening on glass. On zone 2 (defined as the Matrigel from 4 to 14 μm in thickness), the cells caused marked clearing of the Matrigel by creating pericellular zones of lysis around the trophoblast aggregates (Fig. 6C). Scanning electron microscopy (Fig. 6A,B) revealed that the trophoblast groups on zone 2 progressively eroded the Matrigel until the glass surface was exposed. Zone 2 lysis was particularly pronounced after the Matribeaches were immunocytochemically stained for mouse laminin, which resulted in multiple areas of negative staining around

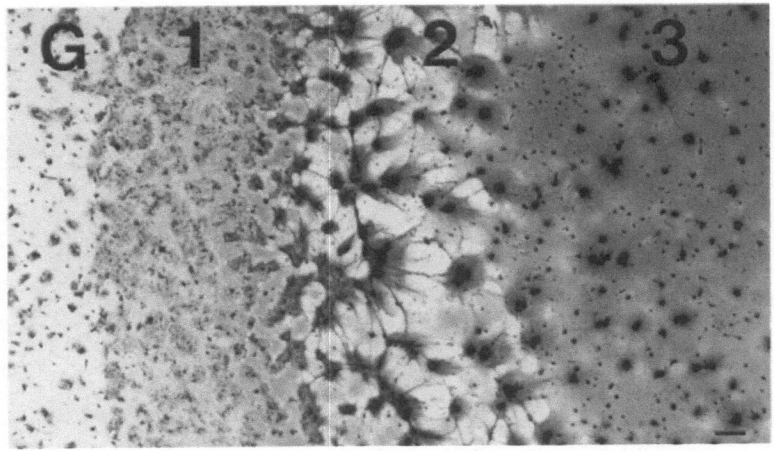

Fig. 5. Trophoblast morphology on a Matrigel slope (Matribeach). Term human trophoblasts were plated onto Matribeach, cultured for 48 h, fixed, and immunocytochemically stained for type IV collagen. Panoramic view of the Matribeach with clearly defined zones 1, 2, and 3. *G*, glass surface; *scale bar*, 200 μm

the cells. The same effect was observed when first-trimester trophoblasts were used. The cells on zone 3 – Matrigel thicker than 14 μm – remained spherical, mostly single, and exhibited little or no zones of lysis. These results were confirmed with multiple primary cultures of trophoblasts. Time course studies revealed that the basic separation of zones occurred by 24 h and continued to be observed through 72 h of culture. The trophoblasts in all three zones were viable, as assayed by their positive staining for α- and β-hCG.

Effect of 8-Bromo-cAMP on Matrigel Degradation

Cytotrophoblast responsiveness to 8-bromo-cAMP led us to examine the effect of this agent on trophoblast-Matribeach interactions. We found that 8-bromo-cAMP completely eliminated zone 2 matrix degradation by normal trophoblasts [34]. The cells appeared uniformly spherical, with little interaction between other cells or the matrix at all thicknesses. In trophoblasts, 8-bromo-cAMP promotes differentiation towards a noninvasive villous syncytial trophoblast phenotype, with induction of the synthesis and secretion of hCG [21, 29]. Additionally, this cyclic nucleotide dramatically downregulates fibronectin production by cytotrophoblasts [86a], thus potentially altering their interaction capabilities with the ECM. Interestingly, JEG-3 choriocarcinoma cells did not alter their morphology or degradative capabilities on zone 2 Matribeach when treated with 8-bromo-cAMP [34], suggesting that their invasive phenotype does not respond to regulatory agents in the same way as normal trophoblasts.

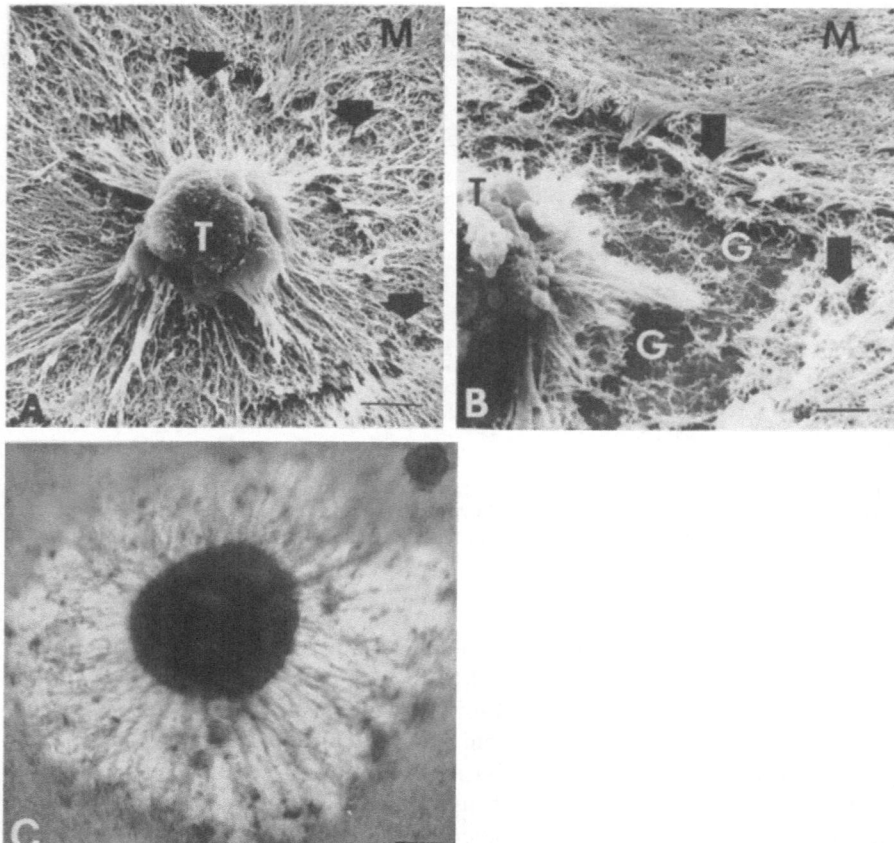

Fig. 6A–C. Trophoblasts degrade Matrigel. **A** Scanning electron microscopy of an aggregate of trophoblasts (*T*) at 48 h with a circumferential area of Matrigel lysis (*arrows*). Thin cytoplasmic processes emanate from the cells and merge into the surrounding Matrigel. Note the flat filamentous nature of the undigested Matrigel (*M*). **B** Now at 72 h, this trophoblast aggregate (*T*) has completely degraded the Matrigel (*arrows*) so that the surface of the underlying cover slip is exposed (*G*). Note again the flat surface of the undegraded Matrigel (*M*). **C** Light-microscopic appearance of an aggregate of trophoblasts on zone 2 of Matribeach after 48 h of culture showing an almost complete clearing of the pericellular Matrigel. *Scale bars*, 10 μm. (Portions of this figure from [34])

What Is the Role of the Plasminogen Activator System in Human Trophoblast Invasion?

The biologic repertoire necessary for the unique trophoblast-ECM interactions that occur during human gestation is unknown. Studies in nonhuman systems have proposed a critical role for trophoblast-secreted plasminogen activator (PA) during implantation and placentation. In mice, trophoblast production of PA correlates temporally with blastocyst invasion [82, 84], and implantation-defective mouse embryos elaborate diminished amounts of PA [3]. A role for PA in human nidation is suggested by the work of several investigators [50, 70], who demonstrated that trophoblasts in culture produce active PA. More recently, the results of Cajot et al. [9] lend support to the concept that any cell type which produces active u-PA can harbor an invasive phenotype. These workers transfected noninvasive mouse L cells with a cosmid containing the complete human u-PA gene. Those cells which expressed human u-PA could both degrade and invade the ECM, suggesting that u-PA expression alone is sufficient to initiate these processes.

PA activity in vascular and extracellular spaces is modulated by PA inhibitors (PAIs), glycoproteins of the serine protease inhibitor (SERPIN) family that covalently bind to and inhibit u-PA [49, 89]. Therefore, it is likely that PA-PAI interactions modulate trophoblast invasion in vivo and control fibrinolysis within the intervillous spaces of the placenta. Until recently, evidence for trophoblast elaboration of PAIs during pregnancy has been indirect. Two well-characterized PAIs, PAI-1 and PAI-2, were isolated from total placental extracts [41, 89, 93]. In addition, plasma PAI-1 and PAI-2 levels increase with advancing gestation, but decrease dramatically soon after delivery [41]. Altered plasma levels of PAIs have been reported in preeclampsia of pregnancy [11, 25], a disease in which abnormalities in fibrinolysis and trophoblast function often occur. Because of the potential role of the PAIs in controlling trophoblast invasion, we initiated studies to examine the synthesis and regulation of PAI-1 and PAI-2 in normal human cytotrophoblasts. PAI-1 and PAI-2 mRNA and protein are produced by cultured cytotrophoblasts, whereas only PAI-1 is found in JEG-3 cells, a malignant trophoblast cell line [18, 19]. PAI-2 is localized by immunocyto-chemistry to villous syncytiotrophoblasts, whereas PAI-1 is present primarily in invasive trophoblasts of implantation sites [17]. These findings suggest an important physiologic role for these proteins in vivo and may indicate that PAI-1 is specifically required for limiting the trophoblast invasive process.

Cyclic AMP and Protein Kinase C Modulation of the PA System

Cyclic AMP agonists have profound effects on cytotrophoblasts in culture by affecting their intracellular morphology, endocrine function, and synthesis of fibronectin [21, 86a]. As cytotrophoblasts undergo morphologic and biochemical differentiation in culture, both hCG and progesterone pro-duction are dramatically stimulated by 8-bromo-cAMP and forskolin [21, 29]. The increase in hCG secretion is the consequence of increased synthesis of α- and β-hCG protein subunits following increased transcription of their corresponding mRNAs [86a]. Cyclic AMP also affects u-PA gene transcription

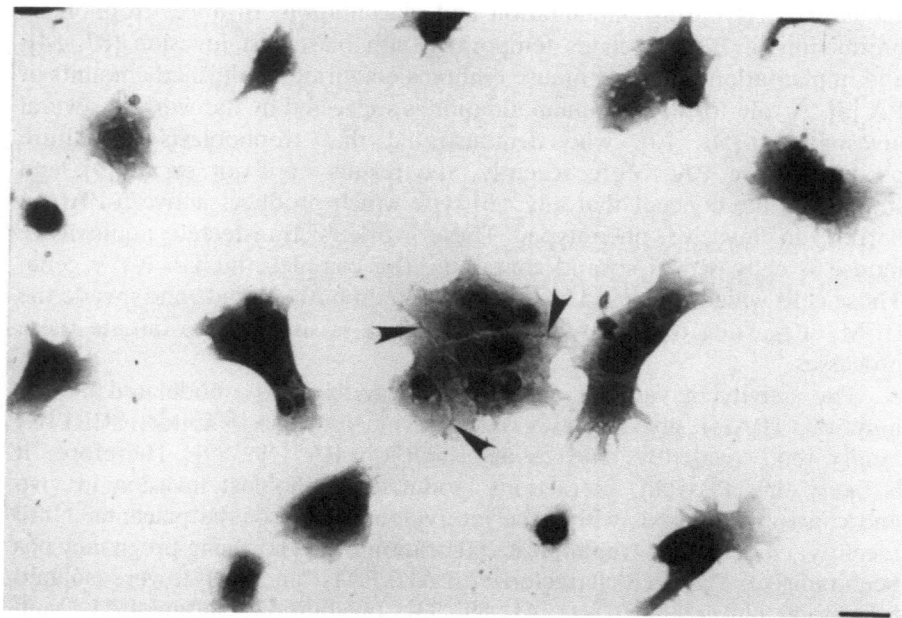

Fig. 7. Urokinase immunocytochemistry. Cytotrophoblasts were cultured for 24-h, fixed for 15 min with Bouin's solution, and immunocytochemically stained with affinity-purified rabbit anti-human u-PA antibody (a gift from Dr. R. Pittman, University of Pennsylvania). This low-power field reveals a mixture of single flattened trophoblasts and forming aggregates; u-PA staining can be identified within the cytoplasm of all of the cells. In addition, fine surface staining can be identified at the contact points of several of the aggregated trophoblasts (*arrow heads*). *Scale bar*, 20 µm

and proteolytic activity [70] as well as PAI production in trophoblastic cells [16, 19].

The PA System in Human Trophoblasts

We have utilized immunocytochemical, immunoblot, Northern blot, and zymographic analysis of cultured human trophoblasts as methods for studying PA and PAI production. We have identified immunoreactive u-PA in normal trophoblasts utilizing affinity-purified rabbit anti-human u-PA antibody (Fig. 7) and have confirmed, by protein immunoblotting, the presence of u-PA antigen in trophoblast cultures. As in other cell systems, trophoblast invasion and remodeling of the ECM may be controlled by a balance between trophoblast elaboration of u-PA and PAIs. We recently determined that human trophoblasts in culture produce both PAI-1, the principal endothelial cell PAI, and PAI-2, a macrophage and placental-associated PAI [17]. Immunoblots of cellular protein extract and conditioned

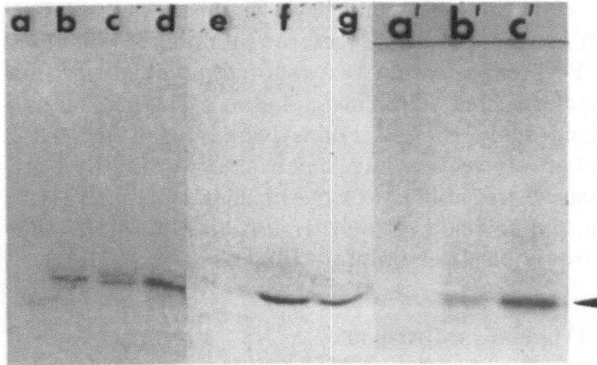

Fig. 8. Immunoblots demonstrating the presence of PAI types 1 (*a–g*) and 2 (a'–c') at 0 h (*a,e,a'*), 24 h (*b,f,b'*), 48 h (*c,g,c'*), and 72 h (*d*) in trophoblast cell extract (*a–d, a'–c'*) and conditioned media (*e–g*). *Arrow*, 46 kDa

media indicate that freshly purified cytotrophoblasts contain barely detectable levels of u-PA, PAI-1, or PAI-2. As the cytotrophoblasts differentiate in culture, the maximal rate of production occurs in the first 24 h. Interestingly, u-PA production appears to peak at 24 h, whereas PAI-1 and PAI-2 production continues at a steady state (Fig. 8). Queenan et al. [70] utilized trophoblasts prepared by the Kliman method and demonstrated secretion of a 50-kDa active u-PA species on plasminogen-dependent zymography of conditioned culture media. These workers also showed maximal activity of this secreted u-PA after 24 h of culture and de novo synthesis of the protein. By pulse-labeling cultured cytotrophoblasts with ^{35}S-methionine, followed by immunoprecipitation with specific anti-PAI antibodies, we have determined (H.J. Kliman, R.F. Feinberg, unpublished results) that both PAI-1 and PAI-2 are synthesized de novo in our cell cultures.

Performing immunocytochemistry (using highly specific rabbit anti-human PAI-1 and PAI-2 antibodies provided by Dr. T.-C. Wun of Monsanto Co.), we have demonstrated [17] that PAI-1 was localized to the trophoblast cell surface, as well as to the cytoplasm, whereas the cellular localization of PAI-2 was only cytoplasmic. Northern blot hybridization analysis using cDNA probes for PAI-1 and PAI-2 (obtained from Dr. T-C Wun [89, 93]) revealed two trophoblast PAI-1 transcripts (2.3 and 3.0 kb) and one 2.3-kb PAI-2 transcript, consistent with findings in other cultured human cells [16, 17]. Interestingly, the cAMP agonist 8-bromo-cAMP appears to coordinately regulate the plasminogen activator system in normal trophoblasts by down-regulating PAI mRNA levels twofold [16] and by up-regulating u-PA mRNA threefold in the first 24 h of culture. In zymographic analysis of serum-free trophoblast media we found secretion of an active u-PA at ~50 kDa (H.J. Kliman, R.F. Feinberg, unpublished results), in agreement with the findings of Queenan et al. [70].

Although trophoblasts elaborate u-PA and PAIs in culture, correlation with trophoblast biology in vivo is critical. Therefore, we undertook studies to characterize PA and PAI expression by trophoblasts in situ within placental villi and implantation sites. Utilizing a monoclonal anti-human u-PA (gift of E. Barnathan, University of Pennsylvania) and the PAI-1 polyclonal antibody described above, we have been able to identify both PAI-1 and u-PA in the invasive trophoblasts of the placental bed [17] (H.J. Kliman and R.F. Feinberg, unpublished results). It appears, therefore, that both in vitro and in vivo, trophoblasts elaborate u-PA and the PAIs.

Role of Other Proteases in Trophoblast Invasion

Matrigel clearing around the trophoblasts on zone 2 of Matribeach appears to be promoted by proteases which degrade both type IV collagen and laminin. Proteolysis of these proteins occurs as a function of time, with immunohistochemical analysis and scanning electron microscopy demonstrating progressively larger areas of negative staining around zone 2 cells from 24 to 72 h (Fig. 6) [34]. The biochemical basis of this trophoblast-mediated proteolysis is unknown, but, as in other cell systems, may be initiated by u-PA [53, 71]. However, the potential role of trophoblast-secreted collagenases [22] or cathepsins [42] in this process is not known. PAs are proposed to be activators of procollagenases [53, 71], and therefore u-PA activation of collagenases may have a role in trophoblast proteolysis and invasion. In gelatin zymographic analysis of serum-free trophoblast media we found, in addition to an active u-PA, secretion of an active gelatinase in the 90-kDa range (H.J. Kliman and R.F. Feinberg, unpublished results). This gelatinase was present in both the plasminogen-containing and plasminogen-free gels, in agreement with the findings of Fisher et al. [22]. However, when Matrigel was substituted for gelatin in our zymograms, only u-PA could be detected, and only in the plasminogen-containing gel. This suggests that the major secreted trophoblast protease capable of initiating Matrigel degradation is probably u-PA, and that this proteolysis is accomplished primarily through the localized activation of plasminogen to plasmin within the ECM.

The Polyphonic Hypothesis of Trophoblast Differentiation

We have noted a paradox about cultured trophoblasts. Cultured trophoblasts are capable of synthesizing and secreting a large array of placental products simultaneously, while trophoblasts in vivo seem to make only select products as defined by their location. For example, examination of tissue sections reveals that villous syncytiotrophoblasts immunochemically

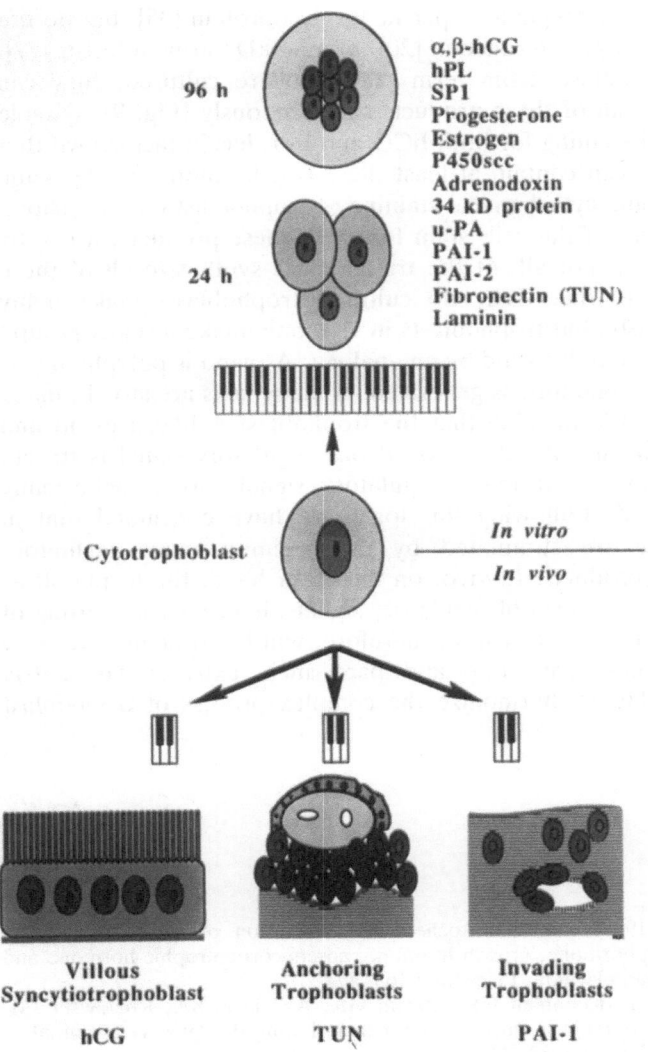

96 h

24 h

α,β-hCG
hPL
SP1
Progesterone
Estrogen
P450scc
Adrenodoxin
34 kD protein
u-PA
PAI-1
PAI-2
Fibronectin (TUN)
Laminin

Cytotrophoblast

In vitro
_____
In vivo

Villous
Syncytiotrophoblast

Anchoring
Trophoblasts

Invading
Trophoblasts

hCG

TUN

PAI-1

Fig. 9. Polyphonic hypothesis of trophoblast differentiation. In vivo, undifferentiated cytotrophoblasts can differentiate into either endocrine villous syncytiotrophoblasts (e.g., hCG secreting), into anchoring cell column trophoblasts [trophouteronectin (*TUN*) synthesizing], or into intramyometrial invading intermediate trophoblasts (containing PAs and PAIs, e.g., PAI-1). Specific products are only expressed by specific trophoblast types (i.e., TUN is not found in villous or invading trophoblasts), suggesting that the local environment dictates the pathway of trophoblast differentiation with a few select regulatory signals (represented by a few *keys from the piano keyboard*). Control may be mediated through autocrine, paracrine, extracellular matrix-mediated, and release of inhibition mechanisms. In vitro, cultured cytotrophoblasts go through a series of morphologic changes, culminating in the production of syncytiotrophoblasts over a 96-h period. During this time, these cells contain and secrete a wide assortment of products simultaneously, suggesting that cultured trophoblasts are stimulated by a variety of regulatory signals at once (represented by a *full piano keyboard*)

stain for hCG, hPL, and pregnancy-specific β_1-glycoprotein [35], but do not stain for PAI-1 [17], trophouteronectin [20], or a 34-kDa growth factor [77]. Yet, when cytotrophoblasts from term placentae are cultured, they can apparently synthesize all of these products simultaneously (Fig. 9). Double immunocytochemical staining for both hCG and hPL has in fact shown that cultured trophoblasts can contain at least these two hormones at the same time [36]. Since immunocytochemical staining of trophoblast cultures shows that the large majority of the cells stain for all of these products, it can be concluded that most, if not all, of the trophoblasts synthesize all of these products at the same time. Why do cultured trophoblasts make many products simultaneously, but trophoblasts in vivo only make a select group?

Our answer can be understood by an analogy. A piano is polyphonic: as each key is struck, a single tone is generated. If many keys are struck, many tones are generated. We imagine that the trophoblast is like a piano and that regulatory signals are like the keys. If one regulatory signal is struck, one product is produced. If many regulatory signals are struck, many products are produced. Following this logic, we have concluded that in culture, trophoblasts are stimulated by many simultaneous regulatory signals, hence many products. In vivo, on the other hand, the trophoblasts are stimulated by a discreet set of regulatory signals, hence a select group of products. The goal of future research, therefore, will be to define precisely how specific hormones (endocrine and paracrine), extracellular matrix proteins, and other factors harmonize the complex process of trophoblast differentiation.

References

1a. Al TA, Fox H (1986) Immunohistochemical localization of follicle-stimulating hormone, luteinizing hormone, growth hormone, adrenocorticotrophic hormone and prolactin in the human placenta. Placenta 7:163–172
2. Albini A, Iwamoto Y, Kelinman HK, Martin GR, Aaronson SA, Kozlowski JM, McEwan RN (1987) A rapid in vitro assay for quantitating the invasive potential of tumor cells. Cancer Res 47:3239–3245
3. Axelrod HR (1985) Altered trophoblast functions in implantation-defective mouse embryos. Dev Biol 108:185–190
4. Bax CM, Ryder TA, Mobberley MA, Tyms AS, Taylor DL, Bloxam DL (1989) Ultrastructural changes and immunocytochemical analysis of human placental trophoblast during short-term culture. Placenta 10:179–194
5. Belisle S, Bellabarba D, Gallo PN, Lehoux JG, Guevin JF (1986) On the role of luteinizing hormone-releasing hormone in the in vitro synthesis of bioactive human chorionic gonadotropin in human pregnancies. Can J Physiol Pharmacol 64:1229–1235
6. Belisle S, Petit A, Bellabarba D, Escher E, Lehoux JG, Gallo PN (1989) Ca^{2+}, but not membrane lipid hydrolysis, mediates human chorionic gonadotropin production by luteinizing hormone-releasing hormone in human term placenta. J Clin Endocrinol Metab 69:117–121

7. Blay J, Hollenberg MD (1989) The nature and function of polypeptide growth factor receptors in the human placenta. J Dev Physiol 12:237–248

7a. Boyd JD, Hamilton WJ (1970) In: The human placenta. MacMillan, London, pp 140–145

8. Branchaud C, Goodyer CG, Guyda HJ, Lefebvre Y (1990) A serum-free system for culturing human placental trophoblasts. In Vitro Cell Dev Biol 26:865–870

9. Cajot JF, Schleuning WD, Medcalf RL, Bamat J, Testuz J, Liebermann L, Sordat B (1989) Mouse L cells expressing human prourokinase-type plasminogen activator: effects on extracellular matrix degradation and invasion. J Cell Biol 109:915–925

10. Conley AJ, Mason JI (1990) Placental steroid hormones. Baillieres Clin Endocrinol Metab 4:249–272

11. DeBoer K, Lecander I, ten Cate JW, Borm JJJ, Treffers PE (1988) Placental-type plasminogen activator inhibitor in preeclampsia. Am J Obstet Gynecol 158:518

12. Diczfalusy E, Troen P (1961) Endocrine functions of the human placenta. Vitam Horm 19:229–311

13. Dodeur M, Malassine A, Bellet D, Mensier A, Evain BD (1990) Characterization and differentiation of human first trimester placenta trophoblastic cells in culture. Reprod Nutr Dev 30:183–192

14. Earl U, Estlin C, Bulmer JN (1990) Fibronectin and laminin in the early human placenta. Placenta 11:223–231

15. Evain BD, Alsat E, Mirlesse V, Dodeur M, Scippo ML, Hennen G, Frankenne F (1990) Regulation of growth hormone secretion in human trophoblastic cells in culture. Horm Res 33:256–259

16. Feinberg RF, Kao L-C, Ringler G, Murray S, Queenan J, Kliman H, Cines D, Wun T-C, Strauss JF3 (1988) Coordinate regulation of urokinase and plasminogen activator inhibitors in human cytotrophoblasts (Abstr). Proc Soc Gynecol Invest 35:231

17. Feinberg RF, Kao LC, Haimowitz JE, Queenan JTJ, Wun TC, Strauss JF3, Kliman HJ (1989) Plasminogen activator inhibitor types 1 and 2 in human trophoblasts. PAI-1 is an immunocytochemical marker of invading trophoblasts. Lab Invest 61:20–26

18. Feinberg RF, Strauss JFIII, Wun T-C, Kliman HJ (1989) Plasminogen activators (PAs) and plasminogen activator inhibitors (PAIs) in human trophoblasts: markers of trophoblast invasion (Abstr). Proc Soc Gynecol Invest 36:487

19. Feinberg RF, Kao LC, Wang C-L, Bui L, Kliman HJ, Strauss JFIII (1990) Plasminogen activator inhibitor (PAI) expression in normal and malignant human trophoblasts: regulation by 8-bromo-camp and phorbol esters (Abstr). Proc Soc Gynecol Invest 37:465

20. Feinberg RF, Kliman HJ, Lockwood CJ (1991) Oncofetal fibronectin: a trophoblast "glue" for human implantation? Am J Pathol 138:537–543

21. Feinman MA, Kliman HJ, Caltabiano S, Strauss JFIII (1986) 8-Bromo-3',5'-adenosine monophosphate stimulates the endocrine activity of human cytotrophoblasts in culture. J Clin Endocrinol Metab 63:1211–1217

22. Fisher SJ, Cui TY, Zhang L, Hartman L, Grahl K, Zhang GY, Tarpey J, Damsky CH (1989) Adhesive and degradative properties of human placental cytotrophoblast cells in vitro. J Cell Biol 109:891–902

23. Fisher SJ, Sutherland A, Moss L, Hartman L, Crowley E, Bernfield M, Calarco P, Damsky C (1990) Adhesive interactions of murine and human trophoblast cells. Troph Res 4:115–138

24. Fournet DN, MacLusky NJ, Leranth CZ, Todd R, Mendelson CR, Simpson ER, Naftolin F (1987) Immunohistochemical localization of aromatase cytochrome P-450 and estradiol dehydrogenase in the syncytiotrophoblast of the human placenta. J Clin Endocrinol Metab 65:757–764

25. Gore M, Eldon S, Trofatter KF, Soong S-J, Pizzo SV (1987) Pregnancy-induced changes in the fibrinolytic balance: evidence for defective release of tissue plasminogen activator and increased levels of the fast-acting tissue plasminogen activator inhibitor. Am J Obstet Gynecol 156:674

26. Handwerger S, Quarfordt S, Barrett J, Harman I (1987) Apolipoproteins AI, AII, and CI stimulate placental lactogen release from human placental tissue. A novel action of high density lipoprotein apolipoproteins. J Clin Invest 79:625–628
27. Hertig AT, Rock J (1956) A description of 34 human ova within the first 17 days of development. Am J Anat 98:435–494
27a. Heyderman E, Gibbons AR, Rosen SW (1981) Immunoperoxidase localisation of human placental lactogen: a marker for the placental origin of the giant cells in "syncytial endometritis" of pregnancy. J Clin Pathol 34:303–307
27b. Hoshina M, Hussa R, Pattillo R, Camel HM, Boime I (1984) The role of trophoblast differentiation in the control of the hCG and hPL genes. Adv Exp Med Biol 176:299–312
28. Jara CS, Salud AT, Bryantgreenwood GD, Pirens G, Hennen G, Frankenne F (1989) Immunocytochemical localization of the human growth hormone variant in the human placenta. J Clin Endocrinol Metab 69:1069–1072
29. Kao LC, Caltabiano S, Wu S, Strauss JFIII, Kliman HJ (1988) The human villous cytotrophoblast: interactions with extracellular matrix proteins, endocrine function, and cytoplasmic differentiation in the absence of syncytium formation. Dev Biol 130:693–702
30. Kato Y, Braunstein GD (1990) Purified 1st and 3rd trimester placental trophoblasts differ in in vitro hormone secretion. J Clin Endocrinol Metab 70:1187–1192
30a. Kato Y, Braunstein GD (1989) Discordant secretion of placental protein hormones in differentiating trophoblasts in vitro. J Clin Endocrinol Metab 68:814–20
31. Kawagoe K, Akiyama J, Kawamoto T, Morishita Y, Mori S (1990) Immunohistochemical demonstration of epidermal growth factor (EGF) receptors in normal human placental villi. Placenta 11:7–15
32. Khodr GS, Siler KT (1978) Localization of luteinizing hormone-releasing factor in the human placenta. Fertil Steril 29:523–526
33. Kleinman HK, McGarvey ML, Hassell JR, Star VL, Cannon FB, Laurie GW, Martin GR (1986) Basement membrane complexes with biological activity. Biochemistry 25:312–318
34. Kliman HJ, Feinberg RF (1990) Human trophoblast-extracellular matrix (ECM) interactions in vitro: ECM thickness modulates morphology and proteolytic activity. Proc Natl Acad Sci USA 87:3057–3061
35. Kliman HJ, Nestler JE, Sermasi E, Sanger JM, Strauss JFIII (1986) Purification, characterization, and in vitro differentiation of cytotrophoblasts from human term placentae. Endocrinology 118:1567–1582
36. Kliman HJ, Feinman MA, Strauss JFIII (1987) Differentiation of human cytotrophoblasts into syncytiotrophoblasts in culture. Troph Res 2:407–421
37. Kliman HJ, Coutifaris C, Babalola GO, Soto EA, Kao LC, Queenan JTJ, Feinberg RF, Strauss JFIII (1989) The human trophoblast: homotypic and heterotypic cell-cell interactions. Prog Clin Biol Res 294:425–434
38. Kliman HJ, Feinberg RF, Haimowitz JE (1990) Human trophoblast-endometrial interactions in an in vitro suspension culture system. Placenta 11:349–367
39. Kliman HJ, Strauss JFIII, Kao L-C, Caltabiano S, Wu S (1991) Cytoplasmic and biochemical differentiation of the human villous cytotrophoblast in the absence of syncytium formation. Troph Res (in press)
40. Krieger DT (1982) Placenta as a source of "brain" and "pituitary" hormones. Biol Reprod 26:55–71
41. Kruithof EKO, Tran-Thang C, Gudinchet A, Hauert J, Nicoloso G, Genton C, Welti H, Bachmann F (1987) Fibrinolysis in pregnancy: a study of plasminogen activator inhibitors. Blood 69:460–466
41a. Kurman RJ, Main CS, Chen HC (1984) Intermediate trophoblast: a distinctive form of trophoblast with specific morphological, biochemical and functional features. Placenta 5:349–369
42. Lah TT, Buck MR, Honn KV, Crissman JD, Rao NC, Liotta LA, Sloane BF (1989) Degradation of laminin by human tumor cathepsin B. Clin Exp Metastasis 7:461–468

43. Liotta A, Osathanondh R, Ryan KJ, Krieger DT (1977) Presence of corticotropin in human placenta: demonstration of in vitro synthesis. Endocrinology 101: 1552–1558
44. Liotta LA, Rao CN, Barsky SH (1983) Tumor invasion and the extracellular matrix. Lab Invest 49:636–649
45. Lobo JO, Bellino FL (1989) Estrogen synthetase (aromatase) activity in primary culture of human term placental cells: effects of cell preparation, growth medium, and serum on adenosine 3',5'-monophosphate response. J Clin Endocrinol Metab 69:868–874
46. Loke YW (1983) Human trophoblast in culture. In: Loke YW, Whyte A (eds) Biology of trophoblast. Elsevier, Amsterdam, pp 663–701
47. Loke YW (1990) New developments in human trophoblast cell culture. Colloq INSERM 199:10–16
48. Loke YW, Gardner L, Grabowska A (1989) Isolation of human extravillous trophoblast cells by attachment to laminin-coated magnetic beads. Placenta 10:407–415
49. Loskutoff DJ, Ny T, Sawdey M, Lawrence D (1986) Fibrinolytic system of cultured endothelial cells: regulation by plasminogen activator inhibitor. J Cell Biochem 32:273
50. Martin O, Arias F (1982) Plasminogen activator production by trophoblast cells in vitro: effect of steroid hormones and protein synthesis inhibitors. Am J Obstet Gynecol 142:402–409
51. Maruo T, Matsuo H, Oishi T, Hayashi M, Nishino R, Mochizuki M (1987) Induction of differentiated trophoblast function by epidermal growth factor: relation of immunohistochemically detected cellular epidermal growth factor receptor levels. J Clin Endocrinol Metab 64:744–750
52. Maruyama I, Bell CE, Majerus PW (1985) Thrombomodulin is found on endothelium of arteries, veins, capillaries, and lymphatics, and on syncytiotrophoblast of human placenta. J Cell Biol 101:363–371
53. Mignatti P, Robbins E, Rifkin DB (1986) Tumor invasion through the human amniotic membrane: requirement for a proteinase cascade. Cell 47:487–498
54. Moll UM, Lane BL (1990) Proteolytic activity of 1st trimester human placenta-localization of interstitial collagenase in villous and extravillous trophoblast. Histochemistry 94:555–560
55. Mulder GH, Maas R, Arts NF (1986) In vitro secretion of peptide hormones by the human placenta: I. ACTH. Placenta 7:143–153
55a. Nestler JE (1990) Insulin-like growth factor-II is a potent inhibitor of the aromatase activity of human placental cytotrophoblasts. Endocrinology 127:2064–2070
56. Nishino E, Matsuzaki N, Masuhiro K, Kameda T, Taniguchi T, Takagi T, Saji F, Tanizawa O (1990) Trophoblast-derived interleukin-6 (IL-6) regulates human chorionic gonadotropin release through iL-6 receptor on human trophoblasts. J Clin Endocrinol Metab 71:436–441
57. Nulsen JC, Silavin SL, Kao LC, Ringler GE, Kliman HJ, Strauss JFIII (1989) Control of the steroidogenic machinery of the human trophoblast by cyclic AMP. J Reprod Fertil [Suppl] 37:147–153
58. Odagiri E, Sherrell BJ, Mount CD, Nicholson WE, Orth DN (1979) Human placental immunoreactive corticotropin, lipotropin, and beta-endorphin: evidence for a common precursor. Proc Natl Acad Sci USA 76:2027–2031
59. Ohlsson R, Larsson E, Nilsson O, Wahlstrom T, Sundstrom P (1989) Blastocyst implantation precedes induction of insulin-like growth factor II gene expression in human trophoblasts. Development 106:555–559
60. Ohtani H, Maruyama I, Yonezawa S (1989) Ultrastructural immunolocalization of thrombomodulin in human placenta with microwave fixation. Act Hist Cy 22:393–395
61. Oike N, Iwashita M, Muraki T, Nomoto T, Takeda Y, Sakamoto S (1990) Effect of adrenergic agonists on human chorionic gonadotropin release by human trophoblast cells obtained from 1st-trimester placenta. Horm Metab Res 22:188–191

61a. Petit A, Guillon G, Tence M, Jard S, Gallo PN, Bellabarba D, Lehoux JG, Belisle S (1989) Angiotensin II stimulates both inositol phosphate production and human placental lactogen release from human trophoblastic cells. J Clin Endocrinol Metab 69:280–286

62. Petraglia F, Sawchenko P, Lim AT, Rivier J, Vale W (1987) Localization, secretion, and action of inhibin in human placenta. Science 237:187–189

63. Petraglia F, Vaughan J, Vale W (1989) Inhibin and activin modulate the release of gonadotropin-releasing hormone, human chorionic gonadotropin, and progesterone from cultured human placental cells. Proc Natl Acad Sci USA 86:5114–5117

64. Petraglia F, Sutton S, Vale W (1989) Neurotransmitters and peptides modulate the release of immunoreactive corticotropin-releasing factor from cultured human placental cells. Am J Obstet Gynecol 160:247–251

65. Petraglia F, Calza L, Garuti GC, Giardino L, De RB, Angioni S (1990) New aspects of placental endocrinology. J Endocrinol Invest 13:353–371

66. Petraglia F, Vaughan J, Vale W (1990) Steroid hormones modulate the release of immunoreactive gonadotropin-releasing hormone from cultured human placental cells. J Clin Endocrinol Metab 70:1173–1178

67. Petraglia F, Garuti GC, Deramundo B, Angioni S, Genazzani AR, Bilezikjian LM (1990) Mechanism of action of interleukin-1-beta in increasing corticotropin-releasing factor and adrenocorticotropin hormone release from cultured human placental cells. Am J Obstet Gynecol 163:1307–1312

68. Pijnenborg R (1990) Trophoblast invasion and placentation in the human – morphological aspects. Troph Res 4:33–47

68a. Pijnenborg R, Robertson WB, Brosens I, Dixon G. (1981) Trophoblast invasion and the establishment of hemochorial placentation in man and laboratory animals. Placenta 2:71–92

69. Posner BI (1974) Insulin receptors in human and animal placental tissue. Diabetes 23:209–217

70. Queenan JT Jr, Kao L-C, Arboleda CE, Ulloa-Aguirre A, Golos TG, Cines DB, Strauss JFIII (1987) Regulation of urokinase-type plasminogen activator production by cultured human cytotrophoblasts. J Biol Chem 262:1093–1096

71. Reich R, Thompson EW, Iwamoto Y, Martin GR, Deason JR, Fuller GC, Miskin R (1988) Effects of inhibitors of plasminogen activator, serine proteinases, and collagenase IV on the invasion of basement membranes by metastatic cells. Cancer Res 48:3307–3312

72. Ren SG, Braunstein GD (1991) Decidua produces a protein that inhibits chorio-gonadotropin release from human trophoblasts. J Clin Invest 87:326–330

73. Ringler GE, Strauss JFIII (1990) In vitro systems for the study of human placental endocrine function. Endocr Rev 11:105–123

74. Ringler GE, Ulloa-Aguirre A, Kao L-C, Nulsen JC, Kallen CB, Kliman HJ, Strauss JFIII (1988) Control of chorionic gonadotropin (hCG) by cyclic AMP: lessons from primary cultures of cytotrophoblasts. In: Mochizuki M, Hussa R (eds) Placental protein hormones. Elsevier, Amsterdam, p 184

75. Ringler GE, Kallen CB, Strauss JFIII (1989) Regulation of human trophoblast function by glucocorticoids: dexamethasone promotes increased secretion of chorionic gonadotropin. Endocrinology 124:1625–1631

75a. Roberts JM, Taylor RN, Musci TJ, Rodgers GM, Hubel CA, Mclaughlin MK (1989) Preeclampsia – an endothelial cell disorder. Am J Obstet Gynecol 161:1200–1204

76. Robinson BG, Emanuel RL, Frim DM, Majzoub JA (1988) Glucocorticoid simulates corticotropin releasing hormone gene expression in human placenta. Proc Natl Acad Sci USA 85:5244–5248

77. Roy CS, Sen MA, Murthy U, Mishra VS, Kliman HJ, Nestler JE, Strauss JFIII, Das M (1988) Biosynthesis and turnover of a 34-kDa protein growth factor in human cytotrophoblasts. Eur J Biochem 172:777–783

78. Saijonmaa O, Laatikainen T, Wahlstrom T (1988) Corticotrophin-releasing factor in human placenta: localization, concentration and release in vitro. Placenta 9:373–385

79. Sakbun V, Koay ES, Bryant GGD (1987) Immunocytochemical localization of prolactin and relaxin C-peptide in human decidua and placenta. J Clin Endocrinol Metab 65:339–343
80. Sakbun V, Ali SM, Lee YA, Jara CS, Bryantgreenwood GD (1990) Immunocytochemical localization and messenger ribonucleic acid concentrations for human placental lactogen in amnion, chorion, decidua, and placenta. Am J Obstet Gynecol 162:1310–1317
81. Sakbun V, Ali SM, Greenwood FC, Bryantgreenwood GD (1990) Human relaxin in the amnion, chorion, decidua-parietalis, basal plate, and placental trophoblast by immunocytochemistry and northern analysis. J Clin Endocrinol Metab 70:508–514
82. Sherman MI, Strickland S, Reich E (1976) Differentiation of early mouse embryonic and teratocarcinoma cells in vitro: plasminogen activator production. Canc Res 36:4208–4216
82a. Shorter SC, Jackson MC, Sargent IL, Redman CW, Starkey PM (1990) Purification of human cytotrophoblast from term amniochorion by flow cytometry. Placenta 11:505–13
82b. Siiteri PK, MacDonald PC (1966) Placental estrogen biosynthesis during human pregnancy. J Clin Endocrinol Metab 26:751–761
83. Sirinathsinghji DJ, Heavens RP (1989) Stress-related peptide hormones in the placenta: their possible physiological significance. J Endocrinol 122:435–437
84. Strickland S, Reich E, Sherman MI (1976) Plasminogen activator in early embryogenesis: enzyme production by trophoblast and parietal endoderm. Cell 9:231–240
85. Truman P, Ford HC (1986) The effect of substrate and epidermal growth factor on human placental trophoblast cells in culture. In Vitro Cell Dev Biol 22:525–528
86. Truman P, Pomare L, Ford HC (1989) Human placental cytotrophoblast cells: identification and culture. Arch Gynecol Obstet 246:39–49
86a. Ulloa-Aguirre A, August AM, Golos TG, Kao LC, Sakuragi N, Kliman HJ, Strauss JFIII (1987) 8-Bromo-adenosine 3',5'-monophosphate regulates expression of chorionic gonadotropin and fibronectin in human cytotrophoblasts. J Clin Endocrinol Metab 64:1002–1009
87. Uzumaki H, Okabe T, Sasaki N, Hagiwara K, Takaku F, Tobita M, Yasukawa K, Ito S, Umezawa Y (1989) Identification and characterization of receptors for granulocyte colony-stimulating factor on human placenta and trophoblastic cells. Proc Natl Acad Sci USA 86:9323–9326
88. Vartio T, Laitinen L, Narvanen O, Cutolo M, Thornell L-E, Zardi L, Virtanen IJ (1987) Differential expression of the ED sequence-containing form of cellular fibronectin in embryonic and adult human tissues. J Cell Sci 88:419–430
89. Wun TC, Reich E (1987) An inhibitor of plasminogen activation from human placenta. Purification and characterization. J Biol Chem 262:3646–3653
90. Yagel S, Casper RF, Powell W, Parhar RS, Lala PK (1989) Characterization of pure human first-trimester cytotrophoblast cells in long-term culture: growth pattern, markers, and hormone production. Am J Obstet Gynecol 160:938–945
91. Yagel S, Lala PK, Powell WA, Casper RF (1989) Interleukin-1 stimulates human chorionic gonadotropin secretion by first trimester human trophoblast. J Clin Endocrinol Metab 68:992–995
92. Yamada T, Isemura M, Yamaguchi Y, Munakata H, Hayashi N, Kyogoku M (1987) Immunohistochemical localization of fibronectin in the human placentas at their different stages of maturation. Histochemistry 86:579–584
93. Ye RD, Wun TC, Sadler JE (1987) cDNA cloning and expression in Escherichia coli of a plasminogen activator inhibitor from human placenta. J Biol Chem 262:3718–3725
94. Yuen BH, Moon YS, Shin DH (1986) Inhibition of human chorionic gonadotropin production by prolactin from term human trophoblast. Am J Obstet Gynecol 154:336–340

2 The Endometrium and Implantation

J. Hustin and P. Franchimont

It is evident that, in the human species, the period during which a blastocyst can implant is extremely limited and is almost totally dependent on the establishment of an equilibrium between the hormonal milieu of the woman and the degree of development of the blastocyst together with the signals it sends to the maternal organism. In the woman, the ovarian cycle is associated with important changes in the morphology and physiology – biochemistry of the uterine mucosa. After ovulation, there is a short period, "a window," during which endometrial receptivity is maximum. It is called the *nidation period*. During these few days, a blastocyst travelling to the uterine cavity can establish a *physical* contact with the maternal organism and eventually implant. The ideal implantation window is situated between the fifth and seventh postovulatory days [i.e., after the luteinizing hormone (LH) surge]. It has been recently characterized from the study of induced ovulatory cycles with subsequent in vitro fertilization and embryo transfer [1].

Implantation means that the conceptus has completely left the "external milieu," i.e., the uterine lumen, and is definitely embedded in the uterine mucosa. This implies that crucial factors exist which permit the penetration of the blastocyst. The first is that the ovarian cycle has followed a physiologic evolution, with corpus luteum development and an adequate production of progesterone. This hormone will induce important changes, which are described as secretory, in a mucosa primarily conditioned by estradiol [2]. That progesterone production must be high and continuous during the conception cycle and thereafter is evident. Occurrence of spontaneous menstruation would quite clearly interrupt the on-going process of early gestation.

We will devote this chapter to the demonstration of endometrial changes which are thought to be necessary for successful nidation. At the beginning, they are clearly linked to the secretory changes induced by progesterone. However, it soon appears that the conceptus itself will modulate this preparation by sending early signals (see Chap. 1) or by inducing endometrial changes via paracrine effects. This chapter will be divided into two major sections: one dealing with the endometrium and its changes within each compartment, and the other discussing the inter-relations between endometrium and early conceptus. We will address two main questions in each of the major sections. With regard to the *preparation of the endometrium to implantation*:

1. Has the human endometrium a particular appearance during the implantation window?
2. What are the properties – morphologic, biochemical – of the different compartments of the endometrium?

and with regard to *implantation and trophoblast endometrium relationship*:

1. How does implantation occur?
2. What are the immediate relations between maternal cells and blastocyst (i.e., primitive trophoblast)?

While in the recent past, most of these points could be answered only in terms of strict morphology, recent immunohistochemical and biochemical studies have brought considerably more knowledge on the physiology and biochemistry of receptive endometrium.

Preparation of the Endometrium for Implantation

The Endometrium During the Implantation Window

The appearance of the endometrium is fairly characteristic. Mucosal thickness is 2.5–4.0 mm. Glands are widened, tortuous, and lined by a simple columnar epithelium. Cytoplasmic vacuoles are no longer present, but there is a copious secretion in the lumina (Fig. 1).

At the electron-microscopic level [3], glycogen particles are still present throughout the cytoplasm. Secretion has begun and follows the pattern of apocrine secretion. The giant mitochondria of the early secretory phase are still conspicuous but irregular. The intercellular spaces of epithelial cells become more dilated though they remain sealed by apical intercellular junctional complexes. The nuclei become more rounded with a small decrease of volume [4]. There is an obvious increase in number and complexity of the nucleoli. Lastly, there is an increase and dilatation of the endoplasmic reticulum and Golgi apparatus. Some lipid droplets are also present. The endometrial stroma is markedly edematous. This dramatic change in stromal cellular density has been beautifully demonstrated by Dockery et al. [5] using morphometric techniques. Predecidual differentiation becomes apparent in superficial stromal cells during this phase. The pattern of vascularization is quite simple. No spiral arteries are present.

The surface epithelium does not participate in the overall secretory pattern. During this period, it is clearly both basophilic and metachromatic (toluidine blue metachromasia). This metachromasia will progressively decrease afterwards.

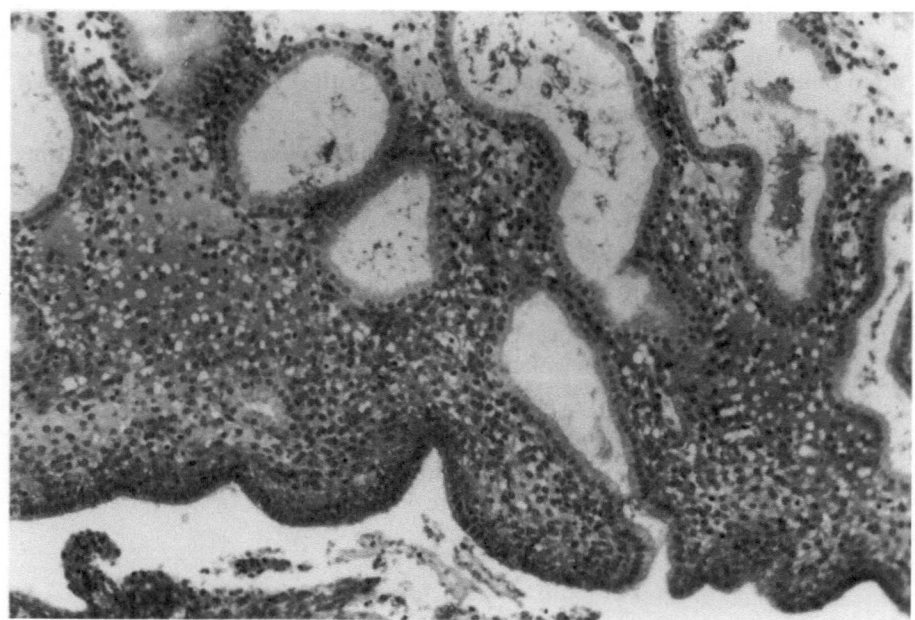

Fig. 1. Normal endometrium. Mid-secretory phase. Note the stromal edema, the dilated glands, and the somewhat darker surface epithelium. Hematoxylin and eosin, ×160

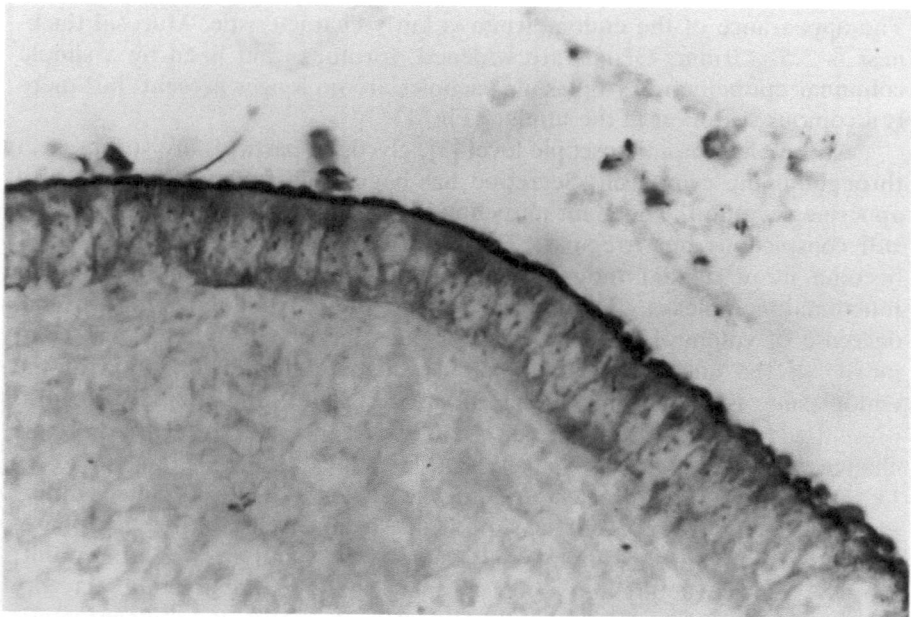

Fig. 2. Surface epithelium. Mid-secretory phase. Note the thick deposit at the apical pole. Colloidal iron stain for acidic mucosubstances, ×400

Artificial conditioning of the endometrium by hormone supplementation is usually feasible. Davies et al. [6] and Critchley et al. [2] have recently pointed to the discrepancies between normal mature endometrium and that obtained either after ovarian stimulation or after treatment regimens defined for medically assisted procreations.

Under sequential estrogen and progesterone, the uterine mucosa is characterized, at the time of embryo transfer, by subnuclear glycogen vacuoles in small round glands. The stroma has a cell density varying from dense cellularity to moderate amounts of edema. Judging from physiologic events, stromal and epithelial maturation are usually not synchronous. Though the percentage of successful implantations is reduced in such conditions, it is clear that they are not completely prevented by the asynchrony of maturation. It is in fact possible that some implantation delay can occur.

It is thus not clear whether a given morphologic pattern of the endometrium is truly a prerequisite for successful nidation. The ease with which a blastocyst can implant elsewhere, in a fallopian tube for instance, demonstrates clearly that this is not the case (see Chap. 17).

The Different Compartments of the Endometrium

Surface Epithelium

It is clear that the surface epithelium is the sole barrier between the external milieu, i.e., the uterine cavity, and the mucosa. Curiously enough, very few studies have been devoted to this important frontier. The surface epithelium is made up of a single layer of cylindrical cells with some intermingled ciliated cells. As a rule, this epithelium does not exhibit as marked changes as the gland cells under hormonal cyclic influences. In particular, secretory changes are scant, and thus epithelium retains its basophilia during the luteal phase. Metachromatic properties are maximal during the nidation window. The colloidal iron-positive material which is present at the cell surface thickens (Fig. 2). Simultaneously, there is a reduction of surface electronegativity as proved by the variation of the pH of the staining solutions [7].

Several sugar moieties are present at the cell surface. They can be defined by their specific lectin binding: during the implantation phase, there is an important L-fucose (Fig. 3) expression, while the presence of sialic acid and 3-fucosyl-N-acetyl lactosamine is reduced (Tables 1, 2); N-acetyl-D-galactosaminyl residues appear at ovulation and are expressed throughout the entire luteal phase. Interest in the mapping of lectin binding of the surface epithelium has been further amplified by the study of possible changes induced by oral contraceptive agents or intrauterine devices (IUD). Preliminary data from our group (Hustin et al., to be published) suggest that both for IUD-bearing patients and women under oral contraception, the presence of fucose residues is strongly reduced, if not totally abolished,

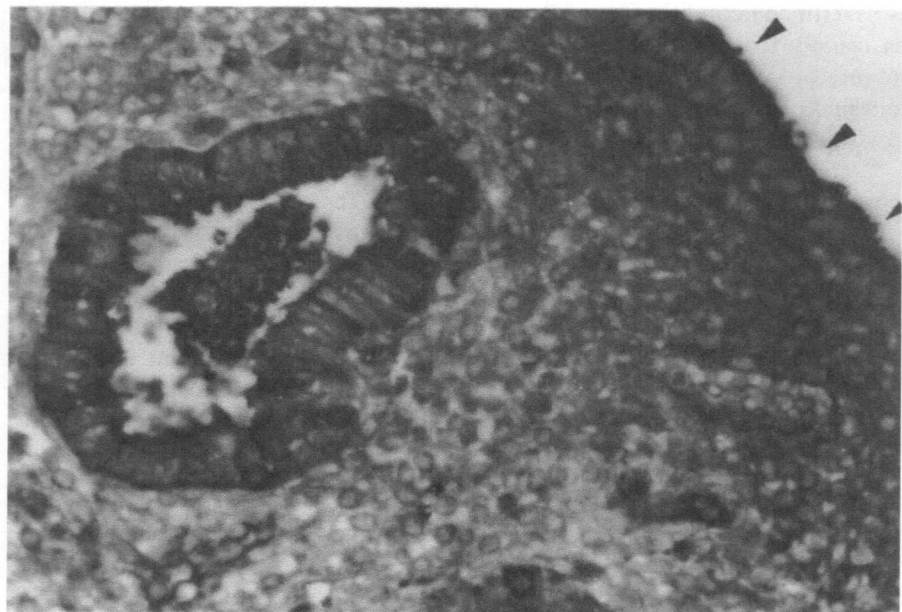

Fig. 3. L-Fucose residues expressed by *Ulex europaeus* binding at the gland cell, surface epithelium (*arrows*), and endothelial cell levels (mid-secretory phase). Lectin binding followed by immunostaining with specific anti-UEA antibody. Avidin, biotin, peroxidase with 3,3'-diaminobenzidine, ×250

Table 1. Lectins or antibodies used and their sugar moiety expression

Lectin or antibody	Main carbohydrate specificity
Concanavalin A (con A)	α-D-Mannosyl
	α-D-Glycosyl
Dolichos biflorus (DBA)	N-Acetyl-α-D-galactosaminyl
Peanut lectin (PNA)	β-D-galactosyl
Soybean lectin (SBA)	N-Acetyl-D-galactosaminyl
	α-D-Galactosyl
Ulex europaeus lectin (UEA)	α-L-Fucosyl
Wheat germ lectin (WGA)	N-Acetyl-β-D-glucosaminyl
Leu M1 (Ab)	3-Fucosyl-N-acetyllactosamine (CD15)

during the nidation window. The significance of this finding obviously awaits further confirmation. It could point to a defective status of epithelial preparation for blastocyst adhesion. Concanavalin-A-binding identifies glucosyl and mannosyl residues and is expressed at the surface epithelium level in the late luteal phase (Hustin et al., in press). Simultaneously, the presence of a heat shock 24K protein is maximally expressed at the surface epithelium level, while it decreases dramatically within the glands [8].

Table 2. Lectin binding to surface epithelium during the secretory phase

Sugar moiety	Lectin	Early secretory phase	Implantation	Late secretory phase and gestation
α-D-Mannosyl α-D-Glycosyl	Con A	−	−	+ +
N-Acetyl-α-D-galactosaminyl	DBA	+	+(+)	+ +
β-D-Galactosyl	PNA	+	+	+
N-Acetyl-D-galactosaminyl α-D-Galactosyl	SBA	+	+	+
α-L-Fucosyl	UEA	−	+ +	+
N-Acetyl-β-D-glucosaminyl	WGA			
(CD15) 3-Fucosyl-N-acetyllactosamine	Leu M1	+	+ −	+

Several cell surface-associated epitopes are currently being investigated [9–11]. They are of a considerable molecular weight, with sialic acid as a major part of the carbohydrate moiety. The presence of sialic acid at the apex of surface cells decreases during the luteal phase; this finding is consistant with our observation of diminished wheat germ lectin (WGA; *Triticum vulgaris*) binding at the surface epithelium level during mid and late secretory phases (WGA binds to N-acetyl-β-D-glucosaminyl residues, most of which belong to sialic acid molecules). It thus appears that during the early and mid secretory phase, the surface epithelium presents with subtle but important apical changes, with maximal expression of 24K protein and fucose residues, while other molecules are reduced or disappear.

Gland Tissue

Except for the synthesis of glycogen, which is the hallmark of early secretory phase, it is intriguing to note that most epithelial productions (secretions) are maximally or only expressed during the late secretory phase, that is, after implantation has occurred. There is one exception, however. During a very limited period (days 4 and 5 post ovulation), the epithelial cells of the tubes display a marked immunopositivity at the lateral and basal cell domains for an antibody which has been shown to recognize the enzyme γ-glutamyl transpeptidase [12] (Fig. 4).

Lectin binding to gland cells follows a pattern somewhat different from that of surface epithelium. If galactose and N-acetyl-galactosaminyl residues are similarly demonstrated during the implantation period, fucose is usually absent.

In the late secretory phase, concanavalin A is markedly bound to the cytoplasm of epithelial cells. True secretory processes are of course present. It seems that glandular epithelium proceeds via a synthetic and storage phase with subsequent release of products during the late secretory phase.

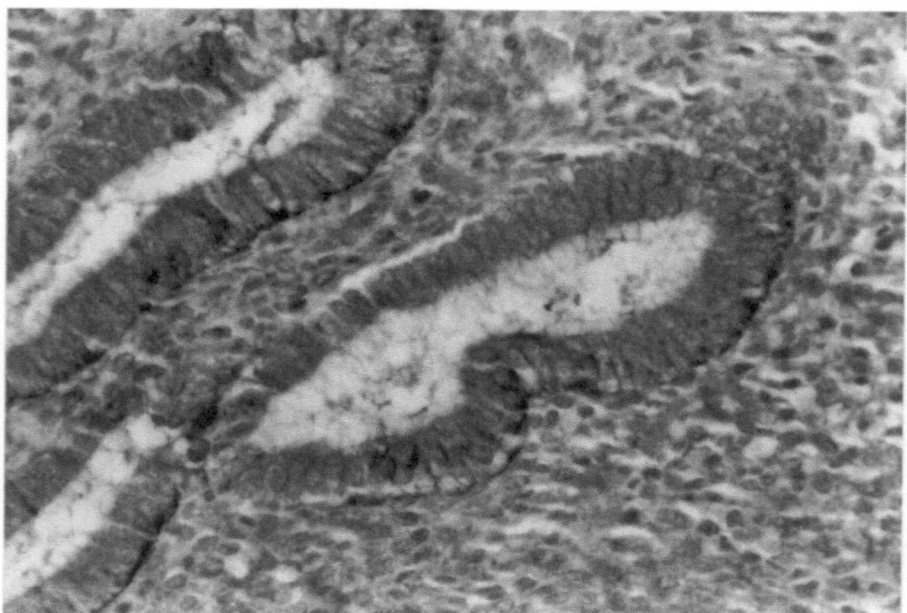

Fig. 4. Positive immunostaining for an antigen related to γ-glutamyl transpeptidase at the lateral and basal cell domains. Early secretory phase. Polyclonal antibody followed by Avidin, biotin, peroxidase with 3,3'-diamino-benzidine, ×250

Apart from glycoproteins such as those defined by the monoclonal antibody D9B1 [11] and which contain large amounts of sialic acid, pregnancy-associated endometrial α_2 globulin (α_2-PEG) is the major secretory product which has been characterized so far [13]. It is a dimeric glycoprotein [14], most probably similar to PP14. Its secretion begins at the time of implantation, increases rapidly, and remains important during the first 12 weeks of gestation. It has a strong homology with the β-lactoglobulin family and possibly also with uteroglobulin. It could exert a local immunoprotection of the embryo and also, like the β-lactoglobulin, it could bind retinol and thus be involved in the transport of this molecule from the endometrium to the trophoblast and to growing embryo, where it is essential for cell growth and organ differentiation [14].

Endometrial Stroma

Endometrial stroma was recognized early by its unique transformation, i.e., decidualization, under progesterone influence. This event begins in the late luteal phase and is not complete until several days later. Decidual cells which are obviously not epithelial (they possess only vimentin and desmin intermediate filaments) not only have the ability to synthetize matrix proteins, but are also able to excrete true secretory products.

A very interesting property of decidual cells concerns the production around them in a "nest fashion" of different matrix proteins (laminin, fibronectin, collagen IV, entactin, heparan sulfate proteoglycan, etc.) together with the surface expression of receptors for those matrix proteins especially laminin [10]. At a distance from individual cells, one can identify collagens I, III, and V, but it appears that collagen VI progressively disappears during the luteal phase. As this molecule is thought to link major fibrils, its absence would allow more water access (edema) and easier migratory cell penetration. Given this observation, free passage of water, ions, and various molecules through the decidua is quite possible.

Decidual cells produce prolactin and α_1-PEG. Decidual prolactin never appears before days 23–24 of the cycle. Its production is considerably increased with extension of the decidualization process [15]. Recent studies have demonstrated, however, that this prolactin is essentially in a glycosylated form which seems to be much less potent than "normal human prolactin (hPRL)." It has in fact reduced lactogenic activity and poor binding to PRL receptors [16]. Decidual prolactin is obviously linked to the presence of amniotic fluid prolactin.

Passage of decidual PRL to amniotic fluid most probably involves a free interstitial passage. It has been suggested that decidual PRL is involved in autocrine and paracrine effects. We have demonstrated that pretreatment with magnesium chloride at high molarity solubilized a majority of decidual surface receptors for PRL and decreased accordingly the intensity of peripheral (membrane bound) PRL immunopositivity [17].

α_1-PEG is produced, albeit in small quantities, by mid and late secretory endometrium. The secretion of this protein increases dramatically during the process of decidualization and particularly during the first weeks of pregnancy [14]. It has been demonstrated to be an insulin growth factor-binding protein (IGF-BP) probably similar to that found in amniotic fluid (PP12). During the first trimester of pregnancy, it is the major endometrial product. IGF-BP must act in autocrine and paracrine manners. In particular [18], it might compete with IGF-1 at the trophoblast receptor level and inhibit the IGF-1 action. Of considerable interest is the real possibility that IGF-BP production could be delayed until largely after implantation. This decidual protein could then be involved in the regulation of trophoblastic extension within the gestational endometrium.

Lastly, it must be noted that decidual cells produce diamine oxidase (polyamine oxidase) well after implantation. This enzyme might be produced as a protection against potentially harmful polyamines coming from the growing embryo [14].

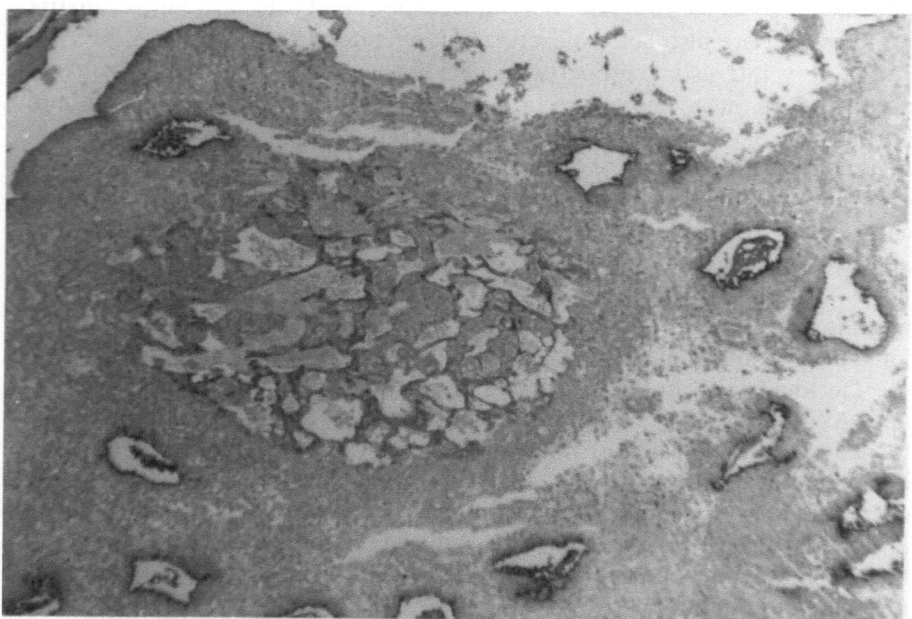

Fig. 5. Early implantation (day 12 post-ovulation). PNA binding demonstrating galactose residues at the surface epithelium and gland level on the one hand, and at the primary lacunar trophoblast on the other. Lectin binding followed by immunostaining with specific anti-PNA antibody. Avidin, biotin, peroxidase with 3,3'-diaminobenzidine, ×65

Implantation – Trophoblast-Endometrium Relationship

The Process of Implantation

Implantation is said to have occurred when the blastocyst has established a close contact with the maternal organism. In the preparation to implantation, three important factors will not be discussed. They concern the transport of the blastocyst, its orientation with the trophoectoderm pole facing the uterine luminal epithelium, and the hatching process (dissolution of the zona pellucida) [19].

The nidation process can be divided in three stages: it begins with fixation-adhesion of the blastocyst, then penetration of the trophoectoderm cells between epithelial cells occurs, and lastly intrusion in the subepithelial zone is effective. Fixation-adhesion of the blastocyst to the surface epithelium is in some way a biologic paradox. There is no other example of epithelial cells shown to adhere by their apical poles. As we discussed earlier, there is a reduction in the electronegative charge of the apex of surface cells during the nidation window [7]. The presence of fucose, galactose, and acetylgalactosaminyl residues at the epithelial cell level surely plays a role. Primitive trophoblast is essentially coated with galactose residues (Fig. 5). Up to now, nothing is known concerning the real mechanisms of adhesion. Just as enzymatic activation is probably needed for blastocyst hatching, the adhesion phenomenon is probably also dependent on an enzymatic molecular rearrangement at the surfaces of both cell types.

After adhesion, the earliest penetration events begin. They have been beautifully illustrated by Lindenberg et al. [20, 21] in elegant in vitro experiments of endometrial epithelial cell and blastocyst cocultures. The first phenomenon is a displacement of epithelial cells by trophoblasts. The trophoblastic cells send long protrusions between epithelial cells. There is an obvious rupture of lateral tight junctions. Interestingly enough, the adjacent endometrial cells show increased cellular activity. One may postulate that this finding could reflect endocytosis and possibly transfer of embryonic signals. Immediately afterwards, the contact between primary trophoblast and the subepithelial zone is established.

Now it must be borne in mind that one of the earliest modifications of the endometrium after fecundation is a tremendous increase in vascular permeability in response to a signal produced by the blastocyst. There is some evidence that platelet activating factor (PAF) is one of such messengers with unique vasodilating properties [22]. It is, however, not clear whether PAF acts alone or also triggers the local production of prostaglandins by the endometrial epithelial cells [23].

Early embryos are fully capable of producing collagen IV protease and of degrading basement membrane [24]. This together with the local edema induced by prostaglandin activation allows motile cells such as trophoblasts to enter the endometrial stroma completely, gliding in a first stage along the widely separated fibrils. Strong evidence exists that trophoblasts attach preferentially to laminin-coated surfaces [25, 26], hence to basement membrane components surrounding decidual cells [10]. Queenan et al. [27] have shown that trophoblasts express proteolytic activities. Various protease inhibitors have been simultaneously demonstrated both in extravillous trophoblasts and endometrial cells [28–30]. A true possibility exists that the depth of trophoblast penetration is regulated, at least partly, by the state of equilibrium between trophoblastic proteases and decidual antiproteases.

Trophoblast-Endometrium Relationship

Not including the endometrial blood vessels which have a distinct temporal relationship with growing trophoblast (this is discussed in Chap. 6), the uterine mucosa comprises four types of cells which are eventually involved in the contact with the early embryo-placental unit and, in fact, almost exclusively with the primary trophoblast which completely surrounds the

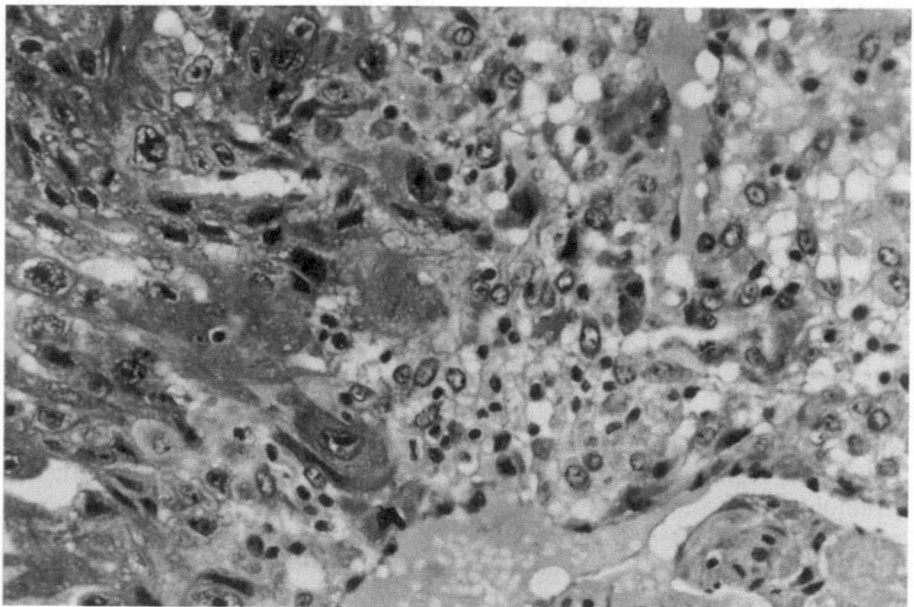

Fig. 6. Early implantation (day 12 post-ovulation). The primary trophoblast (*left*) is in close contact with the different cells of the stroma. No gland is present in this field. Hematoxylin and eosin, ×400

embryo (Fig. 6). These cells are the epithelial cells of the surface epithelium, the epithelial gland cells, the stromal (decidual) cells, and the stromal lympoid cells. The latter are discussed in Chap. 9. According to all the evidence, it seems that surface epithelial cells have the important but limited task to:

1. Present a surface suitable for blastocyst adhesion
2. Separate from each other during the initial stage of implantation and to multiply in order to cover tightly the area where the blastocyst is growing

Probably too, they must carry messages (signals) from the adhering blastocyst to the different compartments of the mucosa. Epithelial gland cells are apparently not modified in their structure even when gland lumina have been opened by growing trophoblasts. Their secretory products will serve as fuel for the growing embryo (glycoproteins) or as carrier proteins for important molecules (α_2PEG for retinol) [14].

More important and complex relations seem to exist between trophoblast and decidual cells. Both PRL and IGF-BP, which are produced by decidual cells, seem to act in a paracrine manner. Handwerger et al. [31] suggest that IGF-1 may exert a positive control on decidual PRL production together with relaxin and a hypothetical placental PRL release factor. Arachidonic acid (originating from fetal membranes) could inhibit this production.

Decidual IGF-BP could compete with IGF-1 for its specific trophoblastic receptors and thus function as a regulator of trophoblastic function [14] and growth. Bell [14] suggests in this respect that true decidualization which controls the highest production of IGF-BP is probably delayed until after complete implantation. Careful histologic studies of early implantation seem to confirm this opinion: in a review of hysterectomy specimens with early pregnancy in situ from the collection of the late Professor Boyd (Cambridge, UK) (this study has been possible thanks to the kindness of Prof. G. Burton, Department of Anatomy, Cambridge, UK), we were impressed by the fact that, under the actively growing trophoblast, the mucosa was still not really decidualized, but that uterine glands were packed against each other with copious amounts of secretion (Fig. 7). This apparent delay in (local?) decidualization could mean that trophoblast growth is controlled only at a later date when penetration within the decidua is sufficient and IGF-BP production has reached its secretory peak.

On the trophoblastic side, hCG has been demonstrated very early. Conflicting results have been presented as to its paracrine action. It appears [32] that in vitro hCG can lower the overall level of decidual protein synthesis. There does not seem, however, to be any effect on the release of decidual prolactin.

Conclusion

The human endometrium displays characteristic morphologic changes during the postovulatory phase. The nidation window is situated between day 5 and day 7 following the LH surge. During this period, important modifications take place at the surface epithelium level in order to facilitate adhesion of the blastocyst.

Most endometrial secretions begin at the end of this period. Glycoprotein production by gland cells is associated with overall endometrial secretions; α_2-PEG is more unique in that it is produced continuously by glands from the late secretory period to the end of the first trimester and is excreted via gland lumina to the amniotic fluid. It seems to be involved in the transfer of retinol which is needed for growth and differentiation.

The most important changes occur, however, after implantation and affect essentially the endometrial stroma. The decidualization process is associated with a remodeling of the matrix fibrillar network, allowing free access and exit of molecules via the interstitial route and of migratory (trophoblastic) cells. In addition, decidualization brings two important products: PRL, the function of which is not totally elucidated (osmoregulation

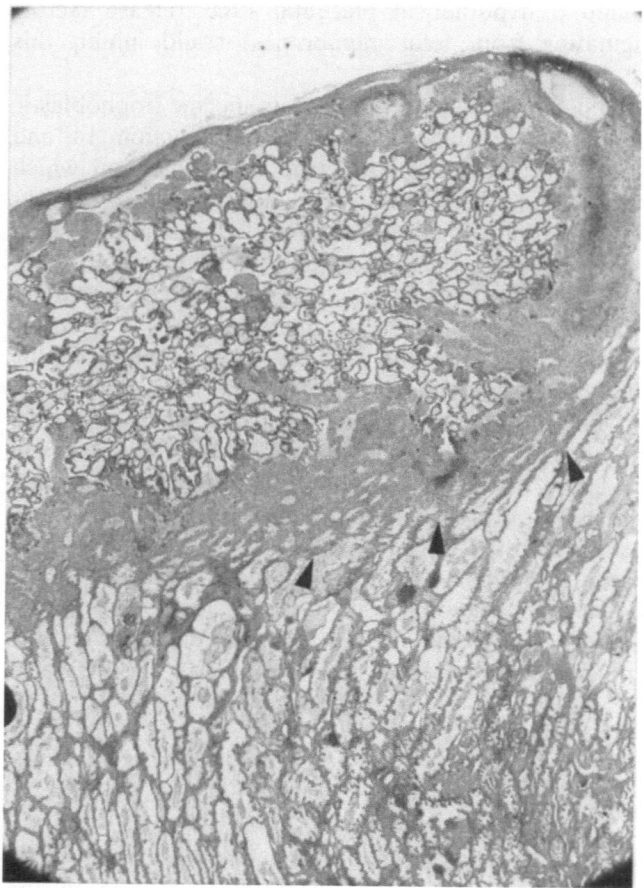

Fig. 7. Specimen H710 (from the collection of the late Prof. Boyd, by courtesy of Prof. G. Burton (Cambridge, UK). Hysterectomy with pregnancy in situ (4 weeks). Lateral section with the edge of the placenta. It is noteworthy that the decidualization process is limited to the close vicinity of the trophoblastic shell (*arrows*). Underneath, endometrial glands predominate and are filled with secretion. Hematoxylin and eosin, ×25

of the amniotic fluid?); and α_1-PEG, which is maximally produced during the first trimester and which is supposed to play an important (paracrine) role in the control of penetration of the endometrium by the implanting trophoblast. A summary of our present knowledge is tentatively detailed in Table 3.

Lastly, it must be stressed that the finding of several endometrial proteins in the amniotic fluid suggests that they are probably essential to the embryo. Aplin [10] has recently suggested that it could be evidence that during the 1st trimester, human placentation would be at least partly

Table 3. Endometrial productions

	Early secretory phase and implantation	Late secretory phase

Surface epithelium	Glycocalyx changes
	                     N-Acetyl-galactosamine
	  • Fucose
	  • ↓ Electronegative charge
	γ-Glutamyl transpeptidase
		
		  • Mannose
		  • Glucose

Gland epithelium	γ-Glutamyl transpeptidase α_2-PEG PP5 Glycoproteins (+) Prostaglandins	 α_2-PEG PP5 Glycoproteins (D9B1) (+++) Prostaglandins
Stroma	Prostaglandins Protease inhibitors Growth factors (?) Matrix proteins Fibrillar components (collagens I, III, V)	(Prostaglandins) Protease inhibitors α1-PEG (IGF-BP) Prolactin (glycosylated) PAPP-A Growth factors Collagen IV – protease Matrix proteins and receptors Fibrillar components

deciduochorial; we would strongly support this hypothesis (see Chap. 6). The importance and duration of this deciduochorial period are, however, still not precise.

Implantation must be regarded as a physiologic event. It is therefore not indicated to speak of trophoblastic invasion. The term "invasion" refers to destructive and out-of-control capabilities of cells (almost always cancer cells) to infiltrate and colonize a given tissue in all planes of space. The arrival of the conceptus within the mucosa proceeds through successive well-recognized steps. Early penetration through the surface epithelium proceeds by simple dislodgement of epithelial cells. These will eventually multiply and entirely cover the penetration site. Once in the mucosa, the conceptus is enclosed in a tissue where connective components have been modified in such a way as to offer:

1. A maximum of anchoring sites (i.e., laminin coating of stromal cells) constituting a specialized matrix structure resistant to breakdown
2. A reduced fibrillar network allowing this progressive entrance of trophoblast

It thus seems that the first stages of nidation do not involve important maternal cell death. When the early embryo begins to grow and occupies more and more space, and only then, do some destructive processes occur. Proteolytic activity is expressed by the trophoblast and probably also some direct molecular transfer takes place to stromal and gland cells which will

Table 4. Control of trophoblast penetration at implantation

Facilitation	Inhibition
Glycocalyx changes at the apical pole of surface epithelium (early and mid-secretory phase)	Glycocalyx changes at the apical pole of surface epithelium (late secretory phase – early gestation)
Hydrolysis of collagen IV (by blastocyst)	
Increased water content of decidua (PAF + PG)	
Laminin coating of decidual cells (attachment)	
Hydrolysis of collagen VI (within decidua)	Competition of IGF-BP with IGF-1 for trophoblastic receptors (growth control)
Proteolytic properties of trophoblast	Decidual antiproteases (α_1-antitrypsin) (α_1-antichymotrypsin)
Lack of placental alkaline phosphatase expression on early trophoblast (?)	
Presence of lymphoid suppressor cells at the implantation site	

die. At this stage, some venous lakes are opened, resulting in the formation of trophoblastic lacunae which contain maternal blood cells.

It is obvious that in opposition to true invasive processes, some control must exist through a harmonious endometrium-trophoblast relationship: once it is reached, the depth of trophoblastic penetration is limited. Clearly, in the vast majority of gestations, trophoblast does not extend further (except for its intravascular component) than the narrow rim of basal layer of endometrium. Some speculations could be made in this respect as to the pathogenesis of placenta accreta.

It is also obvious that in ectopic pregnancies there is no preparation whatsoever of the implantation site. The relative ease with which a conceptus can implant in such abnormal places might suggest that there are few real requirements (see Chap. 17). It might possibly indicate that the human endometrium is perhaps not ideally suited for this essential function and that local conditions would preclude success unless special conditioning is provided. Trophoblastic penetration must thus be under positive or negative control. Most endometrial and trophoblastic properties facilitate implantation (Table 4). After it has occurred, the surface epithelium properties are modified and will probably resist any new apposition. The expression of antiproteases within the endometrium (the decidua?) modulate the proteolytic activites of primary trophoblast.

Lastly, there are the indispensable growth factors: one is IGF-1 which is regulated by the competition of endometrial IGF-BP for trophoblastic IGF receptors. Tumor necrosis factor (TNF) has recently been identified in supernatants of placental and decidual cultures [33]. As this cytokine is able to promote production of growth factors as well as to stimulate fibroblasts and endothelial cell stimulation, it is postulated that TNF might also play a role in the endometrial-trophoblast mutual control. Villous formation is initiated from the lacuna stage onwards. The early placental bed will extend only laterally: this phenomenon is completed at 40 weeks. It is made possible only by the simultaneous and tremendous growth and hypertrophy of the underlying myometrium.

References

1. Edwards RG (1988) Human uterine endocrinology and the implantation window. In: Jones HW, Schrader CH (eds) In vitro fertilization and other assisted reproduction. Ann NY Acad Sci 541:445–454
2. Critchley HOD, Buckley CH, Anderson DC (1990) Experience with a physiological steroid replacement regimen for the establishment of a receptive endometrium in women with premature ovarian failure. Br J Obstet Gynaecol 97:804–810
3. Cornillie FJ, Lauweryns JM, Brosens IA (1985) Normal human endometrium. An ultrastructural survey. Gynecol Obstet Invest 20:113–129
4. Dockery P, Li TC, Rogers AW, Cooke ID, Lenton EA, Warren MA (1988) An examination of the variation in timed endometrial biopsies. Hum Reprod 3:715–720
5. Dockery P, Warren MA, Li TC, Rogers AW, Cooke ID, Mundy J (1990) A morphometric study of the human endometrial stroma during the periimplantation period. Hum Reprod 5:112–116
6. Davies MC, Anderson MC, Mason BA, Jacobs HS (1990) Oocyte donation: the role of endometrial receptivity. Hum Reprod 5:862–869
7. Jansen RPS, Turner M, Johanisson E, Landgren BM, Diczfalusy E (1985) Cyclic changes in human endometrial surface glycoproteins: a quantitative histochemical study. Fertil Steril 44:85–91
8. Manners CV (1990) Endometrial assessment in a group of infertile women on stimulated cycles for IVF: immunohistochemical findings. Hum Reprod 5:128–132
9. Aplin JD, Seif MW (1987) A monoclonal antibody to a cell surface determinant in human endometrial epithelium: stage-specific expression in the menstrual cycle. Am J Obstet Gynecol 156:250–253
10. Aplin JD (1989) Cellular biochemistry of the endometrium. In: Wynn RM, Jollie WP (eds) Biology of the uterus, 2nd edn. Plenum, New York, pp 89–119
11. Smith RA, Seif MW, Rogers AW, Li TC, Dockery P, Cooke ID, Aplin JD (1989) The endometrial cycle: the expression of a secretory component correlated with the luteinizing hormone peak. Hum Reprod 4:236–242
12. Francois C, Calberg-Bacq CM, Gosselin L, Kosma S, Osterrieth PM (1979) Identification, partial purification and biochemical characterization of γ-glutamyltranspeptidase present as a membrane component in skinned milk and multi fat-globule membranes and in mammary tumour virus from the milk of infected mice. Biochem Biophys Acta 567:106–115
13. Waites GT, Bell SC, Walker RA, Wood PL (1990) Immunohistological distribution of the secretory endometrial protein, pregnancy-associated endometrial α_2-globulin, a glycosylated β-lactoglobulin homologue, in the human fetus and adult employing monoclonal antibodies. Hum Reprod 5:105–111
14. Bell SC (1988) Secretory endometrial/decidual proteins and their function in early pregnancy. J Reprod Fertil [Suppl] 36:109–125
15. Maslar IA, Riddick DH (1979) Prolactin production by human endometrium during the mentrual cycle. Am J Obstet Gynecol 135:751–754
16. Heffner LJ, Iddenden DA, Lyttle CR (1986) Electrophoretic analyses of secreted human endometrial proteins: identification and characterization of luteal phase prolactin. J Clin Endocrinol Metab 62:1288–1295

17. Hustin J, Pereira-Leite L, van Cauwenberge JR, Franchimont P (1986) La prolactine dans l'unité foeto-placentaire: données immunohistochimiques. Acta Med Port 7:7–13
18. Bell SC (1989) Decidualization and insulin-like growth factor (IGF) binding protein: implications for its role in stromal cell differentiation and the decidual cell in haemochorial placentation. Hum Reprod 4:125–130
19. Sathananthan H, Bongso A, Ng SC, Ho J, Mok H, Ratnam S (1990) Ultrastructure of preimplantation human embryos co-cultured with human ampullary cells. Hum Reprod 5:309–318
20. Lindenberg S, Hyttel P, Lenz S, Holmes PV (1986) Ultrastructure of the early human implantation in vitro. Hum Reprod 1:533–538
21. Lindenberg S, Hyttel P, Sjogren A, Greve T (1989) A comparative study of attachment of human bovine and mouse blastocysts to uterine epithelial monolayer. Hum Reprod 4:446–456
22. Alecozay AA, Casslen BG, Riehl RM, Deleon FD, Harper MJ, Silva M, Nouchi TA, Hanahan DJ (1989) Platelet-activating factor in human luteal phase endometrium. Biol Reprod 41:578–586
23. Kennedy TG (1983) Embryonic signals and the initiation of blastocyst implantation. Aust J Biol Sci 36:531–543
24. Puistola V, Rönnberg L, Martikainen H, Turpeeniemi-Hujanen T (1989) The human embryo produces basement membrane collagen (type IV collagen) – degrading protease activity. Hum Reprod 4:309–311
25. Loke YW, Gardner L, Burland K, King A (1989) Laminin in human trophoblast-decidua interaction. Hum Reprod 4:457, 453
26. Earl U, Estlin C, Bulmer JN (1990) Fibronectin and laminin in the early human placenta. Placenta 11:223–231
27. Queenan JT Jr, Kao LC, Arboleda CE, Ulloa-Aguirre A, Golos TG, Cines DB, Strauss JF III (1987) Regulation of urokinase type plasminogen activator production by cultured human cytotrophoblasts. J Biol Chem 262:10903–10906
28. Marshall RJ, Braye SG (1987) Immunohistochemical demonstration of α-1-antitrypsin and α-1-antichymotrypsin in normal human endometrium. Int J Gynecol Pathol 6:49–54
29. Earl U, Morrison L, Gray C, Bulmer JN (1989) Proteinase and proteinase inhibitor localization in the human placenta. Int J Gynecol Pathol 6:132–139
30. Feinberg RF, Kao LC, Haimowitz JE, Queenan JT Jr, Wun TC, Strauss JF III, Kliman HJ (1989) Plasminogen activator inhibitor types 1 and 2 in human trophoblasts. Lab Invest 61:20–26
31. Handwerger S, Golander A, Richards R, Thrailkill K, Jorgensen V, Harman I, Grundis A (1989) Paracrine and autocrine factors involved in the regulation of the release of human decidual prolactin and human placental lactogen. In: Genbacev O, Klopper A, Beaconsfield R (eds) Placenta as a model and a source. Plenum, New York, pp 129–140
32. Vicovac L, Vuckovic M, Genbacev O (1987) Does human trophoblast affect decidual cell function during gestation? Troph Res 2:85–94

Anatomy and Physiology
of Placental and Embryonic Development

3 Early Human Placental Morphology

E. Jauniaux, G.J. Burton, and C.J.P. Jones

Introduction

The first detailed descriptions of placental anatomy were published by Leonardo da Vinci and by Vesalius at the beginning of the sixteenth century [1]. Since that time and until the beginning of the twentieth century, the most significant contribution to knowledge of human implantation, placental development and vascularization was provided by William and John Hunter during the eighteenth century (Fig. 1). Their experiments also demonstrated clearly that maternal and fetal bloodstreams are separated by the placental barrier and are, therefore, not anastomosed end to end inside the placenta.

During the last one hundred years, the placental morphology from an early stage of gestation to term has been studied by many workers using light microscopy and, more recently, transmission and scanning electron microscopy (TEM and SEM, respectively). The explosion of new knowledge in the field of antenatal diagnosis in the last decade has been supported by application of these basic science discoveries to clinical research. An increased understanding of placental anatomy has provided the pathophysiologic principles of many diseases affecting pregnancy. Conversely, the advent of new techniques in perinatal medicine, such as ultrasonography, magnetic resonance imaging, or Doppler velocimetry, has allowed in vivo investigations of placental structures from an early gestational stage.

This chapter illustrates and summarizes the various aspects of early placental development and morphology. Peri-implantation trophoblastic development will not be considered in this chapter. Its properties and differentiation, however are described in Chap. 1.

Development of the Early Villous Tree

General Architecture

The formation of chorionic villi begins between 13 and 15 days after ovulation, corresponding to stage 6 of embryonic development and to the end of the 4th week after the last menstrual period [1–3]. Histologically,

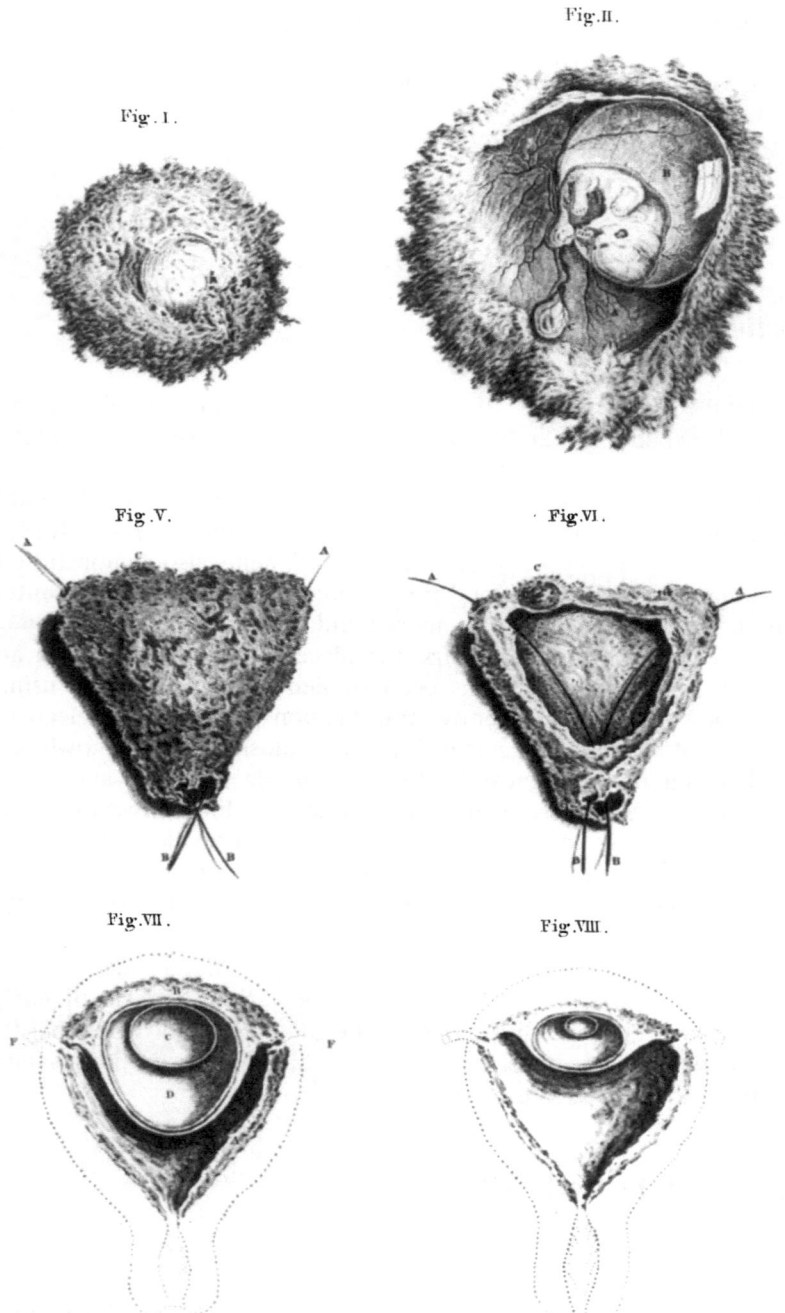

Fig. 1. Plate of William Hunter's *The Gravid Uterus* (Birmingham, 1774) illustrating early implantation in humans. (By courtesy of the King's College Library, London)

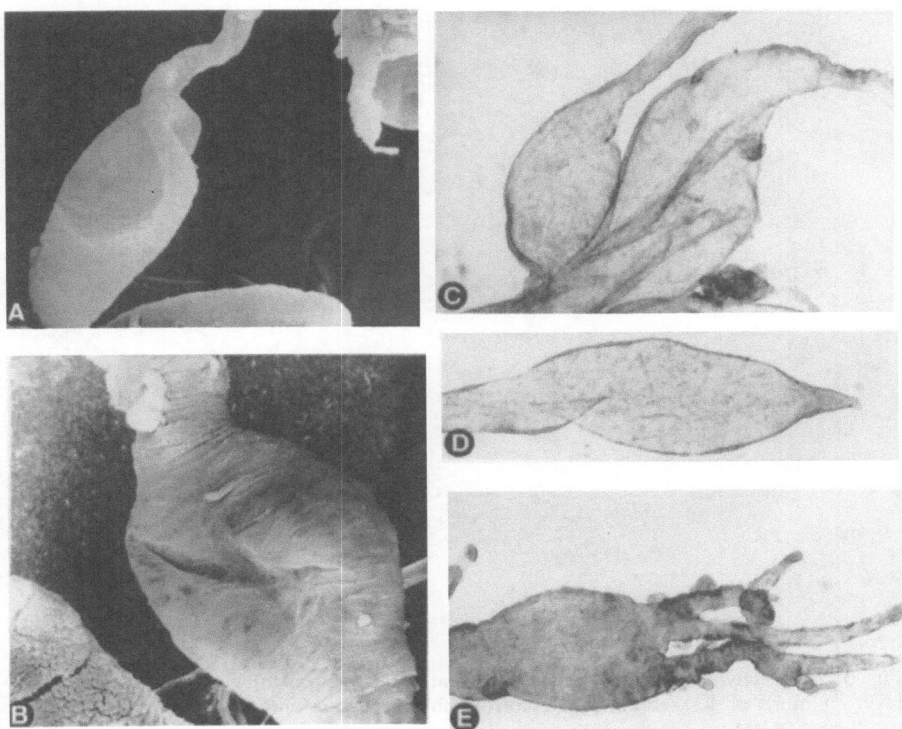

Fig. 2.A–E SEM and phase-contrast micrographs of placental villi at 6 weeks of gestation showing large trunks (**A, B**) with a few developing (branching) villi and syncytial sprouts (**C–E**). (From [7]) **A** ×100; **B** ×120; **C–E** ×25

they are classified into primary, secondary, and tertiary villi. The primary villi are composed of a central mass of cytotrophoblast surrounded by a thick layer of syncytiotrophoblast. During the following week of gestation, they acquire a central mesenchymal core from the extraembryonic mesoderm and become branched, forming the secondary villi. The appearance of embryonic blood vessels within their mesenchymal cores transforms the secondary villi into tertiary villi. At the end of the 5th gestational week, all three primitive types of placental villi can be found, but tertiary villi progressively predominate.

Reconstruction of serially sectioned villi and three-dimensional studies by SEM [4–7] have shown that by 6 weeks of gestation, the villous tree is essentially composed morphologically of long immature trunks, branching mesenchymal villi, and syncytial sprouts (Fig. 2). The mesenchymal villi are characterized by a thick syncytiotrophoblastic layer, a more or less continuous cytotrophoblastic layer, a villous core, and capillaries with lumina containing nucleated red blood cells (Fig. 3).

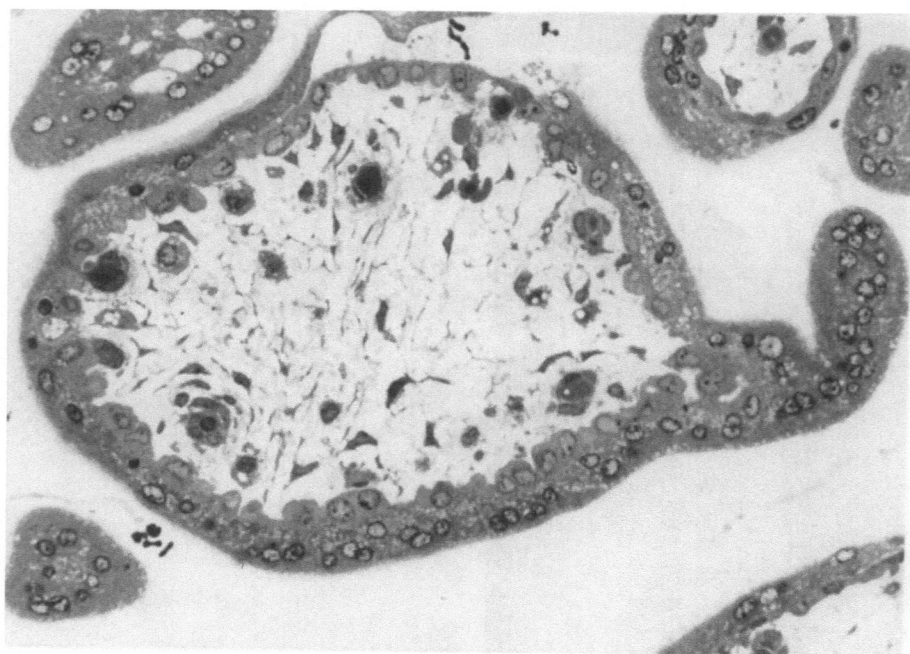

Fig. 3. Mesenchymal villi at 8 weeks of gestation showing a thick syncytiotrophoblastic layer, a more or less continuous cytotrophoblastic layer, and a villous core with few capillaries. ×100

The early villous core is composed of loosely packed mesenchymal cells organized in a delicate network of collagen fibers separated by ill-defined cavities, free of fibers, and sometimes containing macrophages or Hofbauer cells (Fig. 4). Towards the end of the first trimester, the collagen fibres become condensed, creating an intercommunicating network of fluid-filled channels within the stroma. At this stage, the Hofbauer cells are more numerous, adherent to the walls of these channels or free within (Fig. 5). These villi are known as immature intermediate villi and they represent the precursors of stem villi. They reach maximal development by the 16th week.

Up to the 10th week post-menstruation, which corresponds to the last week of the embryonic period (stages 19–23), villi cover the entire surface of the chorionic sac (Fig. 6). As the gestational sac grows during fetal life, the villi associated with the decidua capsularis, surrounding the amniotic sac, become compressed and degenerate forming an avascular shell known as the chorion laeve or smooth chorion [1, 3]. Conversely, the villi associated with the decidua basalis proliferate forming the chorion frondosum or definitive placenta. The stimulus producing the regression of an important portion of the early placenta is not known, but it has been suggested that it may be due to a difference in nutritional supply after implantation [1].

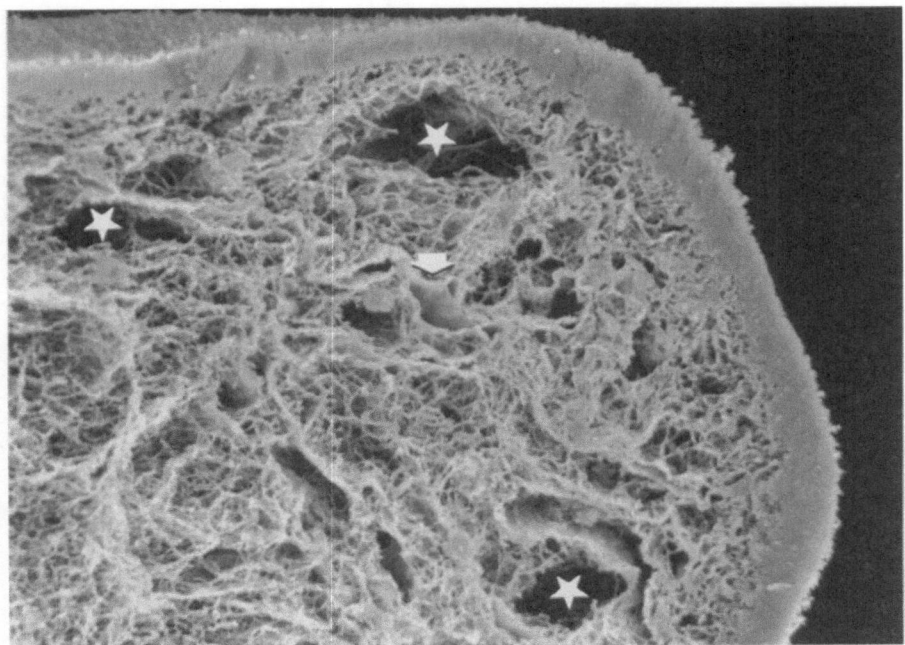

Fig. 4. Cryo-fractured preparation of a mesenchymal villus at 8 weeks of gestation showing the stromal core composed of loosely packed mesenchymal cells organized in a delicate network of collagen fibres. Spaces free of fibres (*stars*) and small blood vessels (*arrow*) can be seen at this stage of development. (From [9]) ×440

The syncytial sprouts are the most numerous villous offshoots of the early placenta [1]. They measure 10–20 µm in diameter and consist of densely packed syncytial nuclei (Fig. 7). The sprouting activity of the trophoblast as revealed by phase-contrast examination is maximal at the end of the embryonic period when the chorion frondosum forms (Fig. 8) and then decreases towards term [7]. However, syncytial sprouts can still be observed in the mature placenta, and it has been suggested that these sprouts are an indicator of placental growth capacity and well-being [1, 8]. During villous development, the sprouts are invaded first by cytotrophoblast and then by mesenchymal elements [7]. For unknown reasons, a proportion of syncytial sprouts are not invaded by cytotrophoblast [9], and their attachment to the villous surface becomes attenuated (Fig. 9). Their stalk may break and the sprout become detached from the villous tree into the intervillous space. Syncytiotrophoblastic sprouts are often deported to the pulmonary vascular bed, where they elicit neither inflammatory changes nor a cellular immune response [1, 10]. The physiologic phenomenon of sprout detachment may play an important role in the immunologic process of pregnancy tolerance.

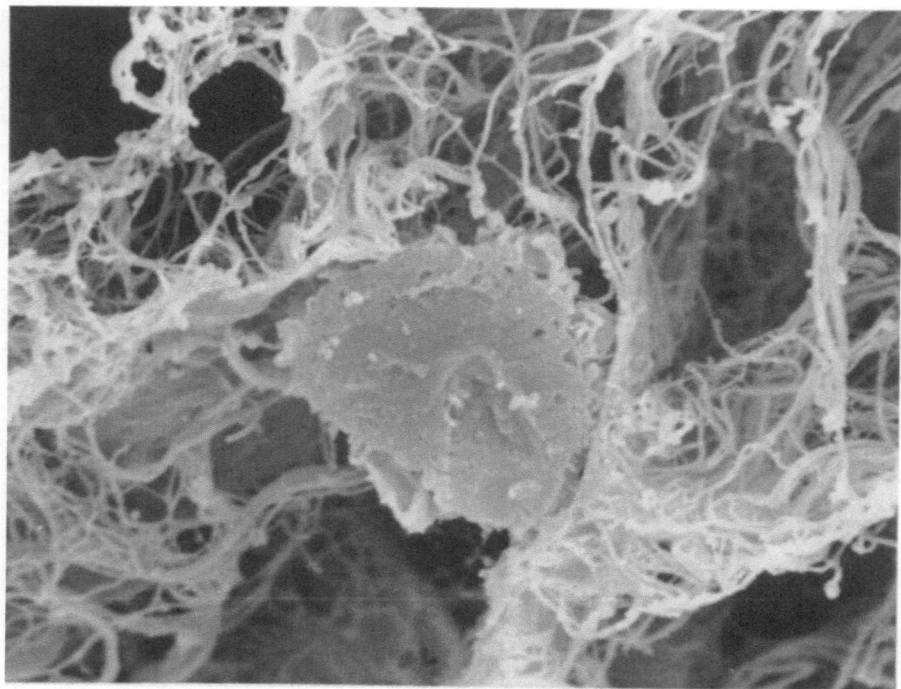

Fig. 5. Hofbauer cell lying within a stromal channel of a mesenchymal villus at 8 weeks of gestation. ×3600

    Mesenchymal villi are continuously formed from trophoblastic sprouts throughout gestation and are the basis for the growth and development of all villous types [6]. Only at the start of the third trimester do the mesenchymal villi transform into mature intermediate villi, which marks the end of the differentiation stage and corresponds to the beginning of placental maturation. Mature intermediate villi possess a more condensed stromal core, and stromal channels are notably absent. The majority of terminal villi arise from the surface of these villi.

Development of the Villous Mesenchyma

The origin of endothelial cells and the formation of fetal villous capillaries have been a matter of controversial debate between anatomists for more than a century. In his classical work describing angiogenesis in the chick embryo, His [11] concluded that angioblasts divide to form cords or syncytial masses, the centre of which liquefies to form embryonic capillaries. Hertig [12] suggested that angioblasts differentiate from the chorionic cytotrophoblast in the presomite stage and also regarded these newly formed cells as

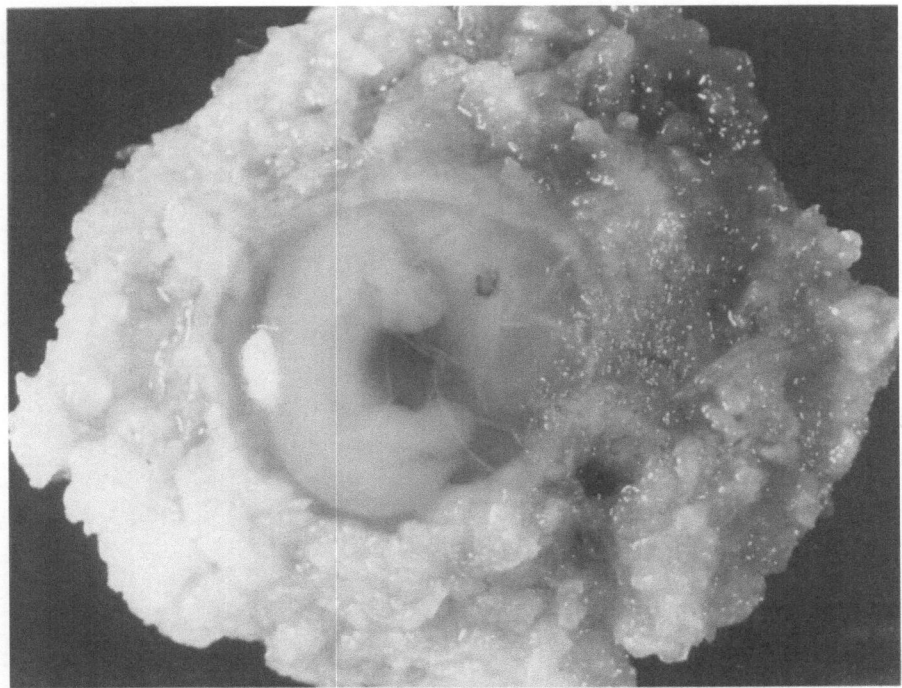

Fig. 6. Photograph of a gestational sac at 8 weeks of gestation. The chorionic villi are well developed and surround the whole periphery of sac

forerunners of vascular plexuses. According to his views, the villous circulation is established when the human embryo reaches the seven-somite stage (22 days after ovulation). These theories prevailed in general until the early 1970s.

If angioblasts were derived from the cytotrophoblast, it could only happen in the previllous stage and later differentiation would not be possible. The trophoblast layer of secondary villi is already separated from the mesenchymal core, in which the blood vessels form, by a complete basal lamina [1, 13]. In very early specimens, at a time when villi are lacking in any vascular structures, only one type of cell is present in the villous mesoblastic core. Recent studies [13, 14] have shown that mesenchymal or stellate cells progressively differentiate into stromal cells and the haemangioblastic cell cords which are the forerunners of the capillary endothelium, pericytes, and hematopoietic stem cells. The first grouping of mesenchymal cells forming haemangioblastic cords is accompanied by the formation of scattered junctions resembling early desmosomes and more extended ones resembling tight junctions. According to Dempsey [13], only pericytes show mitotic activity and are the principal progenitors of capillaries. Demir et al.

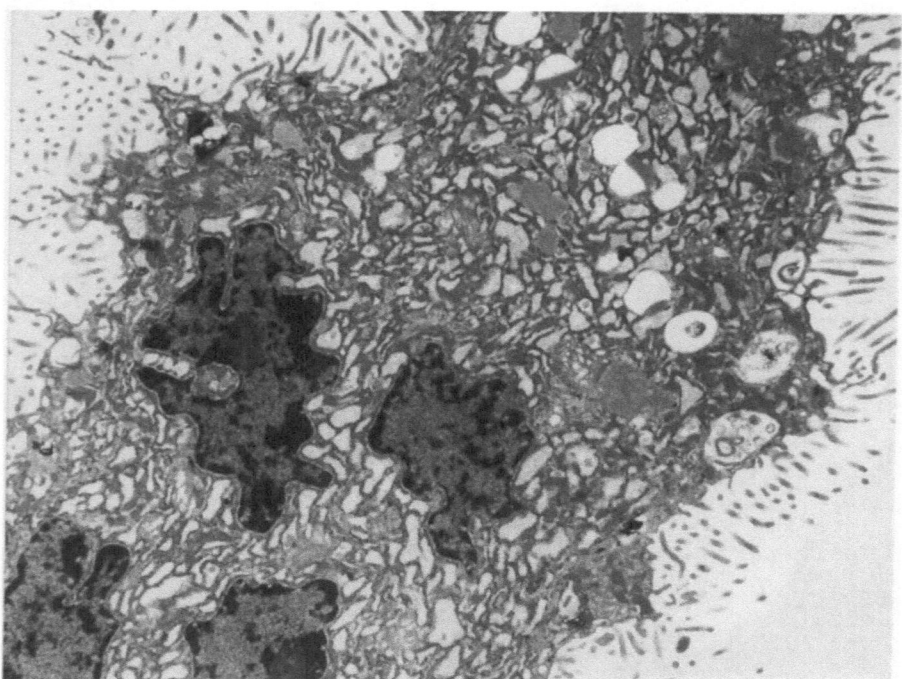

Fig. 7. Electron micrograph of a syncytial sprout at 7 weeks of gestation consisting of densely packed nuclei and organelles. ×8900

[14] have recently reported intermediate stages between mesenchymal and endothelial cells, and between mesenchymal cells and pericytes. The latter being seen not as often as one would expect if they were the main differentiation pathway. Mesenchymal cells also gradually transform into small and large reticulum cells, fibroblast, and macrophages or Hofbauer cells [14]. Intermediate immature macrophages are already present in the villous stroma before haematopoiesis and angiogenesis begin. From these findings, it seems there is little evidence of cytotrophoblastic derivatives migrating into the villous stroma, although Jones and Fox [15] have observed a population of cytotrophoblastic cells at the growing tips of young villi, streaming out from the trophoblast into the stroma to make contact with the developing capillaries. However, the contribution, if any, of these cells to angiogenesis can only be a minor one, as they are seen only whilst new primitive villi are being formed from solid trophoblastic columns [12]. Thus, precursors of all mesenchymal elements derive from extraembryonic mesoderm and infiltrate the villous core from the chorionic plate at the time secondary villi are formed.

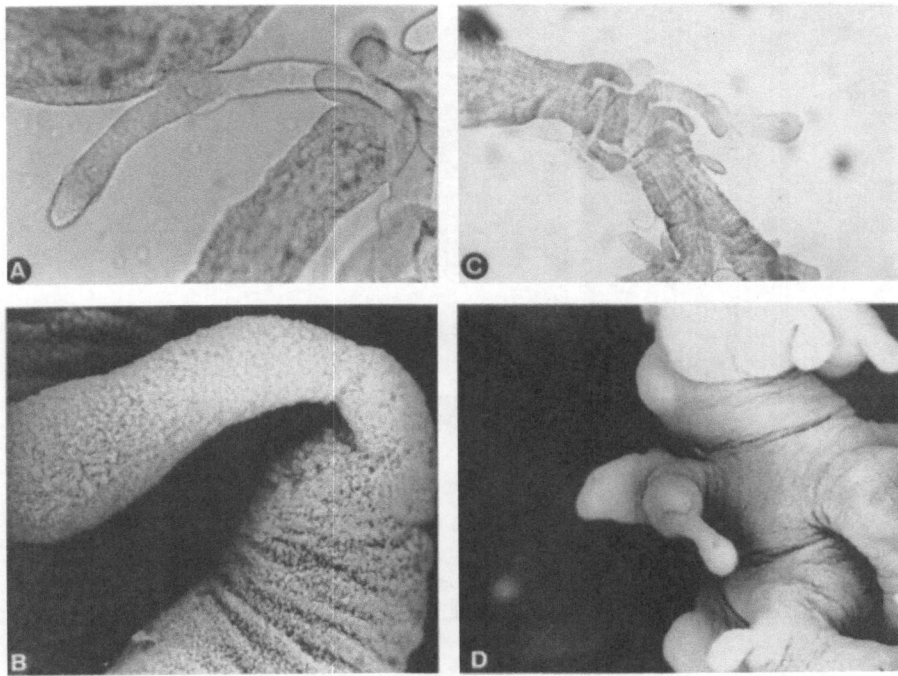

Fig. 8.A–D SEM and phase-contrast micrographs of placental villi at 10 weeks of gestation showing syncytial sprouts. The sprouts are longer (**A, B**) than earlier in gestation (Fig. 2) and more numerous (**C, D**). (From [7]) **A** ×100; **B** ×800; **C** ×100; **D** ×250

Around 21 days after ovulation (stage 10), corresponding to the end of the 5th week after the last menstrual period, the primitive heart begins to beat [3]. However, at that time and until 28 days post-ovulation there is no absolute proof of a continuous villous capillary network [14]. Placental capillary formation is, nevertheless, almost completed by mid-gestation. De novo formation of capillaries from transformation of mesenchymal cells is rare in mature placental tissue and only occurs in the few persisting mesenchymal villi [14]. In the latest stage of pregnancy, capillary growth probably results from division of endothelial cells and/or pericytes, and mitoses can be demonstrated in these cells until term. Around 6 weeks post-ovulation, a basal lamina appears around the capillaries and it is only complete at the beginning of the third trimester [14]. Fetal capillaries rapidly establish contact with the basal trophoblastic surface, protruding deeply in some places and creating primitive vasculo-syncytial membranes similar to those found at term (Fig. 10). Capillaries in intimate contact with the trophoblastic layer can be observed as early as the 6th week of gestation, and there are no hints of a centrifugal shift of the capillaries during development [7, 14].

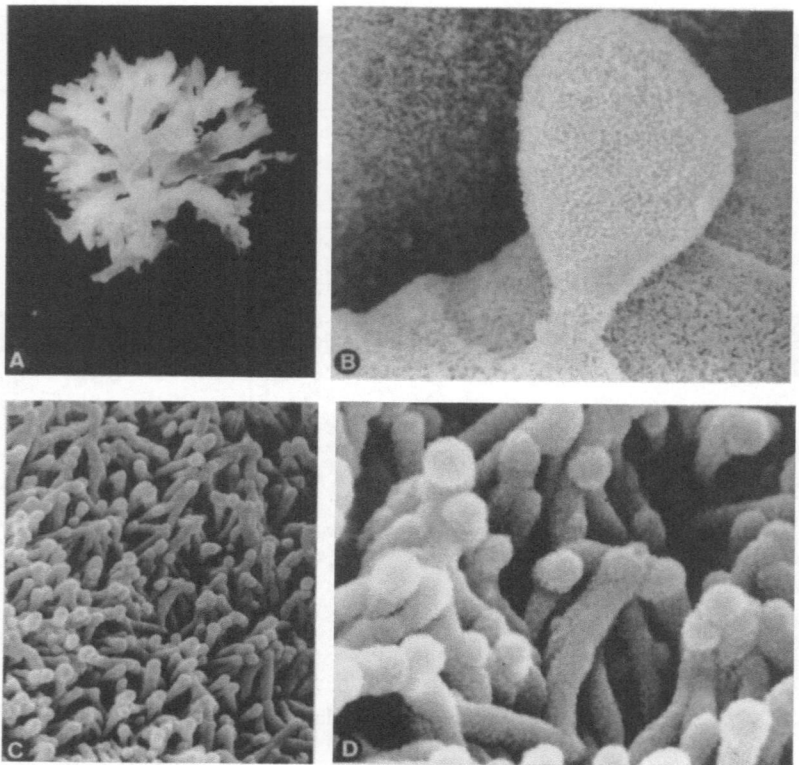

Fig. 9.A–D Macroscopic view of a villous tree at 8 weeks of gestation (**A**) and SEM micrographs of the same specimen showing a syncytial sprout with an attenuated attachment base (**B**) and surface microvilli (**C, D**). **A** ×16; **B** ×2250; **C** ×12 000; **D** ×45 000

Histomorphometric Investigations of the Villous Structures

Histomorphometric techniques allow mapping of the growth of the placental constituents and measurement of the villous area.

Methodological Approach

First-trimester specimens are usually collected after artificial termination of pregnancy performed by means of aspiration and/or curettage of the gestational sac. Consequently, the placenta is fragmented and often incomplete, and the organ volume and, therefore, the absolute quantities cannot be accurately determined during the embryonic phase of human development. The placental volume could be assessed in utero by ultrasonography

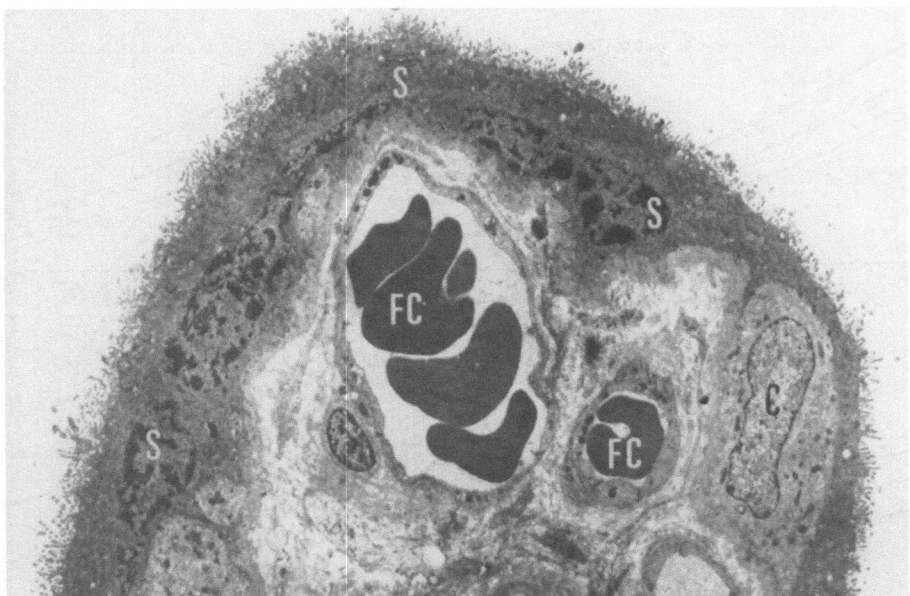

Fig. 10. Electron micrograph of the villous membrane (barrier) at term showing outer syncytiotrophoblast (*S*), occasional cytotrophoblast cells (*C*), and fetal capillary (*FC*). ×3600

before termination of pregnancy. However, because of the degeneration of an important portion of the placenta (chorion laeve) during the second half of the first trimester, the ultrasound measurements may not be reliable. Before 10 weeks of gestation, more accurate measurements of the placental volume could only be obtained from hysterectomy specimens with pregnancy in situ, which are rarely available for scientific investigations.

The morphology of unfixed placental tissue deteriorates rapidly after delivery [16]. Within a few minutes of total ischemia, alterations of the rough endoplasmic reticulum and the mitochondria can be found in the syncytiotrophoblastic layer. The importance of the correct concentration, pH, and osmolality of the fixative solutions cannot be underestimated even for histologic investigations. Every effort should be made to fix the villous material immediately after the termination procedure. Chorionic villus sampling in utero before termination is a possible option, but it may not give a representative picture of the placenta as a whole. Ideally, such studies should be performed longitudinally, but repeated villous sampling would be unsafe and unethical.

Uncertainties over dating early human material have also represented a major problem for previous authors studying placental morphology and development. This problem can now be circumvented by correlating clinical

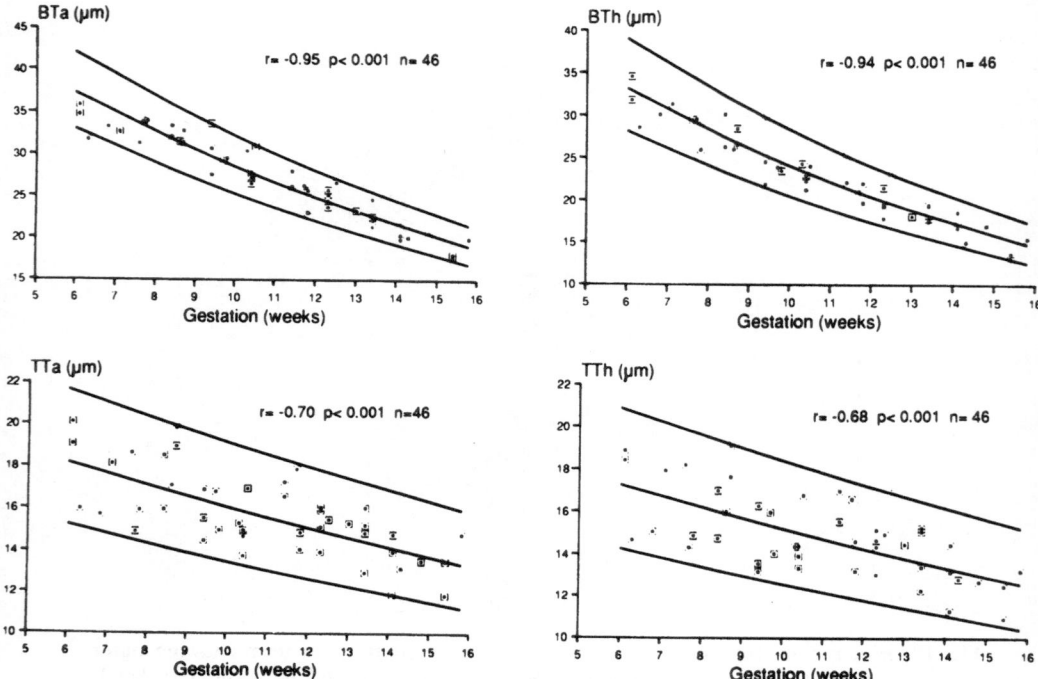

Fig. 11. Scattergrams and regression line (±2 SD) of arithmetic barrier thickness (*BTa*), harmonic barrier thickness (*BTh*), arithmetic trophoblast thickness (*TTa*), and harmonic trophoblast thickness (*TTh*) plotted against gestational age. (From [7])

data with ultrasound measurements in utero of the product of conception. Only placentas from patients with regular menstrual cycles, who are certain of the first day of their last menstrual period, who have never smoked, and who have not used oral contraception for several months before conception should be included in studies dedicated to normal villous development (see Chap. 7).

Histomorphometry of the Early Villous Tree

Few authors have investigated quantitatively the structure of the early placenta [7, 17, 18]. It is well known that in the early stage of gestation the villi are quite large in diameter (±170 μm) and progressively decrease in diameter as term approaches, finally reaching an average size of approximately 40 μm in diameter [19]. Boyd [18] proposes an equation for the villous surface area (m^2) = 0.027 × parenchymal volume (ml) + 0.103 × gestation (weeks) + 0.06 × fetal weight (kg), where parenchymal volume and gestational age account for 94% of the variation in villous surface. At 12

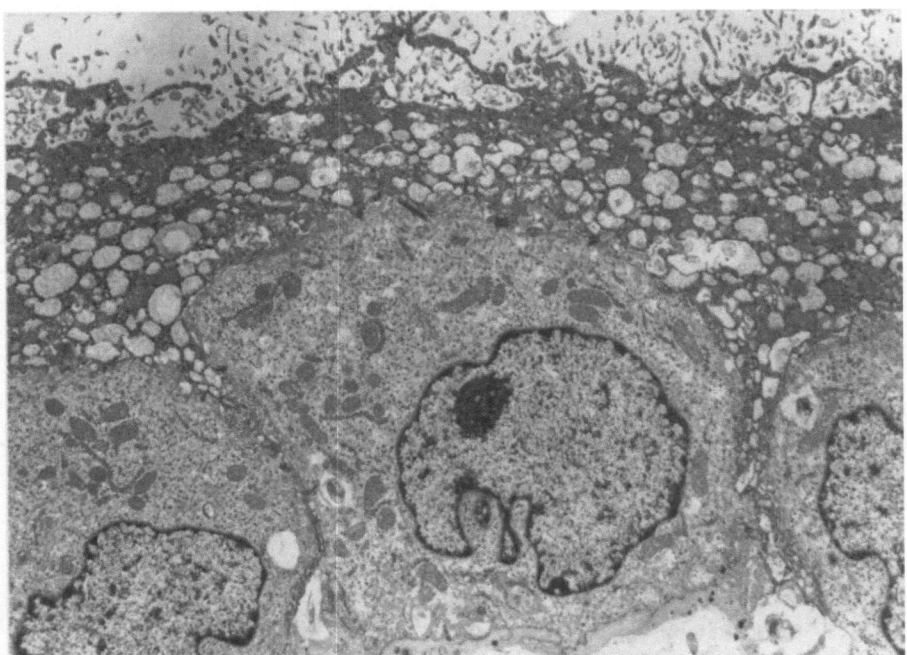

Fig. 12. Electron micrograph of the trophoblast at 11 weeks of gestation showing the continuous layer of cytotrophoblastic cells beneath the syncytiotrophoblast. ×7000

weeks of gestation, the mean value of the villous surface varies between 0.8 and 1.2 m^2 and it increases progressively towards mid-gestation to values between 1.6 and 4.4 m^2 [17].

The villous barrier thickness and trophoblastic layer thickness of branching villi demonstrate an exponential decrease throughout gestation (Fig. 11). Because the barrier thickness includes not only the trophoblastic layer but also the stroma and the capillary endothelium, the magnitude of these changes is greater for the barrier thickness than for the trophoblastic thickness alone [7]. It is generally accepted that initially the cytotrophoblast cells form a continuous layer beneath the syncytiotrophoblast (Fig. 12), but that, as pregnancy advances, these cells become less prominent [1]. This phenomenon must have a major influence on the trophoblastic thickness of the early placenta. The volume fraction for the trophoblast increases until 11 weeks and then subsequently decreases, while the volume fraction for the villous stroma decreases and then increases. The rise and subsequent fall in the volume fraction of trophoblast (Fig. 13) may reflect the growth pattern of the syncytial sprouts [7] which may indicate an attempt to compensate for the loss of two thirds of the placental tissue, occurring at the end of the first trimester. Important changes having a direct influence on the barrier

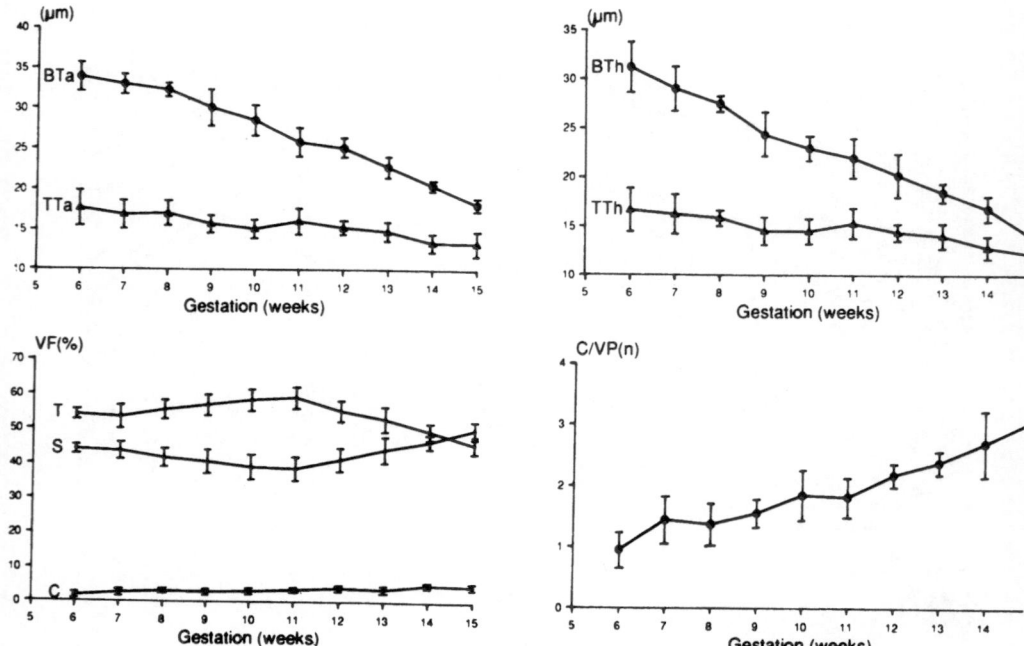

Fig. 13. Moving averages by weeks (±SD) for the arithmetic barrier (*BTa*) and trophoblastic thickness (*TTa*) (**A**); for the harmonic barrier (*BTh*) and trophoblastic thickness (*TTh*) (**B**); for the volume fraction (*VF*) of the villi occupied by trophoblast (*T*), stroma (*S*) and capillaries (*C*) (**C**); and for the number of capillaries per villous profile (*C/VP*) (**D**). (From [7])

thickness also occur in the villous vasculature during early placental development (Fig. 13). The number of capillary profiles per villous profile and the proportion of the villi (volume fraction) occupied by the fetal capillaries increase progressively between 6 and 15 weeks.

Correlation with Biological Function

The Villous Trophoblast

The primitive cytotrophoblast is derived from the cells of the wall of the blastocyst and gives rises to all populations of trophoblastic cells [1, 15, 19]. The evidence bearing on the cytotrophoblastic contribution to the syncytium has been summarized by Boyd and Hamilton [1]. Intermediate and residual cells are often found in the immature trophoblast (Fig. 14), providing additional morphologic arguments supporting the incorporation of the cytotrophoblast into syncytium. The cytotrophoblastic cells are a prominent

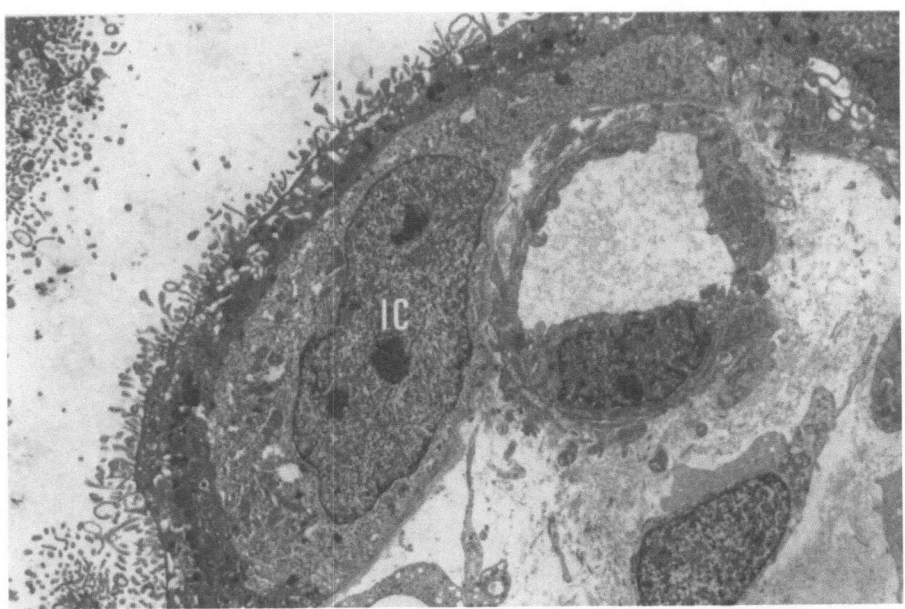

Fig. 14. An intermediate cell (*IC*) from an 18-week villus showing developing endoplasmic reticulum and an increased basophilia compared to undifferentiated cytotrophoblast in Fig. 12. ×6250

feature of the early trophoblast. The presence of large mitochondria and numerous Golgi bodies, as demonstrated by TEM, suggests that these cells have an active metabolism. The cytotrophoblastic layer is not continuous even in very early specimens, and basal infoldings of the syncytiotrophoblast are often seen (Figs. 12 and 15). However, these do not penetrate the trophoblastic basement membrane [15].

The syncytiotrophoblast is the most active component of the human placenta from the embryonic period until the end of gestation. It has been shown by TEM to be the only true syncytial tissue to occur in the human. It has also the highest concentration of organelles, i.e. mitochondria, endoplasmic reticulum, free ribosomes, granules of glycogen, and well-developed Golgi apparatus (Fig. 15), of all villous cells [15, 18, 19]. The first trimester syncytiotrophoblast is rich in electron-dense granules. Immunostaining of electron-microscopy sections has demonstrated that some of these granules are hormone-secretory granules [20, 21]. Clear morphological differences exists between human chorionic gonadotropin (hCG) granules, which predominate during the first trimester, and human placental lactogen (hPL) granules, which are usually found in second- and third-trimester placentas [21]. The syncytiotrophoblast produce many different and unique proteins, mostly of unknown function, which are discussed in Chap. 18.

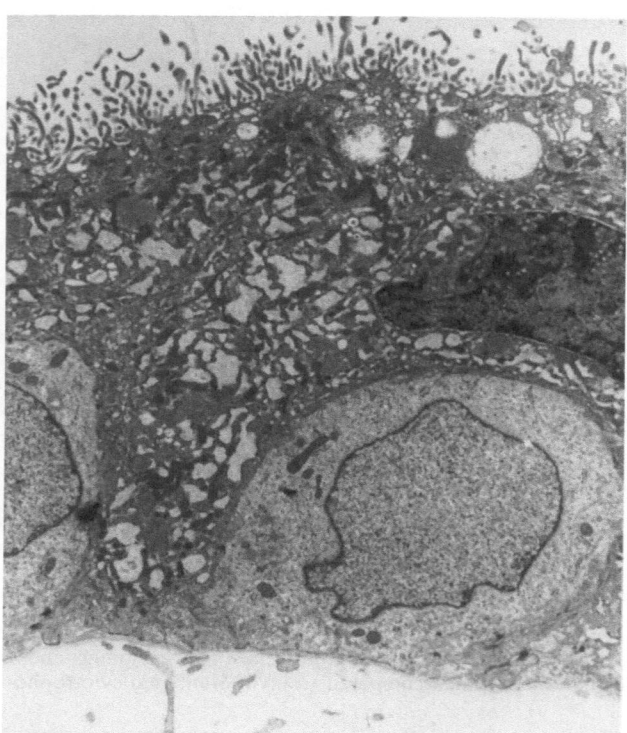

Fig. 15. Electron micrograph of the trophoblast at 12 weeks gestation showing the scalloped brush border of the syncytiotrophoblast. The cytoplasm of the syncytiotrophoblast is infolding in between cytotrophoblastic cells and contains numerous rough endoplasmic reticulum profiles, many mitochondria, and pynocytic vacuoles. ×6600

The hCG levels cannot be correlated with the growth of the placental mass. The trophoblast thickness decreases progressively with gestation (Fig. 13), while hCG levels rise during the second month of gestation, reaching a maximum between 8 and 10 weeks gestation, at a time when most of the placental mass degenerates to form the chorion laeve. The use of specific monoclonal antibodies against hCG subunits shows that α-hCG synthesis is localized to cytotrophoblast while β-hCG synthesis occurs in intermediate cells and the mature syncytium [22]. This finding and correlation of placental histology with levels of hCG suggests that hCG synthesis depends on the rate of differentiation of cytotrophoblast into syncytiotrophoblast and reflects the invasive role of the developing placenta [22].

The trophoblast surface is covered with microvilli (Figs. 9, 15). At 10 weeks, the microvilli are long, measuring approximately 1.5 μm, and become shorter as pregnancy progresses [23]. Microvilli increase the surface area of the villi in contact with maternal blood and, being richly endowed with enzymes and receptors [15], facilitate placental transfer of water, gases, and nutrients to the embryo or fetus. A superficial similarity exists between the distribution of microvilli on the trophoblastic layer microvilli arrangements and those of the brush border of renal and intestinal epithelia, but the villous microvilli are much more pleomorphic and show less morphologic stability [1]. Villous surface areas lacking microvilli have been reported using both TEM and SEM. These smooth areas are randomly distributed and only occupy a small portion of the villous surface [7, 9]. They probably result from villous damage during the collection or fixation process and should not be present on the surface of the normal healthy and well-fixed syncytium.

The Villous Stroma

Early in pregnancy, the villous core is filled by small, spindle-shaped undifferentiated cells with long, thin cytoplasmic processes [5]. These processes give rise to small reticulum cells which, under the SEM, can be seen to possess numerous small processes that resemble "sails" and make close contact with each other. Large reticulum cells and fibroblasts also develop and secrete the collagenous matrix that eventually fills most of the intercellular spaces of the tertiary villous core as well as providing structural support. Myofibroblasts, with well-developed myofilaments, have been described by Feller et al. [24], who suggested that they have a contractile function. Stromal cells form intimate connections with each other and also with the trophoblastic basement membrane [15].

Hofbauer cells, or villous macrophages, are characterized by numerous cytoplasmic vacuoles of variable size, often containing lamellar or granular material (Fig. 16), an eccentric nucleus, numerous Golgi bodies and mitochondria, and an often poorly developed endoplasmic reticulum [15]. They have been observed in the chorionic villi of an 18-day embryo [1] and are present in their highest concentration early in pregnancy. They represent the morphologically obvious members of a much larger mononuclear phagocyte population found in various other fetal tissues [1, 25]. Immature or intermediate Hofbauer cells can be found in the core of secondary villi even before angiogenesis begins and all through the first half of gestation they are the numerically dominant cell type in the villous mesenchyma [14, 26].

Hofbauer cells exhibit avid crystallizable fragment (Fc) receptors, express C3 receptors, and are phagocytic in vitro [25]. Variation in antigen expression has been observed with increasing gestational age. In particular, class II major histocompatibility complex antigens are present in lower density on first-trimester villous macrophages compared with later in gestation [26]. Hofbauer cells also show a strong reactivity to the CD4 antigen that serves as a receptor for the human immunodeficiency virus

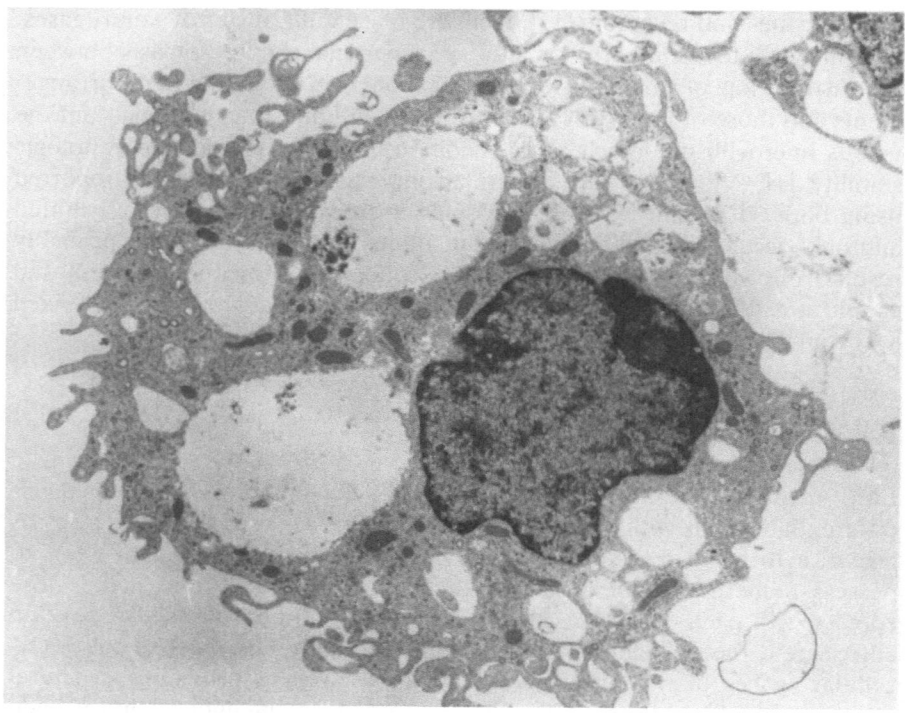

Fig. 16. Typical ultrastructure of a villous macrophage at 7 weeks of gestation, characterized by a vacuolar cytoplasm, few other organelles, surface irregularities, and an eccentric nucleus. ×7600

(HIV). From these findings, it is reasonable to suggest that Hofbauer cells play a critical role in the filtration process of anti-fetal antibodies. A possible role in regulating the early differentiation of the villous mesenchyma has also been proposed for these cells [14].

The Vasculo-Syncytial Membrane

Vasculo-syncytial membranes are points on the villous surface where the syncytiotrophoblast is very thin and separated from the underlying capillaries by only the basal lamina. Although they are believed to be the principal sites of gaseous exchange, comparative studies have shown that they do not occur in all species where gas transport is evidently sufficient [27].

Conclusion

The developing or mesenchymal villi may be considered as the functional villi of the early placenta, where diffusive transport of gases such as oxygen or carbon dioxide should mainly occur. The total thickness of the placental membrane of the developing villi decreases rapidly from 6 weeks to the end of gestation (Fig. 11). However, when the maternal intervillous circulation is established around 12 weeks (see Chap. 6) the mean barrier thickness is still large and varies between 24 and 27 μm. The practical importance of these observations is that during embryogenesis the human gestation lives in an environment poor in oxygen compared with later during pregnancy. This phenomenon may influence other biological events such as the switch of embryonic to fetal hemoglobin and/or may protect the developing embryo from the deleterious effect of high oxygen levels and free radicals.

References

1. Boyd JD, Hamilton WJ (1970) The human placenta. Heffer, Cambridge
2. Wilkin P (1965) Pathologie du placenta humain. Masson Paris
3. Moore KL (1982) The developing human: clinically oriented embryology. Saunders, Philadelphia
4. Kaufmann P (1982) Development and differentiation of the human placental villous tree. Bibl Anat 22:29–39
5. Castellucci M, Kaufmann P (1982) A three-dimensional study of the normal human placental villous core. II. Stromal architecture. Placenta 3:269–285
6. Castellucci M, Scheper M, Scheffen I, et al. (1990) The development of the human placental villous tree. Anat Embryol (Berl) 181:117–128
7. Jauniaux E, Burton GJ, Moscoso JG, et al. (1991) Development of the early placenta: a morphometric study. Placenta 12:269–276
8. Aladjem S (1967) Correlation of pregnancy wastage and hypoplasia of placental syncytium: study by phase contrast microscopy. Obstet Gynecol 30:408–413
9. Burton GJ (1987) The fine structure of the human placental villus as revealed by scanning electron microscopy. Scanning Microsc 1:1181–1828
10. Douglas GW, Thomas L, Carr M, et al. (1959) Trophoblast in the circulating blood during pregnancy. Am J Obstet Gynecol 78:960–969
11. His W (1868) Untersuchungen über die erste aubage des Wirbeltierleibes. Vogel, Leipzig
12. Hertig AT (1935) Angiogenesis in the early human chorion and the primary placenta of the macaque monkey. Contrib Embryol Carnegie Inst Washington 25:37–81
13. Dempsey DE (1972) The development of capillaries in the villi of early human placentas. Am J Anat 134:221–238
14. Demir R, Kaufmann P, Castellucci M, Erbengi T, Kotowski A (1989) Fetal vasculogenesis and angiogenesis in human placental villi. Acta Anat (Basel) 136:190–203
15. Jones CJP, Fox H (1991) Ultrastructure of the normal human placenta. Electron Microsc Rev 4:129–178
16. Jauniaux E. Moscoso JG, Vanesse M, et al. (1991) Perfusion fixation for placental morphologic investigation. Hum Pathol 22:442–449

17. Wilkin P, Bursztein M (1957) Etude quantitative de la grossesse, de la superficie de la membrane d'échange du placenta humain. CR Assoc Anat Leyde 44:830–867
18. Boyd PA (1984) Quantitative structure of the normal human placenta from 10 weeks of gestation to term. Early Hum Dev 9:297–307
19. Fox H (1978) Pathology of the placenta. Saunders, London
20. Morrish DW, Marusyk H, Siy O (1987) Demonstration of specific secretory granules for human chorionic gonadotropin in placenta. J Histochem Cytochem 35:93–101
21. Morrish DW, Marusyk H, Bhardwaj D (1988) Ultrastructural localization of human placental lactogen in distinctive granules in human term placenta: comparison with granules containing human chorionic gonadotropin. J Histochem Cytochem 36:193–197
22. Hay DL (1988) Placental histology and the production of human choriogonadotrophin and its subunits in pregnancy. Br J Obstet Gynaecol 95:1268–1275
23. Tighe JR, Ganod PR, Curran RC (1967) The trophoblast of human chorionic villus. J Pathol Bacteriol 93:559–567
24. Feller AC, Schneider H, Schmidt D, et al. (1985) Myofibroblasts as a major cellular constituent of villous stroma in human placenta. Placenta 6:405–415
25. Wood GW (1980) Mononuclear phagocytes in the human placenta. Placenta 1:113–123
26. Goldstein J, Braverman M, Salafia C, et al. (1988) The phenotype of human placental macrophages and its variation with gestational age. Am J Pathol 133:648–659
27. Steven DH (1983) Interspecies differences in the structure and function of trophoblast. In: Loke YW, Whyte A (eds) Biology of the trophoblast. Elsevier, Amsterdam, pp 111–136

4 Ultrasound Features of the Early Gestational Sac

J.-P. Schaaps

Introduction

Ultrasonography is the only noninvasive method of approaching pregnancy in vivo. The absence of side effects, which is almost certain after 30 years of extensive use in humans, has made it an irreplaceable tool, to be used again and again. Recent advances in ultrasonography, including transvaginal sonography, have significantly improved our ability to investigate the first trimester of pregnancy. First trimester events which used to be described by embryologists can now be visualized in utero with a high degree of accuracy practically from the date of the missed period. The various aspects of normal embryonic development depicted by transvaginal ultrasonography have been recently and remarkably reviewed [1, 2]. Little attention has been paid to the uterine environment of the gestational sac and to the maternofetal interface which are the purpose of the present chapter.

Early Ultrasound Aspects of Singleton Pregnancy

A few days after the expected menstrual period, it is possible to clearly distinguish the presence of an intrauterine gestational sac distinct from the uterine cavity [3–5]. The gestational sac is buried in the uterine mucosa and is always found in an eccentric position with regard to the most echogenic line, which is due to the joining of the two endometrial edges of the cavity (Fig. 1). The differential diagnosis posed by the existence of a pseudogestational sac in ectopic pregnancy is directly solved by endosonography. This intracavitary liquid collection is, of course, median, elongated, and non-intraendometrial. Conversely, the gestational sac appears as a round structure, lacking in echoes, exhibiting the characteristics of a collection of liquid, and surrounded by a region of echogenicity which is more important than the neighboring myometrium [3–5]. This peripheral zone, rich in echoes and surrounding the embryo, is the ultrasound expression of a structure containing a large number of interfaces. It represents a collection of tissues comprising the chorionic villi, the intervillous lakes, the extravillous trophoblast, and the maternal decidua modified by the trophoblastic in-

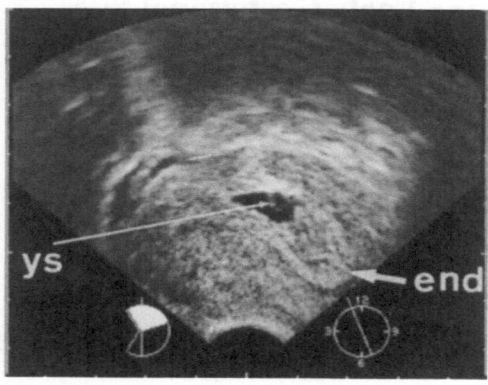

Fig. 1. Gestational sac at 5 weeks' gestation in an eccentric position with regard to the midline of the endometrium area (*end*). The yolk sac (*YS*) is clearly seen

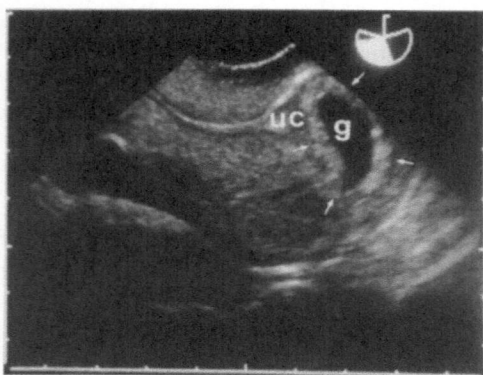

Fig. 2. The gestational sac (*g*), at 7 weeks of gestation, growing into the endometrium. Note that it can be clearly distinguished from the uterine cavity (*uc*) and that it is surrounded by the chorion frondosum (*arrows*)

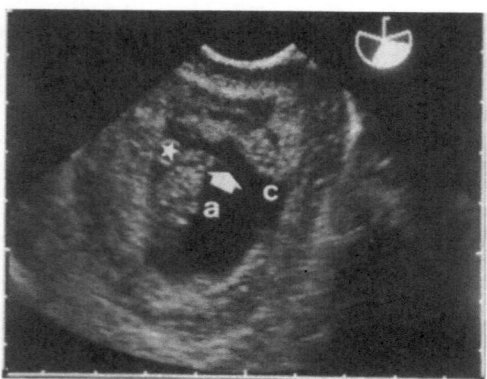

Fig. 3. Sonogram of gestational sac showing the amniotic cavity (*a*) which is very thick and hypoechogenic (*arrow*). The yolk sac (*star*) seems to be close to the embryo, but is in fact located within the external coelom (*c*)

filtration. Therefore, right from the beginning one can verify the existence of two distinct divisions in normal early pregnancy:

- The cavity of the gestational sac
- The peripheral ring and its junction to the uterine tissues

The Cavity of the Gestational Sac

This primitive lacuna was the earliest image of pregnancy supplied by first-generation ultrasonography [6-8]. It appeared to be unique, appearing at 6 weeks under the best examining conditions. The rate at which this amniochorial cavity grows made it one of the first biometric instruments destined to date pregnancy during the first trimester. The biometry of the gestational sac is easy to realize up to 7-8 weeks after the last menstrual period (LMP) because the sac is spherical [3-5]. After this time the sac takes on a beanlike shape, due to two concomitant phenomena:

- The sac grows very quickly into the endometrium. Since the general form of the uterine cavity is elongated, the sac tends to shape itself to the form of the organ which contains it (Fig. 2).
- From 7 weeks onwards, a progressive differentiation between the chorion laeve and the chorion frondosum can be observed. The latter becomes thicker and thicker to form the definitive placental structure. It pushes the boundary of the gestational sac towards the interior, giving it a reniform look.

These modifications to the shape make the acquisition of volume awkward and inaccurate. When dating a pregnancy, the sac measurements are usually only used during the initial critical stages.

Several constituents can be distinguished, whose evolution can be followed during the course of the first trimester.

The Amniotic Cavity

The division between the amniotic cavity and the extraembryonic coelom is visible 5 weeks after LMP. This membrane is very fine and seldom presents itself in cross section as a continuous line [3-5]. Given its poor echogenicity, it can only be seen clearly where the ultrasound beam strikes it perpendicularly. It forms a ring, eccentric in relation to the gestational sac, defining in cross section a residual crescent consisting of the extraembryonic coelom (Fig. 3). Its region of confluence with the wall of the sac indicates the region where the decidua basalis is located. It gradually fills the gestational sac, filling it completely around 11 weeks after LMP.

The Secondary Yolk Sac

The secondary yolk sac can be observed by ultrasonography during the 5th week after LMP (Fig. 4). It is, of course, located outside the cavity, limited

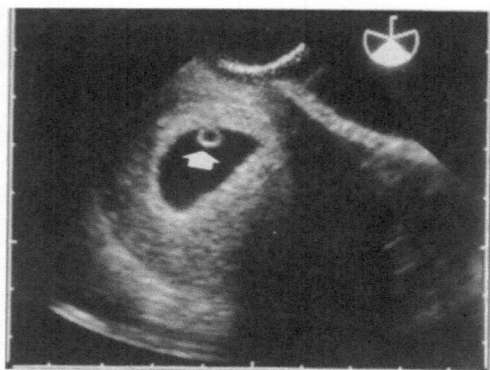

Fig. 4. Sonographic view of the secondary yolk sac (*arrow*)

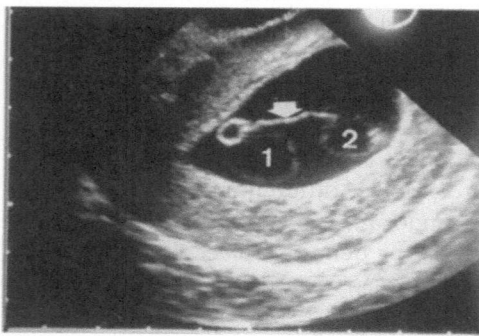

Fig. 5. Monochorial biamniotic pregnancy (*1* and *2*) showing the long pedicule of the yolk sac of the second embryo (*arrow*)

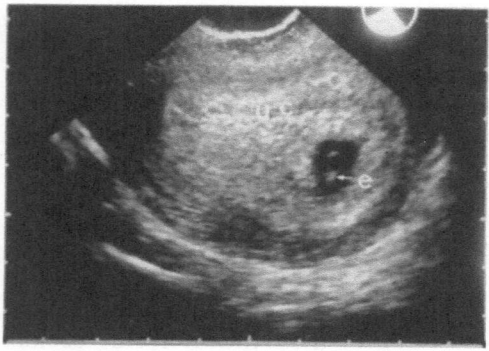

Fig. 6. The gestational sac is located outside the uterine cavity (*uc*) at the end of the 5th week after LMP. The embryo (*e*) is visible close to the yolk sac

by the amniotic membrane and pediculated (Fig. 5). The dimensions of the yolk sac increase slowly throughout the first trimester (see Chap. 11).

The Embryo

At 5 weeks, the embryo measures 2 mm and it can be seen less well than the yolk sac (Fig. 6). At this stage, the embryo is lying next to the wall of the sac, and it can be difficult to distinguish it from parietal echoes [3–5]. This period of uncertainty only lasts a few days as embryonic growth is very rapid (Fig. 4). Much more than the biometry of the gestational sac, the crown-rump length (CRL) of the embryo has been used [9, 10] as a precise indicator of gestational age (Fig. 7). Endosonographic examination is so precise that the CRL can now be used without any correction factor [3–5]. At 5.5 weeks, when CRL is about 3.5 mm, it is possible to determine the embryonic cardiac activity (Figs. 8, 9). This can be detected from the moment when the organ begins to beat (22nd–23rd days of life). At this stage of embryogenesis, the heart consists only of a primitive cardiac tube which begins to bend. The cardiac pulsation is not global, but for around 10 days one can detect a transmitted pulsation similar to a rapid peristaltic movement. This observation, made 5.5–7 weeks after LMP in humans correlates strongly with the findings made by Loeber et al. [11] in guinea pigs. The detection of cardiac activity is therefore a formal criterion for the evolution of the pregnancy [4, 5, 7].

The embryonic cardiac frequency evolves during the 1st trimester. At the moment when it begins to beat, the heart has an average frequency of 100 bpm. At 7 weeks, this frequency reaches an average of 160 bpm. This evolution of cardiac frequency was described in 1975 by Robinson [7, 9], but he placed the maximal frequency at around 10 weeks after amenorrhea. This latter study, having been conducted with less sensitive equipment, did not permit the detection of small oscillations of the embryonic heart.

Using transvaginal ultrasonography, the evaluation of cardiac activity is much more sensitive and does not miss any sort of myocardial activity. In our study of 240 cases, for which gestational age was calculated on the basis of CRL, we established that the acceleration in cardiac frequency was much earlier and quicker than previously described. This increased frequency remains until the end of the first trimester (Fig. 10). The frequency/CRL comparison was chosen in order to exclude a supplementary correlation factor (CRL/gestational age).

The growth of the embryo and embryogenesis can be followed by contact endosonography [3, 4]. The presence of the cerebral ventricules can be asserted at 8 weeks of gestation (Fig. 11) and anencephaly can therefore be excluded at this stage. Embryonic spontaneous activity can also be detected by endovaginal ultrasound as early as 8 weeks' gestation. Total body movements can be observed, which cannot be confused with any transmission of embryonic cardiac activity. Active motoricity is also a

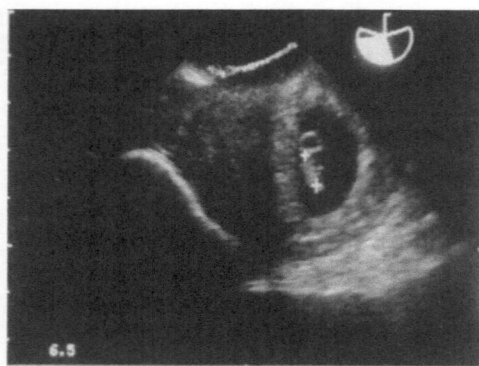

Fig. 7. Sonogram of a 6-week pregnancy with embryonic crown-rump length measuring 6.5 mm (*crosses*)

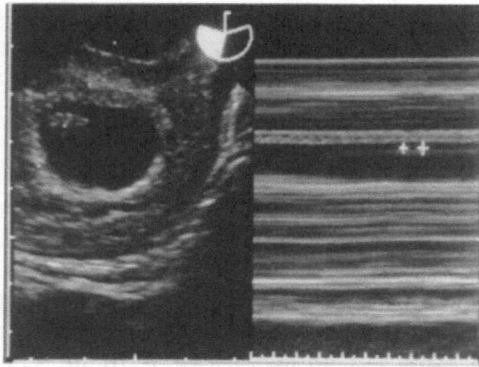

Fig. 8. Cardiac activity detected by the time-motion mode

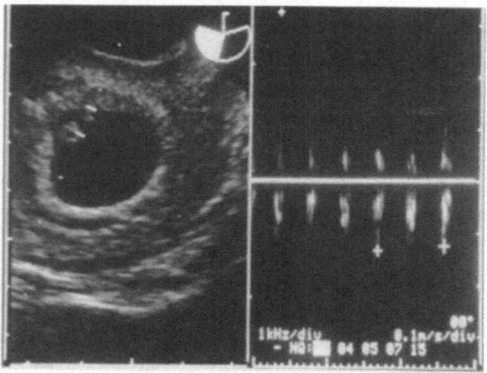

Fig. 9. Pulsed Doppler investigation of the embryonic cardiac activity

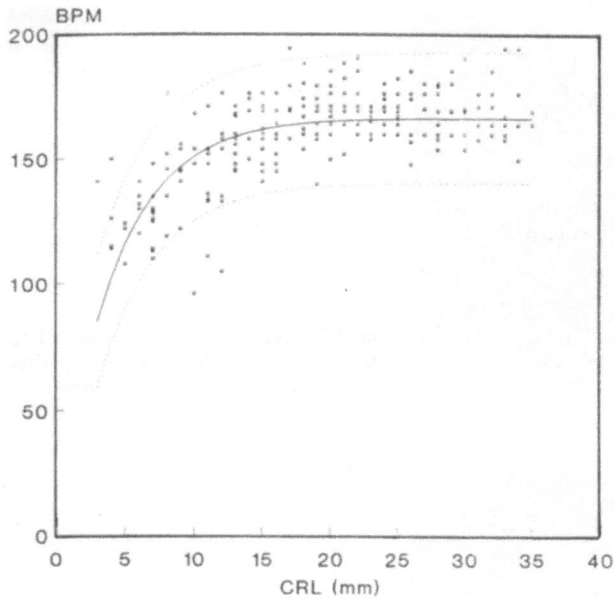

▲

Fig. 10. Evolution of the cardiac rhythm in relation to the CRL. *Squares*, cases; *solid line*, regression curve; *dotted lines*, upper and lower limits. $Y = 166.5 [1 - \exp(-0.238X)]$; $n = 240$

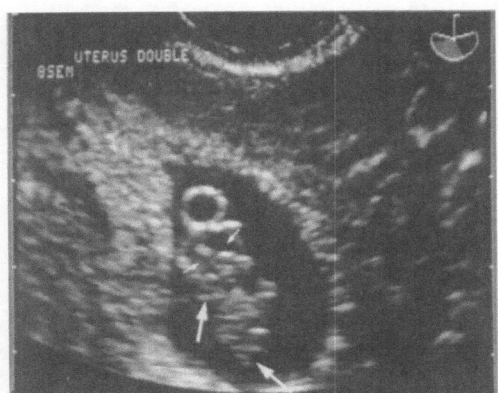

Fig. 11. Pregnancy of 8 weeks. Both cerebral ventricules, one with the choroidal plexus (*short arrows*), both legs and the left arm can be distinguished (*long arrows*)

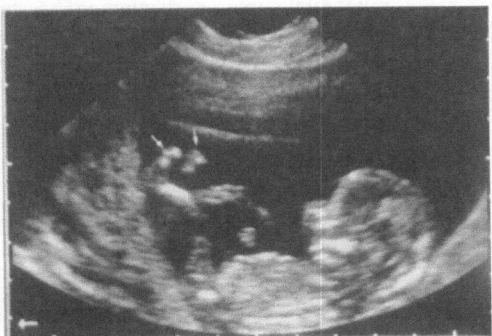

Fig. 12. Pregnancy of 12 weeks. The embryo is completely observed by abdominal ultrasound. The *arrows* indicate the toes

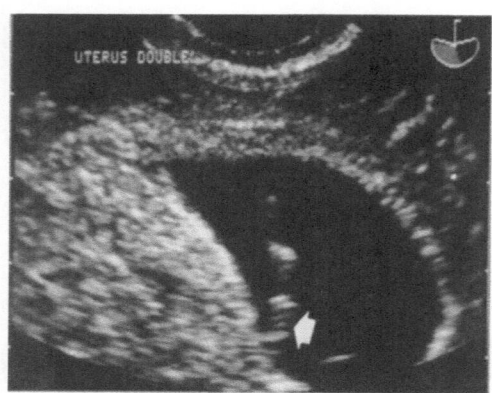

Fig. 13. Pregnancy of 8 weeks. The site of the umbilical cord insertion indicates the decidua basalis (*arrow*)

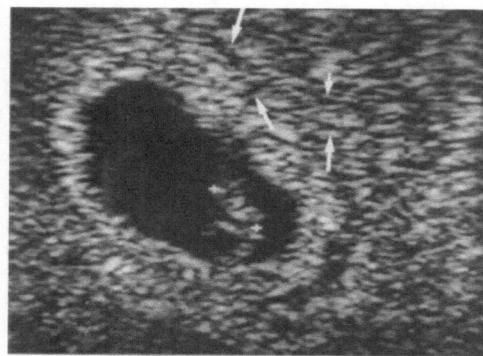

Fig. 14. Pregnancy with 5-mm embryo. The trophoblastic ring is complete. The thickness of the crown is quite uniform all around. *Arrows* indicate maternal blood vessels

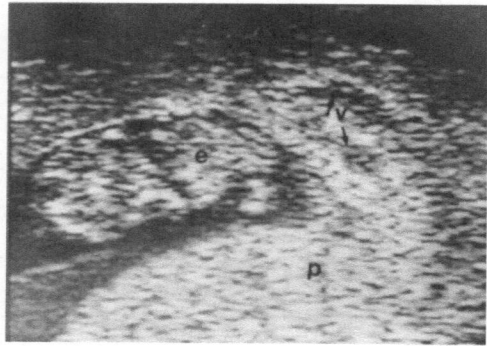

Fig. 15. Pregnancy of 9 weeks. The placental area is beginning to be defined (*p*). The myometrial vessels come closer and closer to the gestational sac (*V*) containing the embryo (*e*)

reliable sign of embryonic wellbeing. Fewer than ten movements detected during a 5-min period constitutes a sign of poor vitality. At 10 weeks after LMP, isolated and individual movements of the members can be detected. They are a contributive help in assuming the presence of four members. The complete normality of the members (three segments) needs 2 more weeks to be asserted (Fig. 12).

The Primitive Umbilical Cord

The primitive umbilical cord can be visualized 6 weeks after LMP. It is the most certain way of defining the site of insertion of the gestational sac (Fig. 13). This parameter is of prime importance in the prognostic evaluation in certain cases of threatened abortion. During the first trimester, the cord is large in proportion to the embryo. In particular, its fetal site of implantation in enlarged due to the presence of primitive intestinal lipple. At this stage, there is no question of parietal defects of the abdomen, and the cord preserves intestinal structures until the end of the first trimester [1, 2].

The Peripheral Ring

The peripheral ring is the site of maternal/embryonic exchanges and is composed on ultrasound by the ring itself and by the modifications of the host organ [2].

The Ring

From the moment when the gestational sac is ultrasonographically evident, the peripheral ring is complete. At this time, the liquid cavities surrounding the embryo are themselves surrounded by the chorion and trophoblastic structures. The echogenicity of this region is the consequence of a number of important interfaces, including the chorionic villi, the intervillous space, and the extravillous trophoblast. The latter progressively infiltrates the maternal decidua and remains uniformly thick all along the periphery of the gestational sac until the age of 7 weeks. At 5 weeks, the ring is 3–4 mm thick and triples within 15 days (Fig. 14). After this time, the topographic distortion between the chorion laeve and the chorion frondosum appears on ultrasound [4].

After 10 weeks of gestation, two-thirds of the trophoblastic ring ceases to grow, while the rest, which turns into the definitive placenta, continues to develop. There is no difference in echogenicity between these two structures, merely a disparity in size. A certain pulsatility resembling a rhythmic scintillation can be seen in the trophoblastic ring after 7 weeks. The rhythm is the same as that of the embryonic heart. From 9 weeks, in certain lateral parts of the trophoblastic ring less rich in echoes, a particularly slow movement can be distinguished. This feature remains limited to the edges of the placental mass [4].

Uterus

One of the first signs of pregnancy is a modification in the ultrasound structure of the uterus. There is a progressive reduction in the intensity of echoes coming from the uterine muscle, clearly visible even by means of conventional abdominal ultrasound [4]. This phenomenon can be related to the important modifications in uterine vascularization occurring from the beginning of pregnancy. The subserous uterine vessels become much more visible along the whole periphery of the organ and vary between 2 and 5 mm in thickness [10]. In sectional views within these structures, small echoes can be detected in motion; these express the circulation of red blood cells. This type of observation can only be made using high frequencies (7.5 MHz), a rapid image frequency (more than 20/s), and maximal enlargement [2, 4]. The position of these peritrophoblastic vessels closely follows the evolution of the trophoblast (Figs. 14, 15). During the formation of the chorion laeve, at the level of the reflected decidua, the vessels are drawn nearer the amniotic cavity, while, at the level of the decidua basalis where the definitive placenta forms, the vascular system is pushed away.

It is therefore not surprising, at around 10–12 weeks, to distinguish a circulatory activity practically next to the cavity of the gestational sac at the opposite pole of the insertion site of the cord. As the pregnancy advances, this vascular network becomes more important, drawing near the tropho-blastic ring again and surrounding it with a compact network of small vessels. Under normal conditions, it is, however, impossible to demonstrate a communication between these vessels and the trophoblastic ring. This movement of particles detected in the vessels is not seen in the trophoblastic ring during the first trimester of pregnancy. During the 3rd month of ges-tation, one can see them in a nonsystematic manner at the edges of the rudimentary placenta [4].

Early Ultrasound Aspects of Multiple Pregnancy

Very early in pregnancy, it is easy to distinguish the presence of two or more gestational sacs. If each sac has its own complete trophoblastic ring, its indicates a bichorial biamniotic pregnancy (Figs. 16–18). Conversely, a single sac containing two distinct amniotic cavities, each with an embryo, represents a monochorial biamniotic pregnancy. In these cases, the external coelom contains two yolk sacs, and the decidua basalis is, of course, shared. If there is only one amniotic cavity, the respective positions and the indi-vidual mobility of these monoamniotic twins allows elimination of the serious pathology of Siamese twins. Later, in a twin pregnancy in which a wall is found, the bichorial nature can only be affirmed if there is in the parietal zone of insertion of the division a sonolucent space proving the confluence of the two chorions. If this sign is not detected, the existence or the

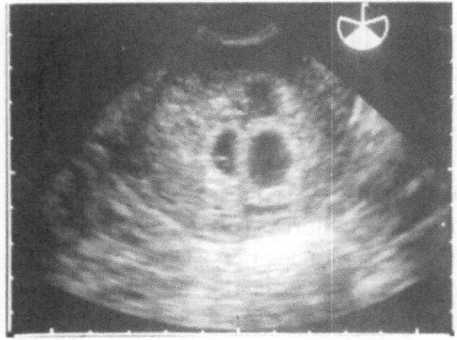

Fig. 16. Triplet pregnancy. Three distinct trophoblastic rings indicate three chorions

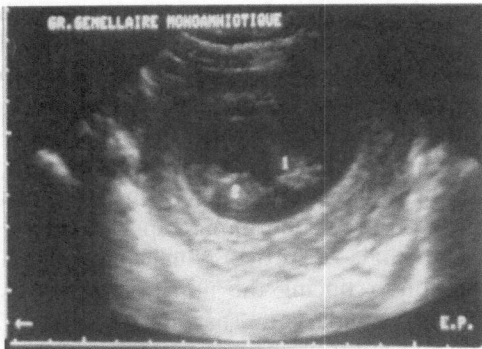

Fig. 17. Monochorial biamniotic twin pregnancy (*1, 2*) at 9 weeks of gestation

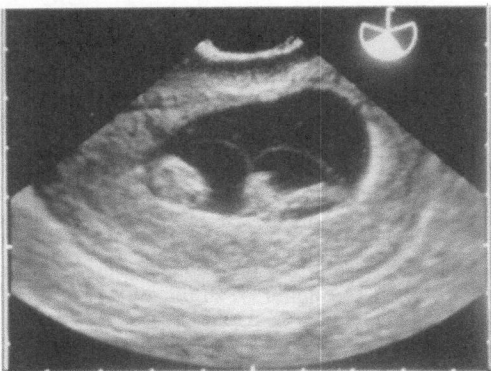

Fig. 18. Monoamniotic twin pregnancy at 7 weeks of gestation showing an excess of amniotic fluid

possibility of further complications due to vascular anastomosis between the two circulations should always be suspected [4].

The development of medically assisted techniques of procreation, and in particular of in vitro fertilization, has led to an increase in multiple pregnancies. As soon as the embryo demonstrates the existence of two pronuclei, the starting point of the pregnancy can be fixed precisely. We have observed biometric discrepancies of up to 7 days between the two embryos issuing from the same procedure, where growth was originally narrowly synchronized. These biometric discrepancies persist throughout the first trimester; they attenuate and disappear towards the middle of pregnancy. It is probably not a recovery phenomenon, but it may be related solely to the margin of error in the technical realization of biometries and the statistical distribution of values which tends to smooth over these small differences. These twin pregnancies evolve completely normally, the growth kinetics of each twin being normal. The only period of pregnancy when all possibility of surveillance is lost is the period between reimplantation (two to four cells) and the moment when the embryo can be detected on ultrasound. It is difficult to think that one of the two embryos is growing more slowly up to this stage followed by a spurt in order to catch up with its twin. This distortion never occurs in monochorionic pregnancy, and it is not impossible that a delay in implantation occurs in the human species similar to that which exists in certain species of bats or in deer. Such phenomena are accounted for by the local conditions reducing the nutritious deposits in the endometrium of these animals, secondary to seasonal or climatic changes. The human endometrium may also not have a totally homogeneous surface, thus slowing embryonic implantation here and there.

References

1. Neiman HL (1990) Transvaginal ultrasound embryography. Semin Ultrasound Comput Tomogr Magn Reson 11:22–33
2. Bernaschek G, Deutinger J, Kratochwil A (1990) Normal early pregnancy. In: Endosonography in obstetrics and gynecology. Springer, Berlin Heidelberg New York
3. Schaaps JP, Lambotte R (1985) Ultrasonic observation of pregnancy during the first trimester using a vaginal approach. In: Fraccaro N, Simoni G, Brambati B (eds) First trimester fetal diagnosis. Springer, Berlin Heidelberg New York, pp 78–79
4. Schaaps JP, Hustin J, Thoumsin H, Foidart JM (1989) La physiologie placentaire. Encycl Med Chir (Paris) Obstétrique, 5505 A 10, 24 pages
5. Schaaps JP, Hustin J, Thoumsin H, Foidart JM (1990) La pathologie placentaire. Encycl Med Chir (Paris) Obstétrique 5037 A 10, 18 pp
6. Schaaps JP, Gillot M, Oosterbosch J (1974) Détermination précoce de l'âge gestationnel par la technique échographique. Rev Fr Gynecol Obstet 69:613–616
7. Robinson HP (1975) The diagnosis of early pregnancy failure by sonar. Br J Obstet Gynaecol 82:849–856
8. Schaaps JP (1983) Gynecological contact ultrasonography: laparoscopic and vaginal way. In: Reichel TZ (ed) Ultraschalldiagnostik. Thieme, Stuttgart, pp 468–470

9. Robinson HP (1973) Sonar measurement of fetal crown-rump lenght as mean of assessing maturity in first trimester of pregnancy. Br Med J 4:28–29
10. Robinson HP, Fleming JEE (1975) A critical evaluation of sonar "crown-rump-lenght" measurements. Br J Obstet Gynaecol 82:702–710
11. Loeber C, Goldberg SJ, Hendrix MJC, Sahn DJ (1983) Dynamic mammalian cardiogenesis investigated by high resolution ultrasound in guinea pig. Circulation 68:841–845

5 Doppler Ultrasound Investigations of Pelvic Circulation During the Menstrual Cycle and Early Pregnancy

D. Jurkovic, E. Jauniaux, and S. Campbell

Introduction

Two-dimensional ultrasonography has become the most common technique for noninvasive investigation of the pelvis of pregnant and nonpregnant women. Continuous and pulsed-wave Doppler ultrasound can add dynamic information about the pelvic circulation which cannot be obtained with imaging alone. The full understanding of Doppler ultrasonography places additional demands on the clinician. In addition to extensive knowledge about both the anatomy and the physiology of the female genital tract, understanding of vascular changes that occur in reproductive life, in pregnancy, or after menopause is now required. The purpose of this chapter is to evaluate the contribution of Doppler ultrasonography to the study of the pelvic circulation during the ovarian and menstrual cycles and in early pregnancy.

The Doppler Principles

Doppler Effect

When an ultrasound acoustic wave is directed against a moving target, the energy which is backscattered undergoes a change in frequency known as the Doppler shift, which is proportional to the relative velocity of the target and to the cosine of the angle between the source of the wave and the displacement axis of the target [1, 2]. This relationship is expressed by the Doppler equation:

$$Fd = 2Fs \, \mathrm{Cos}\theta \, v/c$$

where Fd is the Doppler shift frequency, Fs is the frequency of the source, θ is the angle between the incident ultrasonic wave and the axis of the target (angle of insonation), v is the target velocity, and c is the wave velocity in the medium. Therefore, if the angle θ is maintained, Fd will vary in direct proportion to the target velocity.

Doppler Instrumentation

Doppler instruments are designed to recognize and display the difference in transmitted and received ultrasound frequencies [1–3]. Based on its technical characteristics, Doppler equipment includes three different types of instrument: continuous, pulsed, and color Doppler. In the latter two types, Doppler is combined with ultrasound imaging. This is essential for studies of blood flow in deep pelvic vessels and the developing fetus, and therefore pulsed and color Doppler are nowadays extensively used for this purpose [4–8]. With the pulsed Doppler it is possible to measure velocities selectively at specific locations in the path of the transmitted ultrasound beam. The advantage of color Doppler is that blood flow is displayed in two dimensions and superimposed on the two-dimensional B mode image. It provides a clear picture of vascular anatomy and is used as a guide for velocity measurements that are performed by pulsed Doppler [3–5, 7, 8]. This makes velocimetry studies much faster and more accurate. Another advantage of color Doppler is the identification of blood flow in small vascular branches that are effectively undetectable by traditional Doppler techniques.

Flow Velocity Waveform Analysis

Doppler signals obtained from the blood circulation represent the summation of multiple Doppler shift frequencies backscattered by numerous red cells traveling at different velocities, and the number of cells moving at these speeds varies [1, 2]. The spectrum or flow velocity waveform (FVW) is usually displayed as a real-time ultrasonogram whose vertical axis is the frequency shift and horizontal axis is the time (Fig. 1). The maximal Doppler frequency shift corresponds to the temporal changes in the peak velocity of the red cells during one cardiac cycle.

The analysis of spectral waveforms provides qualitative information such as the presence or the direction of flow and quantitative or semiquantitative

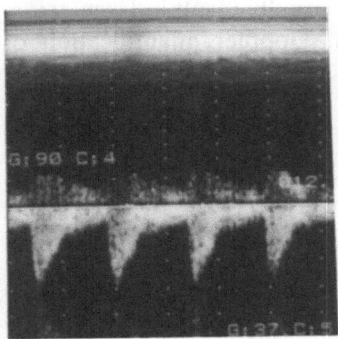

Fig. 1. Flow velocity waveform obtained from the uterine artery in early pregnancy showing moderately high blood velocity and high end-diastolic flow

measurements [1–3]. The shape of the flow velocity waveform indicates the degree of resistance to flow in the arterial vessel under investigation. Absence of frequencies in the diastole is usually found in large arterial vessels (e.g., external iliac artery) supplying a high-resistance vascular bed, while high end-diastolic velocities are usually present in smaller vascular arterial branches supplying fetal parenchymatous organs and in utero-placental circulation.

FVW characteristics may be quantified by calculating an index of impedance to flow. The indices that are most commonly used in clinical practice are A/B ratio, resistance index, and pulsatility index [1–3]. All of them are based on calculation of the ratio between peak systolic and end-diastolic velocity. In this way, index value becomes independent of the angle of insonation. Therefore, FVW analysis may be used for blood flow studies in small nonvisualized arteries with an undefined angle of insonation. Another advantage of FVW analysis, is the relative simplicity of the method and small intra- and interobserver variability. Because of that, FVW analysis is at present accepted as the best way to interpret blood flow changes in different organs and is regularly used for presentation of Doppler data. While the measurement of flow impedance may not always reflect changes in volume flow, in general, a fall of impedance is usually associated with a rise of volume flows.

Quantitative Measurements

Estimates of the mean blood velocity and vessel diameter are required for quantification of blood flow. By multiplying the mean velocity with the cross-sectional area of the vessel, the volume flow can be calculated [1, 2]. Absolute blood velocity as determined by Doppler ultrasound depends mainly on knowing the value of the angle between the ultrasound beam and the vessels of interest. The vascular dimensions and shape may vary during the cardiac cycle, and the assessment of volume flow is limited to large straight vessels. It is therefore important to fully appreciate the anatomy of the vessels and the angle of insonation when making quantitative measurements. Even with a very good measurement technique, quantitative blood flow measurements are poorly reproducible and usually result in large errors in blood volume estimations. For this reason, Doppler quantification of blood flow is rarely used.

Analysis of Blood Flow in Pelvic Vessels

Methodology

A variety of methodological approaches has been developed in attempts to study blood flow in pelvic vessels [9–14]. The common characteristic in all published studies is a primary interest in the assessment of arterial blood flow. Doppler studies of the venous circulation are less accurate, and it is assumed that changes in venous circulation will be a poor predictor of functional changes in organ perfusion.

Studies of blood flow in pelvic vessels require sophisticated Doppler equipment and considerable experience. Therefore, it is not surprising that the first report describing pelvic blood flow characteristics appeared 10 years after the beginning of Doppler studies in obstetrics. Although the number of pelvic circulation studies is relatively small, all available Doppler techniques have been used for this purpose, including continuous, pulsed, and color Doppler. Comparisons between reports is further hampered by the use of both the transabdominal and the transvaginal approach, and different indices of impedance to flow. Some authors have developed their own indices and waveform classification that complicates comparisons even more [12].

The transabdominal approach was abandoned soon after introduction of transvaginal Doppler. The major disadvantage of the transabdominal approach is the need for a distended bladder for the successful visualization of pelvic anatomy. The full bladder displaces pelvic organs backwards, thus increasing the distance between the Doppler probe and the vessels under investigation. The examination is, therefore, limited to the use of low-pulse repetition frequencies that results in decreased measurement accuracy and artifacts. Bladder compression may also cause alterations in blood flow in smaller arteries. Furthermore, patients can rarely tolerate an uncomfortably full bladder long enough to complete what may be a time-consuming Doppler study. Vaginal ultrasonography alleviates the need for bladder filling and is obviously the preferable route for blood flow studies. The probe is located close to the vessel under investigation and offers the advantage of freedom in selecting the optimal pulse repetition frequencies.

The first reports investigated blood flow characteristics in the external and internal iliac arteries, and the uterine and ovarian arteries by a transabdominal route [6, 7]. This was recently followed by transvaginal studies [15-18]. Whichever route is used, the external and internal iliac arteries can be easily identified owing to their relatively large diameter. The internal iliac artery is seen on an oblique scan posteriorly and laterally to the ovaries (Fig. 2). It lies in front of the internal iliac vein, and its distinction is facilitated by the detection of arterial pulsations. The external iliac artery and vein are identified on the lateral pelvic wall as they descend towards the inguinal ligament. It is easy to obtain Doppler signals from both iliac arteries. The FVW show high pulsatility, but, being very different in appearance, they may help in distinguishing between the external and internal iliac arteries [7]. However, it is unlikely that blood flow in these large vessels is measurably affected by physiological changes in reproductive organs during the menstrual cycle and pregnancy, and there have therefore been no attempts to investigate them in detail. In spite of that, it is important to be familiar

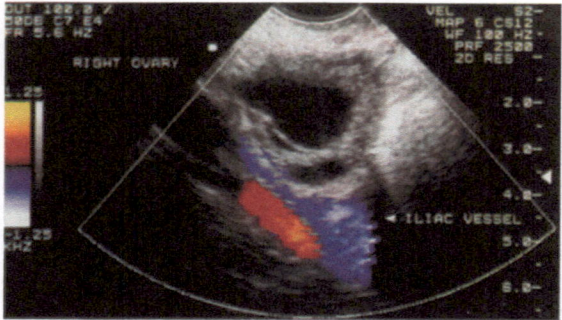

Fig. 2. Demonstration of the internal iliac artery and vein by transvaginal color Doppler. Both vessels are seen postero-laterally to the right ovary that contains the preovulatory follicle

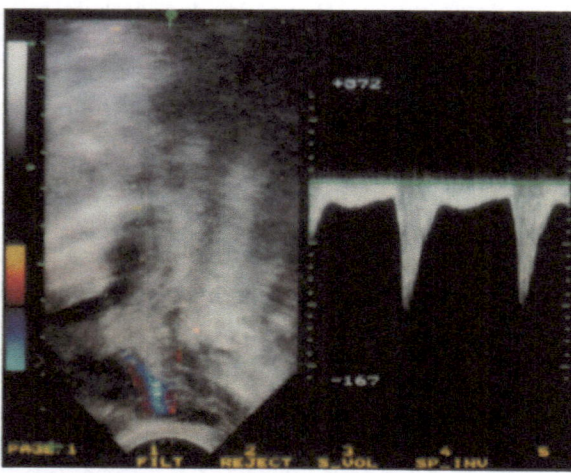

Fig. 3. Color flow mapping of the uterine artery and vein at the level of the uterine isthmus as seen by trans-vaginal color Doppler. The straight part of the vessel is identified as it extends upwards alongside the uterine body. FVW that were obtained from the artery by pulsed Doppler are shown on the *right*

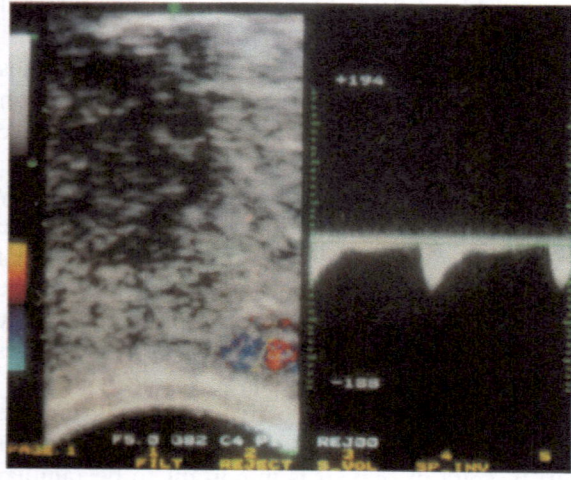

Fig. 4. Demonstration of the ovarian artery and vein by color Doppler (*left*) and FVW showing end-diastolic velocities (*right*)

with their blood flow characteristics. Misinterpretation of signals from the iliac arteries as originating from the ovarian arteries or uterine artery in pregnant patients can easily occur and has important practical implications. This is particularly so with the internal iliac artery because of its close anatomical position to the uterine artery in pregnancy and similarities between the shape of their FVW.

After successful studies of the uterine circulation in the second and third trimesters of pregnancy [8, 9], attempts were made to study uterine and ovarian blood flow in nonpregnant patients. This proved to be more difficult than initially expected. Compared to the iliac arteries, the uterine and ovarian arteries are much smaller and more variable in their anatomical position, which makes their identification on transabdominal B mode scan almost impossible [6, 7]. A high visualization rate of ovarian arteries claimed by some authors was most probably caused by misinterpretation of the internal iliac artery as the ovarian. By using the transvaginal probe [10], the uterine artery can be almost always identified at the level of the uterine isthmus (Fig. 3). Because of its tortuosity, a mixture of uterine arterial and venous vessel loops is seen, rather than the typical straight-looking vessel. Being very thin, the ovarian artery is difficult to identify even by transvaginal scanning (Fig. 4). It is usually covered by bowel loops surrounding the ovary which makes ultrasound and Doppler studies very difficult [11].

A combination of high-resolution B mode image and pulsed Doppler on vaginal probes enables reasonably accurate studies of the uterine artery although it is rarely possible to make an accurate distinction between the main uterine artery and its branches [13]. Even by using the transvaginal probe, the ovarian artery is seldom identified, and in most studies ovarian blood supply is examined by a moving pulsed Doppler sample from the lateral ovarian surface and infundibulopelvic ligament medially through the ovarian tissue until FVW are recorded [14]. Therefore, the results obtained frequently represent a mixture of signals from the ovarian artery, internal iliac artery, terminal branches of the uterine artery, and intra-ovarian flow.

The addition of color Doppler overcomes some of these problems. The main uterine artery can be easily distinguished from its branches, thus increasing reproducibility of measurements (Fig. 5). Even by using color Doppler, the visualization rate of the ovarian artery is not greatly improved. However, identification of other vessels is easier, thus preventing their misinterpretation as the ovarian artery flow. The major advantage of color Doppler is the possibility to study small arterial branches within the reproductive organs. It opens the possibility to study not only differences in blood supply to the whole organs, but also to study differences in perfusion to different tissues within the same organ. What the impact of this recent technique on studies of pelvic circulation remains to be clarified in the future.

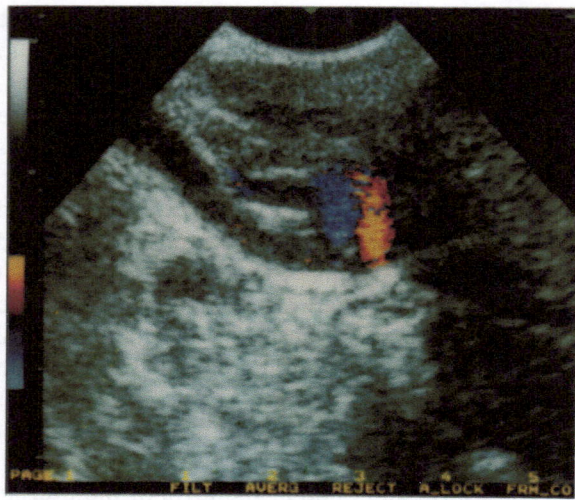

Fig. 5. Color Doppler image of the uterine vessels at the level of the uterine isthmus

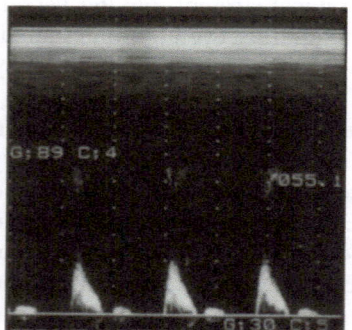

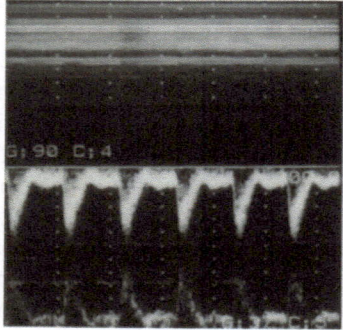

Fig. 6. FVW obtained from the uterine artery in the early follicular phase of the cycle. Note the absence of end-diastolic frequencies indicating high resistance to flow

Fig. 7. Presence of high end-diastolic frequencies in the uterine artery in the luteal phase of the cycle indicates increased uterine perfusion at the expected time of implantation

Physiological Changes During the Menstrual Cycle

Few studies have examined changes in uterine and ovarian perfusion during normal spontaneous menstrual cycles. Goswamy and Steptoe [12], using a transabdominal pulsed duplex system, studied 16 volunteers twice weekly during the menstrual cycle. They reported an increasing diastolic component in the uterine artery waveform associated with rising estradiol levels which increased further following ovulation. They ascribed this to a synergistic effect between estradiol and progesterone. There were some inconsistencies in this study, however. The impedance to flow was highest in mid-cycle when estradiol levels were at their highest, while the lowest resistance index values were on day 18, rising sharply on day 23 when progesterone levels were still rising. Scholtes et al. [17] studied 16 women on four occasions in the spontaneous cycle. They found no significant difference in the pulsatility index of either uterine artery on day 21 compared with day 7, although there was a small, but significantly lower pulsatifility index in the uterine artery on the side of the corpus luteum. Steer et al. [16] used transvaginal color Doppler to identify the uterine arteries on each side in 23 healthy women. Measurements were plotted for each day on either side of the LH peak. The lowest pulsatility index values were found on the day of the luteinizing hormone (LH) peak + 9 days. Their findings are consistent with maximal uterine perfusion at the time of peak luteal function and expected implantation (Figs. 6, 7).

What factors are involved in inducing increased perfusion during the luteal phase has yet to be elucidated [18]. The potential clinical application of these findings was investigated by Goswamy et al. [19]. They found increased resistance to flow in the luteal phase in 48% of women who failed to conceive after three successive embryo transfers. Steer et al. [16] have also defined a pulsatility index greater than 3 on the day of embryo transfer as predictive of failure to implant, and recommended cryopreservation in these cases. This will allow transfer when the uterus is more receptive following hormone therapy. Whether the uterine artery resistance to flow and the uterine receptivity can be manipulated in this way awaits further studies.

Doppler studies of ovarian arteries have shown more consistent results. A high increase in vascular supply to the ovary that contained preovulatory follicle or corpus luteum was documented in all published reports [6, 14, 20]. Differences between the dominant ovarian artery and the contralateral one were observed early in the follicular phase of the cycle, persisting until the late luteal phase [6]. However, blood flow in the dominant ovarian artery showed a steady decrease during the menstrual cycle with absence of any major changes in the periovulatory period. Recent color Doppler studies have confirmed a major role of the developing corpus luteum for increased blood flow in the ovarian artery. Fast development of blood supply to the preovulatory follicle starts after the rise in LH [18]. Over the hours preceding ovulation, there is a progressive increase in the number of blood vessels that supply the preovulatory follicle, accompanied by an increase in blood velocity and overall blood supply (Fig. 8). At the time of follicular rupture, there is a further dramatic increase in blood flow to early corpus luteum (Figs. 9–11). These observations confirm a primary role of the process of neovascularization in the development of blood supply to the corpus luteum. They also indicate an inadequacy of impedance measurements for the description of blood flow changes at the periovulatory period. It seems that velocity measurement and the shape of the waveforms provide more useful information about development of the corpus luteum vascular supply. Newly

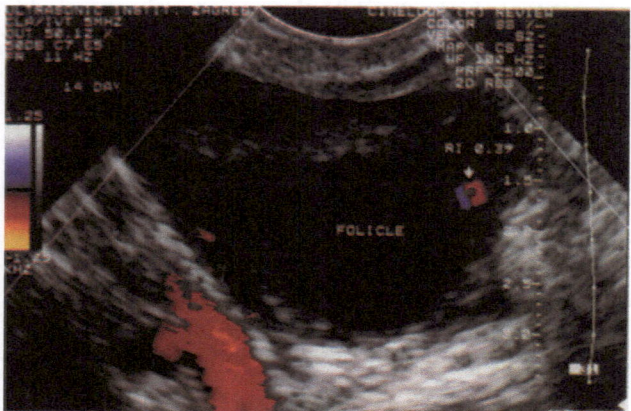

Fig. 8. Color flow mapping of blood supply to the preovulatory follicle in the late follicular phase of the cycle

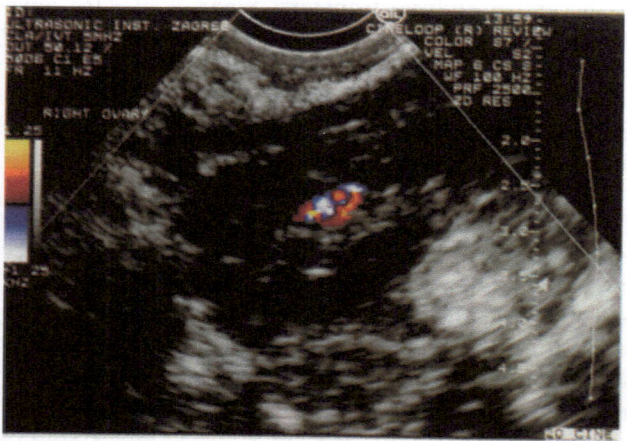

Fig. 9. Transvaginal color Doppler examination showing a highly vascularized area in the right ovary after ovulation. Although the corpus luteum could not be demonstrated on B mode scan, visualization of vascular supply enables its reliable identification

formed vessels lack muscular wall elements and are similar in their organization to blood vessels in tumor growths. This results in low impedance to flow with hemodynamical effects similar to those of the arteriovenous shunt.

The data on intrafollicular flow are novel and may have important implications for the management of women who wish to achieve or avoid conception. More studies are required to determine if blood flow velocity waveforms appearing at the time of the LH rise are necessary quantitatively and temporarily for follicular rupture and the release of a viable oocyte.

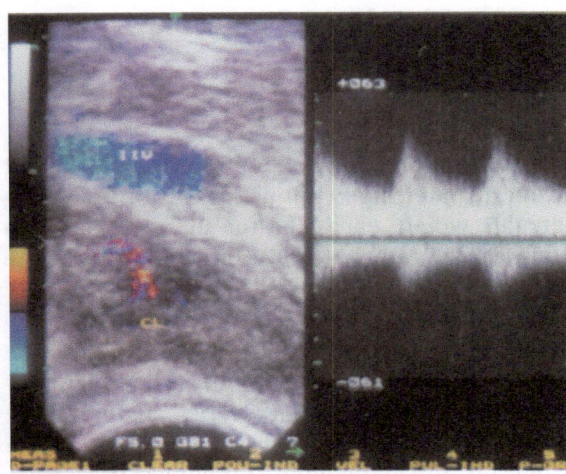

Fig. 10. Corpus luteum FVW show low pulsatility and high end-diastolic frequencies characteristic of low resistance to blood flow

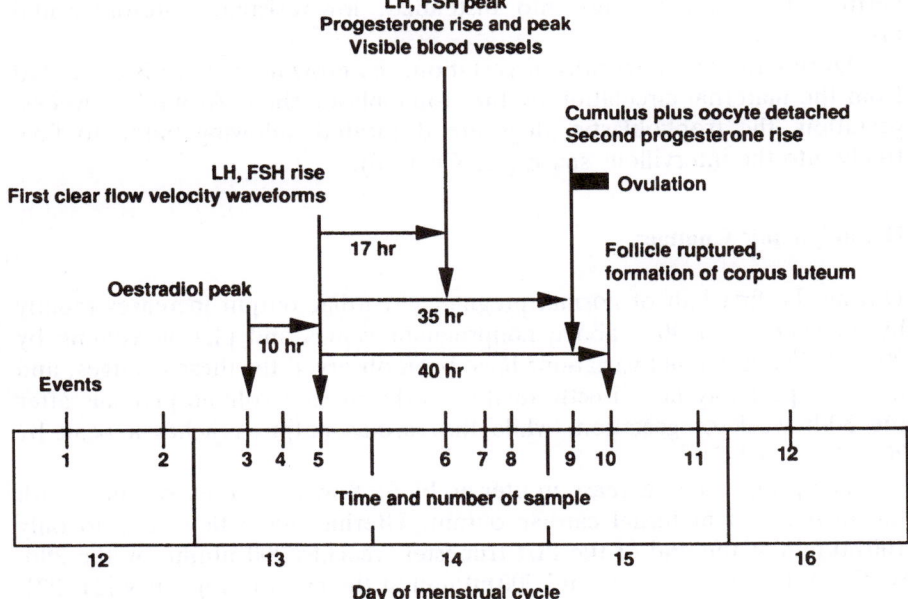

Fig. 11. Temporal relationship between morphological, hormonal, and Doppler parameters in the periovulatory phase of the cycle. There is increase in follicular blood supply at the time of Luteinizing hormone (*LH*) rise and relatively large blood vessels can be seen by color Doppler at the time of ovulation. *FSH*, follicle stimulating-hormone. (From [18])

Physiological Modifications Associated with Pregnancy

Anatomical Changes

The main uterine arteries give off branches which extend inwards for about one-third of the thickness of the myometrium without significant branching and then subdivide into an arcuate wreath encircling the uterus [21]. From this network arise smaller branches, the radial arteries, directed towards the uterine lumen. These arteries become the endometrial spiral arteries as they pass the myometrial-endometrial junction [21].

Human placentation is mainly characterized by diffuse infiltration of the placental bed by extravillous trophoblast (see Chap. 6). The migratory trophoblast infiltrates both the decidua and the spiral arteries, and at 10–12 weeks of gestation it reaches the deciduomyometrial junction. During the first 2 months of the second trimester, the extravillous trophoblast penetrates into the myometrial segments of the spiral arteries to their origins from the radial arteries [21]. These physiologic changes play a crucial role in converting the spiral arteries into distended, low-resistance uteroplacental arteries.

During the first 3 months of gestation, the growing embryo is separated from the maternal circulation by the trophoblastic shell. Around 12 weeks' gestation, the trophoblastic plugs are dislocated, allowing blood to flow freely into the intervillous space (see Chap. 6).

Hemodynamic Changes

During the first half of normal pregnancy, cardiac output increases rapidly by an average of 40% above nonpregnant values and plasma volume by 20%. Wide individual variations have been observed for these changes, and they are probably not directly related as the plasma volume plateaus after the 30th week of gestation, while the cardiac output reaches a peak by 20–24 weeks [22].

The progressive increase in uterine blood flow is also out of phase with the increase of maternal cardiac output. Uterine blood flow rises to only 100 ml/min at the end of the first trimester, reaches 200 ml/min by the 28th week, and is estimated around 500 ml/min at the end of pregnancy [21, 22]. Therefore, until midterm, the rise in uterine blood flow accounts only for a small portion of the increment in cardiac output.

The blood viscosity contributes to the amount of work necessary to push to blood through the system and is directly related to the hematocrit. Maternal total red cell volume increases steadily throughout gestation [22]. The increment varies from 20% to 40% of nonpregnant values, which is considerably less than the increase in plasma volume. As a result hemodilution occurs and is characterized by a decreased hematocrit.

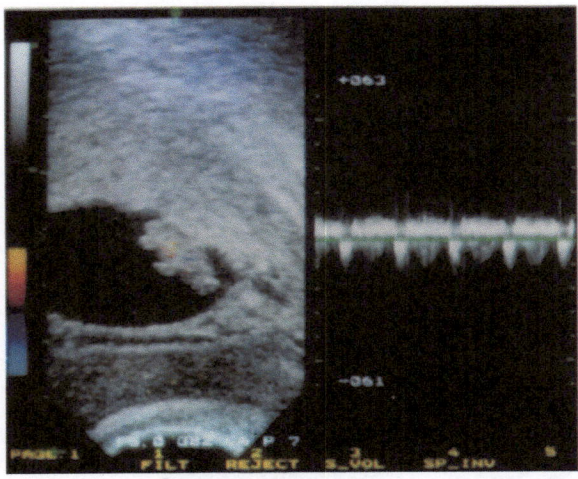

Fig. 12. Demonstration of heart action by pulsed Doppler at 7 weeks' gestation

Doppler Features of Early Pregnancy Circulations

Embryonic Circulation

By using a transvaginal probe, the embryonic heart action can be documented (Fig. 12) as early as 35 days after the beginning of the last menstrual period [23]. In in vitro fertilization (IVF) patients with a certain date of conception, heart action can be detected in all cases 25 days after embryo transfer, i.e., 39 days of menstrual age [24]. From 6 to 9 weeks of gestation there is a rapid increase of the mean heart rate from 113 to 167 beats per minute [25]. The rise in the heart rate parallels the increase in crown-rump length. The increased heart rate is probably required to increase cardiac output and meet the demands of the growing conceptus at this stage of pregnancy. A longitudinal study [24] has established reference ranges for fetal heart rate in early pregnancy. Comparison between normally developing pregnancies and pregnancies that resulted in early miscarriage showed a late onset of cardiac activity in latter group. Another interesting observation was a decreased heart rate in all cases that ended in spontaneous abortion [24]. The significance of this finding needs to be confirmed, and, if proven, the assessment of the fetal heart rate may be used in future as an extremely simple and easily reproducible test for the prediction of pregnancy outcome after infertility treatment and in patients with threatened miscarriage.

Blood flow in the umbilical vessels (Fig. 13) and fetal aorta (Fig. 14) can be visualized by color Doppler how 7 weeks' gestation [15]. The intracranial circulation and renal artery flow can be successfully analyzed from 10 weeks' gestation onwards [26]. Blood flow in the umbilical artery has been most extensively investigated. During the first 12 weeks of gestation the FVW from these vessels (Fig. 15) show high pulsatility and absence of end-diastolic

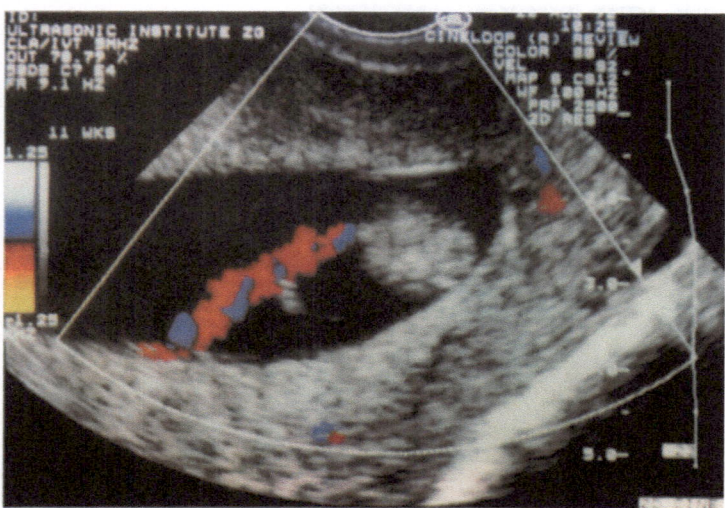

Fig. 13. Examination of the umbilical cord with color Doppler at 11 weeks' gestation shows two arteries (*red*) and vein (*blue*)

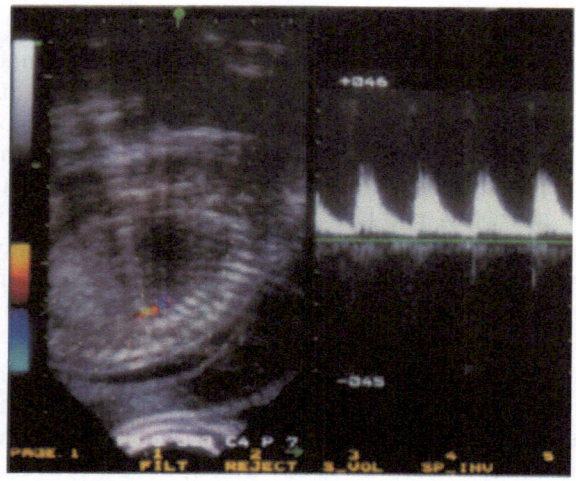

Fig. 14. Demonstration of blood flow in the aorta at 17 weeks' gestation

velocities [27]. Between 12 and 14 weeks there is a progressive development of end-diastolic velocity (Fig. 16), and from 14 weeks it can always be seen in normal pregnancies [28]. It is not clear what the major factors regulating this phenomenon are. It seems that fetal cardiac output, heart rate, and blood pressure do not influence the pulsatility index of the umbilical artery [26]. A more likely explanation could be a fall in vascular resistance in the

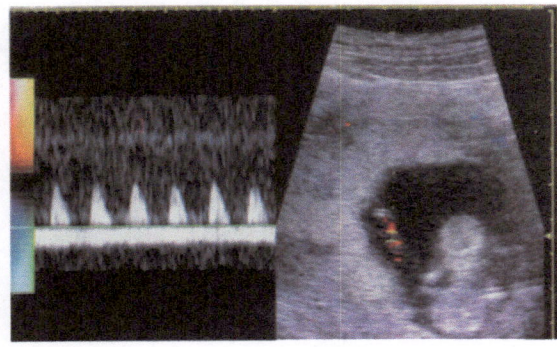

Fig. 15. Doppler examination of the umbilical cord shows absence of end-diastolic velocities at 10 weeks' gestation. (From [28])

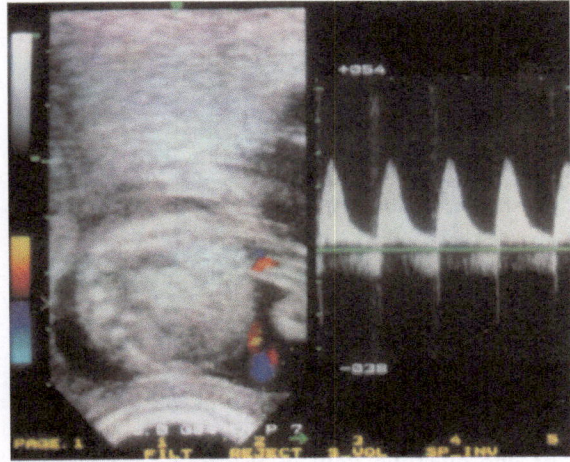

Fig. 16. Examination of blood flows in the umbilical artery at 15 weeks' gestation revealing the presence of end-diastolic velocities

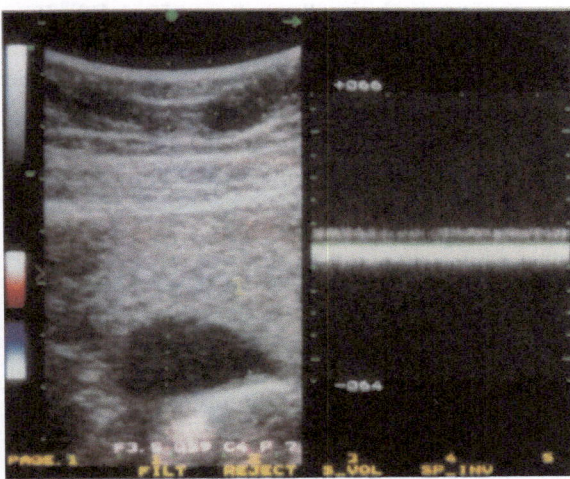

Fig. 17. Color Doppler of the placenta at 16 weeks' gestation showing intervillous flow with a venous flow pattern

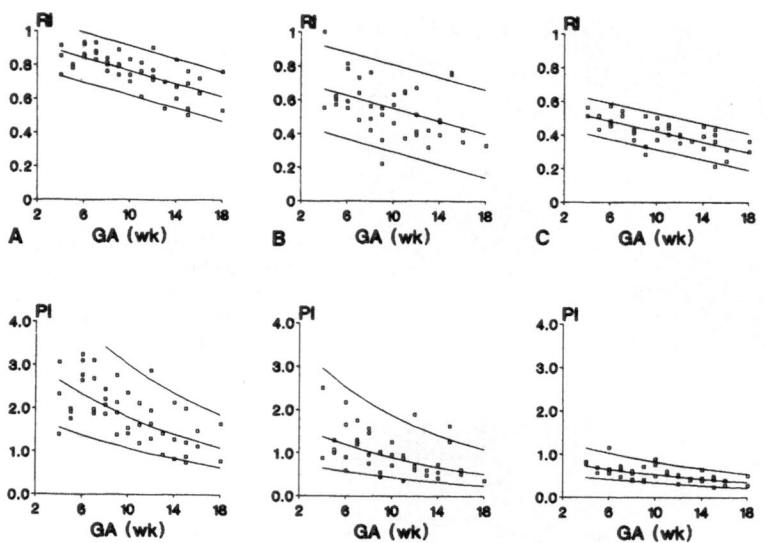

Fig. 18A–C. Reference ranges for the resistance (*RI*) and pulsatility (*PI*) index with gestation in the uterine (**A**), radial (**B**), and spiral (**C**) arteries. Note decreased resistance to flow in all vessels with advancing gestation. (From [33])

villous circulation related to changes in blood flow in the uteroplacental circulation [29].

The anatomical changes in the villous vasculature which are characterized by the progressive increase of the number and the surface area occupied by the fetal vessels must have an important role in the gradual fall in resistance to blood flow in the umbilical circulation. However, in our experience, the appearance of end-diastolic flow in the umbilical arteries coincides exactly with a higher blood velocity in the main uterine arteries and with the presence of a continuous intervillous flow (Fig. 17). The establishment of the intervillous circulation at the beginning of the 2 trimester of pregnancy (see Chap. 6) may be associated with changes in the pressure gradient between the uterine circulation and the intervillous space and/or with alterations in blood gases and metabolites which could influence the appearance of end-diastolic flow in the umbilical circulation [29].

Uteroplacental Circulation

A progressive decrease in resistance to blood flow in the uterine artery during the first trimester of pregnancy has been reported by studies using either continuous wave or pulsed Doppler [10, 29–33]. This decrease continues during the first half of the 2 trimester and then remains steady until term (Fig. 18). The sharp fall in resistance in early pregnancy is most probably caused by the combined effect of hormones secreted by corpus luteum and early trophoblast infiltration [29, 34]. By using color Doppler, it

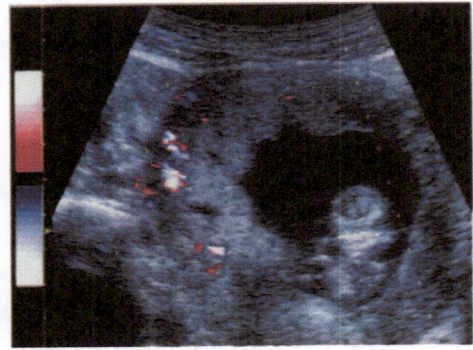

Fig. 19. Color flow mapping of the radial and spiral vessels below the gestational sac at 10 weeks' gestation

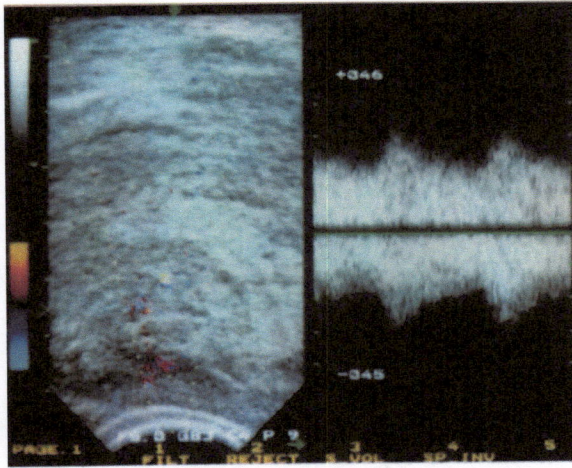

Fig. 20. Spiral artery FVW at 8 weeks' gestation. Note low impedance to flow and spiky shape of the waveform that are characteristic of these vessels

is possible to visualize intrauterine arterial branches, i.e., radial and spiral arteries (Fig. 19) and to analyze their blood flow characteristics by pulsed Doppler [30, 33, 34]. FVW from small arteries show a significantly lower pulsatility compared to the main uterine artery (Fig. 20). This is probably an effect of the branching of the uterine circulation and an increased cross-sectional area resulting in lower impedance to flow [30, 33]. Recognition of this typical flow pattern may be very useful in the early diagnosis of ectopic pregnancy [35].

Studies using conventional Doppler techniques have demonstrated that an increased impedance to flow in the uteroplacental circulation at 18–22 weeks' gestation predicts the development of preeclampsia and intrauterine growth retardation [8]. Transvaginal color Doppler allows the examination of the uteroplacental circulation much earlier in pregnancy (Fig. 21). It remains to be established whether abnormal blood flow indices in the uteroplacental circulation identified in the second trimester are already

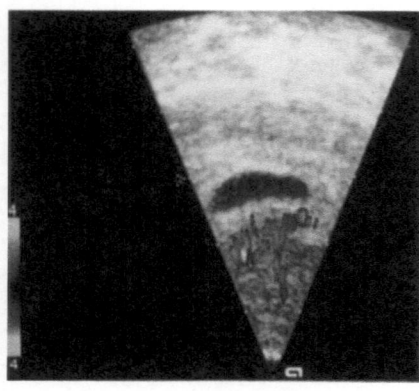

Fig. 21. Color flow mapping of the terminal part of the uteroplacental circulation at 5 weeks' gestation

present in the first trimester of pregnancy. It also remains to be investigated whether abnormal indices of blood flow in the spiral and radial arteries may be used for the diagnosis of abnormal implantation and for prediction of spontaneous abortion in the first trimester.

Recommendations for the Safe Use of Transvaginal Doppler Ultrasound

When the vaginal route is used for assessment of the fetoplacental circulation, the risks of excessive heating of embryonic tissue are higher when compared to the transabdominal route. The transducer itself is a significant heat source. A transducer surface temperature increase of about $0.5°C \, m \, W^{-1} cm^{-2}$ at 5 min in air was reported for commercial Doppler equipment [36]. With endovaginal probes, the increased temperature generated by the transducer will be absorbed into surrounding tissues. Furthermore, the short distance between the transducer and the embryo makes heating from the transducer more likely. The potential for heating is further increased by the use of high frequency transducers (5–10 MHz).

Modern endovaginal transducers usually offer three diagnostic facilities: B-mode, pulsed Doppler and color Doppler. In terms of ultrasonic power intensity output, the pulsed Doppler mode is most likely to expose the embryo to potentially harmful levels. The risk to the embryo is, therefore, substantially reduced if the duration of the pulsed Doppler examination is reduced to a minimum. The ability to visualize vessels of interest by color Doppler simplifies blood flow studies and shortens the use of pulsed Doppler to just a few seconds. At the same time, a reliable and easy identification of the vessels prevents unnecessary exposure of the embryo and other tissues to the "wandering" pulsed Doppler beam [28]. Risks can also be reduced by cooling the ultrasound probe in the antiseptic solution, between examination and switching off the equipment.

Conclusion

Color Doppler will most probably be the dominant technique for the assessment of uteroplacental circulation in the following years. Its introduction opens new and wide areas for blood flow studies in vessels that were inaccessible by traditional Doppler techniques. As experience with color Doppler is still limited, we are very far from understanding its full potential. It is also unclear whether present methods for blood flow analysis based on the use of indices of impedance are appropriate to describe blood flow changes in terminal arterioles or whether new methods have to be developed. The value of blood flow studies of intrauterine, intraovarian, and intraplacental circulation for diagnosis and prediction of various pregnancy abnormalities awaits further studies, although preliminary data suggest a promising role in several areas of clinical importance.

References

1. Maulik D (1989) Basic principles of Doppler ultrasound as applied in obstetrics. Clin Obstet Gynecol 32:628–643
2. Griffin D, Cohen-Overbeek T, Campbell S (1983) Fetal and utero-placental blood flow. Clin Obstet Gynecol 10:565–602
3. Mitchell DG (1990) Color Doppler imaging: principles, limitations and artifacts. Radiology 177:1–10
4. Kurjak A, Zalud I, Jurkovic D, et al. (1989) Transvaginal color Doppler for the assessment of pelvic circulation. Acta Obstet Gynecol Scand 68:131–135
5. Bourne TH, Campbell S, Steer C, et al. (1989) Transvaginal color flow imaging: a possible new screening technique for ovarian cancer. Br Med J 299:1367–1370
6. Taylor KJW, Burns PN, Wells PNT, et al. (1985) Ultrasound Doppler flow studies of the ovarian and uterine arteries. Br J Obstet Gynaecol 92:240–246
7. Kurjak A, Jurkovic D (1986) New ultrasonic technique for assessing circulation in female pelvis. In: Kurjak A, Kossoff G (eds) Recent advances in ultrasound diagnosis 5. Excerpta Medica, Amsterdam, p 198
8. Campbell S, Diaz-Recasens J, Griffin DR, et al. (1983) New Doppler technique for assessing utero-placental blood flow. Lancet 1:675–677
9. Trudinger B, Giles WB, Cook CM (1983) Uterine artery velocity waveforms in pregnancy. J Ultrasound Med 2:10–14
10. Deutinger J, Rudelstrorfer R, Bernaschek G (1988) Vaginosonographic velocimetry of both main uterine arteries by visual vessel recognition and pulsed Doppler method during pregnancy. Am J Obstet Gynecol 159:1072–1076
11. Barber RJ, McSweeney MB, Gill RW, et al. (1988) Transvaginal pulsed Doppler ultrasound assessment of blood flow to the corpus luteum in IVF patients following embryo transfer. Br J Obstet Gynaecol 95:1226–1230
12. Goswamy RK, Steptoe PC (1988) Doppler ultrasound studies of the uterine artery in spontaneous ovarian cycles. Hum Reprod 3:721–726
13. Sterzik K, Hutter W, Grab D (1989) Doppler sonographic findings and their correlation with implantation in an in vitro fertilization program. Fertil Steril 52:825–828
14. Deutinger J, Reinthaller A, Bernaschek G (1989) Transvaginal pulsed Doppler measurement of blood flow velocity in the ovarian arteries during cycle stimulation and after follicle puncture. Fertil Steril 51:466–470

15. Kurjak A, Jurkovic D, Alfirevic Z, et al. (1990) Transvaginal color Doppler imaging. JCU 18:227–234
16. Steer CV, Campbell S, Pampiglione JS, et al. (1990) Transvaginal color flow imaging of the uterine arteries during the ovarian and menstrual cycles. Hum Reprod 4:391–395
17. Scholtes MCW, Wladimiroff JW, van Rijen HJM, et al. (1989) Uterine and ovarian flow velocity waveforms in the normal menstrual cycle: a transvaginal Doppler study. Fertil Steril 52:981–985
18. Collins WP, Jurkovic D, Bourne TH, et al. (1991) Ovarian morphology, endocrine function and intra-follicular blood flow during the peri-ovulatory period. Hum Reprod 6:319–324
19. Goswamy RK, Williams G, Steptoe PC (1988) Decreased uterine perfusion – A cause of infertility. Hum Reprod 3:955–959
20. Bourne TH, Jurkovic D, Waterstone J, et al. (1991) Intrafollicular blood flow during human ovulation. Ultrasound Obstet Gynecol 1:63–69
21. Ramsey EM, Donner NW (1980) Placental vasculature and circulation. Thieme, Stuttgart
22. McAnulty JH, Metcalfe J, Veland K (1982) Cardiovascular disease. In: Burrow GN, Ferris TF (eds) Medical complications during pregnancy. Saunders, Philadelphia, p 145
23. Bernaschek G, Deutinger J, Kratochwil A (1990) Normal early pregnancy. In: Endosonography in obstetrics and gynecology. Springer, Berlin Heidelberg New York, p 41
24. Schats R, Jansen CAM, Wladimiroff JW (1990) Embryonic heart activity: appearance and development in early human pregnancy. Br J Obstet Gynaecol 97:989–994
25. Van Heeswijk M, Nijhuis JG, Hollanders HMG (1990) Fetal heart rate in early pregnancy. Early Hum Dev 22:151–156
26. Wladimiroff JW, Huisman TWA, Stewart PA (1991) Fetal and umbilical flow velocity waveforms between 10–16 weeks' gestation: A preliminary study. Obstet Gynecol 78:812–814
27. Fisk MN, MacLachlan N, Ellis C, et al. (1988) Absent end-diastolic flow in first trimester umbilical artery. Lancet 2:1256–1257
28. Jauniaux E, Jurkovic D, Campbell S (1991) In vivo investigations of anatomy and physiology of early human placental circulations. Ultrasound Obstet Gynecol I: 435–445
29. Jauniaux E, Jurkovic D, Campbell S (1992) Doppler ultrasound features of the developing placental circulations: Correlation with anatomic findings. Am J Obstet Gynecol 166
30. Jauniaux E, Jurkovic D, Campbell S, et al. (1991) Investigation of placental circulations by color Doppler ultrasound. Am J Obstet Gynecol 164:486–488
31. Schulmann H, Fleischer A, Farmakides G, et al. (1986) Development of uterine artery compliance in pregnancy as detected by Doppler ultrasound. Am J Obstet Gynecol 155:1031–1036
32. Thaler I, Manor D, Itskovitz J, et al. (1990) Changes in uterine blood flow during human pregnancy. Am J Obstet Gynecol 162:121–125
33. Jurkovic D, Jauniaux E, Kurjak A, et al. (1991) Transvaginal color Doppler assessment of the uteroplacental circulation in early pregnancy. Obstet Gynecol 77:365–369
34. Jauniaux E, Jurkovic D, Kurjak A, et al. (1990) Assessment of placental development and function. In: Kurjak A (ed) Transvaginal color Doppler. Parthenon, Carnforth, pp 53–65
35. Jurkovic D, Bourne TH, Jauniaux E, et al. (1992) Doppler study of blood flow in ectopic pregnancies. Fertil Steril 58
36. Duck FA, Starritt HC, Haar GR, et al. (1989) Surface heating of diagnostic ultrasound transducers. Br J Radiol 62:1005–1013

6 The Maternotrophoblastic Interface: Uteroplacental Blood Flow

J. Hustin

Soon after implantation, primary trophoblast exerts its proteolytic properties; distended vascular lakes are thus opened, and maternal red blood cells appear within the lacunae at the previllous stage [1]. Knoth and Larsen [2] were surprised to find only one example of such penetration in the ultrastructural study of their 11-day-old implanted ovum. Enders [1], quoting a Carnegie Institute publication by O'Rahilly and Muller, suggests that there is a true blood flow within lacunae of the human placental disk 10–11 days after ovulation, i.e., 4–5 days after implantation. There is now general agreement that the vessels which are tapped are venous capillaries or lakes. It is therefore obvious that only a limited quantity of blood will enter the lacunae through a retrograde effect. Kaufmann [3] has proposed that spiral arteries are tapped around days 28–29. It thus appears that some maternal blood is diverted to the lacunae, but most probably in minute quantities.

Obviously during the initiation of growth of the placenta, several spiral arteries must be encountered. All anatomic studies [3–7] have thoroughly described the focal destruction of the vessel wall by extravillous trophoblast in the course of decidual infiltration. These histologic observations are not an absolute proof of the early establishment of uteroplacental circulation, i.e., of a real blood flow on the intervillous space. On the one hand, it is taken for granted that the early embryo needs important supplies for growth and differentiation. On the other hand, embryonic circulation with proven heart beats is established around days 22–23. The embryo is then about 3 mm long. It is evident that the blood pressure in such embryonic vessels and villous capillaries must be extremely low. The structure of these vessels is quite simple and has no resistance whatsoever to external pressure. Therefore a very low pressure in the intervillous space is mandatory.

Ultrasound studies, especially those using vaginal probes, have demonstrated that, in early gestations, uterine (spiral? myometrial?) vessels could be visualized approximately 1 cm apart from the trophoblastic ring [8, 9] (see also Chap. 4). In these blood vessels, small moving echoes could be identified, synchronous to the maternal pulse.

In later stages, the vascular network becomes more and more conspicuous and grows nearer to the trophoblastic shell which is markedly thickened in the region of the future placenta. However, decidual vessels do not reach the intervillous space, and it is impossible to define any particulate movement in the trophoblastic area during the first 12 weeks. Lastly, with the current practice of chorion villous sampling either through a choriono-scope [10, 11] or with a rigid forceps [9], it has been demonstrated that the villous material so obtained was usually not contaminated with maternal blood.

All these findings led us to reconsider classic theories about early maternoplacental circulation. We will discuss in this chapter all the anatomic data which enabled us to suggest that, during the 1st trimester, the inter-villous space is not directly connected with maternal circulation and is probably not bathed by true maternal blood. We have studied hysterectomy specimens with pregnancy in situ, most often with serial reconstruction of the insertion site, and aspiration material collected from pregnancies voluntarily interrupted between weeks 10 and 12. Lastly, we had the op-portunity to inject radioopaque material into the uterine arteries of three hysterectomy specimens with pregnancy in situ. These specimens were sliced and X-rayed. Later microradiographs of 50-μm sections were also obtained.

Static Morphologic Studies

Hysterectomy Specimens with Pregnancy In Situ

Our early series comprised only three cases [12]. We can now add four new specimens where we could obtain serial blocking and serial sections from the implantation site. Moreover, thanks to the generous cooperation of Prof. G. Burton (Head of Subdepartment of Anatomy, University of Cambridge, UK), we gained access to the collection of the late Professor Boyd [5]. Through a thorough study of the whole collection, we could select 11 specimens which belonged to the first trimester as judged from clinical data and histologic study.

The survey of serial sections from 17 hysterectomies with pregnancy in situ yielded results completely in keeping with previous results [12, 13]. All placentae had a fully developed villous architecture, and most of them had a well-defined division between chorion laeve and chorion frondosum. The interrelations between trophoblast and decidua, from the cellular and bio-chemical points of view, are discussed in Chap. 7.

Nevertheless, whatever the gestational age, the picture was clear. The future fetoplacental unit is like a partially inflated ball completely embedded in the endometrium (Fig. 1). Its outer part is made of the trophoblastic shell. It consists of a thick layer of residual primary trophoblasts mixed together with actively growing cytotrophoblasts and the peculiar extravillous "intermediate" trophoblasts (see Chap. 1). Grossly it appears as an irregular whitish line surrounding the intervillous space (Fig. 2). During the first 12

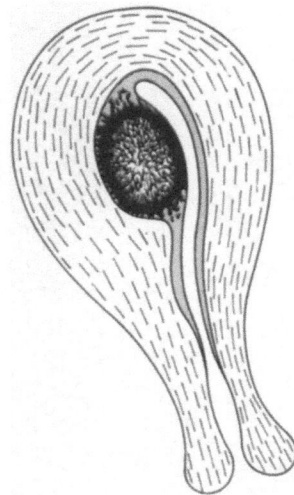

Fig. 1. Concept of early conceptus as a closed system embedded in the gestational endometrium

weeks, the trophoblastic shell is devoid of any loss of continuity. Its inner part is directly linked with the first villous trees by the trophoblastic columns which anchor the villi and provide the necessary supply of proliferative cells (Fig. 3). Its thickness is progressively reduced with time.

During the first weeks, before true partition of the chorion laeve and chorion frondosum, a schematic drawing of the early embryo and placenta would represent the trophoblastic shell as the outer part of a wheel with cytotrophoblastic cell columns and primitive villi as spokes and chorionic plate and embryo as the hub. The trophoblastic shell is, however, not compact. There is a communicating system of lacunae within the cytotrophoblastic masses. These lacunae probably correspond to the early primitive ones of the previllous stages. They are not tight. Intercellular spaces – i.e., between cytotrophoblasts – are directly connected with this labyrinthine network as proven by India ink perfusion (Fig. 4A). Serial sectioning of the implantation site demontrates that the outer part of the trophoblastic shell is not smooth: it is bristling with cell cones. Quite a number of these are strictly located in the openings of spiral arteries. The remainder give rise to the population of wandering interstitial trophoblasts [14].

Like many others, we have been impressed by the so-called physiologic changes occurring at the spiral artery level. These have been described at length [3–5, 7, 15]. Two points are, however, important to note:

1. Spiral arteries are functionally occluded by trophoblastic plugs. These are integral parts of the trophoblastic shell. Cells detach progressively and drip downwards in a "candle-wax" fashion [4]. The cohesivity between trophoblastic cells within the plugs is rather loose. As beautifully

Fig. 2. Hysterectomy with pregnancy in situ (9 weeks). The trophoblastic shell appears as a white compact line merging with the underlying decidua (*arrows*)

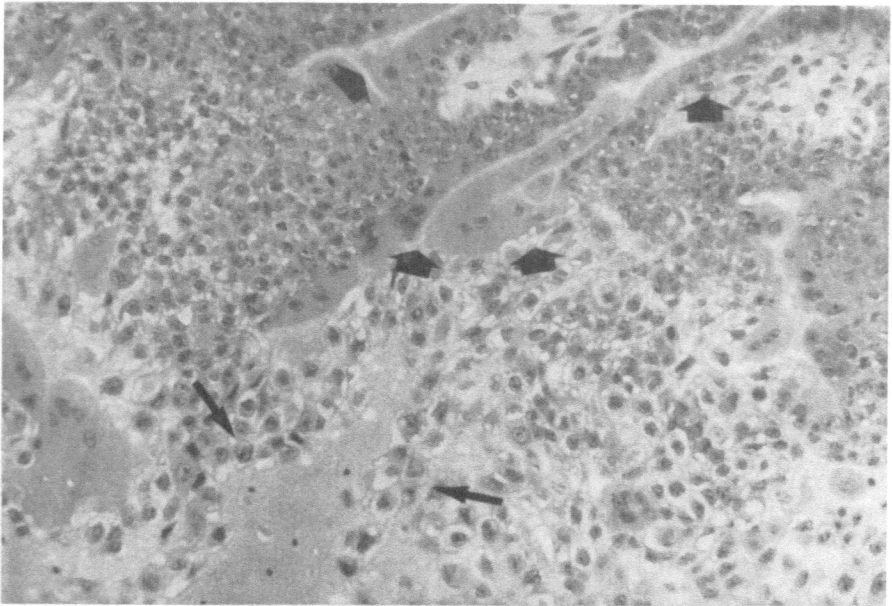

Fig. 3. Normal pregnancy at 7 weeks – voluntarily interrupted. The trophoblastic shell is thick with lacunae and some maternal red blood cells (*small arrows*). The intervillous space is marked by *large arrows*. Hematoxylin and eosin; ×160

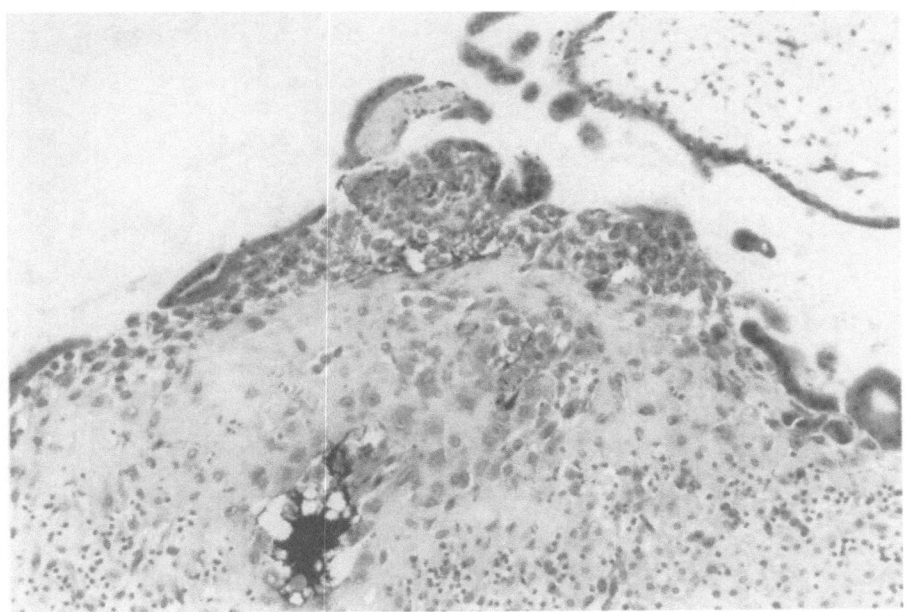

A

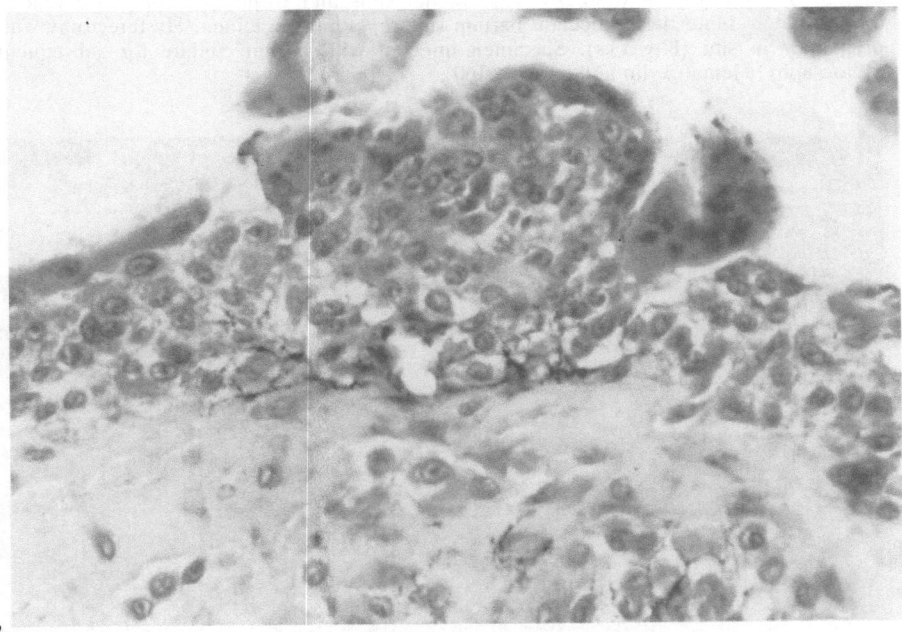

B

Fig. 4. A Specimen H653 (collection of Prof. Boyd, Cambridge University, UK): hysterectomy with 10-week pregnancy in situ. India ink injection prior to fixation. The end of a spiral artery is filled with a trophoblastic plug. Some India ink, however, can be identified between the trophoblastic cells. **B** Higher magnification of the same specimen as Fig. 3. India ink is observed between cytotrophoblasts of the trophoblastic shell. Hematoxylin and eosin; **A** ×160; **B** ×400

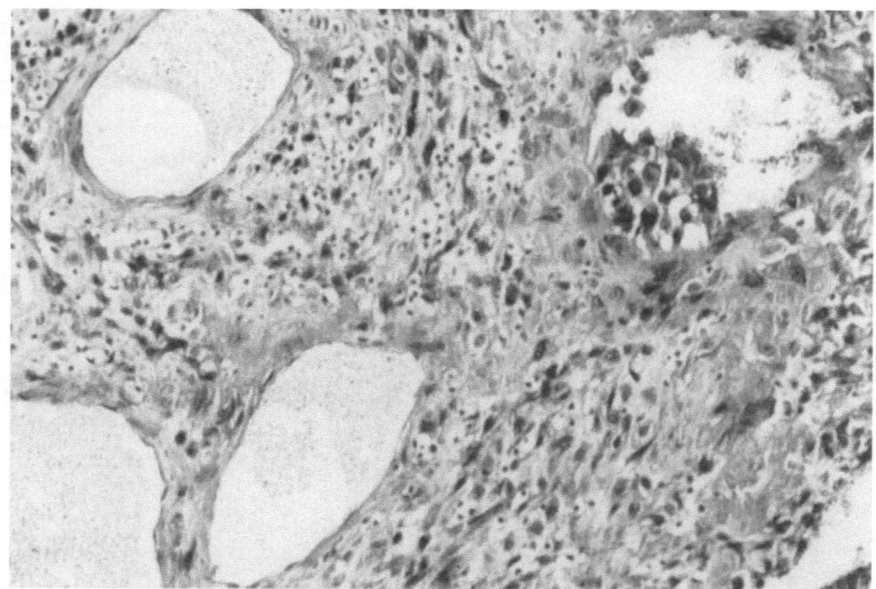

Fig. 5. Spiral artery at 8 weeks. Endothelial cells and trophoblast plug are readily differentiated. Note the particulate barium sulfate within the lumen. Hysterectomy with pregnancy in situ (8 weeks). Specimen injected with barium sulfate for subsequent radiography. Hematoxylin and eosin; ×160

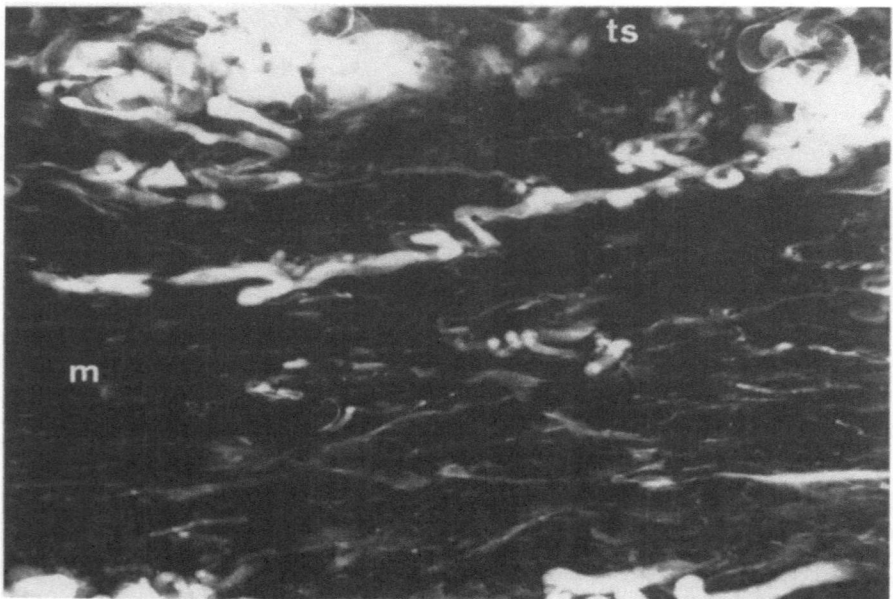

Fig. 6. Hysterectomy with 8-week pregnancy – arterial perfusion with barium sulfate. X-ray of a thick slice. The course of a spiral artery is extremely tortuous and appears very oblique. Note the extreme widening at close contact with the trophoblastic shell (*ts*); *m*, myometrium

demonstrated by the India ink perfusion experiments of Boyd and Hamilton [5] (Fig. 4B), fluid and very small caliber particles can percolate through the plugs and finally reach the labyrinthine system within the trophoblastic shell. Intravascular trophoblastic plugs do not incorporate with the vessel wall nor do they acquire tight junctions with endothelial cells (Fig. 5). This implies that they are loose and must act like valves. Direct observation with the chorionoscope confirms this fact. When the instrument is in place in the intervillous space [9] (see also Chap. 23), vision is perfect and villi appear as whitish threads with some pinkish central discoloration due to the presence of fetal capillaries. In the procedure of obtaining villi for sampling, these are some mandatory movements which modify slightly and temporarily the pressure in the intervillous space and probably lower it. Not infrequently, a very tiny streamer of blood is seen against the trophoblastic shell; it usually subsides rapidly.

2. Spiral arteries run a very tortuous course which is subject to continuous changes. Physiologically, there are always necrotic foci within the decidua. Arterial segments also undergo necrosis. There are, however, new channels which are opened. In serial reconstructions, a given artery may be "opened" several times and occluded by several trophoblastic plugs (Fig. 6). We calculated that the horizontal length of a spiral artery in the decidua was between 500 and 750 µm. Owing to the numerous coils, the real length must be 2–3 mm. This is the accepted length for uteroplacental arteries at term [4–15].

Aspiration Material from Voluntary Interruptions of Pregnancy

We had to insist on the good viability of placenta and embryo in all 75 specimens, as demonstrated by the strictly normal appearance of all histologic structures. These specimens were selected because they all contained the decidua with the implantation site. The inevitable artefacts due to disruption by aspiration or curettage were, of course, almost always present. They were, however, easily recognizable. The most prominent was evidently the presence of blood and decidua intermingled with placental tissue.

The trophoblastic shell was always present though usually fragmented. It appeared thicker than in the hysterectomy specimens with pregnancy in situ. This is perhaps due to the rupture of the shell and some degree of villous contraction towards the chorionic plate. Outer trophoblastic plugs were usually not present in the upper parts of spiral arteries, though isolated trophoblasts could still be identified in many sections.

The inner part of the trophoblastic shell was better preserved in its connection with cytotrophoblastic cell columns and villous development. In most cases, chorion laeve and chorion frondosum could already be differentiated. The operculum, i.e., the endometrial layer covering the chorion

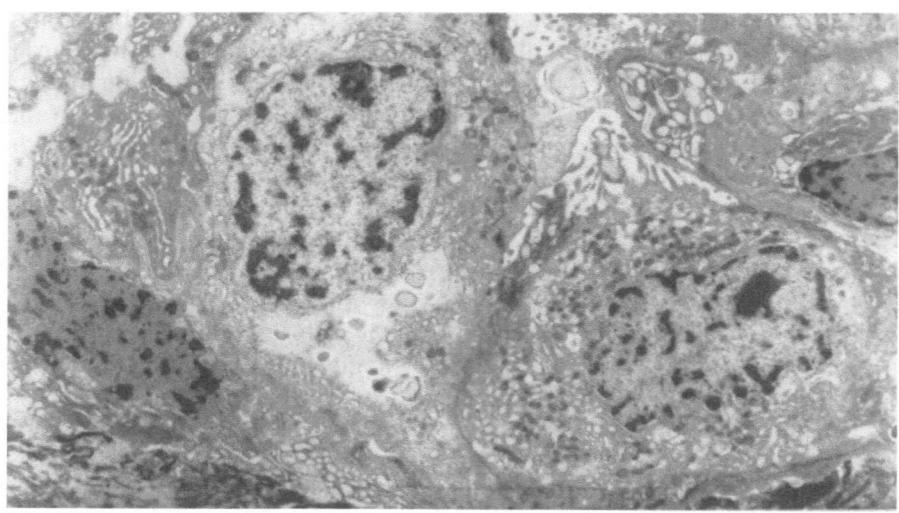

Fig. 7. Intravascular plug. Note the functional (secretory) appearance of trophoblasts Electron microscopy; ×3000

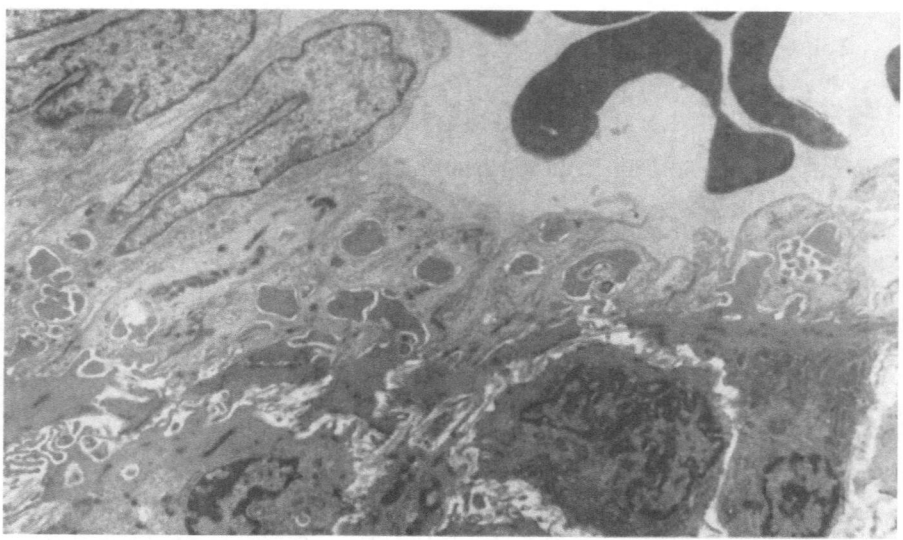

Fig. 8. Spiral artery. Endothelial cells rest on an interrupted basal lamina. Extravascular trophoblasts also stick to the basal lamina. Electron microscopy; ×3000

laeve, was thin with a thick layer of fibrin mixed with decidual cells and trophoblasts. There was no fibrin layer at the implantation site except in very limited areas corresponding to necrotic (ischemic) foci. It was also noted that trophoblast infiltration within the decidua proceeded in two directions: one interstitial, often of greater density around spiral arteries and made of individual cells; and the other still linked to the trophoblastic shell and corresponding to the intravascular plugs.

In selected cases, electron microscopy of the placental site was performed. The intravascular plugs (Fig. 7) were made of cells which differed in their morphology from those of extravascular interstitial trophoblast (Fig. 8). Note in Fig. 8 that endothelial cells look normal, while the basal lamina of the vessel is frequently interrupted.

Angiography of Hysterectomy Specimens

Three cases have been studied so far: two were injected (at 8 and 13 weeks, respectively) by the arterial route, and one (at 21 weeks) by the uterine veins. The specimen with an 8-week pregnancy did not demonstrate any injection of the placental area (Fig. 9), though the perfusion pressure was equilibrated at 100 mmHg at the entrance of uterine arteries. This value is ten times that suggested by Moll et al. [16] for uteroplacental arteries. We have, however, no idea of the pressure which existed at this level during perfusion. In thick 50-μm sections (Fig. 10), the absence of contrast medium between villi was evident. Standard histologic examination confirmed that minute amounts of particulate medium (barium sulfate) were present only in the lacunae of the trophoblastic shell. Uteroplacental arteries appeared, of course, much more distended due to the high perfusion pressure. It ensued that trophoblastic plugs looked somewhat – but not completely – dislodged (Fig. 5).

At 13 weeks, the placental area was immediately and massively injected with contrast medium (Fig. 11), which followed a straight course parallel to the villi. There was no organization pattern in cotyledons. Lastly, the 21-week specimen was injected in retrograde fashion via the uterine veins. Surprisingly, no contrast medium could be observed in the intervillous space. There has not been coherent a explanation for this.

Comments

Obviously, the hypothesis of a delayed blood flow in the intervillous space raises important questions related to transfer from the maternal organism to the embryo. This implies both fuel and oxygen supply. It is logical to assume

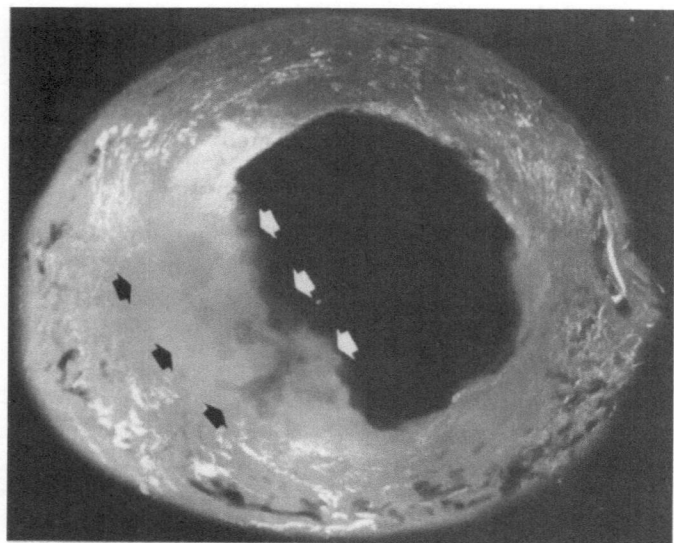

Fig. 9. Hysterectomy with 8-week pregnancy and arterial perfusion with barium sulfate and X-ray. The area comprised between *black* and *white arrows* is the placenta devoid of perfusion medium

Fig. 10. Microradiograph of the same specimen as in Fig. 9. *White arrows* delineate the placental site. No contrast medium is present in the intervillous space

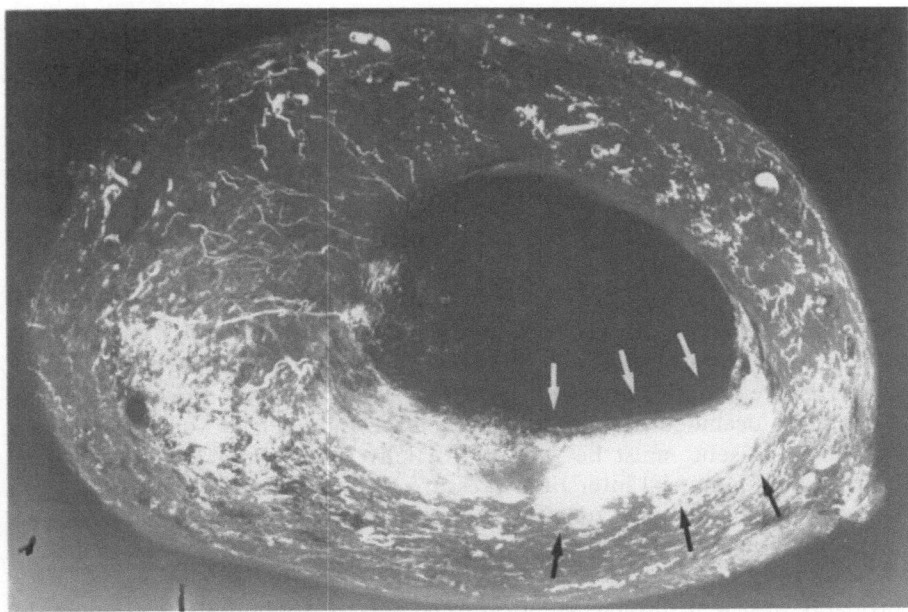

Fig. 11. Hysterectomy with 13-week pregnancy in situ and arterial perfusion with barium sulfate and X-ray. Heavy injection (*arrows*) of the placental area

that intense growth and differentiation of the embryo is associated with important requirements. Some arguments have already been discussed which might suggest that during the first weeks of gestation human placentation is principally deciduochorial [17] (see also Chap. 2). Endometrial secretions and, particularly, glycoproteins seem to play an important part in early embryonic (trophoblastic) nutrition. Bell [18] has suggested that pregnancy-associated endometrial α_1 globulin (α_1-PEG) could be locally used for retinol transfer from mother to conceptus in order to supply a potent stimulus for growth and differentiation. Similarly, decidual cells produce, among others, insulin growth factor-binding protein (IGF-BP) and prolactin. It is also highly probable that other growth factors are produced in order to modulate embryonic growth. We do not pretend that no blood product is present in early intervillous space. The idea is that what is prevented is the early initiation of a true blood flow. On the contrary, some maternal blood will eventually be allowed to percolate between the intra-arterial trophoblastic plugs. When reaching the trophoblastic shell, this small amount of whole blood is probably filtered in a most efficient manner.

It ensues, finally, that a small amount of fluid (plasma) eventually wells out of the inner openings of the trophoblastic shell and fills the intervillous space. It is highly probable that the local pressure is someway between the interstitial intraendometrial pressure and the intravillous pressure, that is to say, very low. Moll et al. [16] demonstrated that in the nonhuman primate, the pressure within spiral arteries of pregnant females was low (more or less 10 mmHg). They suggested that these low values could point to an important placental blood flow. This deduction was made, however, from late pregnancy experiments when uteroplacental arteries open widely in the intervillous space. Boyd and Hamilton [5] postulated that the endovascular trophoblastic plugs could really exist only if the local blood pressure was very low.

The problem of early deciduotrophoblastic circulation is further hampered by the necessary problem of venous return. Endometrial veins are not prominent during the first trimester, while in the myometrium venous channels are readily identified. Just like spiral arteries are more or less completely occluded by trophoblastic plugs, endometrial veins, which are obviously present, must be at least partially closed. One might speculate that local endometrial interstitial pressure could be the closing pressure for veins with only minute quantities of filtered plasma pouring out of the intervillous space.

We postulate that the nutrient transfer to the early embryo must be probably sufficient through the local route and the limited plasma import. In Chap. 3, Jauniaux et al. demonstrate that the thickness of the barrier (trophoblast plus villous core) separating the intervillous space from villous capillaries is higher during the first 12 weeks than during the second trimester. This could be indirect evidence that, early in pregnancy, the nutritional and transfer functions of the placenta are not primordial. The early placenta could rather be an endocrine and growth modulating organ [19]. There remains, however, the important problem of oxygen supply. Given this theory, the only oxygen that can reach the intervillous space is that which is simply dissolved in plasma. Only minute quantities are present and plasma pO_2 is, of course, low. Several facts must, however, be borne in mind:

- Quinn and Harlow [20] have shown, in the animal, that oxygen requirement for optimal blastocyst growth was considerably less than what could be expected.
- In vitro, early animal embryos develop quite well under reduced oxygen tension and exhibit increased glycogen metabolism [21].
- Rodesch et al. (1991, in press) have elegantly demonstrated that pO_2 in the intervillous space increased steadily from week 6 to week 14, and that during early gestation it was lower than that of the underlying endometrium. The authors suggest that the increase of placental pO_2 from first to second trimester is related with the establishment of a continuous maternal blood flow in the intervillous space.
- Lastly, the importance of the hemoglobin moiety in the embryo must be emphasized. During the first 10 weeks, the embryonic hemoglobin is

preeminent. It is a mixture of different hemoglobins which disappear almost completely after 2 months of gestation [22–24]. Embryonic hemoglobins are able to combine with oxygen at the very low tension and pH of interstitial fluids.

It is thus possible that a low oxygen tension is the best choice for embryonic organogenesis. We have already suggested [12] that the first 12 weeks resembled a very short summary of evolution from fish living in water with low oxygen tension to the mammalian embryo depending on its mother's supply.

The concept of a terminal reduced blood flow is necessary for our hypothesis to gain some value. It is well known that one of the earliest consequences of blastocyst formation and apposition is a tremendous vasodilatation and edema probably mediated by platelet-activating factor (PAF) and prostaglandins (see Chap. 2). This is logically associated with an increase of local blood flow. It is, however, essentially mediated by capillary and venous distension without any change in the spiral artery diameter until trophoblastic contact. Our studies of injected uteri with pregnancy in situ suggest that during the first 12 weeks a significant part of the endometrial blood flow is diverted at the base of the mucosa and inner myometrium, probably through the openings of arteriovenous shunts.

It is noteworthy that Ramsey and Donner [4] suggested that a completely free uteroplacental circulation probably did not exist in the first 12 weeks. There must thus be a critical period at the end of the 1st trimester. At that time the fetoplacental unit has increased in size. There is fusion of decidua reflecta (i.e., the operculum) with the decidua parietalis. As a consequence of this increase in size, a significant number of arteries are opened and colonized by plugs. The trophoblastic shell becomes thinner because of the increase of internal volume. Trophoblastic plugs detach progressively from the shell, which is gradually interrupted. Lacunae are no longer present. It is easy to understand how a real blood flow begins in the intervillous space. It is, of course, slow and limited in the beginning. It is tempting to suggest that the establishment of maternoplacental circulation initiates the period of fetal growth when organogenesis is complete.

References

1. Enders AC (1989) Trophoblast differenciation during the transition from trophoblastic plate to lacunar stage of implantation in the Rhesus monkey and human. Am J Anat 186:85–98
2. Knoth M, Larsan JF (1972) Ultrastructure of a human implantation site. Acta Obstet Gynecol Scand 51:385–393
3. Kaufmann P (1981) Entwicklung der Plazenta. In: Becker V, Schiebler TH, Kublif (eds) Die Plazenta des Menschen. Thieme, Stuttgart, p 13
4. Ramsey E, Donner J (1980) Placental vasculature and circulation. Thieme, Stuttgart

5. Boyd JD, Hamilton WJ (1970) The human placenta. Heffer, Cambridge, pp 61–91
6. Gruenwald P (1075) Maternal blood supply to the conceptus. Eur J Gynecol Reprod Biol 5:23–24
7. Pijnenborg R, Dixon G, Robertson WB, Brosens I (1980) Trophoblastic invasion of human decidua from 8 to 18 weeks of pregnancy. Placenta 1:3–19
8. Schaaps JP (1988) Dynamic imaging of the utero placental border in the first trimester of human pregnancy. Troph Res 3:37–45
9. Schaaps JP (1989) Etude de la circulation utéro-trophoblastique au cours du premier trimestre de la grossesse. Thesis, University of Liege
10. Gustavii B (1985) Direct vision technique for chorionic villi sampling in 100 diagnostic cases. In: Fraccaro M, Simoni G, Brambati B (eds) First trimester fetal diagnosis. Springer, Berlin Heidelberg New York, pp 46–50
11. Ghirardini G, Camurri L, Gualerzi C, Fochi F, Foscolu AMS, Spreafico L, Agnelli P (1985) Chorionic villi sampling by means of a new endoscopic device. In: Fracarro M, Simoni G, Brambati B (eds) First trimester fetal diagnosis. Springer, Berlin Heidelberg New York, pp 54–59
12. Hustin J, Schaaps JP (1987) Echographic and anatomic studies of the maternotrophoblastic border during the first trimester of pregnancy. Am J Obstet Gynecol 157: 162–168
13. Hustin J, Schaaps JP, Lambotte R (1988) Anatomical studies of the utero placental vascularization in the first trimester of pregnancy. Troph Res 3:49–60
14. Gosseye S, Fox H (1984) A immunohistological comparison of the secretory capacity of villous and extra-villous trophoblast in the human placenta. Placenta 5:329–348
15. Benirschke K, Kaufmann P (1990) Pathology of the human placenta, 2nd edn. Springer, Berlin Heidelberg New York
16. Moll W, Künzel W, Herberger J (1975) Hemodynamic implications of hemochorial placentation. Eur J Obstet Gynecol Reprod Biol 5:67–74
17. Aplin JD (1989) Cellular biochemistry of the endometrium. In: Wynn RM, Jollie WP (eds) Biology of the uterus, 2nd edn. Plenum, New York, pp 89–119
18. Bell SC (1988) Secretory endometrial/decidual proteins and their function in early pregnancy. J Reprod Fertil [Suppl] 36:109–125
19. Schneider H, Luckhardt M (1989) Entwicklung der Plazenta und des uteroplazentaren Kreislaufes aus morphologischer und funktioneller Sicht. Geburtshilfe Frauenheilkd 49:843–851
20. Quinn P, Harlow GM (1978) The effect of oxygen on the development of preimplantation mouse embryos in vitro. J Exp Zool 206:73–78
21. Khurana NK, Wales RG (1989) Effect of oxygen concentration on the metabolism of [u-14c] glucose by mouse morulae and early blastocysts in vitro. Reprod Fertil Dev 1:99–106
22. Henry RJ, Cannon DC, Winkelman JN (1974) Clinical chemistry: principles and technics. Harper and Row, Hagerstown, pp 1114–1115
23. Kaplan La, Pesce AJ (1984) Clinical chemistry: theory, analysis and correlation. Mosby, St Louis, pp 623–624
24. Salvo G, Samoggia P, Petti S, Guerriero R, Marinucci M, Lazzaro D, Russo G, Mastroberardino G (1985) Haemoglobin switching in human embryos: asynchrony of $\zeta \rightarrow \alpha$ and $\varepsilon \rightarrow \gamma$-globulin switches in primitive and definitive erythropoietic lineage. Nature 313:235–238

7 Placental Biochemistry

E.R. Barnea

Introduction

The placenta is a multifunctional organ which serves as an interface between mother and fetus. Its roles are multiple; they encompass metabolism, transport, and endocrine and immunologic aspects. Metabolism can be multifaceted:

1. It allows local nutrition and growth
2. It is directly committed in transfer and metabolism of nutrients to the embryo
3. It serves as a collecting device for fetal metabolites which are subsequently transported to the maternal organism for elimination
4. It helps to limit the local impact of xenobiotics

In the present chapter, we will discuss essentially local metabolism; the information on transfer is very scant during the first trimester. We will present the significant xenobiotic metabolic activity that takes place in the placenta at this early and critical stage of development. Reference to information dealing with late gestation will be provided only to compare with data available for the first trimester.

Functional Anatomy

Anatomically speaking, two types of embryonic adnexae can be distinguished: one is the yolk sac which is responsible for the embryonal development of blood, germ cells, and primitive gut. This structure is functional only during the first weeks. The other is the chorioallantoic placenta which will be in use during the length of pregnancy as an interface between mother and fetus. Yolk sac functions are discussed in Chap. 11.

Functional Histology

The first contact between the blastocyst and the endometrium occurs at the trophoblastic level. This is the place of anchorage of the blastocyst, the site of materno-embryonic interaction (see Chap. 2). The initial trophoblastic cell is the cytotrophoblast which differentiates into the syncytiotrophoblast very early in development (see Chap. 1).

Though earlier studies suggested that trophoblastic cell properties are similar in all locations, recent work using monoclonal antibodies indicates that the major metabolic/endocrine activity is located in the villous portion [1]. The role of the extravillous trophoblast, principally of the cytotrophoblast type, may serve as an anchor of the placenta to the decidua basalis, and it has a limited metabolic capacity. Other elements of the extravillous trophoblast migrate into the decidua, into the myometrium, or into blood vessels forming trophoblastic plugs (see Chaps. 2, 6).

The fact that most metabolic activity is located in the villous portion of the placenta suggests that maternal environment may condition trophoblastic metabolism. It follows that syncytiotrophoblast is the active component and that it serves as a powerful tool for filtering any information, particularly biochemical, going to or coming from the embryo. The functions of extravillous trophoblasts are discussed in Chaps. 1, 2, and 6. We will focus on villous metabolism.

Information on very early placenta is almost nonexistent. Practically, enough material can be obtained only from the 7th week onward. What is certain is that the first-trimester placenta is metabolically very active. The trophoblast/embryo ratio is 100:1 early in pregnancy (at term, it is roughly 1:6). This implies that the amount of placental tissue is, especially at the beginning, quite capable of multiple biochemical activities which are quantitatively sufficient to protect the developing conceptus.

Certain, but not all, metabolic properties of the trophoblast appear during the 1st trimester. One example is the ability to produce the sex steroids estradiol and progesterone from the 6th or 7th week onward. Placental alkaline phosphatase is expressed by syncytiotrophoblastic cells only later in gestation. This enzyme, which is present at the cell membrane, most probably plays an important role in the control of trophoblastic proliferation which must be down-regulated only after the first 12 weeks.

One of the most controversial subjects regarding metabolic capacity deals with xenobiotic metabolism. Older studies suggested that the capacity of the early placenta in this regard is very limited [2]. Our latest studies demonstrate a substantial activity of various xenobiotic enzymes.

Metabolism

Transfer

Freeze fracture techniques followed by ultrastructural analysis [3] have demonstrated that the maternal and embryonic plasma membranes of the syncytiotrophoblast have different densities in structural proteins. Gap functions are more commonly found early in gestation, while the thickness of the syncytial membrane decreases progressively, from 10 to $1.7\,\mu m$ at term. This implies that placental transfer is facilitated as gestation advances. Placental transfer occurs through four basic mechanisms as follows:

Passive Diffusion. This is the case for oxygen, carbon dioxide, urea, steroids, fatty acids, and fat soluble vitamins, and nutrients [4]. Passive diffusion depends on the concentration gradient. Molecular weight solubility in the lipid layer, degree of ionization and protein binding influence the diffusion coefficient. The uncharged, lipophilic molecules with molecular weight <1000 and those not bound to proteins diffuse passively.

Facilitated Diffusion. Glucose is transferred by this mode. The transfer occurs along the gradient and therefore does not require energy.

Active Transport. Nutrients like L-amino acids, creatine, and electrolytes (Na^+, K^+, Ca^{2+}), and vitamin C are transported. This requires an important consumption of energy [5].

Endocytosis. This mechanism is used in the transfer of immunoglobulins, transferrin, vitamin B_{12}/transcobalamine II. There is binding of molecules to specific sites on the cell membrane followed by the process of internalization. These compounds may be degraded by lysosomes or transported directly to the basal membrane for exocytosis and uptake by villous capillaries [6].

Which mode is prominent in early placenta? All these modes are probably operative and needed for embryonal development, though their relative importance at this period has not been established. Based on recent anatomic and ultrasound data [5], there is evidence to suggest that early placental metabolism is carried out by diffusional processes [5]. Of paramount importance are the moments when embryoplacental circulation and intervillous blood flow are effective (see Chap. 6). During the first 12 weeks, the intervillous space is bathed by a fluid which is probably derived from plasma; its renewal is slow. It seems obvious that a fairly long interaction between various molecules and trophoblasts occurs, thus facilitating transfer.

The uptake of aminoacids represented by α-aminobutyric [6, 7], creatine, and cobalamine were studied in first trimester placenta slices under 100% oxygen. The processes were similar to those at term with regard to energy requirement, metabolism, and specificity; however, the rate of uptake in first-trimester placenta was higher. In this study, it was determined that α-aminobutyric acid and creatine uptake are adenosine triphosphate (ATP) dependent since they were blocked by oubain, a specific Na^+ and K^+ ATPase inhibitor. There are arguments to suggest that these processes are carried out in early placenta by anaerobic pathways. Others have shown that the first-trimester placenta accumulates cobalamine as well as plasma-bound cobalamine and a specific carrier was identified in the tissue [8].

Placental Permeability

Inspite of its large molecular weight, immunoglobulin G (IgG) can cross the placental barrier. Even infectious agents such as viruses (rubella, hepatitis, poliomyelitis), spirochetes (syphilis), and parasites (toxoplasma) can pass the placenta. It has also been recently demonstrated that a vertical transmission of AIDS virus is possible. It appears that the placenta is selective for various medications. Usually, high molecular weight is a transfer-limiting factor: there are, however, exceptions.

First-Trimester Placenta – Metabolic Interaction

The first-trimester placenta interacts with the mother at the trophoblastic shell interface either with the decidua and the uterus or with spiral arteries via trophoblastic plugs (see Chap. 2). In addition, the first-trimester placenta is in direct contact with the embryo and the amniotic fluid.

Metabolic Interaction with the Embryo

The metabolic activities of the embryo are located in the liver, lungs, kidneys, and central nervous system. The importance of the metabolic enzymes in these locations has not yet been investigated. Our preliminary data indicate that several metabolic enzymes are present in these organs (unpublished observations). Others have reported the presence of xenobiotic metabolizing enzymes [9]. We believe that, in addition to the endocrine type of interaction present with the first-trimester placenta (see Chap. 8), a metabolic type of interaction may take place as well. This could correspond to the later concept of a fetoplacental unit.

Interaction with the Mother

There is evidence to suggest that the 1st-trimester placenta is sensitive to maternal influences. These are well documented in advanced gestation, but appropriate data are lacking for the first 12 weeks. Our data on the effect of the serum is the first step in this direction (see below). A metabolic and not only endocrine/paracrine type of interaction of the first trimester placenta, with the adjacent decidua, could be also suggested since the decidua contains several metabolic enzymes.

This discussion was aimed to illustrate the complexity of trophoblastic metabolism in early pregnancy and the differences that may be present between in vitro and in vivo findings, which should be kept in mind in discussing biochemical events. The overall metabolism in the very early stages of pregnancy could well be predominantly anaerobic, especially during the period of embryogenesis and morphogenesis. It seems, in fact, that the local oxygen concentration is minimal during the first 12 weeks (see Chap. 6). This unexpected hypothesis must be kept in mind while looking at selected metabolic pathways in the early placenta.

Metabolic Pathways

Carbohydrates

Glucose metabolism follows two routes: either by conversion to glycogen and by degradation through glycolysis to lactate or by the pentose shunt pathway. The placenta is capable of gluconeogenesis from lactate or pyruvate. The presence of active glucose uptake and lactate production was documented in explants where high doses of nicotine enhanced these activities [10]. The activity of glucose-metabolizing enzymes decreases gradually throughout gestation. This suggests that the enzyme complement once developed has only a limited lifespan without continous replenishment [11].

We have suggested that the metabolism of early placenta is mostly anaerobic. In vitro, the production of progesterone and human chorionic gonadotropin (hCG) is not affected by exposure of placenta explants to D-glucose. However, β-hydroxybutyrate inhibits the production of progesterone and hCG by explants obtained before 9 weeks of gestation. Glucose is important for aerobic metabolism, while β-hydroxybutyrate is involved in anaerobic pathways. These experiments thus point to the relative independence of early placenta from oxygen requirements (Barnea et al, submitted).

Lipids

Lipids are also formed in the placenta. Cholesterol is synthetized locally from acetate, while fatty acids are produced de novo. Triglycerides and phospholipids are formed in the placenta by esterification of fatty acids.

Aminoacids

The placenta receives aminoacids from the maternal circulation. The transport was studied using α-aminobutyric acid as substrate. This process is complex: it requires acetyl choline production, activation of its receptor, membrane phospholipid methylation, and CA^{2+} and aminoacid transport systems. Finally, an organized cytoskeletal system is also needed (5). Membrane-mediated protein synthesis and leucine uptake in inhibited by the absence of K^+ and hyperosmolarity [12]. Transfer of aminoacids can also proceed under anoxic conditions. Term placenta under dual perfusion provides energy requirements through anaerobic glycolysis [13–15].

Energy

The model was the transformation of extracellular dephosphorylated adenine nucleotides to adenosine which was then rephosphorylated within the cell [16]. Mostly adenine nucleotides were formed. The contribution of specific 5'-nucleotidase to the dephosphorylation of extracellular adenosine monophosphate (AMP) was fivefold lower in early placenta than at term. No uptake of nucleotides was observed. The utilization of ^{14}C-adenosine diphosphate (ADP) was higher at term, but no changes were noted in that of ^{14}C-ATP. This suggests that extracellular ATP and ADP are degraded, but are further reutilized for intracellular synthesis [16]. The main pathway for AMP catabolism is deamination, but the trophoblast is capable of producing adenosine during accelerated adenine nucleotide catabolism [17]. Other experiments followed the changes occurring in the catabolism of prelabeled adenine nucleotides and in the adenylate energy charge. Basal energy charge was similar in early and term placenta. Under hypoxic conditions (nitrogen atmosphere) and in the absence of glucose, ATP levels as well as those of energy charge were maintained for 8 h. 2-Deoxyglucose induced a tenfold drop in the ATP level and a 30% decrease in energy charge. Thus, it appears that the trophoblast can tolerate transient hypoxia or lack of exogenous glucose. It was noted that susceptibility to energy deprivation increased with gestational age [17].

Toxicology

Organogenesis is the most vulnerable period in fetal life. Although the mother is exposed to a large variety of environmental pollutants through ingestion or inhalation, the frequency of associated congenital anomalies, embryonal toxicity patterns and death is surprisingly low. Though many compounds are suspected to have a teratogenic potential, this has been clearly demonstrated only for a small number of them.

Such a point suggests that either the embryo is resistant to such exposures or that it does not interact metabolically with the toxin. The embryo may be shielded against such an exposure by intervening tissues separating it from the mother. The first layer is the maternal decidua which is metabolically active and has a significant catechol-O-methyl transferase activity. Then, there is the trophoblastic shell and the whole placenta. Early placenta was considered of little importance as regards embryonic protection. However, our recent studies seem to demonstrate that, on the contrary, first trimester villous tissue plays a very important role in this respect. The barrier it forms against xenobiotics is not physical but metabolic. Term placenta can prevent molecular entry, such as heavy metals (cadmium, lead, mercury) which accumulate locally. The ability of a toxin present in the maternal circulation of affect the embryo depends on:

1. The propensity to develop an adverse reaction (susceptibility) which is genetically determined.
2. Pharmacokinetics of the agent itself: its activity, concentration, mode and duration of administration, liposolubility, and degree of ionization – weak acids are most deleterious since the embryonal milieu is basic [18].
3. The defense mechanisms, barrier and/or metabolism (toxin inactivation) clearance, elimination from the site and conjugation (blocking metabolic activity), and local DNA repair following induced damage.

In order for xenobiotics to interact with trophoblastic cells, specific binding sites have to be present on the cells. The interaction with specific ligands should translate into biologic effects like hormone secretion and/or changes in the local metabolic enzyme activity.

Among the most frequently used medications are tranquilizers. These, especially the group of benzodiazepines, may cause specific congenital malformations. Specific binding sites of the peripheral type have been documented in the embryo (unpublished observations). In addition, and confirming earlier studies, specific clonazepam-nonresponsive sites were demonstrated in early placenta. No major differences were present in the receptor properties between early and term tissues. Exposure of the placenta to benzodiazepine of its potent analogues stimulated both hCG and progesterone secretion. The maximal responses were noted in concentrations which are close to the affinity of the ligand to the binding site.

This strongly suggests that the first trimester placenta may reduce the amount of diazepam reaching the embryo by local binding. By its binding the drug is likely to modify placental hormonal secretion, which may have a local role and perhaps also indirectly affect the embryo. Moreover, benzodiazepine can be metabolized locally. A further confirmation that maternal drug use affects the first-trimester placenta was provided when we found that in vivo exposure to neuroleptics, such as thioridazine, changed the pattern of spontaneous pulsatile hCG secretion in superfused explants of young placentae. These data confirm that the first trimester placenta is very sensitive to neuroleptic drug exposure; in the case of benzodiazepine there is receptor dependency. The nature of these interactions, protective or not, remains to be determined [19, 20].

Another major group of drugs are opiates. Specific binding sites for these compounds were documented at term. In the first-trimester placenta, we have recently provided evidence that both K type receptor agonists, i.e., dynorphins, and μ- and Δ-receptor agonist, i.e., β-endorphin, specifically affect local hormonal secretion [21, 22]. These compounds operate through opiate receptor-dependent mechanisms. At term, a down-regulation of the K-type site was demonstrated after administration of methadone. In the first-trimester placenta, a tenfold higher dose (500 μM) of diacetylmorphine was required to affect ^{14}C-leucine incorporation into total proteins as compared to term. This may be due to differences in the density of opiate receptors. Lactate production and glucose uptake were not modified.

It is clear that opiates interact with the placenta. For reduction of the drug transfer, local binding is very likely, as well as a local opiate degradation. The consequences of the metabolic changes on the embryo must still be investigated.

Xenobiotic Metabolism

Xenobiotic metabolism is a major and recent aspect of placental function. The information on the enzymes involved was well investigated at term while knowledge about the first trimester is limited and mostly based on older studies [23]. Our laboratory has been involved in studies on the activity of metabolic enzymes in the first-trimester placenta with the prospect of seeking embryonal protection or damage. Generally speaking, xenobiotic enzymes can be divided into two groups. Phase I enzymes are considered as the xenobiotic activating pathway. This is carried out by hydroxylation of metabolically inert compounds. Phase II enzymes are regarded as inactivators of reactive metabolites. This is obtained by transformation of reactive compounds to inactive metabolites and by conjugation (glucuronidation, sulphation, gluthatione conjugation). Evidently, a given xenobiotic can activate phase I, phase II, or both pathways (Fig. 1).

Phase I Enzymes

Aryl hydrocarbon hydroxylase is regarded as a major carcinogen/mutagen-activating enzyme in the organism as well as in the placenta. Its characteristics have been previously well described at term [24]. Using chromatographic techniques, it was demonstrated that specific toxins/mutagens are formed by this enzyme. It was also shown that maternal cigarette smoking causes a significant increase in the placental enzyme activity as compared to nonsmokers.

Earlier studies using fluormetric techniques suggested that aryl hydrocarbon hydroxylase activity in the first-trimester placenta was very low.

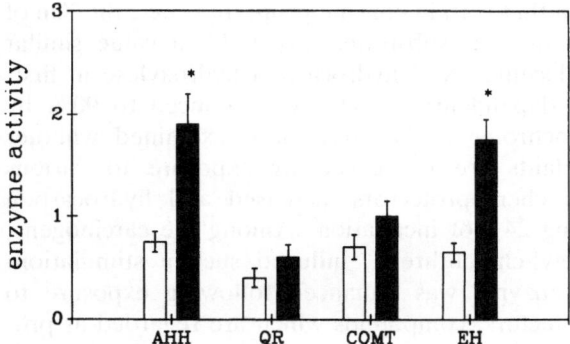

Fig. 1. Enzyme activities of the 1st-trimester placenta before incubation (zero time; *open columns*) and after 16 h of incubation (*solid columns*). The data are expressed as mean ±SEM. *Asterisks*, $P < 0.01$ versus zero time. Enzyme activities are expressed as follows: aryl hydrocarbon hydroxylase (*AHH*): nmol/mg protein per 10 min; quinone reductase (*QR*): µmol/mg protein per 10 min; catechol-*O*-methyl transferase (*COMT*): nmol/mg protein per 30 min; estrogen hydroxylase (*EH*): nmol/mg protein per 30 min

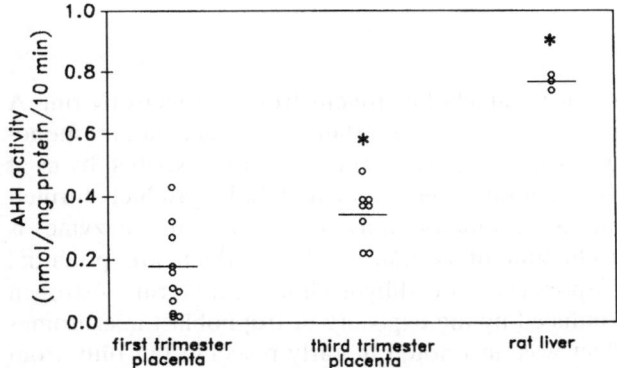

Fig. 2. Aryl hydrocarbon hydroxylase (*AHH*) activity in placentas from patients in the first and third trimesters and in rat liver. The activity in the third trimester was higher than in early gestation. Data depict individual enzyme values in the patients studied. *Asterisks*, significant difference from third trimester

We have recently challenged this statement with the use of a newer radio-enzymatic technique [25]. Contrary to the previous method, this technique measures overall aryl hydrocarbon hydroxylase activity in the placenta by hexane separation of the products formed.

The details of these investigations were recently reported [26]. We have demonstrated that the mean basal activity at term was twofold that of the first-trimester placenta. This is at variance with the usual decrease in activity of many enzymes with advancing gestation. One may speculate that the increase in enzyme activity is linked in some way with the presence of xenobiotics in the environment (Fig. 2).

In addition, it was shown that certain enzyme properties, i.e., the Km of benzo(a)pyrene, which was used as substrate, was $2\,\mu M$, a value similar to that reported for term placenta. Aryl hydrocarbon hydroxylase in first-trimester placenta is P-450 dependent; its activity is reduced to 90% by high concentrations of cytochrome c. We have also examined whether first-trimester placenta explants are influenced by exposure to various xenobiotics: carcinogens or chemoprotectors increased aryl hydrocarbon hydroxylase activity following 24 h of incubation. Among the carcinogens, benzopyrene, but not methyl-cholanthrene, induced such a stimulation. Moreover, activity of the enzyme was enhanced following exposure to various groups of chemoprotectors, compounds which are regarded as protective against cancer or mutagens. The increase was two- to threefold which is usually considered as proof of enzyme induction.

We have also investigated whether aryl hydrocarbon hydroxylase activity could be modified by maternal cigarette smoking. Our preliminary data suggest that both at term and in early tissues, there are significant differences between the mean aryl hydrocarbon hydroxylase activity of smokers and nonsmokers (unpublished observations).

Estrogen Hydroxylase

Estrogen hydroxylase produces catechol estrogens from estrogens by ring A hydroxylation; it is considered as an important P-450-dependent mono-oxygenase. Estrogen hydroxylase properties were recently described by us at term [27]. In this study, we identified the major metabolic products formed by high-performance liquid chromatography (HPLC). This enzyme is also involved in the production of certain products which are potential carcinogens, especially epoxides and dihydrodiols. At term, estrogen hydroxylase activity was induced by the exposure of trophoblast microsomes to benzopyrene. This effect was also noted in early placenta but only from smoking patients. Finally, various chemoprotectors increased markedly (two- to threefold) the enzyme activity after 24-h incubation. The increase was similar in magnitude to that found for aryl hydrocarbon hydroxylase. However, here there were no differences between the basal enzyme activity in smokers and nonsmokers.

The consequences of induced estrogen hydroxylase activity may be multiple: the formation of catechol estrogens is increased. These estrogen derivatives (especially the 2-hydroxy estrogens, which are prominent) are considered as antiestrogens and may counteract estrogen action. Catechol estrogens could also impede the vasodilatory effect of estrogens in both uterus and decidua. In addition, we have already demonstrated that catechol estrogens could influence placental steroidogenesis [28, 29] (Fig. 3).

Thus, the phase I enzymes, namely aryl hydrocarbon hydroxylase and estrogen hydroxylase, are active within the first trimester placenta and are modulated by environmental exposures. Such observations suggest that the

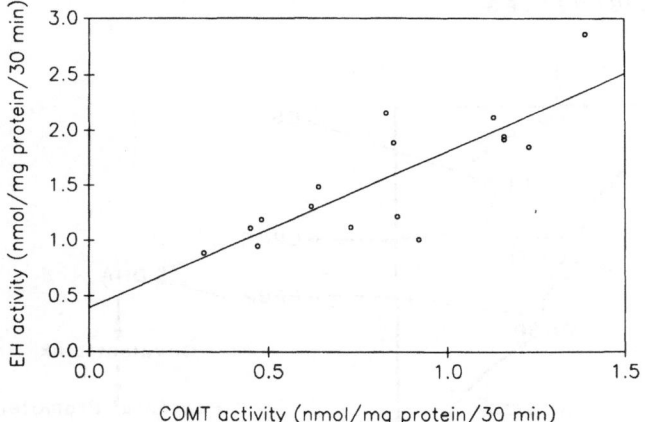

Fig. 3. Correlation between catechol-O-methyl transferase *(COMT)* and estrogen hydroxylase *(EH)* activities in cultured placental explants following treatment with carcinogens, chemoprotectors, or serum ($R^2 = 0.75$; $P < 0.01$)

first trimester placenta is capable of activating various mutagens/carcinogens through hydroxylation processes. Based on this evidence, the 1st-trimester placenta does not protect the embryo. However, local xenobiotic hydroxylation may induce a protective effect since such activation could promote local phase II enzyme activity, thus inactivating the reactive metabolites formed and thereby limiting their passage to the embryo.

Type II Enzymes

Catechol-O-methyl Transferase

We have also examined type II enzyme activity in the first-trimester placenta. The enzyme properties were similar to those reported at term [30]. The preferable substrate for the enzyme was 2-hydroxyestrone, the Km of which was 100-fold superior to that of catecholamines. We found that catechol-O-methyl transferase activity changed during the first trimester. A threefold increase was noted between weeks 7–8 and 11–13. Moreover, a significant (threefold) reduction in the enzyme activity was noted in cigarette smokers at 11–13 weeks. In cultures of isolated trophoblastic cells, $50\,\mu M$ benzopyrene increased the catechol-O-methyl transferase activity two- to threefold. There was no similar enzyme induction in the explants obtained from smokers.

In vitro, catechol-O-methyl transferase activity increased following exposure to various chemoprotectors (Sudan I, *tert*-butyl-4-hydroxyanisole and dicoumarol) for 24 h. The increase in enzyme activity was parallel to that of estrogen hydroxylase [31]. Thus, estrogen hydroxylase forms

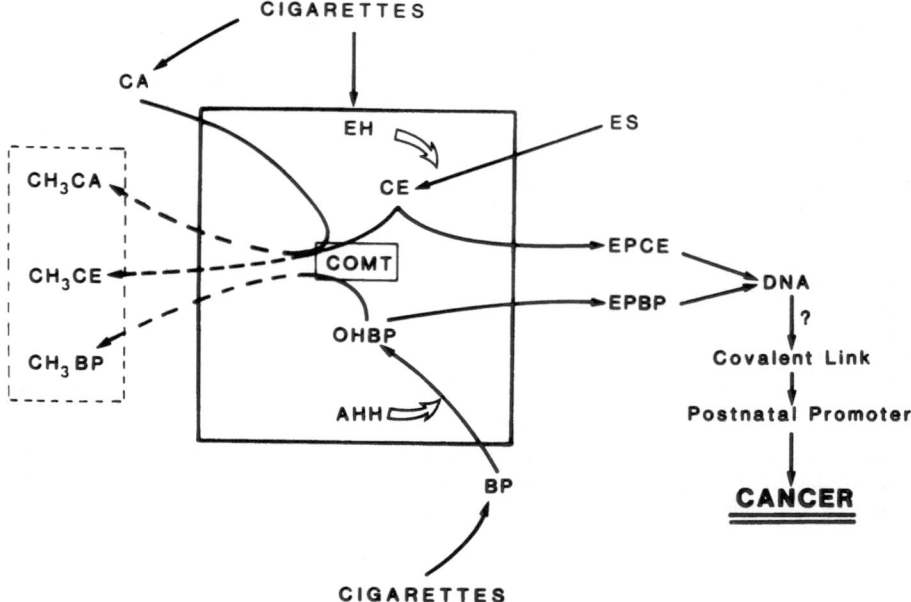

Fig. 4. Proposed hypothesis for the role of catechol-*O*-methyl transferase in cigarette smoking-related toxin inactivation by the human placenta, thus protecting the embryo. The *area within the dotted lines* denotes catechol-*O*-methyl hydroxylase-dependent methylation of catecholamines (CH_3CA), catecholestrogens (CH_3CE) and benzo(a)pyrene (CH_3BP). The *right-hand* side denotes formation of activated carcinogens: epoxides of catechol estrogens (*EPCE*) and benzopyrene (*EPBP*) which may reach the embryo and which, with a postnatal promoter, may lead to development of cancer. *CA*, catecholomines; *EH*, estrogen hydroxylase; *CE*, catecholestrogens; *COMT*, catechol-*O*-methyl transferase; *OHBP*, hydroxylated benzopyrene; *AHH*, aryl hydrocarbon hydroxylase; *ES*, estrogen

catechol estrogens while catechol-*O*-methyl transferase inactivates them to methylestrogens. In vivo, however, this coordination is less evident because of the presence of catecholamines, which are also a substrate for catechol-*O*-methyl transferase. The first-trimester placenta expresses a complex metabolic action in this instance: a cytosolic enzyme (catechol-*O*-methyl transferase) is capable of interacting with a microsomal fraction (estrogen hydroxylase) (Fig. 4).

Quinone Reductase

Using 2,6-dichlorophenol indophenol as substrate, we studied quinone reductase activity in the placenta. This is a typical phase II enzyme which has a protective role against cancer and mutagenesis in other systems [32]. The role of quinone reductase is to inactivate reactive toxic/mutagenic metabolites through a two-electron reduction of quinones to hydroquinones.

Table 1. Effect of 10% pregnant serum fractions above or below 1000 molecular weight and heat inactivation on xenobiotic enzyme activity in first-trimester placenta. Data are expressed as percentage difference from controls

Enzyme	Control (%)	Serum (%)	HIS (%)	MW < 1000 (%)	MW > 1000 (%)
AHH	100	180	210	480	520
EH	100	80	90	130	120
COMT	100	220	210	90	230
QR	100	23	50	85	34

[a] $P < 0.05$ versus control.
HIS, heat inactivated serum; AHH, aryl hydrocarbon hydroxylase; EH, enzyme hydroxylase; COMT, catechol-O-methyl transferase; QR, quinone reductase.

It was of great interest to find that the first trimester placenta has a substantial quinone reductase activity in the micromolar range per milligram of protein. This activity appears to be almost 1000-fold higher than the activity of the phase I enzymes aryl hydrocarbon hydroxylase and estrogen hydroxylase. This strongly suggests that the early placenta has a significant protective metabolic potential.

Properties of quinone reductase were also examined: the enzyme characteristics appeared to be similar to that of the rat liver. Quinone reductase activity is blocked by exposure to high concentrations of dicoumarol, a specific enzyme inhibitor. No differences were found in the basal quinone reductase activity between smokers and nonsmokers. A significant increase was noted in enzyme activity following exposure to various carcinogens and chemoprotectors: this effect was dose dependent and was not affected by smoking. The induction occurred within 6 h with a two- to threefold increase in enzymatic activity, and a return to basal levels after 24 h only in high concentrations [33]. Thus, early placenta responds to xenobiotics in vivo and in vitro. Previous exposure to tobacco in vivo does not reduce the level of enzymatic activity.

Although the degree of carcinogenicity of the compounds tested (polycyclic aromatic hydrocarbons) varied widely, their effect on quinone reductase activity was similar. Thus, quinone reductase may emerge as a major protective mechanism in the early placenta.

The activity of phase I and phase II enzymes was checked against time in cultures of 1st-trimester placenta. An interesting pattern emerged: aryl hydrocarbon hydroxylase and estrogen hydroxylase activities increased two- to threefold after 24-h culture, while quinone reductase and catechol-O-methyl transferase activity did not change significantly (Fig. 1). This suggests that the enzymes that activate xenobiotics are probably under inhibitory influences in vivo, while phase II enzymes express their full effect in the maternal organism. We therefore examined (Barnea et al., submitted) the effect of the maternal environment, i.e., maternal serum, on the activity

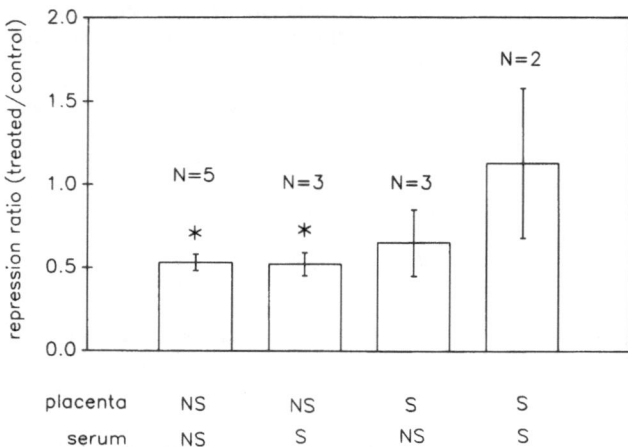

Fig. 5. The effect of 10% serum from smoking and nonsmoking women on aryl hydrocarbon hydroxylase activity in placentas of smokers (*S*) or nonsmokers (*NS*). Incubation was carried out for 16 h. Mean ± SEM. *Asterisks, P* < 0.01 versus control

of these enzymes in vitro. Various concentrations of homologous maternal serum obtained from 7–11-week pregnancies were incubated with placental explants for 24 h. Only a serum concentration of 10% was effective. Differences were observed according to the maternal status (smoker or nonsmoker). Aryl hydrocarbon hydroxylase (Fig. 5) and estrogen hydroxylase activities decreased in a nonsmoking environment. The effect of serum on phase II enzymes was less evident.

The specific factors involved in the modulation of enzyme activity have not been identified so far, but they appear to be multiple. We have examined whether the effect of the serum is direct or indirect. To test this point, we used incubation with placental homogenates. The serum effect appeared to be mostly indirect: the serum did not interact directly with the free enzyme, but the maintenance of membrane integrity was required. The effect was specific since male serum had no effect. The specific factors involved are currently being investigated. Components with different molecular weights are involved (either with a low molecular weight <1000 or with a molecular weight >1000). Their effect on the activity of different enzymes varies. Some factors appear to be heat stable (Table 1).

Thus, the observations on enzyme modulatory serum factors are new and the identification of such factors could be of importance. The importance of maternal and perhaps embryonal influences present in vivo should be taken into account in the estimation of the activity of various local enzymes. We have recently demonstrated [34] that, at term and first trimester, the activity of xenobiotic metabolizing enzymes in vitro is modified after 24-h preincubation with heavy metals, cadmium, mercury. This effect is dose dependent and correlated with intracellular metal concentration (unpublished observations).

Superfusion – A Model for Toxicology

The experimental models of use for the study of placental metabolism and toxicology usually rely on static cultures [35]. The effect of various toxins on placental metabolism is observed on explant cultures following acute exposure. This approach enables examination of a direct but not delayed effect. Therefore, our superfusion model of placental explants mimics the placental environment in situ and could thus be an important addition to screening and determining the effects of xenobiotics on local metabolism (see Chap. 8). The advantages are:

1. Transport of solutes functions via a gradient.
2. There is no effective physical placental barrier.
3. An intimate contact between the solutes in the medium and the trophoblastic cells is always possible.

We have recently demonstrated [36] that preexposure to carcinogens such as benzo(a)pyrene can modify the spontaneous pulsatility of hCG by the superfused placenta. This finding suggests that benzopyrene enters the trophoblast and then modifies the hormone secretion. This aspect was further examined by determining the pulsatile pattern of hCG secretion in superfusion following different degrees of maternal exposure to cigarette smoking. Clear changes in the pattern were noted, with heavy smoking presenting for the first time a good correlation between the degree of smoking and modification of placental hormonal secretion (unpublished observations).

In other studies, we have shown that maternal exposure to neuroleptic drugs can modify the basal pattern of hCG pulsatility in superfusion. This further emphasizes the point that the 1st-trimester placenta is very sensitive to environmental effects, and studies in this direction should be pursued [20]. Chorion villus biopsy has provided a large amount of information regarding various congenital metabolic disorders and it is used extensively for prenatal diagnosis [37]. Several defects can be detected by assaying enzyme activity and the concentration of metabolites present in the tissue. Poenaru [38] has recently reviewed 38 metabolic disorders. This is additional proof of the enzymatic wealth of the early placenta.

Future Directions

1. In-depth investigation of phase I and II enzyme activities, correlation with specific response to xenobiotics in vivo and in vitro.
2. Studies on pathways of local metabolism of various drugs such as tranquilizers and opiates.

3. Metabolic embryoplacental interaction with the use of multiple models: static, dynamic, short- and long-term cultures, co-cultures, etc.
4. Identification of embryo-derived factors affecting placental metabolism.
5. Identification of maternal influences, i.e., serum factors which could affect early placenta.

References

1. Loke YW (1987) Early human trophoblast. In: Beard RW, Sharp F (eds) Early pregnancy loss: mechanisms and treatment. Springer, Berlin Heidelberg New York
2. Juchau MR (1971) Human placental hydroxylation of 3.4-benzopyrene during early gestation and at term. Toxicol Appl Pharmacol 18:665–675
3. Sideri M, de Virgiliis G, Rainoldi A, Remottig (1983) The ultrastructural basis of the nutritional transfer: evidence of different patterns in the plasma membranes of the multilayered placental barrier. Troph Res 1:15–25
4. Faber JJ, Thornburgh KL (1983) Oxygen exchange. In: Faber JJ, Thornburg KL (eds) Placental physiology. Raven, New York, pp 63–74
5. Sastry RBV, Barnwell SL, Moore DR (1983) Factors affecting the uptake of alpha-amino acids by human placental villus: acethylcholine, phospholipid methylation, Ca++ and cytoskeletal organization. Troph Res 1:81–100
6. Faber JJ, Thornburg KL (1983) Placental transfer of nutrients and wastes. In: Faber JJ, Thornburg KL (eds) Placental physiology. Raven, New York, pp 151–160
7. Young M (1982) Placental aminoacids transfer and metabolism. Placenta Suppl 2:177–184
8. Ng WW, Miller RK (1983) Transport of nutrients in the early human placenta: aminoacids, creatine, vitamine B_{12}. Troph Res 1:121–134
9. Pelkonen O (1984) Detoxification and toxification processes in the human feto-placental unit. In: Drugs and pregnancy. Academic, London, pp 63–72
10. DeSoye G, Schmon B, Hartmann M, Dohr G, Flaschka G, Jones CJ (1992) Opposite effects of an antilepolytic hormone (nicotinic acid) on the production/secretion of hCG and hPL by isolated trophoblast cells. Troph Res 6:199–210
11. Shafrir E, Diamant YZ (1979) Regulation of placental enzymes of the carbohydrate and lipid metabolic pathways. In: Beard RW (ed) Pregnancy, metabolism diabetes and the fetus. Ciba Found Ser 63:161–179
12. Genbacev O, Cemerikic B, Cetkovic M (1983) Interaction between membrane function and protein synthesis in the human placenta in vitro. Troph Res 1:197–208
13. Faber JJ, Thornburg KL (1983) Permeability of the placental membrane for hydrophilic substances. In: Faber JJ, Thornburg KL (eds) Placental physiology. Raven, New York, pp 151–160
14. Illsley NP, Penfold P, Bardsley SE, Tracey BM, Aarnoudse JG (1983) The effects of anoxia on human placental metabolism and fetal substrate profiles investigated by an in vitro placental perfusion technique. Troph Res 1:55–70
15. Penfold P, Illsley NP, Purkiss P, Jennings P (1983) Human placental aminoacids transfer and metabolism in oxygenated and anoxic conditions. Troph Res 1:27–36
16. Vettenranta K, Raivio KO (1989) Extracellular adenine nucleotides in human trophoblastic purine nucleotide synthesis. Placenta 10:472
17. Vettenranta K, Raivio KO (1988) Adenine nucleotide catabolism in the human trophoblast early and late in gestation. Pediatr Res 24:373–379
18. Nau H, Scott WJ (1986) Weak acids may act as teratogens by accumulating in the basic milieu of the early mammalian embryo. Nature 323:276–278
19. Barnea ER, Fares F, Gavish M (1989) Modulatory action of benzodiazepines on human term placental steroidogenesis in vitro. Mol Cell Endocrinol 64:155–159

20. Shurtz-Swirski R, Cohen Y, Barnea ER (1991) In vivo exposure to neuroleptics modifies embryo placental relationship in vitro. Hum Reprod (in press)
21. Barnea ER, Ashkenazi R, Kol S, Erlik Y, Sarne I (1991) Effect of beta endorphin on placental hCG secretion in the first trimester. Hum Reprod 6 (in press)
22. Barnea ER, Ashkenazi R, Sarne I (1991) Effect of dinorphin upon pulsatile hCG secretion in vitro. J Clin Endocrinol Metab 73:1093–1098
23. Juchau MR, Namkung MJ, Rettie AE (1987) P-450 cytochromes in the human placenta: oxidations of xenobiotics and endogenous steroids. Troph Res 2:235–263
24. Juchau MR (1976) Drug biotransformation reactions in the placenta. In: Mirkin (ed) Perinatal pharmacology and therapeutics. Academic, New York, pp 71–118
25. DePierre JW, Moron MS, Johannesen K (1975) A reliable sensitive and convenient radioactive assay for benzopyrene monoxygenase. Ann Biochem 63:470–484
26. Barnea ER, Avigdor S (1991) Aryl hydrocarbon hydroxylase activity in the placenta: Induction by carcinogens and chemoprotectors. Obstet Gynecol Invest 32:4–9
27. Barnea ER, McLusky NJ, Purdy N, Naftolin F (1988) Estrogen hydroxylase activity in the human placenta at term. J Steroid Biochem 31:253–255
28. Barnea ER, Fakih H (1985) The role of catechol estrogens in placental steroidogenesis. Steroids 45:427–432
29. Barnea ER, Naftolin F (1987) Estrogens and catecholamine metabolism: possible interaction during pregnancy. J Endocrinol Invest 10:329–340
30. Barnea ER, McLusky NJ, DeCherney AH, Naftolin F (1988) Catechol-O-methyl transferase activity in the human term placenta. Am J Perinatol 5:121–127
31. Barnea ER, Avigdor S (1990) Coordinated induction of estrogen hydroxylase and catechol-O-methyl transferase by xenobiotics in the first trimester placenta in vitro. J Steroid Biochem 35:327–331
32. Talalay P, Prohaska HJ (1987) A quinone reductase with special functions in cell metabolism and detoxification. In: Ernester L, Eastbrook RW, Hochstein P, Orrenius S (eds) DT-Diaphorase. Cambridge University Press, Cambridge, pp 61–66
33. Avigdor S, Zakheim B, Barnea ER (to be published) Quinone reductase activity in the first trimester placenta: effect of cigarette smoking and carcinogens. Repro Toxicol
34. Boady WY, Urbach Y, Barnea ER, Yannai S (1991) In vitro effect of mercury on AHH, COMT, QR and G-6-P dehydrogenase activities in term placenta. Pharmacol Toxicol 68:317–321
35. Beaconsfield R, Cemerikic B, Genbachev O, Sulovic V (1987) The placenta as a model for toxicity screening of new molecules. Troph Res 2:343–356
36. Barnea ER, Shurtz-Svirsky R (1991) Polycyclic aromatic hydrocarbon induced modification of pulsatile hCG secretion in first trimester placental explants. Hum Reprod (in press)
37. Kazy Z, Rozovsky IS, Bakherev VA (1982) Chorionic biopsy in early pregnancy: a method of early prenatal diagnosis for inherited disorders. Prenat Diagn 2:39–45
38. Poenaru L (1987) First trimester prenatal diagnosis of metabolic diseases: a survey in countries from the European community. Prenat Diag 7:333–341

8 Endocrinology of the Placenta and Embryo-placental Interaction

E.R. Barnea and R. Schurtz-Svirsky

Introduction

The human placenta undergoes major changes during pregnancy. Initially, human chorionic gonadotropin (hCG) secretion increases, reaching a plateau at 9–10 weeks. Subsequently, hCG production decreases, while that of sex steroids and other protein hormones such as human placental lactogen (hPL) and *Schwangerschafts protein* 1 (SP1) increase until term. The information on the regulation of hormone secretion by the 1st-trimester placenta is limited and mostly based on data derived from term measurements. The information accumulated by us and others indicates, however, that, in order to understand the first-trimester hormonal regulation, studies must be conducted on that early tissue. In the present chapter, aspects of endocrine and para/autocrine control will be discussed. Information will be restricted to first-trimester placenta, and allusion to term will only be used for comparison. Discussion of hormone levels in biologic fluids is done elsewhere (see Chap. 20).

Hormone Synthesis/Secretion

Human Chorionic Gonadotropin

Human chorionic gonadotropin is a glycoprotein (molecular weight 36000–42000) composed of two dissimilar subunits, α and β, which are joined noncovalently. Both hCG and the pituitary, follicle-stimulating hormone (FSH), thyroid-stimulating hormone (TSH), and luteinizing hormone (LH) have an almost identical α-subunit, but have a specific β-subunit which dictates their function. The microheterogeneity of hCG structure has been stressed in recent years by the finding of variations in its sialic acid content; the lower the content, the higher the metabolic clearance rate. Changes in the O-linked carbohydrates have also been detected in the hCG molecule.

The principal roles of hCG are to maintain progesterone secretion from the corpus luteum and promote dihydroepiandrosterone sulfate (DHEAS) production by the fetal adrenal and steroidogenesis by the fetal testis.

The free subunits have no biologic function; nor are the secreted subunits products of hormone breakdown. Free α-hCG is larger than the combined form and is unable to combine with the β-hCG subunit [1]. The increase in hCG secretion is accompanied by an increase in free β-hCG production, which declines after the hCG peak. In contrast, free α-hCG levels increase until term. Thus, the β-subunit production is the rate-limiting step in hCG biosynthesis. β-Core fragments is believed to be a breakdown product of placental β-hCG subunit by metabolism carried out by the renal parenchyma.

The cytotrophoblast continuously differentiates into syncytium. The process of syncytial fusion appears to be the principal event that leads to hCG and subunit synthesis (see Chap. 1). The accepted scheme is as follows: α-hCG is activated in committed cytotrophoblasts (Langhans cells) with later expression of β-hCG gene. The intermediate cells elaborate both subunits and exist until 8–10 weeks, their number declining only thereafter [2]. At the syncytiotrophoblastic level, hCG is found in the surface microvilli and in the cisternae of the rough endoplasmic reticulum, but not in the Golgi apparatus. Newly synthesized hCG is rapidly secreted by the syncytiotrophoblast by a process which involves attachment to plasma membranes. A significant but limited storage of hCG and free β-hCG in granules is found in the early placenta but not at term. This finding reflects changes in biology during gestation [3]. mRNA coding for hCG subunits is expressed more in the first-trimester than in term placenta. Similarly, the tissue ratio α-hCG/hCG increases steadily until term.

Earlier studies showed that hCG may modulate local hormone secretion. Accordingly, it was recently shown that the placenta contains specific binding sites for hCG, and this is also true for the first trimester [4]. Therefore, an auto/paracrine form of regulation is likely, as it was shown that hCG induced local adenylate cyclase activity.

Human Placental Lactogen

Human placental lactogen is a polypeptide, made of 96 aminoacids, which has a 96% structural homology to growth hormone (GH). It has, however, only 3% of the biological potency of GH. hPL is produced by the syncytiotrophoblast, and its secretion is believed to be mostly constitutional and increases in direct correlation with the trophoblastic mass. The importance of hPL in pregnancy maintenance has been questioned since in some cases no hPL at all is produced. A specific mRNA was detected in the placenta with an increase throughout gestation. The involvement of hPL in glucose and lipid metabolism has been suggested. hPL is principally secreted into the maternal compartment, and therefore the levels in the cord blood are low. A para/autocrine effect of hPL on placental steroid secretion was previously reported in vitro. The early placenta also contains somatostatin, localized in the cytotrophoblast; the highest concentration is found early and decreases, thereafter [5]. The decrease in the production of somatostatin contrasts with the production of hPL which increases throughout pregnancy.

The Luteoplacental Shift of Sex Steroids

In contrast to hCG, which is a pure placental glycoprotein, sex steroids are initially produced by the ovary. The luteoplacental shift occurs as early as 6–7 weeks for estradiol and at 8–9 weeks for progesterone [6]. The declining circulating levels of 17-OH-progesterone after the 6th week reflects the shift in progesterone secretion. The high density receptors for low density lipoproteins (LDL) and acetylated LDL, present as early as the 6th week of gestation, may be the major source of cholesterol needed for local progesterone formation. While the LDL site density is similar both in early and term placenta, that of acetyl LDL is twofold higher in the first trimester tissues [7]. Specific binding sites for progesterone have been detected in the early placenta. That progesterone is necessary for maintenance of pregnancy has been clearly demonstrated by the use of its antagonist RU-486 as an effective abortion-inducing drug [8]. Whether the principal effect of progesterone is exerted on the placenta or on the decidua has not been well established so far (see *Mechanism of Progesterone Action* in Chap. 24).

Prostaglandins

The placenta is capable of synthestizing prostaglandins with arachidonic acid as precursor. The major products obtained by chromatography were found to be HETE and leucotrienes belonging to the lipoxygenase pathway: this is more active in the early placenta than at term. The formation of these prostaglandin metabolites is blocked by SKF525, which suggests that this metabolism is P-450 dependent [9]. In addition to immune and vascular effects, prostaglandins may be involved in hCG regulation as was previously suggested in JEG choriocarcinoma cells [10].

Models Used for Studying the first-Trimester Placenta In Vitro

Several experimental models have been used. These can be divided into two major types, static and dynamic, according to the type of culture.

Static Models

In the static model, the contact with the medium is continuous: this allows cellular secretory products to pass continually through the placental membrane into the medium. However, the placenta is a rich source of a large variety of products which could affect subsequent local secretion. For example, a hormone such as hCG could, when it reaches high levels in the medium, modify its own secretion by negative feedback. It might simultaneously affect the secretion of other products as well. Finally, the qualitative or quantitative changes induced in these products could affect hCG secretion itself. Thus, a whole variety of para/autocrine types of interactions may take place in static cultures making the effect of tested agents less evident.

Cellular interaction is important in the placenta. In tissue explants, cell-to-cell contact and communication are maintained, and both cytotrophoblasts and syncytiotrophoblasts are present and contiguous. In contrast, isolated cell cultures are essentially of the cytotrophoblast type, and only later in culture do syncytial elements become apparent. Differences in response to GH are observed at term when explants or isolated cells are compared [11]. Similarly, differences in the response to progesterone are apparent between the two culture models [12]. In our view, the use of a simple culture medium is more interesting since serum (with a very complex composition) could interfere with local para/autocrine processes.

At term, the hCG concentration of explants is diminished. The secretion of hCG in the medium increases with time, in a manner quite similar to that of the first-trimester tissues. In contrast, isolated trophoblast cells exhibit clear differences according to gestational age: first-trimester cells reduce their hCG production in vitro, while term trophoblasts increase it for a few days. This might suggest that a promoting factor is eliminated by cell separation in the first-trimester tissue: it is recovered at term as cell aggregation and syncytium formation become more prominent [13].

Dynamic Cultures

Our laboratory has been involved in the development of a novel superfusion model to study the dynamics of hormone secretion in the placenta. Our system is based on the hypothesis that the exchanges between placenta and surrounding tissues occur mainly through diffusion which brings nutrients and maternal hormones to the placenta and carries away hormones, metabolites, and waste products from the conceptus (see Chap. 7). Our system mimics this process and provides a near physiologic environment for the explants. It differs from the placental perfusion apparatus which is used for studies at term. hCG levels fluctuate in the maternal circulation in the first trimester [14]. Using the novel superfusion model we have recently con-

Table 1. Properties of endogenous hCG secretion by first-trimester placental explants during superfusion following different sampling intervals

Sampling intervals (min)	6	2.4		1.2		1	0.5	
Frequency of pulse cosinor (min)	46.8 ± 7.6	18.4 ± 1.7	51.6 ± 27.6	27.6	10.8	22	11	8
P	0.01	0.04	0.026	0.088	0.082	0.042	0.066	0.026
Pulse interval (min)	17.8 ± 4	13.8 ± 2.7	35.9 ± 2.5	15.2 ± 1.2	8.4 ± 0.7	1.8 ± 0.9	5 ± 0.5	7 ± 0.9
Peak width (min)	15.2 ± 3	10.2 ± 10.7	10.7 ± 10.9	7.2 ± 0.9	4.8 ± 0.4	5 ± 0.6	3 ± 0.2	4 ± 0.01
Mean peak height (mIU/mg protein)	394 ± 102	17 ± 1.6	43.4 ± 6.8	1.5 ± 1.3	5 ± 0.3	10 ± 0.4	2 ± 0.06	4 ± 0.08
Area under the curve	389 ± 55	29.6 ± 8.15	234 ± 36	27.5 ± 5.2	8.36 ± 0.9	24 ± 2.4	9.4 ± 0.8	5.6 ± 0.02

The data are presented as mean ± SEM.
P, best degree of statistically significant interval which has the best fitting tau as analyzed by the Cosinor program.

firmed that maternal episodic hCG bursts are most likely of placental origin [15]. The lower pulse frequency in vivo (60–180 min) is due to the infrequent (i.e., 10–30 min) collection interval. In superfused explants, using a reduced collection interval (0.5–6 min) two pulse frequencies could be detected [16]. The first occurs every 18–20 min, while the other is more frequent (every 7–8 min). Analysis of the data was carried out by advanced computerized programs (Cosinor or PULSAR) (Fig. 1, Table 1). The higher pulse frequency in superfusion may be explained by the fact that, in order to detect all pulses, medium collection must be at least four times more frequent than the pulse frequency itself.

We have assayed more than 200 placentas in culture: we have noted that the spontaneous pulse frequency is fairly constant in the first-trimester tissues. However, within the same placenta the hCG pulse amplitude is variable: this suggests that multiple pulse generators are involved in the development of hCG pulsatility. Further work has revealed that, even after 4 h of superfusion, hCG pulsatility was maintained without observing a decline towards baseline levels, i.e., without evidence of tissue exhaustion. This was recently confirmed by histologic studies. Moreover, explants which were cultured for 2 days in static cultures, prior to superfusion, also maintained the pulsatile pattern of hCG secretion.

Preincubation with cycloheximide eliminated pulsatility in superfusion, suggesting that hCG pulsatility depends on protein synthesis. Other experiments were carried with ^{3}H leucine-preincubated explants which were subsequently washed off with cold leucine and superfused. The effluent was precipitated with 5% TCA and counted. The profile of the precipitate also appeared to be pulsatile. This strongly suggests that the early placenta operates with off and on periodicity in its secretory products, although there is a basal tonic secretion. The identification of the proteins in question is currently under investigation. Superfused placenta at term revealed a pattern of secretion which was completely different [16]. The pulse frequency was around 45–55 min, and pulse amplitude was only 30%–35% over baseline compared to the major peaks seen with first-trimester tissue (Fig. 2). There was also a clear decrease to the baseline during superfusion. Such observations point to important differences between early and late placenta as regards hCG modulation. Preliminary evidence that pulse frequency decreases with advancing gestation in vivo was recently confirmed by other [15]. At 20–24 weeks it was every 90 min while in the first trimester it was 60 min.

Some preliminary experiments were also carried out with superfused isolated cells from early placenta. In contrast to explants, hCG production was not pulsatile [17]. This may illustrate the need for intact cytotrophoblast-syncytiotrophoblast interaction. The support needed for continued pulsatile hCG secretion might thus depend on cellular interaction and release of local factors. We believe that GnRH may have a major role in promoting this secretion (see below).

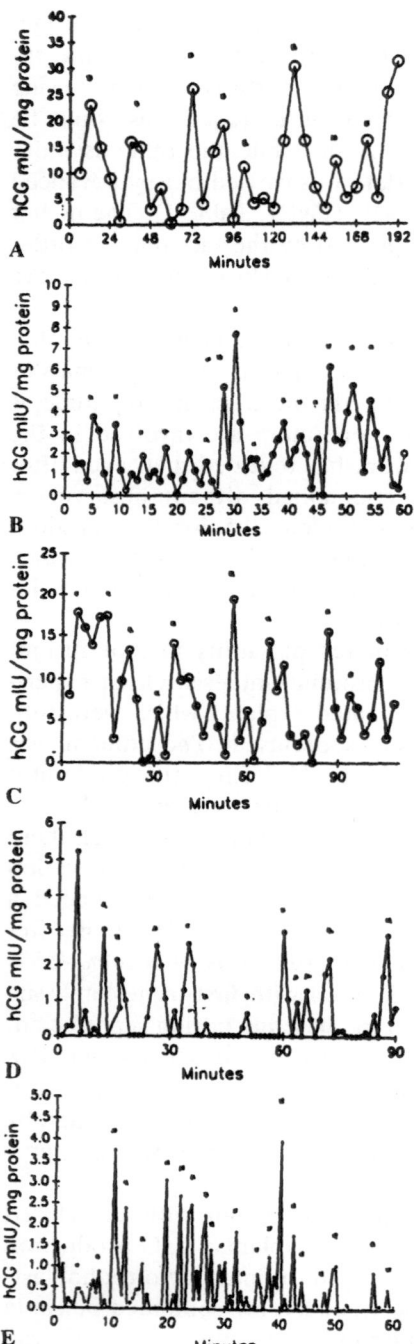

Fig. 1A–E. Spontaneous pulsatile hCG secretion by superfused placental explants (1st trimester and term). Collection interval of the effluent is every 6 min (**A**), 2.4 min (**B**), 1.2 min (**C**), 1 min (**D**), and 0.5 minutes (**E**). No change were noted in baseline levels throughout superfusion. hCG was measured by radioimmunoassay (MAIAclone, Serono). Intra-assay variability was 1.7%. Pulse analysis by computerized program (PULSAR)

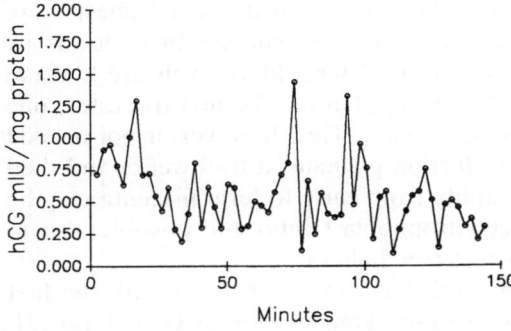

Fig. 2. Spontaneous pulsatile hCG secretion by superfused placental explants at term. Collection interval 2.4 min

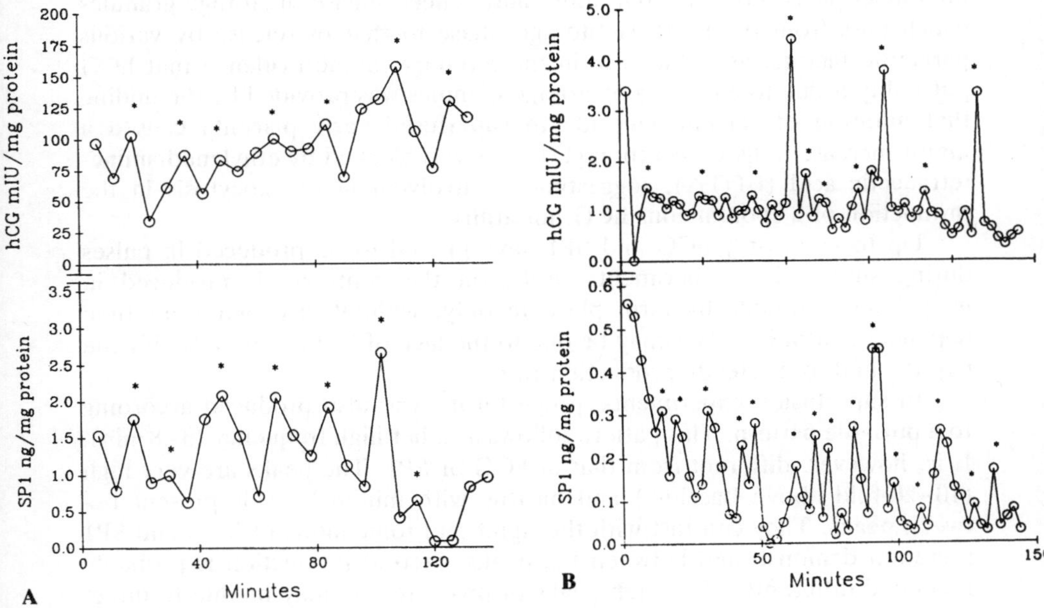

Fig. 3. Patterns of SP1 pulsatility by 1st-trimester (**A**) and term explants (**B**). The sampling interval was 2.4 min in both cases. *Stars*, pulses identified by PULSAR

The patterns of other hormones were also examined in superfusion. SP1, a glycoprotein, is a placental marker, but without a known specific function. In the first-trimester tissues, the pulse frequency of SP1 was 15–16 min, demonstrating a rather good concordance with hCG pulsatility. However, in contrast to hCG at term, SP1 pulsatility increased to 7–8-min frequency (Fig. 3) [16].

This strongly suggests that, since hCG levels in the first trimester are elevated, a high pulse frequency is required to maintain these levels by offsetting metabolic clearance. In contrast, at term hCG levels are low and are correlated with the low pulse frequency. During the first trimester, one role of hCG is to maintain the corpus luteum. This, however, is not needed afterwards, and therefore hCG production plateaus at 8–9 weeks and then declines. The change in hCG pulsatility from early to term placenta may be due to a change in the cyto/syncytiotrophoblast ratio and possibly also to other metabolic and endocrine factors (see below).

According to immunohistochemical data, and contrasting with the first trimester, there are no detectable storage granules for hCG at term [3]. Storage granules may be an important prerequisite for the expression of high hCG pulsatility. This pulsatility may reflect release of storage granules which pass from the synthesis/storage phase to that of release by various paracrine factors, as is the case in the pituitary. Some evidence that hCG pulsatility is due to emptying of storage granules was provided by the finding that addition of calcium chloride to superfused early placenta caused a major increase in hCG secretion [17]. This was blocked by ethylenediamine-tetraacetic acid (EDTA), suggesting an involvement of exocytosis in the production of a spontaneous hCG pulsatility.

Up to now, only hCG and SP1 have proved to be produced in pulses during superfusion experiments. hPL, on the contrary, is produced in measurable amounts by term placenta only, without any significant time dependent variation. This may be due to the lack of storage granules for the peptide and to its steady production rate.

In superfusion experiments, progesterone was also produced according to a pulsatile pattern. This pattern follows a rather high frequency (3–8 min). It is, however, different from that of hCG or SP1. The peaks are very high (10–20-fold above baseline) and narrow with minimal levels present between peaks. They contrast with the significant tonic mode of hCG and SP1 secretion demonstrated between the peaks. Steroid production is probably regulated differently. The high peaks of progesterone may be due to direct production by the early placenta without any significant intracellular storage. This is similar to the pulsatile nature of progesterone secretion by the corpus luteum. A similar pulsatility of progesterone was recently documented in vivo during early pregnancy [18]. The pulse frequency was much less frequent; this is likely to be caused by the infrequent collection intervals, i.e., 30 min [18].

Regulation of hCG Secretion

Gonadotropin-Releasing Hormone

Effect of GnRH in Static Cultures

The early placenta is a target organ for GnRH, and this molecule may be one of the major paracrine regulators of hCG production. Early work has shown that in static cultures and using high concentrations, GnRH stimulates hCG secretion [19]. It has also been demonstrated that GnRH has an effect not only on hCG production but also on that of the free α-subunit, progesterone, estradiol, and estriol [20]. It must be noted that the concentrations of GnRH used were two or three orders of magnitude higher than the local GnRH content $(10^{-8} M)$, and the increase seen in hCG secretion was only 50%–100%. GnRH can also stimulate hCG secretion in patients with gestational trophoblastic disease under treatment. This could point to persistent viability of neoplastic trophoblastic cells.

We have recently examined the effect of both GnRH and of a potent of GnRH agonist (estimated potency 100-fold that of GnRH) on hCG secretion for 24 h. Very low doses $(10^{-11} M)$ of the agonist and $10^{-9} M$ of the native compound increased hCG secretion, while higher concentrations had no effect. In first-trimester explants, stimulatory effects of GnRH on hCG biosynthesis were also demonstrated in experiments where incubation with S-35 methionine was used.

Effect of GnRH

We have reasoned that, since hCG secretion is pulsatile, the most effective mode of decapeptide administration could be episodic as well. It had been previously reported that the high endogenous placental GnRH may saturate and/or down-regulate the local GnRH receptor [21]. We found that at low concentrations $(10^{-10} M)$, the GnRH analogue and GnRH $(10^{-8} M)$, stimulated pulsatile hCG secretion in superfused explants of young placentas. This was noted with 1-min pulses. The response was prompt and evident within 6 min. The exposure increased both pulse amplitutde (two- to sevenfold) and frequency [15, 22]. This closely resembles the GnRH effect on pituitary LH secretion in terms of pulse amplitude only. We later demonstrated that the effect of GnRH agonist and GnRH was dose dependent. The effect of the analogue was of two orders of magnitude higher than that of the native compound. A biphasic type of response was noted, where low and high doses were less effective. The maximally effective concentration, $10^{-8} M$ GnRH, is that which locally supports the proposition of a paracrine role for GnRH. The effect of a low concentration of GnRH agonist was time dependent, and it increased hCG secretion when administered for 1–20 min. The effect of longer pulses was biphasic; an initial increase of the hormone was followed by a decrease close to baseline. However, when stopping GnRH administration there was a major increase in pulsatile hCG secretion. This type of response suggests that a down-regulation of the GnRH receptor may have taken place. The prompt restoration of hCG release after stopping GnRH administration argues against tissue exhaustion. In contrast, at term and using a wide range of concentrations, no effect of GnRH agonist was noted upon hCG secretion by superfused explants.

Receptor Dependence of GnRH Action

There are specific GnRH receptors at term, and the decapeptide binds specifially to isolated trophoblastic cells in order to stimulate hCG secretion [21]. We have shown that both previously described types of binding sites, one with a Ka of $10^{-8} M$ and the other which is a low affinity-high capacity site – Ka $10^{-5} M$ – can also be detected in the first-trimester tissues. We also demonstrated [22] that lower doses of a GnRH antagonist blocked GnRH effect in superfusion. This inhibition effect was reversible, and the refractoriness to GnRH stimulation was suppressed shortly after stopping the antagonist administration. Moreover, the antagonist effect was specific since it did not block the stimulatory effect of epidermal growth factor (EGF) upon hCG secretion in superfusion. In static cultures such a block could only be demonstrated in the second trimester.

Localization and Secretion of GnRH

GnRH is produced by the cytotrophoblast and acts on the syncytium where specific receptors are present on the cell membrane. The placenta contains GnRH: its content is higher in early tissues that later in gestation. Significant amounts of GnRH-like material are secreted by superfused placental explants; this confirms the physiologic role of GnRH in controlling hCG production (unpublished observations). This secretion appears to be modulated by various cations, prostaglandins, β-adrenergic agents, and steroids at term [23].

There are reports which indicate that placental GnRH is immunologically and chemically identical to hypothalamic GnRH and that the decapeptide extracted from the placenta can promote pituitary LH secretion. Recent work has pointed to the presence at term of a prohormone which in vitro appears to be more potent than native GnRH to induce hCG secretion; an encoding gene has been isolated [24].

Why Is There a Lack of GnRH on hCG Secretion at Term?

We found that GnRH agonist added in various concentrations for 1 min to superfused explants at term had no effect on hCG secretion. This may be caused by several factors, but probably not by receptor properties which appear to be similar to those in the first-trimester tissues (unpublished observations). At term, the presence of hCG secretory granules is limited, if it exists, and therefore pulsatility is much less evident [16]. Most probably only that portion of gonadotropins which is readily available for secretion, i.e., which has passed from the synthesis/storage phase to that of secretion, will be released. In addition, a post-receptor defect may be present at term.

The discovery of specific differences between first-trimester and term placenta could be an important start in elucidating the control of hCG secretion. There is a possibility that placental inhibin is involved in suppressing endogenous GnRH, and therefore in lowering hCG secretion at term. In contrast, local FRP, part of the inhibin family, was suggested to have a promoting effect upon GnRH secretion at that time [25].

Mechanism of Action of GnRH

The information on the GnRH mechanism of action in first-trimester placenta is rather limited. We have found that Ca^{2+} pulses stimulates hCG secretion in superfusion. It has been recently demonstrated that at very high concentrations, GnRH increases hCG secretion by isolated cells at term [21]. This action of GnRH is likely to be mediated by the low affinity site which we also found to be present in the early placenta (unpublished observations). Others have found that this effect was blocked by verapamil, a calcium channel blocker, as well as by calmodulin antagonists [26]. There is also a possibility that at term the effect of GnRH is mediated by Ca^{2+} activation through protein kinase C but not by hydrolysis of placental inositol phosphate [27].

Effect of Progesterone on GnRH-Induced hCG Secretion

During the luteal phase, progesterone modifies GnRH secretion, thus inducing a low-frequency and high-amplitude pattern of LH secretion. In early pregnancy, progesterone may be involved in the establishment of a plateau in hCG production at 9–10 weeks with a subsequent decline. Lower levels of free subunits are obtained following incubation of explants with progesterone [28].

Recently we demonstrated that isolated cells under progesterone influence have a reduced hCG secretion in long-term culture, while cellular viability remains good [12]. In superfusion, $10^{-6} M$ progesterone decreased hCG secretion when given in short pulses: lower concentrations proved less effective.

This inhibitory effect concerned both the frequency and the amplitude of the pulse [15]. Co-administration of maximally effective doses of GnRH agonist with progesterone inhibited the GnRH agonist effect rapidly and reversibly (Fig. 4). The concentration of progesterone was similar to that

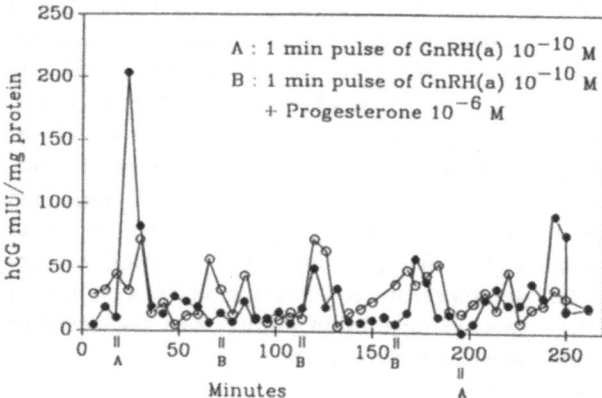

Fig. 4. Effect of GnRH $10^{-10} M$ given for 1 min in superfusion (**A**). A significant increase in the amplitude of the hCG was noted. This was blocked by co-administration of $10^{-6} M$ progesterone (**B**). *Open circles*, control; *solid circles*, treatment

secreted by the explants in superfusion [12]. This steroid effect is rapid and is probably not exerted through the genome; it might prevent the exocytosis of early placenta storage granules induced by GnRH. In contrast, the effect of low concentrations of cortisol and estradiol was stimulatory on hCG secretion in static culture. Wilson and Jawad [29] have demonstrated that cortisol and dexamethasone inhibited native hCG secretion and stimulated free α-subunit production in long term cultures.

In explants to which fetal calf serum was added, the effect of high doses of progesterone, cortisol, or DHEAS was stimulatory, while that of testosterone was inhibitory, and estradiol had no effect after 4 days on hCG secretion [30]. At term, cortisol had no effect. That testosterone of fetal origin may decrease hCG secretion in vivo has been suggested, since hCG levels are lower in pregnancies with a male fetus.

The Role of Opioid Peptides

The pathways leading to the formation of opioid peptides from pro-opiomelanocortin are similar in the placenta and central nervous system. Among the proopiomelanocortin metabolites β-endorphin, enkephalins, adrenocorticotropic hormone (ACTH), and dynorphin have been identified in the placenta [31]. The amniotic fluid is also an important source of such peptides. The involvement of opioid peptides in the modulation of hypo-thalamic endocrine function is well recognized during the luteal phase. Recently, Cemeric et al. [32] demonstratred that high concentrations of morphine stimulated hCG secretion in first-trimester tissues, but not in term placenta. It is not clear, however, whether opioid peptides are directly

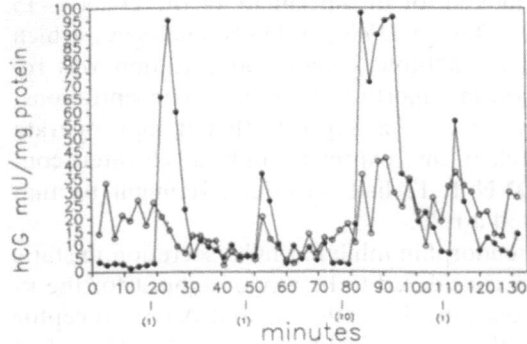

Fig. 5. Effect of DYN 1–13 ($10^{-10}M$) on hCG secretion in superfusion (*solid circles*). The 10-min pulse effect was more pronounced than that of 1 min pulses. The collection interval was 2.4 min. *Open circles*, control

involved in the modulation of hCG production. We have investigated the effect of two classes of peptides, β-endorphin (which has a high affinity to the μ and Δ class of receptors) and dynorphin (which binds preferentially to the κ-receptor) on hCG secretion by the early placenta. In superfusion, dynorphin (DYN) 1–13, a potent dynorphin fragment, stimulated pulsatile hCG secretion (Fig. 5) [33]. This effect was dose dependent and biphasic. The most significant effect was noted at $10^{-10}M$ concentration, that is, within the range of opioid peptide affinity to its respective κ-receptor at term [34]. The properties of the opiate receptor in first-trimester tissues are not known.

At high and low concentrations, the effect of DYN 1–13 was less evident. This type of biphasic response was similar to that seen with GnRH. In addition, the effect of DYN 1–13 seemed to be time dependent: a longer pulse was more effective than a 1-min pulse. The effect of DYN 1–13 was specific and receptor dependent. When naloxone, an opiate receptor antagonist, was added to the medium, at a concentration one order of magnitude higher than the opiate peptide, it blocked all effect. It was of interest to note that, after 10 min of combined administration, there was a delayed but significant increase in hCG secretion. High doses of naloxone alone ($10^{-9}M$) stimulated hCG secretion, though at $10^{-10}M$ no effect could be obtained. The stimulatory effect of naloxone alone suggests that there is a continuous and inhibitory effect of opioid peptides on hCG secretion, and that it can be counterbalanced by naloxone. It is tempting to speculate that part of the local hCG regulation in the first-trimester placenta is mediated by opioid peptides. The tonic inhibitory effect was more evident when DYN 1–13 was added to the medium for 90 min; hCG production increased for the first 20 min and then returned to basal levels. When DYN 1–13 administration was stopped, there was a major increase in hCG secretion.

This may indicate that brief administration of low doses can cause receptor up-regulation, increasing the production of hCG. Later, a down-regulatory effect may appear. It is reversible, and the refractoriness to DYN

1–13 is short. We have further looked for the specificity of the DYN 1–13 effect on hCG secretion. Using Des-tyr-DYN, a DYN analogue, which binds poorly to the κ-receptor, a 1000-fold higher concentration was required to stimulate hCG secretion in superfusion. At high concentrations, Des-tyr-DYN had a stimulatory effect; this suggests that it may operate through nonopiate-mediated mechanisms, expressed only at elevated concentrations. In static cultures DYN 1–13 had no effect. It might be that receptor desensitization has played a role.

In contrast, we found that β-endorphin inhibited hCG secretion in static incubations [35]. This molecule is considered to be a poor agonist for the κ-receptor. Indeed, at very low concentrations, this μ- and Δ-type receptor agonist inhibited hCG secretion following preincubation for 24 h. The effect was dose dependent and was seen in concentrations most likely to interact with the receptor. Though no such type of receptor has been described in the placenta, our preliminary data suggest the presence of such sites in the 1st-trimester tissues. We have examined whether this interaction was carried out at the opiate receptor site. Addition of naloxone abolished the inhibition seen with β-endorphin alone. The lack of effect of N-acetyl β-endorphin, a nonactive analogue of the opiate peptide, confirmed the specificity of the molecule effect. The effect of endorphin appears to vary according to the gestational age. At 7–9 weeks hCG secretion decreased, while at 11 weeks a reversal of the situation was seen and hCG levels increased in the media. Thus, in vivo, β-endorphin might be involved in the initiation of the decrease in hCG secretion following the plateau.

Out preliminary data point to the possible existence of receptors of the μ- and Δ-type in the first trimester placenta which would strengthen the physiologic role of opioid peptides in situ. The opposing effects of β-endorphin and dynorphin on the placenta suggest that the endogenous opioid system has an important and complex role in controlling hCG secretion. It is complex for the following reasons:

1. Different concentrations produce different effects which at low doses could be opiate mediated, while at high doses might not be opioid.
2. The effect is dependent on the model used: dynorphin is effective in superfusion experiments and not in static cultures, while the effect of β-endorphin appears to be just the opposite. Evidently, both models are needed to examine the full range of opioid peptide effects.
3. The effect of β-endorphin differs according to the gestational age.
4. These results illustrate the need to use the first-trimester placenta for investigations instead of relying on information obtained with term tissues.
5. Native opioid peptides are important for the comprehension of the endocrinology/paracrinology of the first trimester.
6. The properties of opioid peptide-binding sites need to examined further in the early tissues.

We have examined the mechanism by which DYN 1–13 might stimulate hCG secretion. We found that at defined concentrations, interesting results were obtained. When hCG levels increased, there was an associated decrease in local progesterone secretion, suggestive of a paracrine/autocrine form of regulation. Our work also suggested that the effect of DYN 1–13 was not modified when it was co-administered with maximally effective concentrations of GnRH agonist. This suggests that the action of dynorphin is not exerted through increasing levels of endogenous GnRH, but probably shares a common second messenger: this hypothesis must, however, be tested. DYN 1–13 may act either by suppressing Ca^{2+} entry to the cell or by blocking adenylate cyclase; the latter could lower local progesterone secretion, thereby causing the increase in hCG [36]. At term, it has been suggested that GnRH effect is Ca^{2+} dependent.

There is thus a distinct possibility that hCG production in the first-trimester tissue is regulated via a paracrine/autocrine control. This intra-placental modulation could induce spontaneous tonic (basal) and clonic (pulses) events, at least partly analogous to the production of LH by the pituitary: it is highly probable that different pathways coexist, and that the mechanisms involved are different in early and term tissues. They necessitate the integrity of the tissues.

Effect of Growth Factors

The early placenta is characterized by its very rapid growth from a few cells to 20–30 g at the end of the first trimester. Thereafter the increase is much slower, reaching only an additional 20 fold increase in weight until term. The major involvement of various growth factors in the regulation of the first-trimester placenta is strongly suspected [37]. These are regarded as stimulating mitosis and/or differentiation. The list is growing (Table 2); these factors have been identified in the circulation, amniotic fluid, and placenta where specific binding sites could also be demonstrated, suggesting the existence of probable paracrine and autocrine effects. Some growth factors stimulate, others inhibit or do not affect placental hormone secretion.

EGF is a peptide which leads to phosphorylation of membrane protein and acts as a mitogen by increasing DNA and RNA synthesis. At term, it increases hCG secretion and enhances trophoblastic differentiation, but in the first trimester early reports were conflicting. We have recently examined the effect of some growth factors on the first trimester placenta under various experimental methods. hCG production increased in explants after 24 h of incubation with EGF [38]. Moreover, exposure of isolated tropho-blastic cells to EGF induced an increase in hCG secretion during the first week. This appears to be related to hCG production de novo since the media were collected at daily intervals. As expected, there was a progressive decrease in hCG secretion during the culture period, and EGF delayed this

Table 2. Presence of growth factors and their receptors in early first trimester human placenta: their effect on hCG production in vitro

Growth factors	Presence	Specific receptors	In vitro influence Cells	Explants	Superfusion
Epidermal growth factor	+	+	+	+	+
Insulin	+/−	+		−	−
Insulin-like growth factor 1	+	+	+	=	
Interleukin-1	+	+	+		
Interleukin-6	+	+	+		
Fibroblast growth factor		+		=	
Parathyroid hormone 1–34		+	+	+	+
Macrophage colony-stimulating factor					
Granulocyte colony-stimulating factor	+	+			
Transforming growth factor β	+				
Interferon α	+				
Tumor necrosis factor α	+				

decline. The addition of EGF to trophoblastic cells did not appear to modity their rate of differentiation or their long-term viability. It has been reported [39] that exposure to EGF could induce an increase in free α-subunit.

In superfusion experiments, EGF had a rapid effect on hCG secretion. This effect was dose dependent, increasing both the pulse frequency from 3 to 5/h and the amplitude. At a 100-ng/ml concentration, the effect became less evident. Prolonged administration of EGF (i.e., 90 min in superfusion) induced an initial increase in hCG, followed by a decrease to below baseline levels, suggesting a down-regulation (Fig. 6). This proved durable since the subsequent pulse of EGF failed to elicit a significant response. There is thus a possible modulation by EGF of the affinity of its own receptors. We conclude that EGF has a dual regulatory effect on hCG secretion. The first is rapid; it involves emptying of storage granules, without hormone synthesis; the delayed effect is obtained by stimulation of hormone production. We have even demonstrated that overnight incubation with EGF leads to a major increase in the pulsatile hormone secretion as evidenced by the increased area under the curve and pulse amplitude [40]. This response was obtained with early tissues, but with 12-week tissues the effect was reversed.

These experiments suggest that the action of EGF is long lasting and that it is present even after the removal of the growth factor from the medium by extensive washing: most probably, specific binding of EGF to its receptor has taken place. Alternately, EGF could have affected the genome or other steps required for hCG biosynthesis and secretion. In the first-

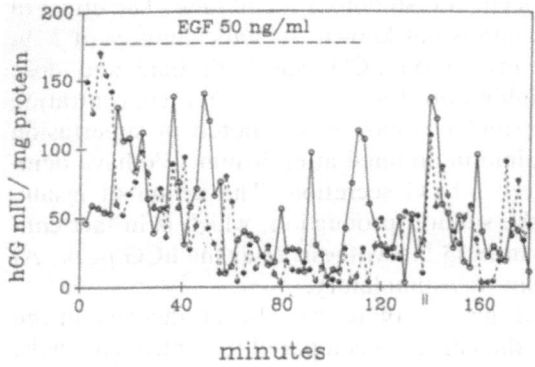

Fig. 6. Effect of continuous administration of EGF (50 ng/ml) for 90 min on hCG secretion from placental explants in superfusion (*solid circles*). A significant decrease in pulsatility was noted following the initial increase; 40 min after stopping EGF administration, a 1-min pulse of 50 ng/ml EGF was, however, ineffective (11), *Open circles*, control

trimester placenta, EGF receptors are located at the syncytial cell membrane and/or in the nucleus.

EGF-like material was found to be secreted by the trophoblast itself [41]. The effect of EGF does not appear to be mediated by GnRH, since continuous administration of an GnRH antagonist in superfusion was incapable of blocking EGF action, though it blocked that of GnRH. In explants of first-trimester placenta, it was suggested that EGF stimulates the secretion of free α-hCG subunit and hPL as well.

The effect of other mitogens, i.e., PTH 1–34 (an active fragment of the parathyroid hormone which is, however, devoid of calciotrophic effects) was examined (unpublished observations). PTH 1–34 added to explants for 24 h at 7–9 weeks increased hCG secretion, in a dose-dependent and biphasic fashion. However, in 11–14-week placenta there was an inhibition of hCG production. The response was similar in superfusion experiments: following overnight preincubation there was either an increase or a decrease of the amplitude and area under the curve. We then tested the interaction between PTH 1–34 and EGF. At 7–9 weeks, we found that there was an additive effect, suggestive of synergistic action. The mechanism of action of PTH 1–34 is not clear. It is partly only adenylate cyclase dependent, while that of EGF appears to be mediated by tyrosine kinase. There might be an interaction between the two growth factors which could activate similar second messengers, possibly tyrosine kinase, or act at the binding sites. In isolated cells an increase in hCG secretion was demonstrated with PTH 1–34 which was potentiated by EGF [42].

Insulin is regarded as the major hormone involved in the control of carbohydrate homeostatis. In recent years this hormone has been found to have endocrine effects on the placenta as well, and in one sense it is considered a growth factor since its structure is similar to that of insulin-like growth factors. We have previously reported [43] that insulin modulates both estradiol and progesterone secretion by term explants. It probably binds on specific sites on the syncytial brush border membrane. The effect of insulin on the 1st-trimester placenta is not known. In static cultures of 7–9-week placentas, the effect of insulin on hCG was both time and dose dependent with the highest inhibition for a 5–50 µU/ml concentration (unpublished observations). A similar response was noted in superfusion experiments with a maximal inhibition attained after 30 min. We have demonstrated that insulin lowers free β-hCG secretion. The action of insulin may be mediated by reducing the subunit production, which is in fact considered as the rate-limiting step in hCG biosynthesis after the hCG peak. At 11–14 weeks, the effect of insulin was stimulatory.

Following preincubation with insulin, there was also an increase in the pulse amplitude and area under the curve, as compared to control channels. In these low concentration experiments, insulin action was mediated solely by binding to insulin receptors and not to insulin-like growth factor I (IGF-I) binding sites to which it attaches at high concentrations. Moreover, we found that IGF-I does not influence the secretion of hCG by the early placenta. In a recent work, it has been shown that 100 nM IGF-I added to isolated cells could stimulate hCG secretion after 4 h of culture, irrespective of the gestational age. The placental concentration of IGF-I decreases towards term. This could be indicative of the time variability of auto/paracrine regulation.

IGF-II is structurally similar to IGF-I; it is produced by the placenta and binds to specific sites [37]. The levels of IGF-II mRNA are higher in second-trimester than first-trimester tissues. Its effect on the placenta is not known, but it has been suggested that it could be involved in cellular proliferation.

In preliminary studies we found that fibroblast growth factor did not stimulate hCG secretion by early placenta explants. The secretion of hCG was increased by incubations with human macrophage colony-stimulating factor (hM-CSF) and granulocyte colony-stimulating factor (G-CSF). Production of these factors was demonstrated in the early placenta in concentrations much lower than in the decidua [44].

Other workers have reported that the cytokines which are compounds involved in cell proliferation and differentiation have also endocrine effects on the early placenta in vitro [45]. Interleukin-6, which is actually derived from first-trimester tissues, stimulated hCG secretion, in a receptor-dependent fashion, independent of the action of GnRH, although their combined effect was synergistic. Interleukin-1 also stimulates hCG production, and it has been suggested that this is obtained by local production of interleukin-6. Such studies well illustrate the complexity of interrelations between growth factors.

Recently there has been much investigation regarding the role of inhibin and related compounds, which were first identified in the ovary [46]. Inhibin inhibits pituitary FSH production in women. Circulating inhibin levels increase with pregnancy, and immunocytochemical data reveal that levels of inhibin α- and β-A-subunits are highest in early placenta and decrease later, while those of β-B-subunits increase steadily to term [47]. It was suggested that activin could stimulate GnRH secretion by cultured cells at term while inhibin would suppress it. There are no data for the 1st trimester.

Transforming growth factors (TGF) are so named because of their ability to promote cell colony formation. TGF were described in the placenta [48]. α-TGF has a structure similar to EGF and binds to the same receptor. Its biologic effect is not known. A related polypeptide, activin, has a structural homology to β-TGF at term.

Platelet-derived growth factor (PDGF) is a glycoprotein which has been identified in the early placenta where specific binding sites were also detected. Its biological function is not known.

Effect of Cellular Mediators

The effect of cyclic adenosine monophosphate (cAMP) in hCG secretion has been examined in several studies; it is not known, however, whether this second messenger is a physiologic mediator of hCG secretion. The stimulatory effect of cAMP analogues on hCG and free α-hCG and pro-gesterone secretion in both early and term cells has been demonstrated. The stimulation of endogenous cAMP by cholera toxin had a similar effect [13]. 8-Bromo-cyclic guanidine monophosphate (8-bromo-cGMP) is also stimulatory on early tissues. Some data are indicative of the hCG being structurally different. This molecular heterogeneity, in its turn, induces variability in the in vitro biologic activity [49].

Embryo-Placental Interaction

The concept of a fetoplacental unit is a well-accepted entity. The placenta is an incomplete steroidogenic organ which needs to receive DHEA from the fetal or maternal adrenals in order to synthesize estrogens. The local trans-formation of progesterone to androgens, as occurs in the adrenal glands and gonads, is very limited. However, the concept of embryo/placental inter-action during the first trimester is new in the human, although it has been proposed in several mammals. Application of this concept to human beings provides new insights into the regulation of the early placenta. It has been shown that decapitation of pig fetuses could alter placental function. In the human species, intrauterine fetal death in the second trimester could be associated, for a period of time, with a persistent placental function [50, 51]. This led some people to think that the human embryo did not play an important role in the control of placental function. There are, however, some observations which contradict this:

1. Trophoblast cultures obtained after recent embryonal death grow poorly compared to controls.

2. Ectopic pregnancies without a demonstrable embryo have low levels of circulating hCG and are less likely to rupture the tube.
3. Recent studies examining the results of surgical embryonal reduction for multiple pregnancy have shown that shortly after the embryo was injected with potassium chloride, hCG levels were reduced [52].
4. Sex-linked differences have been noted with hCG levels being higher when the embryo is female [53].

Of course, all these data provide only circumstantial evidence. In addition, it is well known that the end of embryogenesis coincides with the hCG plateau: clearly, serum hCG levels can be used to separate the embryonal and fetal periods.

In order to investigate the possible embryo/placental interaction during the 1st trimester, we have used co-culture techniques to test the effect of the embryonal tissue on hCG and progesterone secretion by the early tissues [54, 55]. Our recently published data demonstrate that various embryonal organs, i.e., liver, lung, adrenal gland, spinal cord, and brain, have a significant modulatory effect upon the secretion of these hormones. This was dose dependent and was already evident at very low concentrations when sonicated extracts of the organs were incubated with the placental explants. With 7–9-week tissues, an inhibitory effect on hCG secretion was noted, while the effect produced was reversed at 11–14 weeks.

More recently, we have confirmed that the modulation of hCG production could also be noted after tissue co-culture. This strongly suggests that the secretion of the active principles occurs in vivo and that it is not only a structural element of some embryonal cells liberated only during sonication. For spinal cord explants combined with placental explants, the effect was seen both in static co-cultures following overnight incubation and in superfusion. In co-superfusion experiments, the spinal cord explants were placed before the placental explants to obtain the inhibitory effect (Fig. 7). A decrease in the hCG pulse amplitude and area under the curve was noted at 7–9 weeks. In addition, there was also a decrease when the placental explants were placed in superfusion after preincubation of the medium with the spinal cord. Thus, these embryo-derived compounds also exerted a delayed effect which was similar to that seen when the co-superfusion was carried out. An increase both in the pulse amplitude and area under the curve was noted in co-cultures carried out at 11–14 weeks.

The effect on progesterone secretion varied according to the organ tested. Some organs acted similarly on hCG and progesterone secretion, while for others the effect was opposite. A gestational age and dose-dependent effect on progesterone secretion could also be seen. The effect obtained with embryonic adrenal gland is noteworthy. We found that water-soluble extracts had a strong modulatory effect on hCG secretion by explants in static cultures.

Preliminary data in vivo suggested the presence of a putative steroid (possibly DHEAS) which exerts an inhibitory effect on hCG secretion.

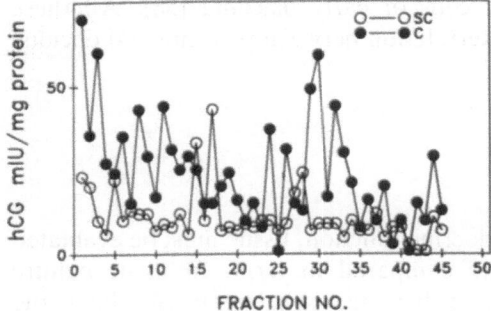

Fig. 7. Reduced hCG secretion by superfused placental explants cultures following overnight pre-exposure to embryonic spinal cord tissue (*open circles*) (8 weeks' gestational age). *Solid circles*, control

However, since the major effect was noted with water-soluble extracts and since with a 11-week adrenal gland it was stimulatory, it does not appear that it is responsible for lowering hCG secretion after the hCG plateau. It seems, on the contrary, that the opposite should be probable since hCG exerts a trophic effect on adrenal function in the fetus.

The biochemical nature of the secreted compounds is under investigation; preliminary results suggest that the compounds in question are water soluble and are of low molecular weight. It may thus be that in vivo the embryo plays a role in the modulation of placental function in order to suit its needs. Before hCG peak, when hCG levels rise, the effect could be inhibitory; after the hCG peak, when gonadotropic stimulation is needed for testicular and adrenal function, hCG production is stimulated, though to a lesser degree than in the early 1st trimester.

We have thus demonstrated that in superfusion experiments, hCG pulsatility (peak to trough differences) is intense, while in vivo, though detectable, it is not as important. In addition to the metabolic clearance that certainly plays a role in shaping the hCG pattern, there may be in vivo a tonic inhibition of hCG pulsatility: this should be eliminated once the influence of the embryo is removed, as seen in placental cultures. It has been shown that in cases of defective fetus (Down's syndrome), the levels of hCG are more elevated in the second trimester.

Placental Interaction with the Decidua

There are some indications that addition of culture media from the early placenta explants decreases decidual protein synthesis but does not affect secretion of prolactin (PRL). Removal of hCG from the media by preincubation with an antiserum abolished this effect [56]. However, this is not supported by other data showing that neither hCG nor progesterone appear to modify PP12, PP14, and PRL secretion by decidual cell cultures [57]. A receptor for PP12 in the first-trimester placenta was recently reported. Preliminary reports indicate that media derived from decidual cultures modify hCG secretion by isolated cells of early placenta [58]. All these experiments point to a paracrine interrelation between placenta and decidua (see Chap. 2).

Conclusions

In order to understand placental endocrine function, tissue must be evaluated at different stages of gestation and compared to term. In tissue culture studies, there is a need for static and dynamic experiments. We favor the use of explants since cellular interaction is maintained, while hCG secretion by isolated cells is not pulsatile. Based on recent work, it appears that the regulation of hCG and progesterone is complex. Various molecules may exert different actions in the pre- and post-plateau phase of hCG synthesis. Moreover, the placental tissue synthesizes several hormones or factors which can produce either a paracrine or even an autocrine action. These aspects can be addressed by dynamic culture techniques.

On the other hand, it appears that in some instances the static cultures are more advantageous (i.e., β-endorphin effect on hCG compared to that of DYN 1–13). The use of choriocarcinoma tissue or cell lines does not provide comparable results, and subsequently normal tissue is to be used.

A very wide range of concentrations of molecules is needed since the observed effect may vary with the same compounds, and the early placenta appears to be exquisitely sensitive to concentration changes, especially in superfusion experiments. Moreover, the effect of various compounds on hCG production can be rapid or slow, addressing either secretory processes (exocytosis of storage granules) or true biosynthesis which obviously needs a longer period. Combinations of static and dynamic techniques help to delineate those two mechanisms and aid in the study of this dual form of regulation. Lastly, the basic properties of first-trimester placenta are beginning to be elucidated from an endocrine point of view. This will allow more knowledge about the pathologic and altered embryo-placental relationships.

Future Directions

1. Comparative studies of endocrine and paracrine effects of all placental molecules at various stages including first trimester before and after hCG peak, using both dynamic and static in vitro models.
2. Attempts to selectively eliminate para- and autocrine influences in order to define precisely specific effects of a given hormone or factor.
3. Identification of the embryonal and decidual origin of given compounds which modulate placental function.
4. Studies devised to understand patterns of hormone secretion and their regulation with the use of the superfusion model.

5. Design of a scheme summarizing hormonal interplay in the early placenta.

References

1. Chen R, Barnea ER, Benveniste R (1986) Characterization of glycoprotein hormone free alpha-subunit from human pituitary and placenta extracts. Horm Res 23:38–49
2. Hoshina M, Hussa R, Pattillo R, Camel MH, Boime I (1984) The role of trophoblast differentiation in the control of the hCG and hPL genes. Adv Exp Med Biol 176:299–311
3. Morrish DW, Marusyk H, Siy O (1987) Demonstration of specific secretory granules for human chorionic gonadotropin in placenta. J Histochem Cytochem 35:93–101
4. Reshef E, Lei ZM, Rao CV, Prodham DD, Chegini N, Luborsky JL (1990) The presence of gonadotropin receptors in non pregnant uterus, human placenta, fetal membranes and decidua. J Clin Endocrinol Metab 70:421–427
5. Kumasaka TN, Nishi N, Jai Y, Kido M, Saito I, Okayasu K, Shimizu S, Hatakeyama S, Sawano S, Kokubu N (1979) Demonstration of immunoreactive somatostatin-like substance in villi and decidua in early pregnancy. Am J Obstet Gynecol 134:39–48
6. Devroey P, Camus M, Palermo G, Smitz J, Vanwaesberghel Wisanto A, Wibjbo I, Van Steirteghem AC (1990) Placental production of estradiol and progesterone after oocyte donation in patients with primary ovarian failure. Am J Obstet Gynecol 162:66–70
7. Alsat E, Mondon F, Malassine A, Rebourcet R, Goldstein S, Cedard L (1987) Identification of distinct receptors for native and acetylated – low density lipoproteins in human placental micrivilli. Troph Res 2:17–28
8. Rodger MW, Baird DT (1987) Induction of therapeutic abortion in early pregnancy with the antigestagens RU 486 in combination with prostaglandin pessary. Lancet ii:1415–1418
9. Rose MP, Elder MG, Myatt L (1987) Arachidonic acid metabolism in the human placenta. Troph Res:71–85
10. Ilekis J, Benveniste R (1985) Effects of EGF, phorbol myristate acetate and AA on choriogonadotropin secretion by cultured human choriocarcinoma cells. Endocrinology 116:2400–2409
11. Barnea ER, Perlman R, Bick T, Hochberg Z (1989) Effects of human growth hormone upon term placental hormone secretion in vitro. J Gynecol Obstet Invest 27:133–136
12. Barnea ER, Feldman D, Kaplan M (1991) The effect of progesterone upon first trimester trophoblastic cell differentiation and human gonadotropin secretion. Hum Reprod 6:905–909
13. Kato Y and Braunstein GD (1990) Purified first and third trimester placental trophoblasts differ in in vitro hormone secretion. J Clin Endocrinol Metab 70:1187–1192
14. Owens OM, Ryan KJ, Tulchinsky D (1981) Episodic secretion of human chorionic gonadotropin in early pregnancy. J Clin Endocrinol Metab 53:1307–1309
15. Barnea ER, Kaplan M (1989) Spontaneous, gonadotropin-releasing hormone induced, and progesterone inhibited pulsatile secretion of human chorionic gonadotropin in the first trimester placenta in vitro. J Clin Endocrinol Metab 69:215–217
16. Kaplan M, Barnea ER, Bersinger N (1991) Varying patterns of hCG, SP1 and hPL secretion in superfused placental explants of first and third trimester. Acta Endocrinol 124:331–337
17. Barnea ER, Shurtz-Swirsky R, Kaplan M (1992) Factors controlling spontaneous hCG pulsatility in superfused first trimester placental explants. Hum Reprod (in press)

18. Nakajima ST, McAuliffe T, Gibson M (1990) The 24 hour profile of the levels of serum progesterone and immunoreactive human chorionic gonadotropin in normal early pregnancy. J Clin Endocrinol Metab 71:345–353
19. Khodr GS, Siler-Khodr TM (1978) The effect of luteinizing hormone releasing hormone on human chorionic gonadotropin secretion. Fertil Steril 30:301–304
20. Siler-Khodr TM, Khodr GS, Valenzuela G, Rhode J (1986) Gonadotropin-releasing hormone effects on placental hormone secretion during gestation: progesterone, estrone, estradiol and estriol. Biol Reprod 34:255–264
21. Belisle S, Guevin JF, Bellabarba B, Lehoux JG (1984) Luteinizing hormone releasing hormone binds to enriched human placental membranes and stimulates in vitro synthesis of bioactive chorionic gonadotropin. J Clin Endocrinol Metab 59:119–126
22. Barnea ER, Kaplan M, Naor Z (1991) Comparative aspects of GnRH, its agonist on pulsatile hCG secretion by the human placenta in superfusion: Reversible inhibition by GnRH antagonist. Hum Reprod 6:1063–1069
23. Petraglia F, Vaughan J, Vale W (1990) Steroid hormones modulate the release of immunoreactive gonadotropin – releasing hormone from cultured human placental cells. J Clin Endocrinol Metabl 70:1173–1178
24. Seeburg PH, Adelman JP (1984) Characterization of cDNA for precursor of human luteinizing hormone releasing hormone. Nature 311:666–668
25. Petraglia F, Vaughan I, Vale W (1989) Inhibin and activin modulate the release of GnRH, hCG and progesterone from cultured human placental cells. Proc Natl Acad Sci USA 86:5114–5117
26. Iwashita M, Watanabe M, Adachi T, Shinozaki Y, Takeda Y, Sakamoto S (1988) The effect of diacyglicerol on hCG release by trophoblast cells: comparison with GnRH. In: Mochizuki M, Hussa R (eds) Placental protein hormones. Exerpta Medica, Amsterdam, 798:287
27. Belisle S, Petit A, Bellabarba D, Escher E, Lehoux JG, Gallo-Payet N (1989) Ca^{++}, but not membrane lipid hydrolysis, mediates human chorionic gonadotropin production by luteinizing hormone – releasing hormone in human term placenta. J Clin Endocrinol Metab 69:117–121
28. Maruo T, Matsuo H, Ohtani T, Hoshina MH, Mochizuki M (1986) Differential modulation of chorionic gonadotropin (CG) subunit messengers ribonucleic acid levels and CG secretion by progesterone in human placenta in normal placenta and choriocarcinoma cultured in vitro. Endocrinology 119:855–864
29. Wilson EA, Jawad MJ (1982) Stimulation of human chorionic gonadotropin secretion by glucocorticoids. Am J Obstet Gynecol 142:344–348
30. Ahmed NA, Murphy BE (1988) The effects of various hormones on human chorionic gonadotropin production by early and late placental explant cultures. Am J Obstet Gynecol 159:1220–1207
31. Liotta AS, Krieger DT (1980) In vitro biosynthesis and comparative postranslational processing of immunoreactive precursor corticotropin/beta endorphin by human placental and pituitary cells. Endocrinology 106:1504–1511
32. Cemeric B, Genbacev O, Sulovic V, Beaconsfield R (1988) Effect of morphine on hCG release by first trimester human trophoblast in vitro. Life Sci 42:1773–1780
33. Barnea ER, Ashkenazi R, Sarne I (1991) Effect of dynorphin upon pulsatile hCG secretion in vitro. J Clin Endocrinol Metab 73:1093–1098
34. Ahmed MS, Cavinato AG (1987) Partial purification of the opioid receptor from human placenta. Troph Res 2:279–287
35. Barnea ER, Ashkenazi R, Kol S, Tal Y, Erlik Y, Sarne I (1991) Effect of beta endorphin on placental hCG secretion in the first trimester. Hum Reprod 6:1327–1331
36. Barnea ER, Check JH, Ashkenazi R, Sarne Y (1992) Evidence for a paracrine interaction between progesterone and hCG secretion in the first trimester placenta (abstract). Soc Gynecol Invest (in press)
37. Baly J, Hollenberg MD (1989) The nature and function of polypeptide growth factor receptors in the human placenta. J Dev Physiol 12:237–248
38. Barnea ER, Kaplan M, Feldman D, Morrish DW (1990) Dual stimulatory effect of epidermal growth factor upon first trimester placental hCG secretion in vitro. J Clin Endocrinol Metab 71:923–928

39. Maruo T, Matsuo H, Oishi T, Hayashi M, Nishino R, Mochizuki M (1987) Induction of differentiated trophoblast function by epidermal growth factor: relation of immunohistochemically detected cellular receptor level. J Clin Endocrinol Metab 64:744–750

40. Barnea ER, Feldman D, Shurtz-Swirsky R, Kaplan M (1992) Gestational age dependent, rapid and delayed effect of EGF upon hCG secretion by the first trimester explants. Troph Res 6:173–187

41. Morrish DW, Laborde NP, Bhardwaj D, Dabbagh LK (1989) Localization of EGF in human placenta. 8th International Congress of Endrocrinology, Kyoto (abstr 15–19–020)

42. Alsat E, Mirlesse M, Dodeur M, Evain-Brion D (1989) Regulation of epidermal growth factor receptors in human trophoblastic cells in culture: effect of parathyroid hormone. Trophoblast Conference, Dourdan, 1989. Placenta 10:509

43. Lavy G, Barnea ER, De Cherney AH (1987) The effect of insulin on oestradiol and progesterone release by normal and diabetic placentae in vitro. Placenta 8:443–448

44. Shigeru M, Saito Y, Motoyoshi K, Saito M, Ichijo M (1990) Localization and production of hM-CSF and G-CSF in human placental and decidual tissues. Proceedings of the International Conference on Placenta, Tokyo

45. Nishino E, Matsuzaki N, Masuhiro K, Kameda T, Taniguchi T, Takagi T, Siji F, Tanizawa O (1990) Trophoblast-derived interleukin-6 (IL-6) regulates human chorionic gonadotropin release through IL-6 receptor on human trophoblasts. J Clin Endocrinol Metab 71:436–441

46. McLachlan RI, Healy DL, Robertson DM, Burger HG, De Kretzer DM (1987) Circulating immunoreactive inhibin levels during the luteal phase and early pregnancy. Fertil Steril 48:1011

47. Petraglia F, Sawchenko P, Lim A, Rivier J, Vale W (1987) Localization secretion and action of inhibin in human placenta. Science 237:187

48. Sporn MB, Roberts AB, Wakefield LM, Assoian RK (1987) Transforming growth factor beta: biological function and chemical structure. Science 233:532–534

49. Hilf G, Merz WE (1985) Influence of cyclic nucleotides on receptor binding, and microheterogeneity of human chorionic gonadotropin synthesized in placental tissue culture. Mol Cell Endocrinol 39:151–159

50. Hustin J, Gaspard U (1977) Comparison of histological changes seen in placental tissue cultures and in placenta obtained after fetal death. Br J Obstet Gynaecol 84:210–216

51. Davies J, Glasser SR (1967) Light and electron microscopic observations on a human placenta two weeks after fetal death. Am J Obstet Gynecol 98:1111–1112

52. Wapner RJ, Davies GH, Johnson A, Weinblatt VJ, Ficher RL, Jackson LG (1990) Selective reduction of multifetal pregnancies. Lancet 335:8681

53. Danzer H, Braunstein GD, Rasor J, Forsythe A, Wade M (1980) Maternal serum human chorionic gonadotropin concentrations and fetal sex predictions. Fertil Steril 34:336

54. Barnea ER, Simon RJ, Kol S (1989) Human embryonal extracts modulate placental function in the first trimester: effects of visceral tissues upon chorionic gonadotropin and progesterone secretion. Placenta 10:331–334

55. Shurtz-Swirski R, Cohen Y, Barnea ER (1991) Modulatory effect of embryonal neural tissue upon hCG secretion by the first trimester placenta. Placenta 12

56. Vicovac L, Vuckovic M, Genbacev O (1987) Does human trophoblast affect decidual cells functions during gestation? Troph Res 2:85–94

57. Song Guang R, Braunstein GD (1990) Progesterone and hCG do not stimulate placental proteins 12 and 14 or prolactin production by human decidual tissue in vitro. J Clin Endocrinol Metab 70:983–989

58. Abramovich DR, Page KR, Pearson CK (1989) Does decidua influence first trimester trophoblast hCG secretion. Placenta 10:494

9 The Gestational Uterine Environment

J.N. Bulmer

Introduction

Considerable interest has recently been focused on local intrauterine immunoregulatory mechanisms during pregnancy in both humans and experimental animals. The availability of monoclonal antibodies (mAbs) directed against leucocyte subsets has stimulated interest in the function in normal pregnancy of the leucocytes which form a substantial cellular component of human decidua throughout gestation. Immunological and molecular biological techniques are gradually revealing clues to the potential in vivo roles of decidual lymphocytes and macrophages. The aim of this chapter is to review recent developments in the immune status of human decidua, focusing primarily on the first trimester.

Morphological Aspects

In early pregnancy human endometrium undergoes characteristic morphological changes under the influence of progesterone to form decidua. Similar changes noted in the late secretory phase of the menstrual cycle have been termed "predecidua", whilst stromal decidualisation in association with progesterone treatment is referred to as "pseudodecidua". Endometrial stromal cells enlarge, forming polygonal cells with a large pale vesicular nucleus and pale cytoplasm containing abundant glycogen and lipid. Three anatomical areas of decidua can be identified: decidua parietalis (vera) lines the uterine cavity away from the implantation area; decidua capsularis overlies the developing embryo and blends with decidua parietalis as the uterine cavity is obliterated after the third month of gestation; decidua basalis lies between the chorionic sac and the basal endometrium and forms the maternal part of the placenta.

The complexity of human decidua has been documented in both light- and electron-microscope studies. Various cell types have been described as components of human decidua: these include small decidual cells, metachromatic decidual cells, lymphocytes and so-called endometrial stromal granulocytes [1]. These latter cells, also termed Körnchenzellen or K cells, have been characterised by the presence of variable numbers of phloxinophilic cytoplasmic granules and a small rounded or reniform nucleus [2, 3]. Although present in the stroma of proliferative endometrium, "endometrial stromal granulocytes" increase in number in the late secretory phase of the menstrual cycle and reach maximal frequency in first-trimester decidua. They were formerly considered to derive from endometrial stromal cells and, in common with their proposed analogue in the rat, the granulated metrial gland (GMG) cells, to secrete relaxin [2]. More recent evidence indicates that human endometrial stromal granulocytes and their proposed rodent analogues are leucocytes, most probably a type of granulated lymphocyte [3–5]. A more appropriate terminology such as "endometrial granulated lymphocyte" (eGL) may therefore be proposed.

In early pregnancy, columns of cytotrophoblast proliferate from the tips of chorionic villi forming a cytotrophoblast shell around the developing conceptus. Extravillous trophoblast from this shell invades uterine decidua basalis and inner myometrium, and migrates up spiral arteries replacing endothelial cells with endovascular trophoblast. Individual mononuclear cytotrophoblast cells ultimately fuse to form multinucleate giant cells so that syncytial giant cells are the predominant trophoblast cell within the placental bed at term. Considerable difficulty may be encountered in trying to distinguish between maternal cells and fetal extravillous trophoblast cells at the light microscope level [1]. However, immunohistochemical techniques using trophoblast-specific mAbs have allowed immunological identification of fetal cells within maternal tissues and have highlighted their abundance [6].

Decidualisation is a feature of haemochorial placentation, and proposed functions have included nutrition of the embryo, endocrine secretion and immunoregulation. Suggested immunoregulatory roles include both protection of the mother from excessive invasion by extraembryonic fetal trophoblast cells and protection of the embryo from maternal immune rejection. Interest in local intrauterine immunoregulatory mechanisms during pregnancy has thus focused attention on potentially immunocompetent cells within the decidualised endometrium.

Leucocyte Populations in Human Decidua

Leucocytes are a major component of human decidua throughout pregnancy, although the relative proportion of different leucocyte populations differs between first trimester and full-term tissues. In the first trimester, 26%–36% of stromal cells in human decidua parietalis and decidua basalis are leucocytes expressing CD45 (CD45$^+$); these form three major groups: macrophages, granulated lymphocytes and T lymphocytes. B lymphocytes, plasma cells, polymorphonuclear leucocytes and classic CD16$^+$ CD57$^+$ natural killer (NK) cells are rare in normal early pregnancy decidua [3].

Macrophages

Macrophages account for nearly 40% of leucocytes in 1st-trimester human decidua. Although macrophages are present in endometrial stroma throughout the menstrual cycle, in first trimester decidua they increase both in absolute number and as a proportion of the total leucocyte population. They are present in both decidua parietalis and decidua basalis where they are often closely associated with extravillous trophoblast cells. Decidual macrophages can be detected with mAbs directed against CD14 and CD68, but most are unreactive for CD11b. The majority express class II major histocompatibility complex (MHC) gene products, although fewer cells express human leucocyte antigen (HLA)-DP and HLA-DQ than HLA-DR, particularly in first-trimester decidua [7]. Most also express the cell adhesion molecule CD11c. Antigen expression may provide clues to function. For example, macrophages in term decidua basalis express a surface antigen associated with down-regulatory stages of inflammation, providing some support for an immunoregulatory, non-inflammatory role [8]. However, human decidual macrophages also contain lysosomal enzymes such as acid phosphatase, non-specific esterase, α1-antitrypsin and α1-antichymotrypsin which indicates phagocytic capacity.

Granulated Lymphocytes

Lymphocytes with an unusual antigenic phenotype are a major component of human first-trimester decidua, accounting for up to 75% of leucocytes in human first-trimester decidua [3]; they are uncommon in the second half of pregnancy. They were initially detected as a cell population which expressed T lineage markers such as CD2 and CD7, but failed to express other T cell antigens such as CD3, CD4 and CD8 or the classical NK cell antigens CD16 and CD57. They were also positive for CD38 which is expressed by immature and activated cell populations. More recently, it has been shown that the cells are intensely reactive for CD56, a marker of NK cells and other large granular lymphocytes. Single and double immunohistochemical studies indicate that there are two phenotypic groups of CD56+ lymphocytes in human decidua. Expression of CD38 and CD56 overlaps, but only 50% – 60% of CD56+ cells also express CD2. To date there have been no studies examining separately the function and distribution of CD56+ CD2+ and CD56+ CD2− subgroups in normal and pathological pregnancy.

CD56+ cells are scattered throughout decidua basalis and decidua parietalis forming aggregates adjacent to endometrial glands and arterioles (Fig. 1). Phenotypically similar cells can be identified in late secretory phase endometrium, particularly in areas which show predecidual stromal changes. The distribution of CD56+ cells matches that of the endometrial stromal granulocytes, and it has become generally accepted that these cells are granulated lymphocytes eGL (Fig. 2). The characteristic cytoplasmic

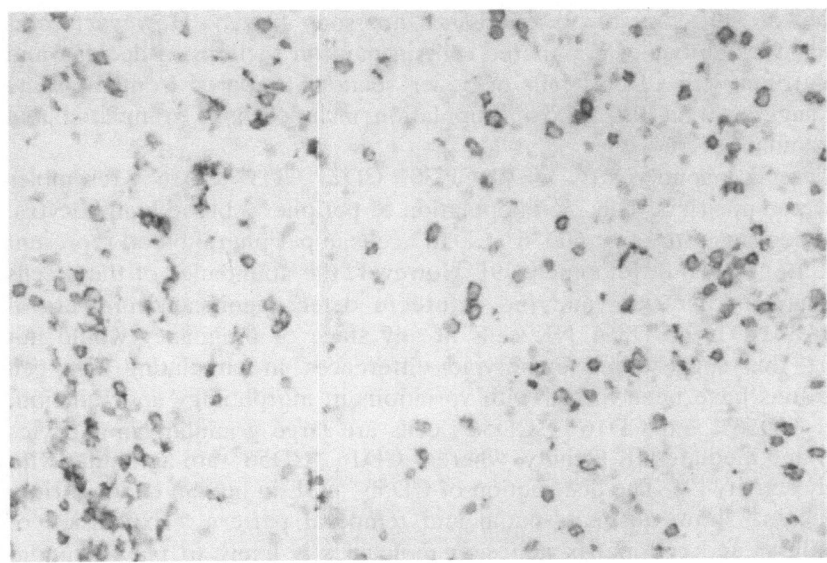

Fig. 1. Frozen section of first-trimester human decidua labelled by an indirect immunoperoxidase technique with NKH1 (anti-CD56) showing numerous positive cells. ×160

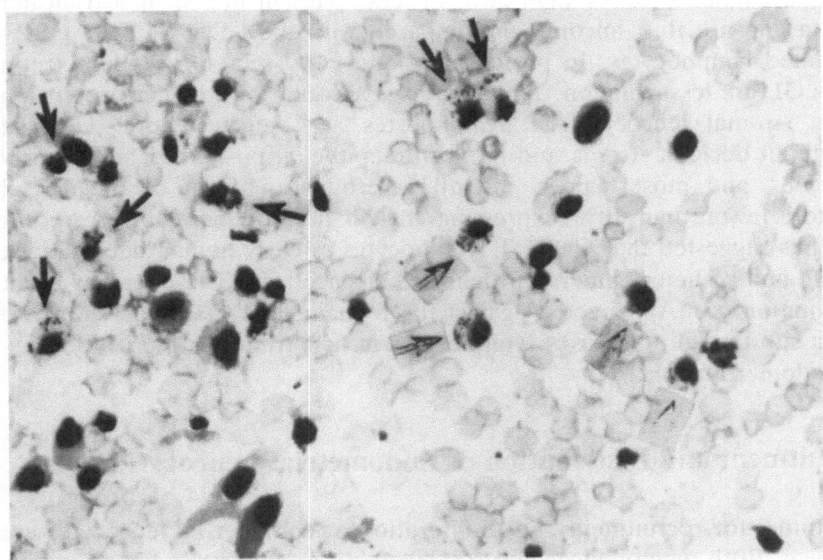

Fig. 2. Imprint preparation of first-trimester human decidua stained with Giemas showing numerous cells with cytoplasmic granules (*arrows*). ×400

granules of eGL are lost when tissues are snap frozen. However, comparison of a number of granulated cells in paraffin sections of decidua and endometrium with CD56$^+$ cells in frozen sections prepared from the same tissue suggests that the CD56$^+$ population includes both granulated and non-granulated cells.

The eGL phenotype (CD56^{++} CD38$^+$ CD2$^\pm$ CD3$^-$ CD16$^-$) resembles that of a minor (less than 2%) population of peripheral blood lymphocytes. It has been suggested that CD56$^+$ CD16$^-$ cells in peripheral blood represent a stage in NK cell development [9]. However, the abundance of these cells in human endometrium and the failure to detect significant numbers of "mature" CD16$^+$ CD56$^+$ NK cells at any stage of pregnancy would not support this suggestion. Phenotypic differences in circulating NK cell populations have been linked with variations in morphology and function. CD16$^+$ CD56$^+$ and CD16^{dim} CD56$^+$ cells are large granular lymphocytes (LGL) and mediate NK activity, whereas CD16$^-$ CD56$^+$ are agranular with low NK activity [9]. The distribution of CD56$^+$ eGL in human endometrium and decidua shows distinct spatial and temporal patterns. Expression of intercellular and cell-matrix adhesion molecules is likely to play a fundamental role in determining the distribution of eGL within non-pregnant and gestational endometrium (see below).

T Lymphocytes

T cells account for fewer than 20% of leucocytes in human first-trimester decidua and are thus uncommon compared with the CD56$^+$ CD3$^-$ CD2$^\pm$ granulated lymphocytes. In proliferative and early secretory endometrium when eGL are less common, T lymphocytes account for a higher proportion of the stromal leucocytes. T lymphocytes are detected both scattered throughout decidua stroma and in an intraepithelial position. The majority are CD8$^+$ and most express the $\alpha\beta$ heterodimeric form of the T cell receptor, fewer than 10% expressing the $\gamma\delta$ heterodimer [10]. A recent study has suggested that CD3$^+$ T lymphocytes in early human decidua lack both $\alpha\beta$ and $\gamma\delta$ heterodimers on their cell surface [11]. These studies await confirmation; a mAb directed against the β chain, βF1, has been shown to label a substantial number of lymphocytes in first trimester human decidua and endometrium [12].

Recruitment and Distribution of Endometrial Leucocytes

The stimuli for recruitment and proliferation of endometrial leucocytes are unknown. Both macrophages and eGL increase in number in late secretory phase endometrium as well as in first-trimester decidua: an embryo is therefore not essential for their recruitment and proliferation, although an embryonic signal may "activate" cell function for implantation.

Cell Proliferation

The increase in numbers of eGL in the late secretory phase may be partly due to influx from blood, but local proliferation plays an important role. Although it was formerly claimed that eGL were terminally differentiated cells, there is clear evidence of eGL proliferation within late secretory phase endometrium. eGL are observed in mitosis, particularly premenstrually, and CD56$^+$ cells express the nuclear proliferation antigen Ki67 [13]. Nevertheless, eGL have proved difficult to maintain in vitro, and clones of CD56$^+$ CD3$^-$ lymphocytes prepared from first-trimester decidua show a low proliferative frequency [14]. The stimulus for eGL proliferation in premenstrual endometrium may be due to a hormonal or cytokine signal which has not yet been mimicked in vitro.

Adhesion Molecules

Adhesion molecules play a vital role in regulation of lymphocyte migration and cellular interactions in the immune response [15]. LFA-1 (CD11a) is a β_2 integrin which is expressed by T lymphocytes; it reacts with a counter-receptor ICAM-1 (CD54) or ICAM-2, which are members of the immunoglobulin family. LFA-2 (CD2) reacts with LFA-3 (CD58) on targets and both are members of the immunoglobulin family. The β_1 integrins (VLA-1–6) include receptors which bind to the extracellular matrix components fibronectin, laminin and collagen. β_1 integrins are expressed on non-haematopoietic and leucocytic cells and have been designated VLA (very late activation) antigens since VLA-1 and VLA-2 are expressed on lymphocytes 2–4 weeks after antigen stimulation in vitro.

The distribution of leucocytes in human endometrium and decidua may be due to adhesion molecules. In our studies, stromal reactivity for ICAM-1 was associated with aggregates of CD11a$^+$ lymphocytes in non-pregnant endometrium. In early pregnancy decidua, aggregates of CD11a$^+$ lymphocytes are associated with decidualised stromal cells showing surface ICAM-1 reactivity. LFA-3 is expressed by glandular epithelium, vessels and scattered stromal lymphocytes. A large proportion of CD56$^+$ lymphocytes in decidua also express CD11a and CD2, and CD56$^+$ CD2$^+$ often form a halo of cells surrounding endometrial glands (J.N. Bulmer, unpublished). Aggregation of endometrial lymphocytes around glands and vessels may be due to expression of LFA-3 by glands associated with CD2$^+$ cells or to expression of LFA-3 or ICAM-1 by endothelial cells associated with CD2 and CD11a, respectively.

Fibronectin forms an important component of the pericellular matrix around stromal cells in non-pregnant endometrium and decidua, outlining the cells; the distribution does not alter markedly with menstrual cycle stage. In contrast, laminin is scanty in proliferative endometrium, but gradually increases in the secretory phase, and in first-trimester decidua laminin forms a capsule surrounding the decidual cells, suggesting a distribution within the external lamina [16-18]. Explants of decidua produce laminin in vitro, but do not secrete it into the culture medium. It has been suggested that production of laminin by decidual cells may play a major role in facilitating trophoblast invasion [19]. The expression of VLA antigens in relation to alterations in endometrial extracellular matrix around the time of implantation and early placental development are worthy of further consideration.

Functional Studies of Leucocytes in Human Decidua

The in vivo role of the various decidual leucocyte populations remains to be established, but in vitro studies may provide information concerning the function of leucocytes in normal pregnancy. The three major leucocyte populations in first-trimester decidua will be considered in turn.

Macrophages

As well as classical phagocytic and anti-inflammatory roles, macrophages play a central role in the immune response acting as antigen-presenting cells. Antigen expression may provide clues to cell function. Macrophages in first-trimester decidua express class II MHC antigens and CD11c, suggesting an immunological role. At term they express an antigen associated with down-regulatory stages of inflammation, leading to the suggestion that decidual macrophages may be involved in inhibition of specific cellular responses [8].

Accessory Cell Function

Antigen-presenting capacity has been reported in both murine and human decidua. Fibronectin-adherent cells from first-trimester human decidua can act as accessory cells for mitogen-induced lymphoproliferation, for presentation of antigen to primed T cells, for activation of suppressor cells, for development of cytotoxic reactions and in a primary lymphoproliferative response [20, 21]. The precise type of cell responsible for antigen presentation has not been determined, but anti-class II MHC antibodies lead to a decrease in accessory cell function. Dorman and Searle [21] noted class II MHC-positive lymphoid and non-lymphoid cells in cell suspensions; CD14[+] macrophages were scarce and occasional CD1[+] cells were detected. Macrophages appear to be prime candidates as decidual antigen-presenting cells in early pregnancy. However, scanty CD1[+] cells which could function as potent accessory cells have also been detected in first-trimester decidua. The role of accessory cells in induction of immunosuppression is of interest since antigen-presenting cells extracted from human decidua can induce generation of $CD5^+$ $CD8^+$ suppressor T cells in peripheral blood lymphocytes exposed in vitro to fetal cells [20].

Immunosuppression

Decidual macrophages may also play a role in immunosuppression. It is well established that supernatants produced by first-trimester human decidua suppress mitogen-induced lymphocyte proliferation and the mixed lymphocyte reaction (MLR), but the cell type responsible remains uncertain; stromal cells, lymphocytes, macrophages and epithelial cells have all been implicated. Macrophages in human decidua have been reported to produce prostaglandin E_2 (PGE_2), which mediates immunosuppressive activity which can be partly inhibited by indomethacin [22]. This immunosuppressive activity has also been shown to block activation of decidual lymphocytes with potential anti-trophoblast activity [23]. PGE_2 has also been reported to be produced by decidualised stromal cells and the relative importance of PGE_2 production from these two sources is not known.

Cytokine Production

Macrophages are also able to produce various cytokines which may have an immunoregulatory role. Receptors for macrophage colony-stimulating factor (M-CSF), granulocyte CSF (G-CSF), macrophage-granulocyte CSF (GM-CSF) and tumour necrosis factor α (TNFα) have been reported on human placenta, and it has been proposed that cytokines play an important role in placental growth regulation [24]. Macrophages could thus contribute to production of growth and immunoregulatory cytokines in human uteroplacental tissues. Production of interleukin-1 (IL-1), TNFα and G-CSF by human decidua has been reported, but the cellular site of origin has not been precisely determined.

Non-specific Anti-inflammatory and Phagocytic Roles

Decidual macrophages are also equipped for phagocytic and non-specific anti-inflammatory function. It is possible that their principle in vivo role is in the non-specific defence of the fetoplacental unit to infective agents. It is notable that T cells are scarce in human decidua and B cell and plasma cells rarely detected. Furthermore, the extensive invasion of maternal uterine tissues by extravillous trophoblast, which is a feature of normal pregnancy, is likely to lead to tissue debris which may require removal by phagocytosis.

Granulated Lymphocytes

The dramatic increase in number of $CD56^+$ granulated lymphocytes, which occurs in the late secretory phase of the menstrual cycle and in early pregnancy decidua, suggests that these cells play a role in implantation and placentation. The presence of large numbers of granulated cells in decidua in other species with haemochorial placentation, including the GMG cells of rat and mouse, provides further support for a pregnancy-related function.

NK Activity

LGL in peripheral blood are associated with NK activity. CD56[+] eGL of varying purity extracted from human decidua by panning, flow cytometry and density gradient centrifugation have all been shown to mediate NK activity against the NK-sensitive target K562 [25–28]. However, eGL purified to >90% purity show lytic levels which are comparable with unfractionated peripheral blood lymphocytes which contain only 10%–20% NK cells. This suggests that eGL are poor effectors compared with peripheral blood NK cells; this may be an inherent property of the cells, but could also result from enzyme digestion of tissues and lengthy purification techniques. In our laboratory, NK activity was not enhanced by IL-2 or interferon γ (IFNγ), and eGL did not function as lymphokine-activated killer (LAK) cells [27]. The enhancement of NK activity with IL2 noted in other studies may be due to differences in the precise populations studied.

Christmas et al. [14] have produced clones of CD3[−] lymphocytes from first-trimester human decidua. CD3[−] CD16[+] CD56[+] clones showed high lytic activity against NK and LAK targets, but CD3[−] CD16[−] CD56[+] clones showed low cytotoxic activity. These results have led to the suggestion that the NK activity of eGL reported in other studies may be due to the presence of a small population of CD16[+] CD56[+] cells within the isolated CD56[+] cells. The function of the two phenotypically distinct subsets of CD2[+] and CD2[−] eGL has not so far been examined separately, but Christmas et al. did not note functional differences between CD2[+] and CD2[−] clones.

It is interesting to note that very low or negligable NK activity was detected in cell suspensions of purified eGL prepared from five samples of non-pregnant endometrium from various stages of the menstrual cycle [29], thus supporting the proposal that an embryonic signal may be required for functional activation.

Anti-trophoblast Cytotoxicity

Detection of NK activity by eGL has led to the suggestion of a role in the control of trophoblast invasion, although there is no direct evidence for this in tissue sections of decidua basalis. Human first-trimester cytotrophoblast is resistant to lysis by decidual and peripheral blood NK cells in a chromium release assay, apparently lacking the relevant NK target structure [30]. However, trophoblast does appear to be lysed by NK cells which have been pretreated with IL-2, presumably leading to LAK cell generation. The resistance and susceptibility of trophoblast cells to NK lysis may be an important factor in pregnancy loss (see below).

Mouse GMG cells fail to lyse NK targets or trophoblast cells in chromium release assays [31]. However, time lapse video studies of co-cultures have demonstrated that GMG cells can directly lyse occasional labyrinthine trophoblast cells in vitro [32]. This technique allows prolonged direct visualisation of co-cultures: chromium release assays reflect killing over a short period and may not be sufficiently sensitive to detect killing of occasional individual cells. In preliminary studies we have examined co-cultures of CD56$^+$ eGL and human trophoblast by scanning electron microscopy: eGL consistently showed close associations with trophoblast cells. The interactions of CD56$^+$ eGL with normal trophoblast is thus worthy of further study.

Immunosuppression

A population of small granulated non-T non-B lymphocytes isolated from murine decidua apparently mediate immunosuppression by secretion of a transforming growth factor β_2 (TGFβ_2) [33]. A similar suppressor mechanism has been proposed for "small" suppressor cells in human decidua [34]. The prominence of eGL in human decidua at the time of implantation led to suggestions that they may function as suppressor cells. In vitro studies do not support this proposal. Supernatants from unfractionated decidual suspensions showed higher suppression of mitogen-induced lymphocyte proliferation than those from purified eGL. Furthermore, eGL supernatants from some specimens led to stimulation of the mixed lymphocyte reaction [29]. Thus, although eGL may mediate a component of decidual immunosuppression, other mechanisms must play an important role.

Cytokine Secretion

Various cytokines may be produced by peripheral blood NK cells, and it has been proposed that cytokines play an important role in control of placental growth. Supernatants of purified eGL may stimulate a mixed lymphocyte reaction and also cause proliferation of placental cells (Ritson and Bulmer, unpublished). This raises the possibility that eGL secrete cytokines which maintain normal placental development. Supernatants from >90% purified eGL contain various cytokines on Western blot analysis, including IL-1, TNFα, TGFβ and M-CSF. Christmas et al. [14] analysed cytokine production by decidual lymphocyte clones and detected TNFα, IFNγ and TGFβ; CD3$^-$ clones (proposed to be eGL clones) generally produced higher levels of TGFβ. The role of eGL in normal pregnancy remains uncertain. Further studies of cytokine production are required as are more investigations of the effect of cytokines on normal and pathological trophoblast cells. The possibility that these cells play a role in the trophoblast invasion of maternal uterine tissues also merits further study.

T Lymphocytes

T lymphocytes form a low proportion of leucocytes in first trimester human decidua which have not yet been isolated for functional studies. T cells may be detected in an intraepithelial position as well as in stroma throughout the menstrual cycle and early pregnancy. This distribution may reflect diverse functions. Most T lymphocytes in human decidua are CD8[+] and hence could function as suppressor cells. Brierley and Clark [35] reported large hormone-dependent suppressor cells in murine decidua expressing surface antigens characteristic of suppressor T cells. CD8[+] T cells may also secrete cytokines, including IL-2, IL-3, IFNγ and GM-CSF, which may play a role in control of placental growth [24]. Although GM-CSF has been reported to cause proliferation of murine placental cells, there are as yet no comparable data for humans, and Hill et al. [36] have suggested that high doses of GM-CSF may be detrimental to embryo survival.

Although decidual T cells lack the p55 IL-2 receptor, in non-pregnant endometrium T cells express class II MHC and VLA-1 antigens, suggesting long-term activation [37]. Reports that decidual T cells lack both αβ and γδ forms of T cell receptor [11] await confirmation; if confirmed, alternative non-classical pathways of activation must be considered.

Role of Endometrial Leucocytes in Pregnancy Loss

Non-MHC-restricted cytotoxicity may play a role in pregnancy loss, although there are as yet few studies in humans. Murine trophoblast is not susceptible to lysis by NK or cytotoxic T cells, but is killed by LAK cells [38]. Resorption rates in the CBA/J × DBA/2 mouse model are increased by enhancement of NK activity with polyIC and decreased with anti-asialo GM1 antibody, and the implantation site is infiltrated with asialo-GM1[+] cells [39]. Accumulation of lymphocytes has been noted at the implantation site of pregnancies destined to fail, but cells were not phenotyped. Michel et al. [40] reported lymphocytes with small (<1 μm) granules in women destined to abort and lymphocytes with large (>1 μm) cytoplasmic granules in normal pregnancy. These results await confirmation, but do raise the possibility that granulated lymphocytes may mediate pregnancy loss in humans. It can be proposed that loss of the IL-2 blocking essential for normal pregnancy could lead to generation of LAK cells either from the resident eGL population or from NK cells newly recruited from peripheral blood. These questions remain to be answered, but studies to date should stimulate interest in this area.

Endometrial Stromal Cells

Involvement of stromal cells in secretion of the extracellular matrix in decidua has been noted (see above). Decidualised stromal cells also have endocrine functions such as prolactin production. However, considerable attention has been focused on the role of decidualised stromal cells as suppressor cells. As well as decidual macrophages, decidual cells have also been reported to mediate immunosuppression which can be abrogated with indomethacin or anti-PGE$_2$ antibody [22]. As noted before, this PGE$_2$-mediated immunosuppression suppresses potential anti-trophoblast activity by inhibiting IL2-receptor generation and IL-2 production in situ [23]. It should be noted, however, that both glands and stromal cells participate in production of PGE in non-pregnant endometrium [41].

A recent study [42] indicates that decidualised stromal cells mediate immunosuppression apart from PGE$_2$ production. First-trimester decidual cells purified on a Percoll gradient and free from macrophages and T lymphocytes produced a soluble factor which blocked lymphocyte responses to mitogens in the mixed lymphocyte reaction and cytotoxic T cell generation. Supernatants from the purified decidual cells inhibited IL-2 and IFNγ production by peripheral blood lymphocytes and inhibited IL-2 receptor expression. Low molecular weight factors including prostaglandins were removed by dialysis, and the molecular weight of the immuno-suppressive factor was 43–67 kDa. Thus, decidualised endometrial stromal cells may play a vital immunoregulatory role in normal human pregnancy.

Endometrial Epithelial Cells

Antigen Expression

Endometrial glands persist throughout gestation though they may be attenuated and difficult to distinguish without immunohistochemistry. Endometrial glands appear to lose or substantially decrease expression of class I MHC antigens during pregnancy and also rarely express class II MHC antigens [6]. Leucocytic aggregates related to glands in early pregnancy may reflect increased susceptibility to immune attack, and activated T cells have been identified within this infiltrate.

Functional Observations

Lectin Binding

Lectins can be used as histochemical probes to determine carbohydrate composition of cell membranes and secretions. Although lectins have been used to examine normal non-pregnant, hyperplastic and malignant endo-metrium, there have been few studies of decidua. However, a recent study of endometrium in primary infertility demonstrated an abnormal lectin binding profile in the peri-implantation phase, suggesting that delayed or deficient secretion of glycoproteins may render the uterus "unfavourable" for implantation [43]. This approach provides another tool for investigation of pregnancy pathology.

Cytokine Secretion

Endometrial epithelial cells in murine pregnancy have been shown to be the source of cytokines such as M-CSF, which may play a role in placental growth [44]. Comparable studies are in progress for human tissues: our data from immunohistochemistry and in situ hybridisation indicates that epithelial cells in human decidua also produce M-CSF. Endometrial epithelial cells also produce pregnancy proteins, including PP14, which may play a role in local immunosuppression.

Intraepithelial Lymphocytes

Intraepithelial lymphocytes (IEL) in human endometrium have attracted less attention than those in the gastrointestinal tract. Macrophages, eGL and T cells may occur in an intraepithelial position. The relative proportions of T cells and eGL vary according to the menstrual cycle stage. CD56$^+$ IEL increase in number from proliferative to late secretory phase accounting for 50% IEL in early pregnancy. The remainder are CD3$^+$ CD8$^+$ T lymphocytes. The proportion of IEL with cytoplasmic granules also increases in first-trimester decidua, and results indicate that at least a proportion of the CD3$^+$ CD8$^+$ IEL must be granulated [45]. IEL function in human endometrium is not known. Separation of IEL for functional studies would pose technical problems, but further in situ analysis of IEL in abnormal endometrium and pathological pregnancy may provide clues to their function.

Conclusions

There has been considerable progress in our understanding of the decidualised uterine endometrium which forms a favourable environment for the developing human conceptus. In vitro studies and investigation of animal models are providing clues to the functions of the many cell types constituting human decidua, but their in vivo role generally remains uncertain.

References

1. Pijnenborg R, Dixon G, Robertson WB, Brosens I (1980) Trophoblast invasion of human decidua from 8–18 weeks of pregnancy. Placenta 1:3–18
2. Dallenbach-Hellweg G (1987) The normal histology of the endometrium. In: Histopathology of the endometrium, 3rd edn. Springer, Berlin Heidelberg New York, pp 25–92
3. Bulmer JN, Morrison L, Longfellow M, Ritson A, Pace D (1991) Granulated lymphocytes in human endometrium: further histochemical and immunohistochemical studies. Hum Reprod 6:791–798
4. Bulmer JN, Hollings D, Ritson A (1987) Immunocytochemical evidence that endometrial stromal granulocytes are granulated lymphocytes. J Pathol 153:281–288
5. Peel S (1989) Granulated metrial gland cells. Springer, Berlin Heidelberg New York (Advances in anatomy, embryology and cell biology, vol 115)
6. Bulmer JN (1988) Immunopathology of pregnancy. Baillière's Clin Immunol Allergy 2:697–734
7. Bulmer JN, Smith JC, Morrison L (1988) Expression of class II MHC gene products by macrophages in human uteroplacental tissues. Immunology 63:707–714
8. Mues B, Langer D, Zwadlo G, Sorg G (1989) Phenotypic characterisation of macrophages in human term placenta. Immunology 67:303–307
9. Nagler A, Lanier LL, Cwirla S, Phillips JH (1989) Comparative studies of human FCRIII positive and negative natural killer cells. J Immunol 143:3183–3191
10. Bulmer JN, Morrison L, Longfellow M, Ritson A (1991) Leucocytes in human decidua: investigation of surface markers and function. In: Chaouat G (ed) Maternofetal relationship: molecular and cellular biology. Libbey, Paris, pp 189–196
11. Dietl J, Horny HP, Ruck P, Marzusch K, Kaiserling E, Griesser H, Kabelitz D (1990) Intradecidual T lymphocytes lack immunohistochemically detectable T-cell receptors. Am J Reprod Immunol 24:33–36
12. Yeh C-JG, Bulmer JN, Hsi B-L, Tian W-T, Rittershaus C, Ip S (1990) Monoclonal antibodies to T cell receptor γδ complex react with human endometrial glandular epithelium. Placenta 11:253–261
13. Pace D, Morrison L, Bulmer JN (1989) Proliferative activity in endometrial stromal granulocytes throughout the menstrual cycle and early pregnancy. J Clin Pathol 42:35–39
14. Christmas SE, Bulmer JN, Meager A, Johnson PM (1990) Phenotypic and functional analysis of human CD3⁻ decidual leucocyte clones. Immunology 71:182–189
15. Springer TA (1990) Adhesion receptors of the immune system. Nature 346:425–434
16. Kisalus LL, Herr JC, Little CD (1987) Immunolocalisation of extracellular matrix proteins and collagen synthesis in first trimester human decidua. Anat Rec 218:402–415
17. Aplin JD, Charlton AK, Ayad S (1988) An immunohistochemical study of human endometrial extracellular matrix during the menstrual cycle and first trimester of pregnancy. Cell Tissue Res 253:231–240
18. Earl U, Estlin C, Bulmer JN (1990) Fibronectin and laminin in the early human placenta. Placenta 11:223–231
19. Loke YW, Gardner L, Burland K, King A (1989) Laminin in human trophoblast-decidua interaction. Hum Reprod 4:457–463
20. Oksenberg JR, Mor-Yosef S, Ezra Y, Brautbar C (1988) Antigen presenting cells in human decidual tissue. III. Role of accessory cells in the activation of suppressor cells. Am J Reprod Immunol Microbiol 16:151–158
21. Dorman PJ, Searle RF (1988) Alloantigen presenting capacity of human decidual tissue. J Reprod Immunol 13:101–112
22. Parhar RS, Kennedy TG, Lala PK (1988) Suppression of lymphocyte alloreactivity by early gestational human decidua. I. Characterisation of suppressor cells and suppressor molecules. Cell Immunol 116:392–410

23. Parhar RS, Yagel S, Lala PK (1989) PGE$_2$-mediated immunosuppression by first trimester human decidual cells blocks activation of maternal leukocytes in the decidua with potential anti-trophoblast activity. Cell Immunol 120:61–74
24. Wegmann TG (1988) Maternal T cells promote placental trophoblast growth and prevent spontaneous abortion. Immunol Lett 17:297–302
25. King A, Birkby C, Loke YW (1989) Early human decidual cells exhibit NK activity against the K562 cell line but not against first trimester trophoblast. Cell Immunol 118:337–344
26. Manaseki S, Searle RF (1989) NK cell activity of first trimester human decidua. Cell Immunol 121:166–173
27. Ritson A, Bulmer JN (1989) Isolation and functional studies of granulated lymphocytes in first trimester human decidua. Clin Exp Immunol 77:263–268
28. Ferry BL, Starkey PM, Sargent IL, et al. (1990) Cell populations in the human early pregnancy decidua: natural killer activity and response to interleukin-2 of CD56-positive large granular lymphocytes. Immunology 70:446–452
29. Bulmer JN, Longfellow M, Ritson A (1991) Leukocytes and resident blood cells in endometrium. Ann NY Acad Sci 622:57–68
30. King A, Kalra P, Loke YW (1990) Human trophoblast cell resistance to decidual NK lysis is due to lack of NK target structure. Cell Immunol 127:230–237.
31. Croy BA, Kassouf SA (1989) Evaluation of the murine metrial gland for immunological function. J Reprod Immunol 15:51–69
32. Stewart IJ, Mukhtar DDY (1988) The killing of mouse trophoblast cells by granulated metrial gland cells in vitro. Placenta 9:417–426
33. Clark DA, Flanders KC, Banwatt D, et al. (1990) Murine pregnancy decidua produces a unique immunosuppressive molecule related to transforming growth factor β-2. J Immunol 144:3008–3014
34. Clark DA, Falbo M, Fowley RB, et al. (1988) Active suppression of host-vs-graft reaction in pregnant mice. IX. Soluble suppressor activity obtained from allopregnant mouse decidua that blocks the cytolytic effector response to IL-2 is related to transforming growth factor-β. J Immunol 141:3833–3840
35. Brierley J, Clark DA (1987) Characterisation of hormone dependent suppressor cells in the uterus of mated and pseudopregnant mice. J Reprod Immunol 10:201–218
36. Hill JA, Haimovici F, Anderson DA (1987) Products of activated lymphocytes and macrophages inhibit mouse embryo development. J Immunol 139:2250–2254
37. Tabibzadeh S (1990) Evidence of T-cell activation and potential cytokine action in human endometrium. J Clin Endocrinol Metab 71:645–649
38. Head JR (1989) Can trophoblast be killed by cytotoxic cells? In vitro evidence and in vivo possibilities. Am J Reprod Immunol 20:100–105
39. Gendron RL, Baines MG (1988) Infiltrating decidual natural killer cells are associated with spontaneous abortion in mice. Cell Immunol 133:261–267
40. Michel M, Underwood J, Clark DA, et al. (1989) Histologic and immunologic study of uterine biopsy tissue of women with incipient abortion. Am J Obstet Gynecol 161:409–414
41. Smith SK, Kelly RW (1987) The effect of estradiol 17β and actinomycin D on the release of PGF and PGE from separated cells of human endometrium. Prostaglandins 34:553–561
42. Matsui S, Yoshimura N, Oka T (1989) Characterisation and analysis of soluble suppressor factor from early human decidual cells. Transplantation 47:678–683
43. Klentzeris LD, Bulmer JN, Li T-C, Morrison L, Warren A, Cooke ID (1991) Lectin binding in women with unexplained infertility. Fertil Steril 56:660–667
44. Arceci RJ, Shanahan F, Stanley ER, Pollard JW (1989) Temporal expression and localisation of colony stimulating factor 1 (CSF-1) and its receptor in the female reproductive tract are consistent with CSF-1 regulated placental development. Proc Natl Acad Sci USA 86:8818–8822
45. Pace DP, Longfellow M, Bulmer JN (1991) Intraepithelial lymphocytes in human endometrium. J Reprod Fertil 91:165–174

10 Functional Aspects of Embryology

J.G. Moscoso

Molecular Maps and Signals

To achieve normal development, the human embryo relies on a time-dependent interaction of highly specific cell functions, including cell migration, localization, cell division, differentiation and programmed cell death [1]. Accumulating evidence indicates that patterns of migration are regulated by a transiently expressed family of genes known as homeobox. These modulate the generation of molecules which "prime" embryonic cells and mesenchyme, therefore pre-determining the spatial domain of specific cell lineages. Thus, primitive cells, following these molecular cues, migrate and localize in specific regions of the embryo [2]. Furthermore, homeobox-containing genes participate in the segmentation of organs such as the brain of vertebrates [3] and the body itself in insects [4, 5]. Another set of morphoregulatory molecules assists cells in "communicating" or interacting with one another as they migrate or aggregate. These are known as cell adhesion molecules (CAM), cell junction molecules (CJM) and cadherins. The latter is a family of integral cell membrane glycoproteins which mediate cell-cell adhesion using Ca^{2-} [6]. A different set of molecules known as substrate adhesion molecules (SAM) has been identified in the extracellular matrix and is responsible for mediating cell-substrate adhesion [7]. Thus, most embryonic cells are influenced by a complex array of time-dependent molecular expressions and interactions, which, if adequately balanced, should determine normal development (Fig. 1). It is therefore obvious that during this delicate phase the embryo is particularly vulnerable to negative epigenetic influences [8].

Development of the External Form

Following predictable patterns, the external morphological features of the human embryo change rapidly during development. Thus, their correct time sequence can be correlated with other fundamental changes of form at organ level or cytochemical differentiation at cell level. In this section, a brief

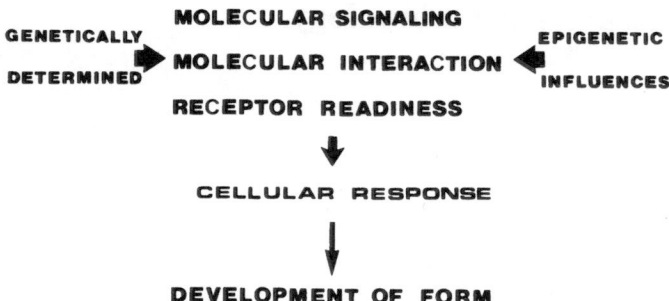

Fig. 1. Molecular basis of organogenesis, a schematic view

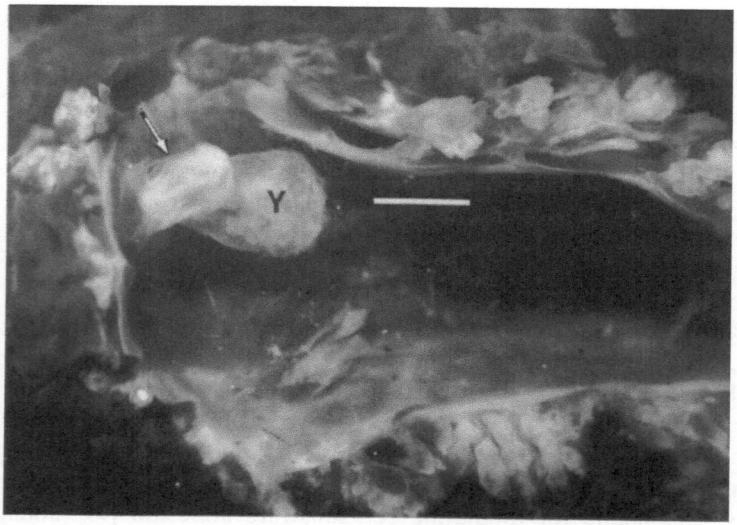

Fig. 2. Human gestational sac at 17 days post-fertilization (Streeter's stage 7). It contains a human embryo (*arrow*) attached to the yolk sac (*Y*) which shows a constriction zone beneath the embryonic plate. *Scale bar*, 3 mm

account of the various changes observed in the external form of the human embryo is presented as a baseline for comparison with the development of other internal organs.

A two-cell stage conceptus can be observed anytime after 22 h following fertilization. Thereafter, series of cell divisions transform the shape of the developing embryo into "morula" and "blastocyst". Implantation occurs during the 6th–7th days. Seventeen days later the embryo appears as an ovoid structure attached to the developing secondary yolk sac (Fig. 2). On close examination, the embryonic plate shows lobulated borders. The

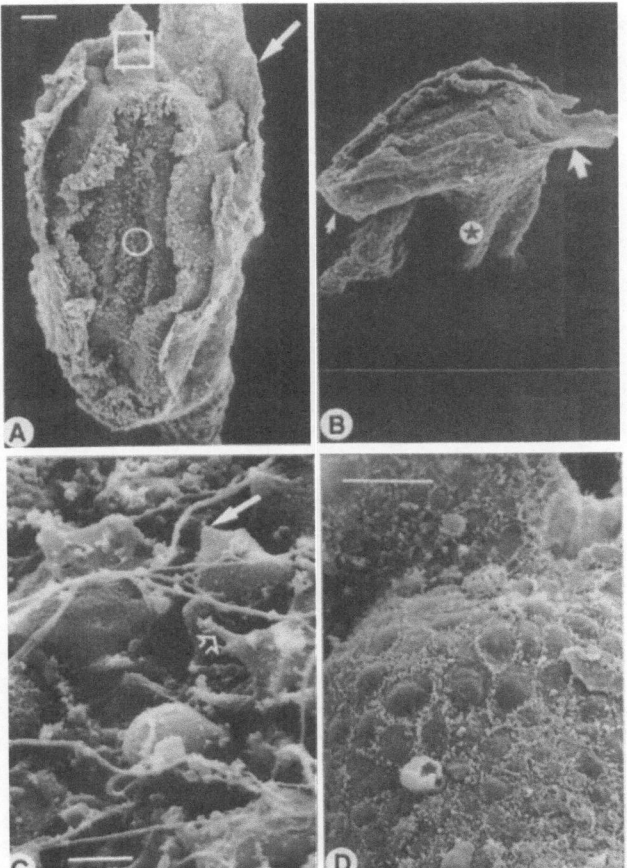

Fig. 3A–D. Human embryo from Fig. 2. **A** Dorsal view. There is partial loss of mid-line ectoderm revealing the primitive streak (*circle*). Note the lateral lobulations (*square*) around the embryonic disk. These are in immediate continuity with the developing amnion (*arrow*). Scanning electron microscopy (SEM), ×120. **B** Right lateral view. *Large arrow*, cephalic end; *small arrow*, curved caudal end; *star* subembryonic zone of constriction of the yolk sac SEM, ×99. **C** Detailed view from the encircled area in **A** showing cells from the primitive streak. These appear loose, with stellate external forms (*solid arrow*). Some cells have short dendritic projections (*open arrow*). Note the numerous slender, long filaments crossing the area. SEM, ×7200. **D** Close-up view of the *framed zone* in **A**. The ectodermal cells covering the lateral lobulations show a convex smooth external surface. Numerous slender and short microvilli are present in intercellular boundaries. SEM, ×870. *Scale bar*, 20 μm

cephalic and caudal poles together with the primary streak can be recognized (Fig. 3). Gradually, the embryo adopts a tubular shape as the neural canal deepens and the primary yolk sac transforms into the definitive secondary yolk sac. The first somites appear and fuse at about the 20th day

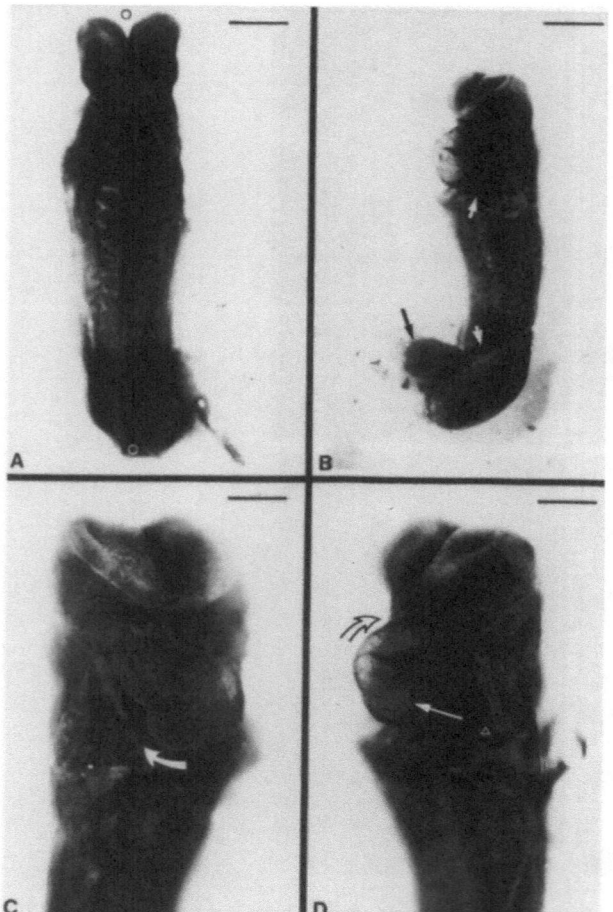

Fig. 4A–D. Human embryo at 24 days post-fertilization (Streeter's stage 11). **A** Dorsal view. Twelve to thirteen somites are readily apparent. The anterior (*black circle*) and posterior (*white circle*) neuropores are still patent. *Scale bar*, 0.4 mm. **B** Left oblique view. The coelomic space is between the *two white arrows*. The *black arrow* points to the developing umbilical bud. *Scale bar*, 0.3 mm. **C** Right oblique close-up view of the outlet segment (*arrow*) of the cardiac loop. *Scale bar*, 0.2 mm. **D** Left oblique close-up view. The *arrows* point to the orientation of inlet (*white arrow*) and outlet segments (*open black arrow*) of the cardiac loop. *Triangle*, pericardial space; *scale bar*, 0.2 mm

post-fertilization. At 24 + 1 days, up to 16 somites can be identified on direct examination. The anterior and posterior neuropores remain patent (Fig. 4A). Thus, the closing neural tube maintains a direct contact with the amniotic fluid. The septum transversum or putative diaphragm is clearly defined. Beneath it there is a large coelomic cavity, and the developing

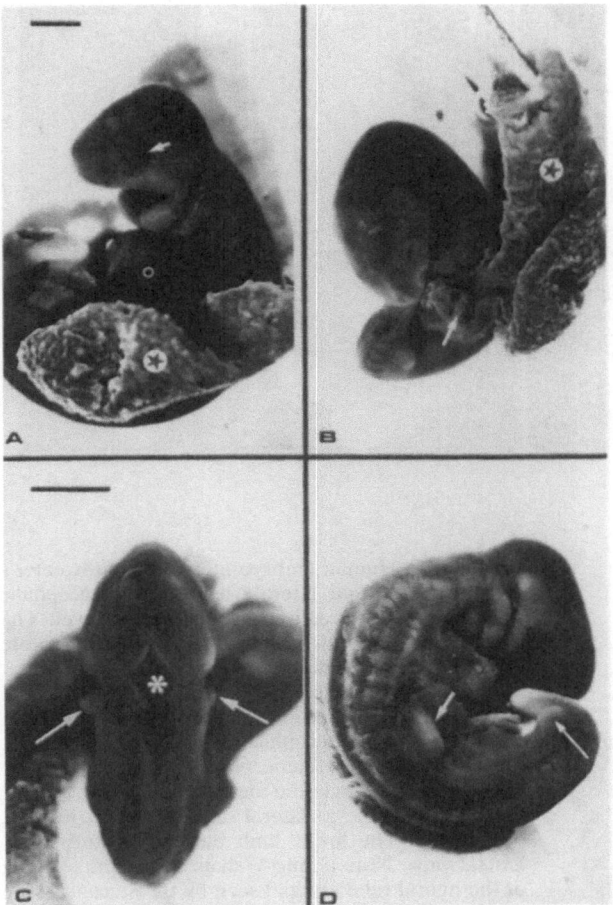

Fig. 5A–D. Human embryo at 28 + 1 days (Streeter's stage 12). **A** Left lateral view. Ophthalmic placode (*white arrow*), cardiac eminence (*circle*), yolk sac (*star*). Note the pharyngeal arches. *Scale bar*, (for **A, B, D**), 1 mm. **B** Fronto-cephalic view. The umbilical cord shows three vessels (*white arrow*). **C** Coronal view. Note the tadpole shape of the head and neck segments. The second pharyngeal arches show lateral prominences (*arrows*). Developing pontine flexure (*asterisk*) and the putative fourth ventricular space. *Scale bar*, 1 mm. **D** Right lateral view. Four branchial arches and somites are clearly seen. The upper limb bud (*short arrow*) is developing at a faster pace. The tail is prominent. *Long arrow*, lower limb

umbilical cord is at the caudal end of the embryonic tube. Neither pharyngeal arches or limb buds are visible on the epiblastic surface (Fig. 4B–D). At 27 days (Streeter's stage 12) (Fig. 5) the embryo has adopted a "C" shape. The neuropores close at 28 days [9]. The lens placode and three pharyngeal arches can be recognized. The superior and inferior limb buds

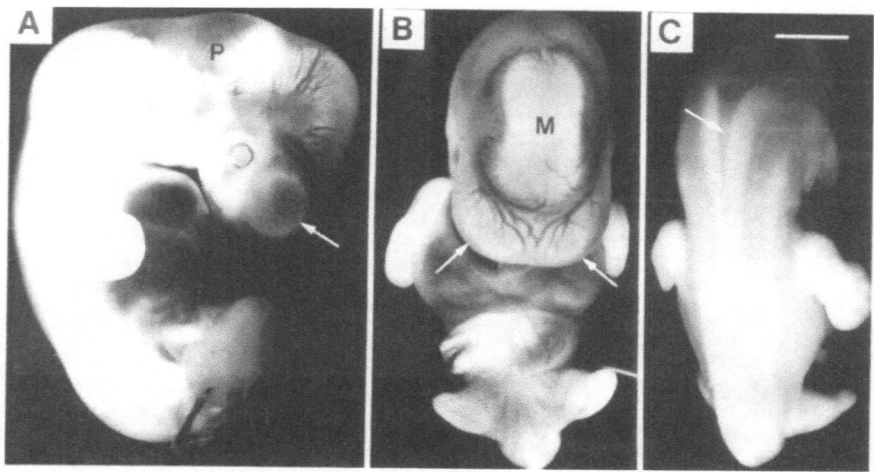

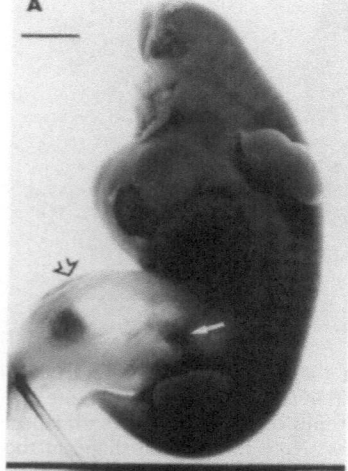

Fig. 6A–C. Human embryo at 36.5 days (Streeter's stage 16). **A** Right lateral view. The telencephalic vesicles (*arrow*) have not passed the eye level. The eye is open, deposition of pigment has started. Note the prominent vasculature over the developing brain. The pontine flexure shows a 90° angle (*P*). **B** Frontal view. The head and liver are prominent organs. Note the size of the mesencephalon (*M*) in relation to that of the developing telencephalic vesicles (*arrows*). Large vessels encircle the mesencephalon. The upper limb buds have a lateral and slight downward orientation. The lower limb buds are upturned. **C** Dorsal view. Note a long V-shaped cephalic opening of the neural tube (*arrow*) seen by translucency. The neural tube is closed along the dorsal line. *Scale bar*, 1 mm

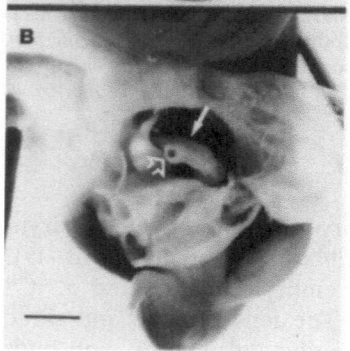

Fig. 7A,B. Human embryo at 35 days (Streeter's stage 15). The head has been previously excised. **A** Left lateral view. A small loop of hind gut (*white arrow*) is entering a well-developed physiological hernial sac (*open arrow*). *Scale bar*, 1.4 mm. **B** Frontal view. The opened physiological hernial sac shows a loop of hind gut (*white solid arrow*) and the cut end of the vitelline artery (*open arrow*) beneath the putative antimesenteric border (*tenuous white line*). The tail is prominent

are visible. The former develop ahead of the latter (Fig. 5D). The vitelline vessels and duct together with the umbilical cord are seen attached between the second and third distal segments of the embryo's body.

At 36.5 days (Streeter's stage 16) (Fig. 6) the embryo is unfolding from its previous "C" shape. The head is bent over the abdomen, and the cervical and lumbar flexures can be recognized. The eyes are seen on lateral views only and show a ring-like distribution of pigment (compare with Fig. 9). At this stage, the pharyngeal arches are fused. The anterior chest wall is in close apposition to the developing heart. The limb buds are paddle-shaped, and the putative upper limbs show early digital ray formation. The elbow flexure is not present. The lower limb buds show an upward turn. The liver is prominent and occupies most of the developing abdominal cavity. The physiological hernia has formed and contains a small loop of developing hind gut. The tail is clearly visible (Fig. 7). At 43 days (Streeter's stage 18) (Fig. 8), the embryo has distinct human features. The opened eyes contain abundant pigment. The cervical and lumbar flexure are clearly seen. The elbow flexure is present. The hands show early lobulation and are "drumming" the abdomen. The head and liver remain prominent while most of the anterior chest wall is obscured by the head and upper limb buds. The lower limbs almost face each other and show well-defined digital rays. The tail is significantly shorter and has a blunt tip (compare with Fig. 7A). At 11 weeks of gestation (Fig. 9), apart from the external genitalia, all the external features of a well-developed human are self-evident. The head is prominent, the eye lids have developed and therefore the eyes are closed. The abdominal circumference is slightly larger than the chest circumference. The physiological hernia is reduced to a shallow "cup".

Organ System Development

The Nervous System

The central nervous system derives from the ectodermal layer and can be recognized as a neural plate during the 18th day after fertilization (stage 8). The neural groove appears on the neural plate at day 18 (stage 9). The adjacent ectoderm elevates and forms neural folds just before the first pair of somites become visible. Fusion of neural folds at this level takes place between days 21 and 22 (stages 9–10) when the embryo is about 3.3 mm (crown-rump) in length [10]. Other points of fusion at the cephalic end of the human embryo have been described [11]. Formation of the neural tube (neurulation) is complex [12, 13]. Observations of neurulation in animal models suggest that adjacent developing structures, such as the notochord, may play important roles in determining the morphology of the spinal cord and of specific brain segments [14]. At 28 days post-fertilization, there are 42–45 pairs of somites and the neural tube is closed (Fig. 5).

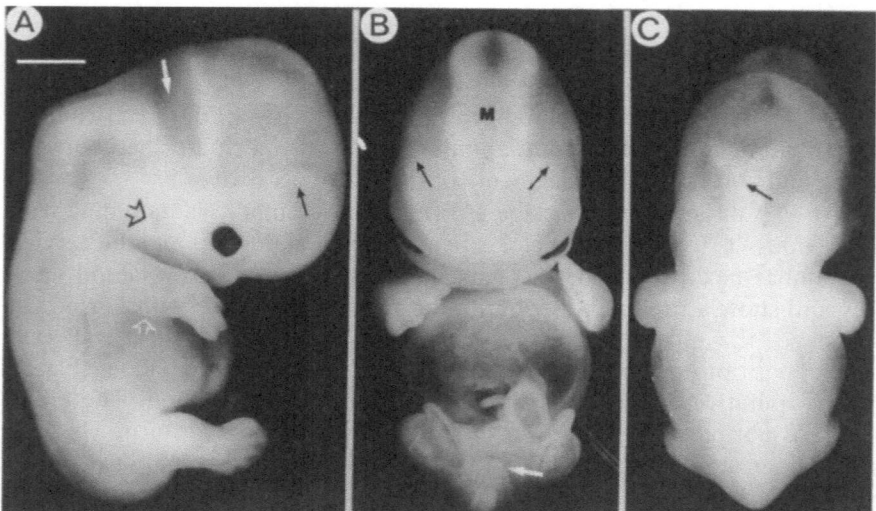

Fig. 8A–C. Human embryo at 43 days (Streeter's stage 18). **A** Right lateral view. The telencephalic vesicle (*arrow*) is just beyond the eye level. The latter contains abundant pigment. The pontine flexure (*white arrow*) is closing (compare with Fig. 6). Developing auricular hillock (*open black arrow*). The right elbow is clearly apparent (*open white arrow*). There is obvious lobulation of digits in both hands. The lower limb buds show digital rays only. *Scale bar*, 1.8 mm. **B** Frontal view. The mesencephalon (*M*) is prominent. There is symmetric growth of the telencephalic vesicles (*black arrows*). Observe the position and orientation of the hands and feet in relation to the abdomen. A short. blunt tail is still present (*white arrow*). **C** Posterior view. The neural tube is fully covered by epiblast. The *arrow* points to the distal end of the developing fourth ventricle seen by translucency (compare with Fig. 6)

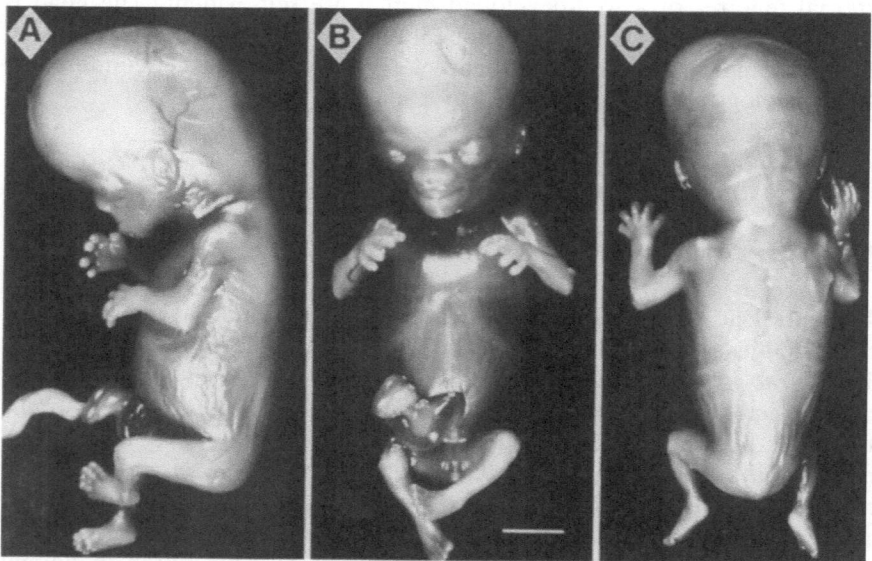

Fig. 9A–C. Human fetus at 11 weeks of gestation (foot length: 6.1 mm). **A** Left lateral view. The head is prominent, and there is predominance of the fronto-parietal region. The developing ear is at eye level. There is subtotal reduction of the physiological hernial sac. **B** Frontal view. The frontal region of the head is prominent. The face is formed. The eyes are closed. *Scale bar*, 8 mm. **C** Dorsal view. The base of the neck is broad. The dorsal region of the spine is unremarkable. The slightly wrinkled appearance of the epidermal surface is due to a fixation artefact and storage. *Scale bar*, 2 mm

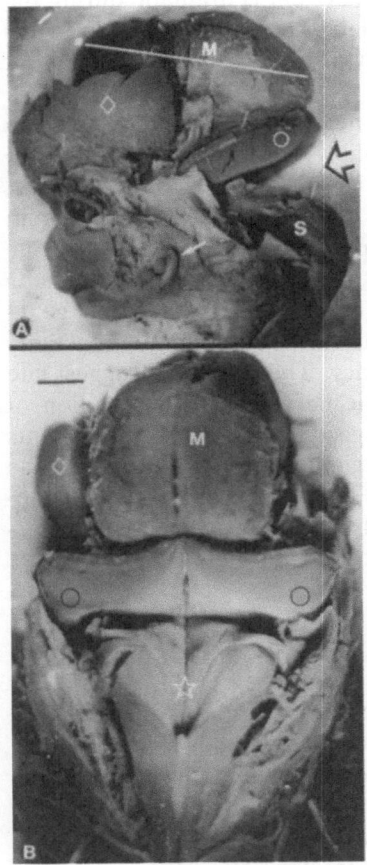

Fig. 10A,B. Head of a human fetus at 9.5 weeks of gestation. **A** Left lateral view. The posterior margin of a partially excised left brain hemisphere (*diamond*) does not reach the level of the left ear. Note the predominating mesencephalon (*M*). The developing cerebellar hemisphere (*circle*) shows a smooth convex surface. The brain stem (*S*) is in a horizontal position. The pontine flexure (*open arrow*) is still present; *small arrow*, ear. **B** Dorsal view. The cerebellar hemispheres (*circles*) resemble the roof top of an oriental temple. *Star*, fourth ventricle. *Scale bar*, 0.9 mm

The brain develops from the fourth somite headwards. At 32 days (Streeter's stage 14), three brain "vesicles" can be identified; the hindbrain or rhombencephalon, the midbrain or mesencephalon and the forebrain or prosencephalon [9, 15]. The putative brain hemispheres appear as two small symmetrical vesicles on either side of the mid-line, but do not reach the eye level. The brain vasculature is apparent (Fig. 6). At 43 days of gestation (Streeter's stage 18), the developing brain hemispheres are beyond the eye level (Fig. 8). At 9 weeks, the pontine flexure is still apparent (Fig. 10). Neither the insula, sulci nor gyri have formed. In the developing cerebellum, the rhombic lips or putative cerebellar hemispheres resemble the roof top of an oriental temple if observed from a dorsal view. The floor of the developing fourth ventricle is well exposed (Fig. 10B). Early development of the brain stem and fourth ventricle observed under scanning electron microscopy has been previously described (see [16] for review). At 12 weeks of gestation, the brain hemispheres have partially covered a prominent midbrain.

The rate of growth of the central nervous system is dramatic during the first half of pregnancy. According to Mikhailets [quoted by 17], it has over 400% volume compared with that of the body of the embryo. At the end of gestation the volume is about 42% only.

Away from the central nervous system, nerve fibres in the periosteum and synovium of the developing knee joint show clear immunoreactivity to protein gene product 9.5 at 8 weeks of gestation. Active peptides such as substance P can be observed at 11 weeks in nerve fibres of the periosteum and synovium. However, the C-flanking peptide of neuropeptide tyrosine and tyrosine hydroxilase immunoreactivity (associated with sympathetic nerve fibres) is absent until the 13th week of gestation. Calcitonin gene-related peptide and vasoactive intestinal peptide nerve fibre immunoreactivity is absent in the knee joint throughout pregnancy. These findings suggest late cytochemical cell differentiation in autonomic nerves [18]. Myelination of nerve fibres can be observed at 14 weeks of gestation in many components of the spinal cord [19]. Thus, from the beginning of organogenesis, the central nervous system, in particular, appears to be concerned with developing its various anatomical components, whilst signs of specific cell differentiation are observed towards the end of the 1st trimester.

Cardiac Development

The heart is among the first organs to develop early in embryonic life, and the functional competence of the cardiovascular system is obtained long before other organ systems. The study of the complex anatomical changes in this four-chamber organ has been greatly improved by the combined use of microdissection and scanning electron microscopy [20, 21].

Early during the 3rd week of gestation (Streeter's stage 9), a plexus of primitive vessels, beneath the cephalic end of the embryonic plate, fuses to form the cardiac tube. This important event marks the start of true cardiac morphogenesis. Both the right and left dorsal aortas connect with the primitive cardiac tube. Owing to an accelerated growth, it bends forwards forming a loop at about the 26th day (Streeter's stage 12) (Fig. 4B–D). Thus, descending inlet and ascending outlet segments can be recognized. Internally, the cardiac tube is lined by putative endocardial cells, its thick walls are made up of cardiac "jelly" containing a few migrating endocardial cells. At 32 days (Streeter's stage 15), both atria, the primitive left and right ventricles and the truncus arteriosus can be identified on external examination (Fig. 11). Internally, the valves of the sinus venosus and the septum primum approach the atrioventricular junction. At this level, the superior and inferior atrioventricular cushions are unfused (Fig. 12). The foramen interventriculare is wide, and the first embryonic trabeculae can be observed in the apical region of both developing ventricles. The truncus arteriosus shows developing ridges (Fig. 12). Fusion of the atrioventricular cushions

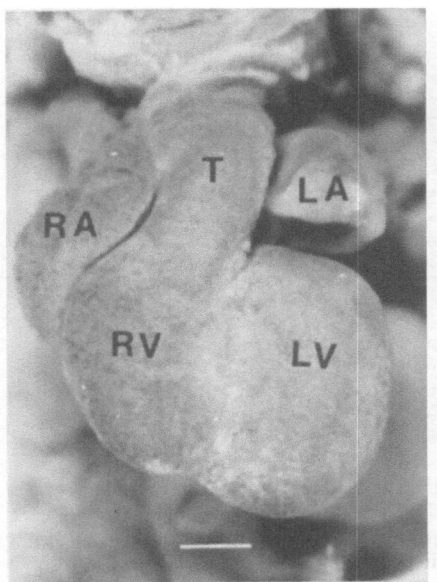 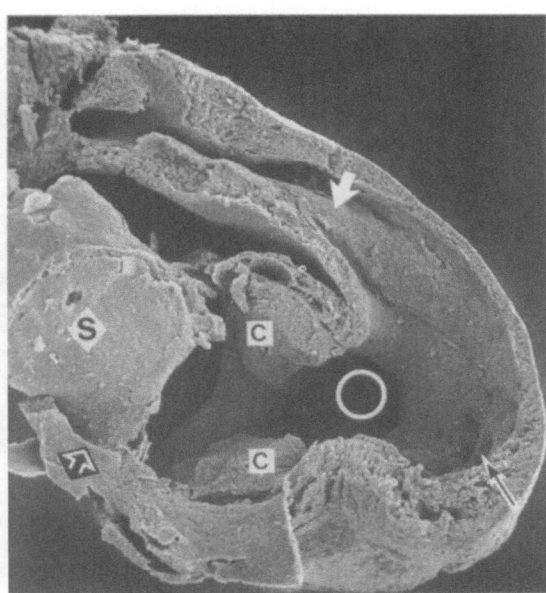

Fig. 11. Embryonic heart at 32 days of gestation. Neither the aorta nor the pulmonary arteries have developed. *T*, truncus arteriosus *RA*, right auricle; *LA*, left auricle; *RV*, right ventricle; *LV*, left ventricle. *Scale bar*, 1.3 mm

Fig. 12. Right lateral aspect of an embryonic heart at 32 days of gestation. The developing septum primum (*S*) and the valve of the right sinus venosus (*open arrow*) are at a distance from the unfused artrioventricular cushions (*C*). The latter appear as two clearly defined structures at the level of the atrioventricular junction. The foramen interventriculare is widely patent (*circle*). *Solid black arrow*, embryonic trabeculae; *white arrow*, conal ridge. SEM, ×500

and truncal ridges occurs before the 40th day of gestation. At 42 days (Streeter's stage 18) (Fig. 13A) the septum primum divides both atria. The valve of the sinus venosus is prominent. The anterior leaflet of the tricuspid valve (not shown) and both the anterolateral and posteromedial leaflets of the mitral valve have delaminated (Fig. 13B). The interventricular septum continues developing, and the right and left ventricular walls show prominent embryonic trabeculae. The aorta is connecting to the left ventricle, and both the aortic and pulmonary valves show developing cusps (Fig. 13A).

At this stage, the subaortic infundibulum is closely related to the crista supraventricularis and the anterior leaflet of the tricuspid valve (Fig. 14). This is a previously undescribed feature which may contribute to the understanding of the anatomical relationships of the aorta in cases of tricuspid atresia. Closure of the foramen interventriculare occurs during the 7th week of gestation. Here, the aorta loses its relationship with right ventricular structures and connects to the left ventricle. As development continues,

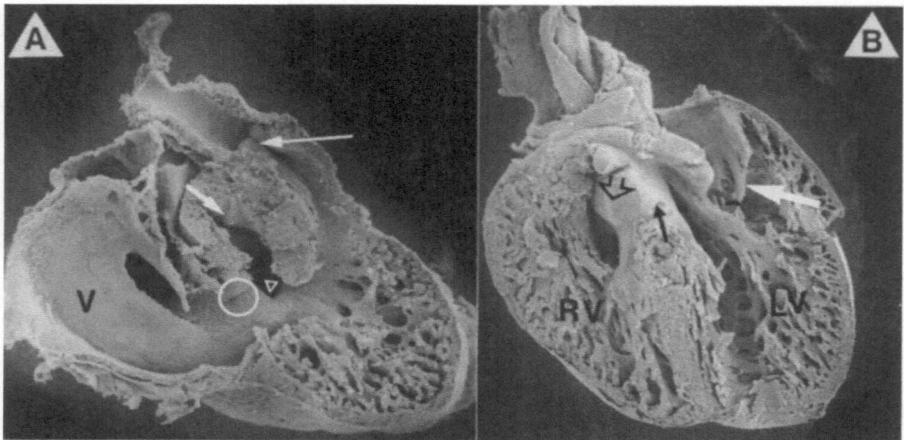

Fig. 13. A Right lateral aspect of an embryonic heart at 42 days of gestation. *V*, valve of the right sinus venosus. The atrioventricular cushions are fused (*circle*). The foramen interventriculare appears reduced in diameter (*triangle*). The aorta (*short arrow*) is connecting to the left ventricle. Note the embryonic trabeculae on the septal aspect of the right ventricle. *Long arrow*, pulmonary valve. **B** Four-chamber view of a heart at 42 days of gestation. There is no septal leaflet of the tricuspid valve (*open arrow*). The leaflets of the mitral valve can be recognized (*white arrow*). The *black solid arrow* marks the lower margins of the foramen interventriculare. Note at this level the smooth endocardial surface between the *black arrows*. *RV*, right ventricle; *LV*, left ventricle. SEM, ×50 (for **A, B**)

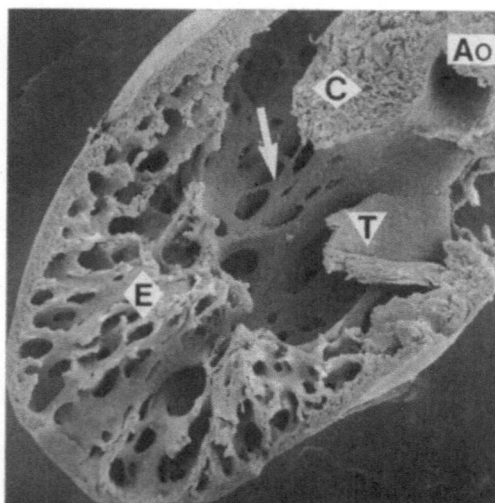

Fig. 14. Parietal aspect of the right ventricle before closure of the foramen interventriculare (42 days). The subaortic infundibulum (*Ao*) is directly related to the crista supraventricularis (*C*) and the anterior leaflet of the tricuspid valve (*T*). Note the embryonic trabecula (*E*) towards the apical region. *Arrow*, definitive trabeculae. SEM, ×56

the pulmonary, aortic and tricuspid valves show a typical alignment and relationships. The embryonic trabeculae are gradually replaced by definitive trabeculae which develop in a base-to-apex direction following the longer ventricular axis (Fig. 15). Delamination of the septal leaflet of the tricuspid valve starts after closure of the foramen interventriculare, spreading antero-superiorly from the diaphragmatic aspect of the septum (Fig. 16). Gradual development of the membranous portion of the interventricular septum starts at 8–9 weeks and can be observed after 10–11 weeks of gestation. It is the last morphological feature to develop and marks the end of cardiac organogenesis. Thus, the external (Fig. 17) and internal (Fig. 18A) morphology of the human heart resembles that observed later in gestation or after birth.

The left ventricular mass predominates over that in the right ventricle and the weight of both ventricles increases rapidly from 50 mg at 9 weeks to over 150 mg at 14 weeks (Fig. 18B). The development of form in the heart is paralleled by cytochemical differentiation of some cardiac cells. For example, atrial natriuretic peptide can be identified in both atria and ventricles as early as 7 weeks of gestation [22, 23]. When examined using molecular probes, at 12 weeks of gestation there is highly selective gene expression of heavy chain β myosin in the ventricular myocardium [24] (Fig. 19). Further-more, the expression of neural cell adhesion molecules (N-CAM) is greater in early than in late fetal life. Interestingly, significant re-expression occurs in pathological conditions such as hypertrophy or following cardiac trans-plant [25].

Respiratory System

The primordium of the lower respiratory system develops during the 4th week of gestation. At 32 days (Streeter's stage 14), a short trachea and both right and left lungs in early lobation can be recognized (Fig. 20). The primitive respiratory epithelium remains apparently uncommitted until the end of the 7th week. Thereafter, ciliation of the epithelial surface, starting on the membranous (posterior segment) trachea spreads in a caudal direction. Ciliation of the carinal angle takes place at 8 weeks (Fig. 21). At 12 weeks, the cartilaginous trachea (anterior segment) appears uncommitted when examined under the scanning electron microscope (Fig. 22). During a transitional period between the 12th and 13th weeks, neuroepithelial body-like structures differentiate at the level of the carinal angle (not shown). This is followed by a stage of pseudociliation in which cells destined to become ciliated cells show proliferation of long and slender microvilli (Fig. 23). The ciliary escalator is, however, established in the posterior trachea well before 10 weeks, and ciliated cells beat following the beating pattern observed after birth (Fig. 24) [26–28]. Mucin-secreting cells differentiate during the 12th–13th weeks of gestation and are first seen in the mem-

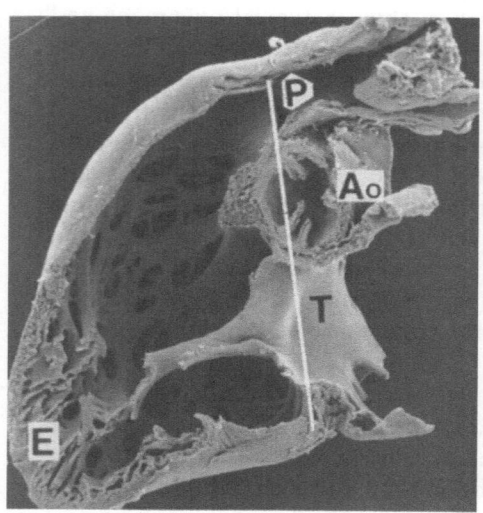 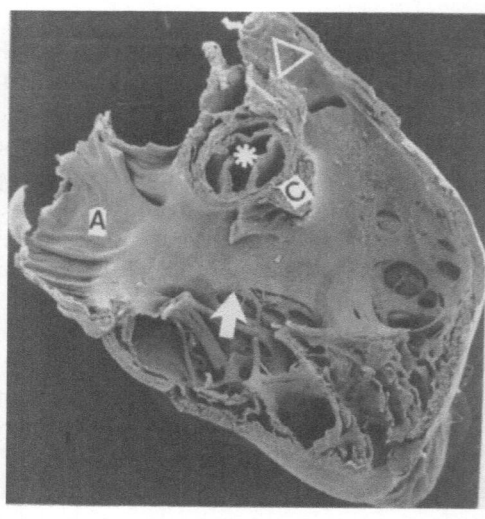

Fig. 15. Parietal aspect of the right ventricle at 10 weeks of gestation. Note the axial alignment that exists between the pulmonary (*P*), aorta (*Ao*) and anterior leaflet of the tricuspid valve (*T*). The embryonic trabeculae (*E*) have been displaced towards the apex of the ventricle. SEM, ×500

Fig. 16. Septal aspect of the heart in Fig. 15. The septal leaflet of the tricuspid valve shows early delamination (*arrow*). *A*, right atrium; *C*, crista supraventricularis; *asterisk*, aortic valve orifice; *triangle*, pulmonary trunk. SEM, ×500

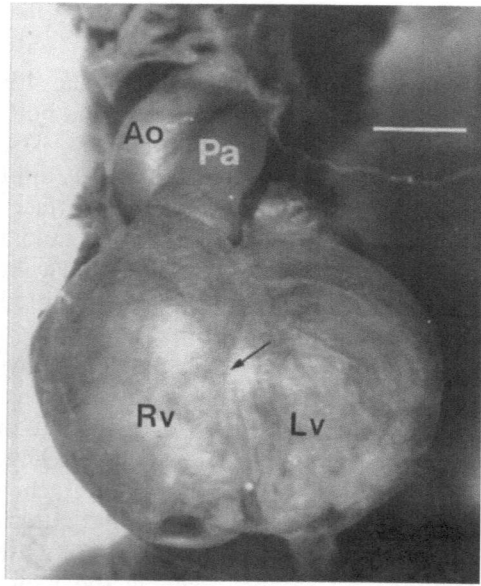

Fig. 17. External form of the heart at 7.5 weeks of gestation. The aorta (*Ao*) and pulmonary (*Pa*) arteries are clearly shown (compare with Fig. 11). *RV*, right ventricle; *LV*, left ventricle; *arrow*, left anterior descending coronary artery; *scale bar*, 1.6 mm

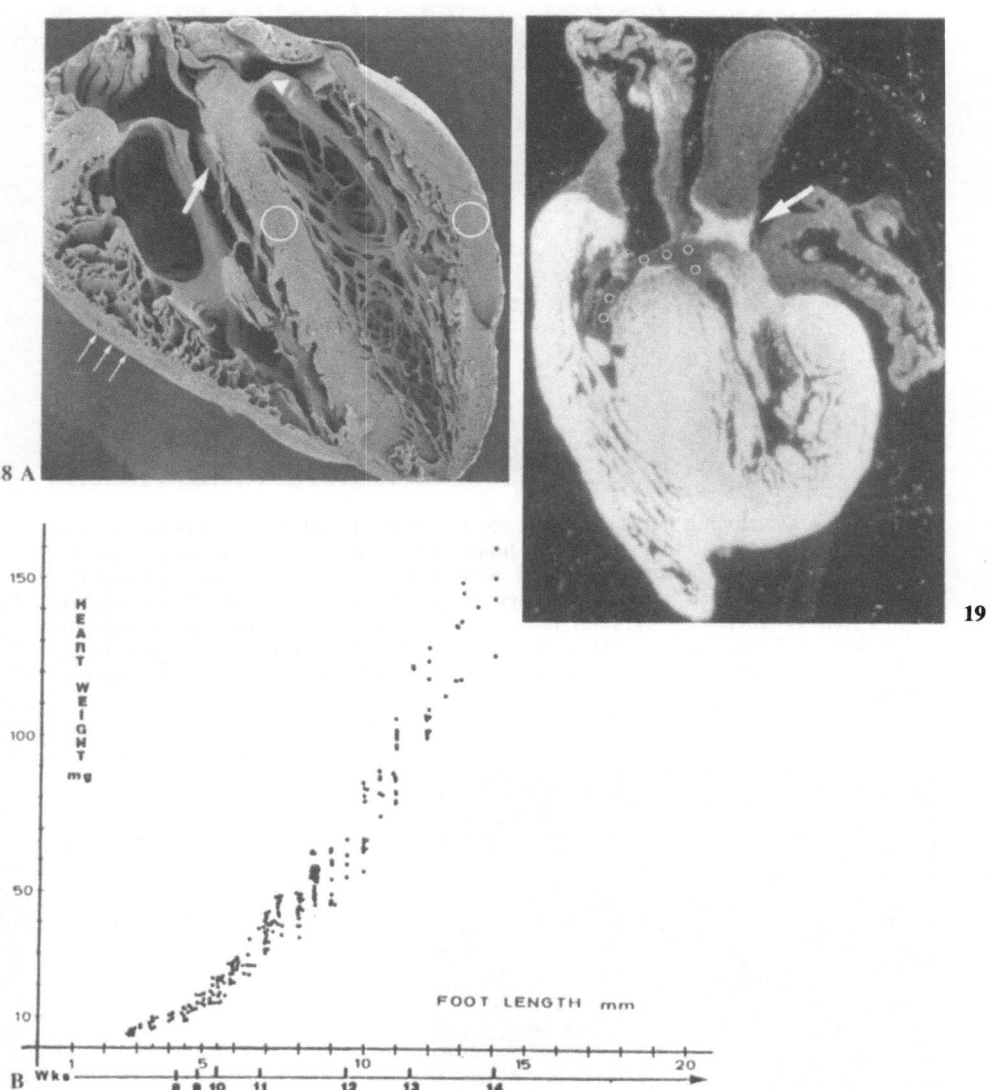

Fig. 18. A Four-chamber view of a heart at 11 weeks of gestation. The septal leaflet of the tricuspid valve (*large arrow*). Note the different levels of septal attachment of both atrioventricular valves. *Triangle*, mitral valve. The walls of the left ventricle and septum (*circles*) are thicker than that of the right ventricle (*small arrows*). **B** The weight of both ventricles is plotted against the foot length and the gestational age. Note the rapid increase of the ventricular mass after the 7th week of gestation. **A** SEM, ×500

Fig. 19. A heart at 11 weeks of gestation shows selective gene expression of heavy chain β myosin in both ventricles. Note some gene expression at the base of the aorta (*arrow*). The atrioventricular valve region gives no signals *open circles*, no signal in atrioventricular region. (Specimen prepared by Dr. P. Barton at the Royal National Heart and Lung Institute, London, UK)

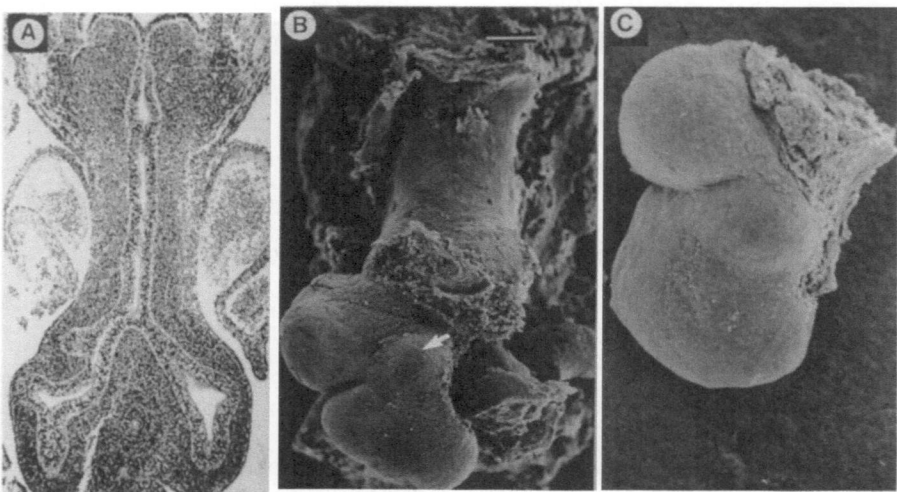

Fig. 20A–C. Developing human lungs at 35 + 2 days of gestation. **A** Coronal section showing primitive respiratory epithelium lining the larynx, trachea and lung buds. No cartilage rings are yet present. **B** Frontal view of an embryonic lung of a comparable gestational age to that shown in **A**. The right lung shows early upper and lower lobation with a small eminence in between (*arrow*), presumably representing the developing middle lobe. C Posterior aspect of the embryonic left lung from **B**. Its lobation pattern is a right lung mirror image (compare with **B**). **A** ×100; **B** SEM, ×●; C SEM, ×●; *scale bar*, 200 µm

Fig. 21. The posterior carinal angle shows early ciliation (*arrows*) of the respiratory epithelium. SEM, × 480

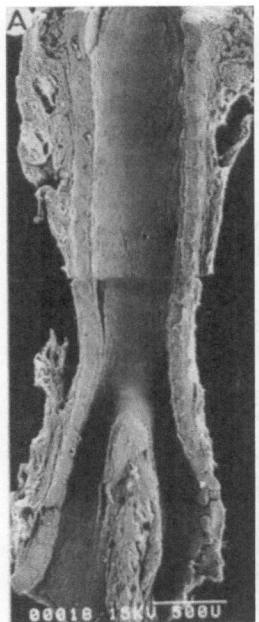
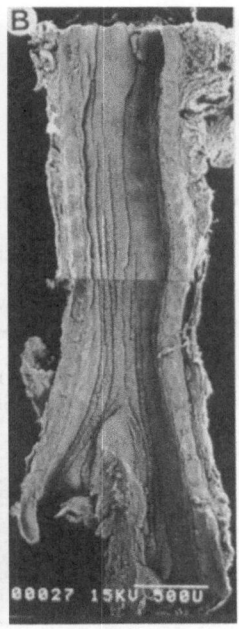
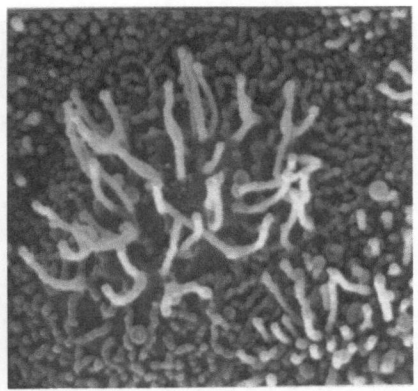

Fig. 23. Pseudociliated cells from the respiratory epithelium of the cartilaginous tracheal segment. SEM, ×12 000

Fig. 22A,B. Composite picture of the trachea at 12 weeks of gestation. Note the apparently uncommitted respiratory epithelium on the cartilaginous trachea (**A**). The white peppering on the membranous segment (**B**) represents ciliating cells distributed along its full length

branous trachea. More distal segments of the respiratory airway show an increased rate of proliferation of air sacs. Their epithelial lining is made up of cells with polygonal boundaries. These appear uncommitted when observed at the epithelial surface level (Fig. 25). Active peptides are expressed in lung cells towards the end of the first trimester. Messenger RNA-encoding gastrin-releasing peptide has been identified in human lungs at 9–10 weeks and remains active for a short period of time [29]. These findings correlate well with the period of canalicular development, when the peptide may operate as a growth factor.

Fetal Development Between 9 and 14 Weeks of Gestation: An Overview

During a transitional stage between 9 and 14 weeks of gestation, organs from other systems show a sudden leap in their developmental pattern, this being either at organ level or at the level of cell differentiation. In the

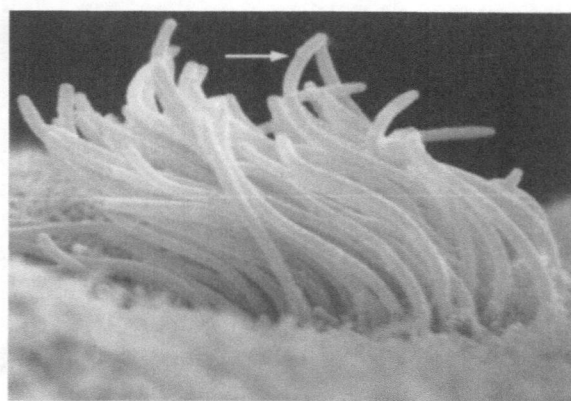

Fig. 24. Beating ciliated cell at 10 weeks of gestation. Most ciliary shafts are at the end of the recovery phase. A few are entering the effective stroke phase as their distal third curve (*arrow*) before the forward beat. SEM, ×13 000

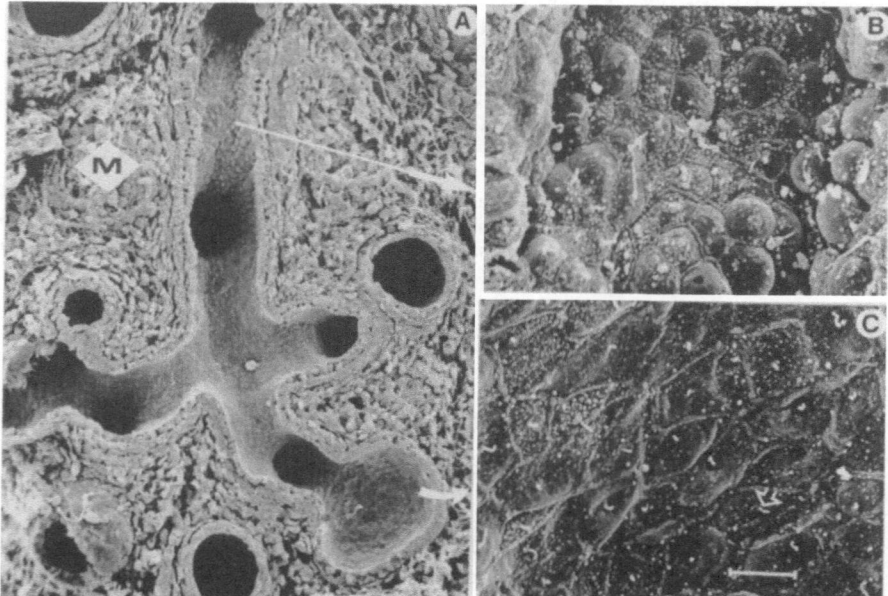

Fig. 25A–C. Fetal lung at 14 weeks of gestation. **A** Some developing air ducts branch at 90° from the respiratory bronchiole. Note the amounts of mesenchymal tissue (*M*) in relation to the space occupied by the air ducts and terminal air sacs. *Long arrow*, detailed view of the respiratory epithelium is shown in **B**; *curved arrow*, close-up view of the air sac epithelium is shown in **C**. SEM, ×260 **B** Most epithelial cells have a convex luminal surface. The shallow intercellular ridges follow a geometrical pattern. SEM, ×1800 **C** The epithelial cells appear flattened and polygonal. Most of them have a centrally placed primary cilium (*arrow*). The intercellular spaces also have a geometrical profile. SEM, ×2100; *scale bar*, 10 μm

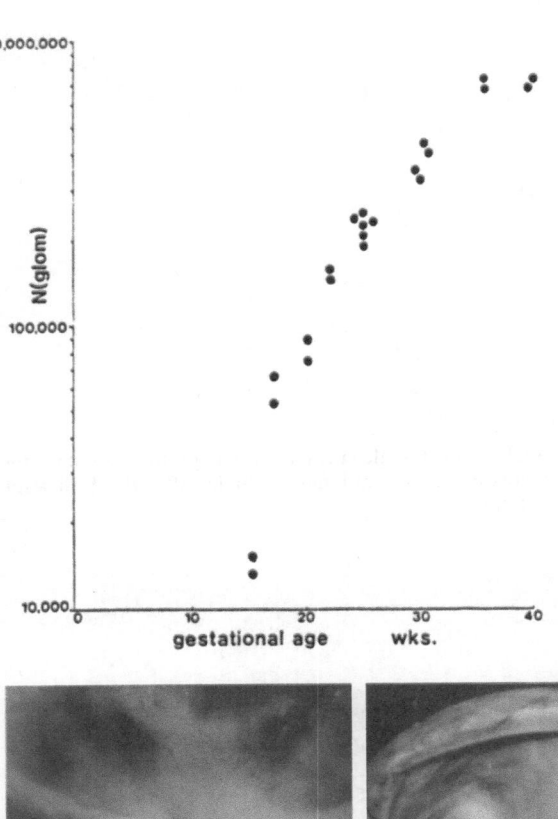

Fig. 26. The absolute numbers of glomeruli [*N(glom)*] have been plotted against the gestational age in humans. There is a rapid increase of glomeruli at the start of the 2nd trimester. (From [30])

Fig. 27. A The verum montanum (*circle*) clearly indicates the gender of the male fetus at 10 weeks of gestation. **B** The bladder of a female fetus at a similar gestational age for comparison. *Arrows*, ureteric orifices. *Scale*, 3 mm
▼

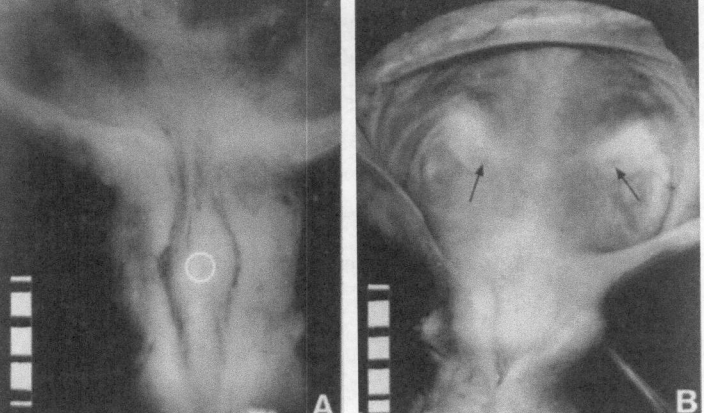

fetal kidney, the total number of developing glomeruli "jumps" from just over 10000 at 12–13 weeks to nearly 250000 halfway through the second trimester and under one million before birth (Fig. 26) [30]. In the bladder, the verum montanum is well defined in males by the 10th week of gestation (Fig. 27), well before sex differences can be established by external examination. The interureteric distance changes at the same rate in both sexes. However, in males the length of the lower urinary tract grows at a faster pace from the 12th week onwards (Fig. 28) [31].

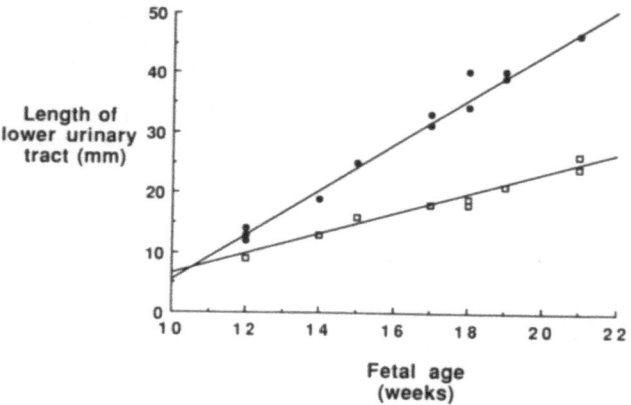

Fig. 28. Length of the human male and female urinary tract plotted against the gestational age in weeks. There is a steady increase in length in males after the 12th week of gestation. *Circles*, male; *squares*, female

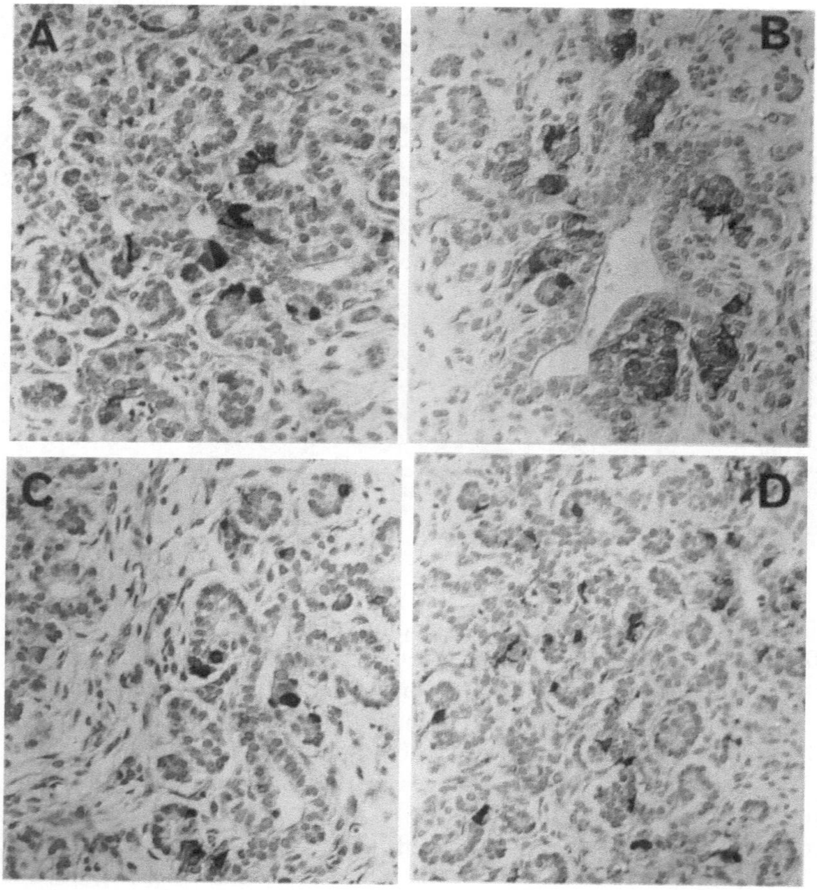

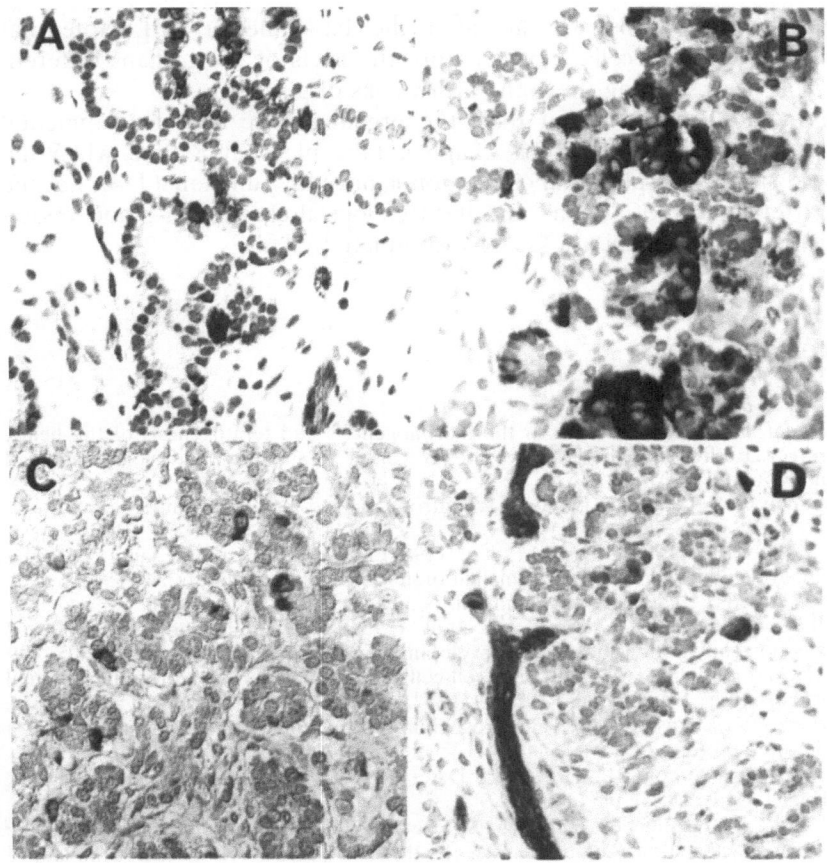

Fig. 30A–D. Positive immunohistochemical response of pancreatic tissue to synaptophysin (**A**), chromogranin (**B**), Gly, ala, trp, lys (GWAK) (**C**) and protein gene product 9.5 (**D**) in a fetus at 9 weeks of gestation. **A–D** ×160

In the fetal skin, the onset of expression of peanut lectin-binding glyco-proteins starts at 12 weeks of gestation and correlates well with stratification of keratinocytes [32]. In the pancreas, immunoreactivity to native pancreatic hormones (Fig. 29) and general neuroendocrine markers (Fig. 30) is observed between the 9th and 10th weeks of gestation (J.G. Moscoso, personal observations). However, islet cell amyloid polypeptide typically

◀ **Fig. 29A–D.** Positive immunohistochemical response of pancreatic epithelium to insulin (**A**), glucagon (**B**), pancreatic polypeptide (**C**) and somatostatin (**D**) in a fetus at 9 weeks of gestation. **A, B** ×160; **C, D** ×40

associated with β cells, appears after the 1st trimester [33]. Note that pancreatic endocrine cells appear before the islets of Langerhans differentiate. This occurs during the 10th–11th weeks of gestation.

It should be emphasised that, the organogenetic and developmental events briefly described in this chapter, take place at a time when the maternal side of the placental circulation is not fully functional [34], i.e. the spiral arteries are not "open" and the topological changes of blood vessels within the chorionic villi are far from complete [35].

References

1. Pexieder T (1975) Cell death in the morphogenesis and teratogenesis of the heart. Springer, Berlin Heidelberg New York (Advances in anatomy, embryology and cell biology, vol 51) (3)
2. Edelman GM (1988) Topobiology. An introduction to molecular embryology. Basic, New York
3. He X, Treacy MN, Simmons DM, et al. (1989) Expression of a large family of POU-Domain regulatory genes in mammalian brain development. Nature 340:35–42
4. Lewis J (1989) Genes and segmentation. Nature 341:382–383
5. Wilkinson DG, Bhatt S, Cook M, et al. (1989) Segmental expression of Hox-2 homeobox-containing genes in the developing mouse hindbrain. Nature 341:405–409
6. Takeichi M (1988) The cadherins: cell-cell adhesion molecules controlling animal morphogenesis. Development 102:639–655
7. Parks S, Wieschaus E (1991) The drosophila gastrulation gene concertina encodes a G alfa-like protein. Cell 64:447–458
8. Charnes ME, Simon RP, Greenberg DA (1989) Ethanol and the nervous system. N Engl J Med 321:442–454
9. O'Rahilly R, Muller F, Hutchins GM, et al. (1984) Computer ranking of the sequence of appearance of 100 features of the brain and related structures in staged human embryos during the first 5 weeks of development. Am J Anat 171:243–257
10. O'Rahilly R (1975) A color atlas of human embryology. Saunders, Philadelphia
11. O'Rahilly R, Gardner E (1979) The initial development of the human brain. Acta Anat (Basel) 104:123–133
12. Heuser CH (1957) Corner GW. Developmental horizons in human embryos. Contrib Embryol 36:31–41
13. Streiter A (1951) Ein menschlicher Keimling mit 7 Urwirbelpaaren (Keimling Ludwig). Z Mikrosk Anat Forsch 57:181–248
14. Schoenwolf GC (1982) On the morphogenesis of the early rudiments of the developing central nervous system. Scanning Electron Microsc 1:209–308, 1982
15. Muller F, O'Rahilly RO (1988) The first appearance of the future cerebral hemispheres in the human embryo at stage 14. Anat Embryol (Berl) 177:495–511
16. Tanaka O, Otani H, Fujimoto K (1987) Fourth ventricular floor in human embryos: scanning electron microscopy observations. Am J Anat 178:193–203
17. Blinkov SM, Glezer II (eds) (1968) The human brain in figures and tables. Plenum, New York, pp 126, 334
18. Hukkanen M, Mapp PI, Moscoso G, et al. (1991) Inervation and neuropeptide containing nerves in human fetal knee joint (Abstr). J Pathol 163:181–A
19. Friede RL (1989) Developmental neuropathology, 2nd edn. Springer, Berlin Heidelberg New York
20. Pexieder T (1986) Standardized method for study of normal and abnormal cardiac development in chick, rat, mouse and dog (Abstr). Teratology 33:91C–92C

21. Moscoso G, Pexieder T (1990) Variations in microscopic anatomy and ultrastructure of human embryonic hearts subjected to three different modes of fixation. Pathol Res Pract 186:768–774

22. Wharton J, Anderson RH, Springall D et al. (1988) Localization of atrial natriuretic peptide immunoreactivity in the ventricular myocardium and conduction system of the human fetal and adult heart. Br Heart J 60:267–274

23. Takemura G, Fujiwara H, Mukoyama M, et al. (1991) Expression and distribution of atrial natriuretic peptide in human hypertrophic ventricle of hypertensive hearts and hearts with hypertrophic cardiomyopathy. Circulation 83:181–190

24. Barton PJR, Moscoso G, Thompson R (1991) Detection of myosin gene expression in cardiac muscle using probes derived by polymerase chain reaction (PCR). Int J Cardiol 30:116–118

25. Gordon L, Wharton J, Moore SE, et al. (1990) Myocardial localization and isoforms of neural cell adhesion molecule (N-CAM) in the developing and transplanted human heart. J Clin Invest 86:1293–1300

26. Moscoso GJ, Driver M, Whimster WF (1987) Ciliogenesis of the human respiratory epithelium during the pre-natal period (Abstr). Pediatr Pathol 3:354

27. Moscoso GJ, Driver M, Codd J, Whimster WF (1988) The morphology of ciliogenesis in the developing fetal human respiratory epithelium. Pathol Res Pract 183:403–411

28. Moscoso GJ, Nandra K, Driver M (1989) Ciliogenesis and ciliation of the respiratory epithelium in the human fetal cartilaginous trachea. Pathol Res Pract 184:161–167

29. Spindel ER, Sunday ME, Hofler H, et al. (1987) Transient elevation of messenger RNA encoding gastrin-releasing peptide, a putative pulmonary growth factor in human lungs. J Clin Invest 80:1172–1179

30. Hinchliffe S, Sargent PH, Howard CV, et al. (1991) Human intrauterine renal growth expressed in absolute numbers of glomeruli assessed by the "direct" method and Cavalieri's principle. Lab Invest (in press)

31. Cutner A, Cardozo LD, Driver M, Moscoso G (1990) The development of the human fetal lower urinary tract (Abstr). Int Urogynecol J 1:75

32. Watt FM, Keeble S, Fisher C, et al. (1989) Onset of expression of peanut lectin-binding glycoproteins is correlated with stratification of keratinocytes during human epidermal development in vivo and in vitro. J Cell Sci 94:355–359

33. Rindi G, Terenghi G, Westermark G, et al. (1991) Islet amyloid polypetide in proliferating pancreatic B cells during development, hyperplasia and neoplasia in man and mouse. Am J Pathol 1991 (in press)

34. Hustin J, Schaaps JP (1987) Echographic and anatomic studies of the maternotrophoblastic border during the first trimester of pregnancy. Am J Obstet Gynecol 157:162–168

35. Jauniaux E, Burton GJ, Moscoso G, et al. (1991) Development of the early human placenta: A morphometric study. Placenta 12:269–276

11 Morphology and Significance of the Human Yolk Sac

E. Jauniaux and J.G. Moscoso

Introduction

The yolk sac is the most representative organ of the first trimester of human pregnancy. Its disappearance at the end of the third month of gestation represents a turning point in human development in utero. The human yolk sac is an extremely small and fragile structure which makes its investigation difficult. However, during the past 50 years, several studies have demonstrated the complexity and importance of the yolk sac function in early pregnancy. This chapter is a brief review of the anatomy and the potential role of the human yolk sac in relation to embryonic and fetal development during the first 12 weeks of gestation.

Embryology of the Yolk Sac

Around day 9 after ovulation, corresponding to the beginning of the 4th week after the last menstrual period (stage 5c of embryonic development), the implanted blastocyst is composed of two cavities separated by the bilaminar embryonic disk, i.e., the amniotic cavity and the primary yolk sac, and is surrounded by the extraembryonic mesoderm and the trophoblastic ring [1, 2]. The largest cavity, which corresponds to the primary yolk sac is isolated from the extraembryonic mesoderm by the exocoelomic membrane (Fig. 1A). Both structures originate from cells delaminated from the cytotrophoblast [2].

The extraembryonic mesoderm progressively increases, and by the end of embryonic stage 5 (12 days after ovulation or 26th menstrual day) it contains isolated spaces which rapidly fuse to form the extraembryonic coelom [1, 2]. As the latter forms, the primary yolk sac decreases in size and the secondary yolk sac arises from cells growing from the embryonic disk inside the primary yolk sac [1, 2]. At the end of the 4th week of gestation, the developing coelom splits the extraembryonic mesoderm into two layers: the somatic mesoderm lining the trophoblast, and the splanchnic mesoderm covering the secondary yolk sac and the embryo (Fig. 1B).

At the beginning of the 5th week of gestation, the secondary yolk sac grows rapidly and by the 23rd embryonic day (37th menstrual day) is larger than the amniotic cavity [1]. The yolk sac communicates by a large implantation basis to the ventral part of the flat trilaminar embryo surrounded by its amniotic cavity [1, 2]. At this stage of embryonic development, the caudal part of the embryo is attached to the chorion by a thin bridge of extra-embryonic mesoderm called the connecting stalk, containing the allantois (Fig. 1C). During the 6th week of gestation, successive foldings of the embryo constrict the yolk sac basis creating the yolk stalk, which connects the definitive yolk sac to the ventral part of the embryo, near the connecting stalk (Fig. 1D). During the folding process, a portion of the secondary yolk sac is enclosed within the developing embryonic area to form the primitive gut [1]. The yolk stalk is progressively reduced to a relatively small duct called the vitelline or omphalomesenteric duct. With subsequent growth of the amniotic sac and elongation of the vitelline duct, the yolk sac is removed from the body wall (Fig. 2). The embryonic coelom is gradually obliterated, and the amnion sheathes the connecting stalk, forming the epithelial covering of the definitive umbilical cord [2].

At the end of the first trimester, the secondary yolk sac begins to shrink and it has usually disappeared by 20 weeks of gestation [1, 2]. Vestigial yolk sacs of various sizes and forms may be found throughout gestation and are recognizable beneath the amnion near the placental end of the umbilical cord (Fig. 3). The vitelline duct is then obliterated and detaches from the apex of the original midgut loop before the latter has returned to the abdominal cavity [2]. In nearly 2% of adults, the proximal intraabdominal part of the vitelline duct persists as a diverticulum of the ileum, known as Meckel's diverticulum [2]. Umbilical cord remnants of the vitelline duct are found in 2.4% of the umbilical cords examined at term (Fig. 4). They are generally located at the fetal end of the umbilical cord and are not associated with any particular perinatal complications [3].

Anatomy of the Secondary Yolk Sac

Macroscopic Features

The definitive secondary yolk sac appears externally as a spherical structure with a honeycomb pattern and is covered by numerous superficial small vessels merging at the basis of the vitelline duct (Fig. 5). The vascularization of the yolk sac wall takes place in the mesoderm surrounding the vitelline duct and communicates with the primitive cardiovascular system of the embryo by means of paired vitelline veins and arteries [1].

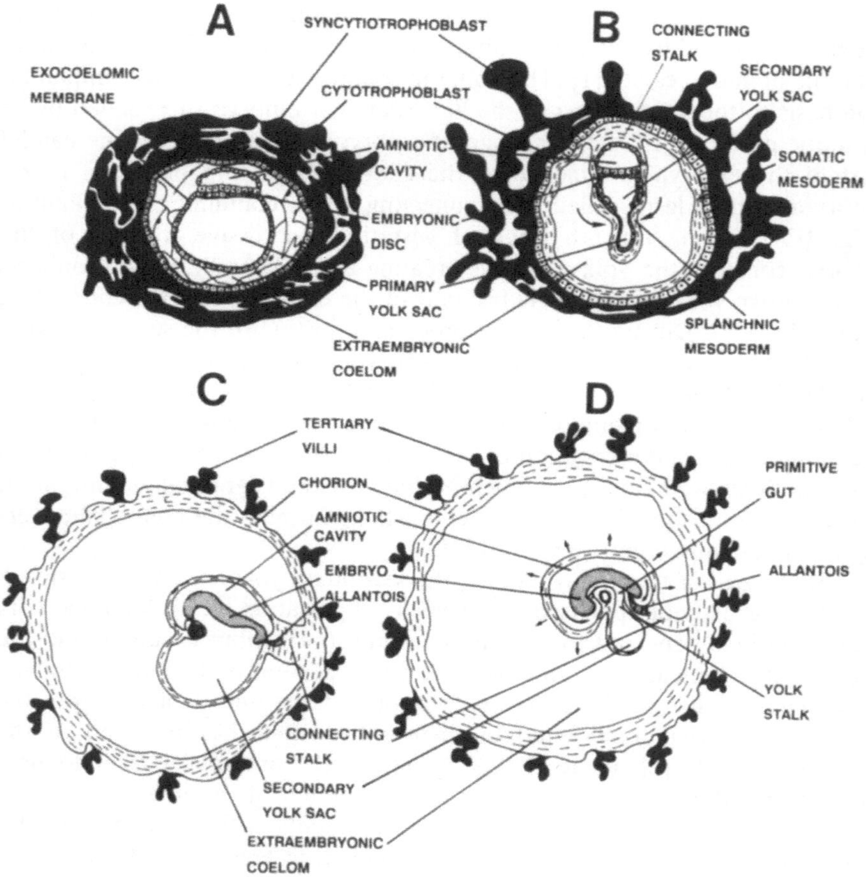

A

EXOCOELOMIC
MEMBRANE

SYNCYTIOTROPHOBLAST

CYTOTROPHOBLAST

AMNIOTIC
CAVITY

EMBRYONIC
DISC

PRIMARY
YOLK SAC

EXTRAEMBRYONIC
COELOM

B

CONNECTING
STALK

SECONDARY
YOLK SAC

SOMATIC
MESODERM

SPLANCHNIC
MESODERM

C

TERTIARY
VILLI

CHORION

AMNIOTIC
CAVITY

EMBRYO

ALLANTOIS

CONNECTING
STALK

SECONDARY
YOLK SAC

EXTRAEMBRYONIC
COELOM

D

PRIMITIVE
GUT

ALLANTOIS

YOLK
STALK

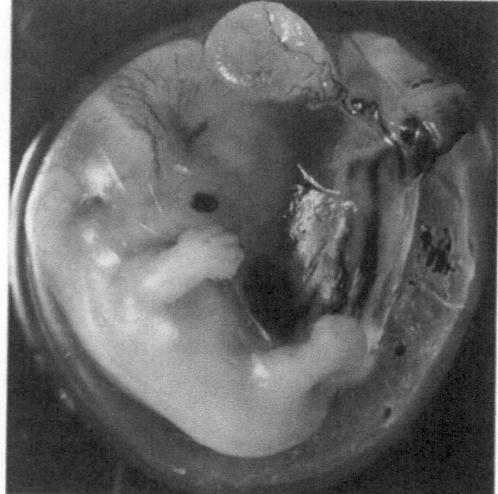

▲

Fig. 1A–D. Human pregnancies at
the beginning (**A**) and at the end
(**B**) of the 4th menstrual week and
during the 5th (**C**) and the 6th (**D**)
menstrual week. (From [20])

Fig. 2. Embryo and amniotic sac
at 9 weeks of gestation, showing
a well-vascularized yolk sac located
outside the amniotic cavity

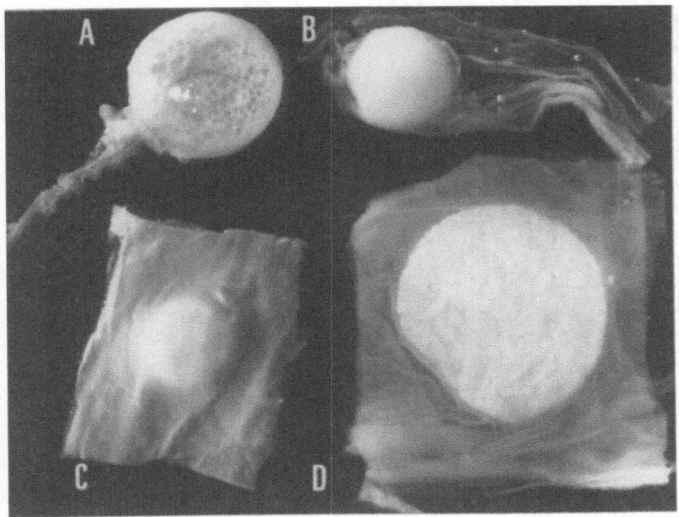

Fig. 3A–D. Macroscopic views of the human yolk sac at 9 (**A**), 12 (**B**), 16 (**C**), and 32 (**D**) weeks of gestation. Note the progressive flattening of the yolk sac structure

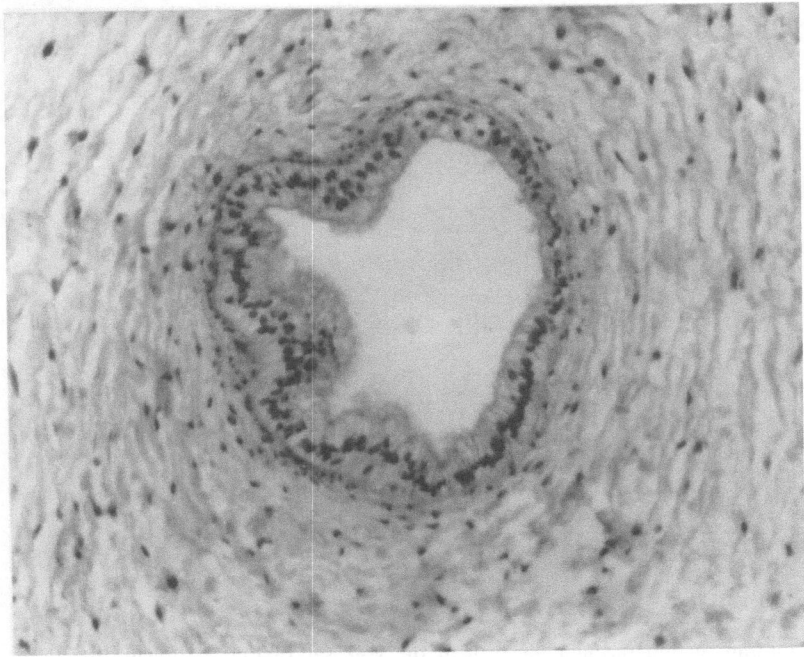

Fig. 4. Microscopic section of the umbilical cord at 40 weeks of gestation showing a well-preserved vitelline duct lined by columnar epithelium containing mucus. Hematoxylin and eosin, ×200

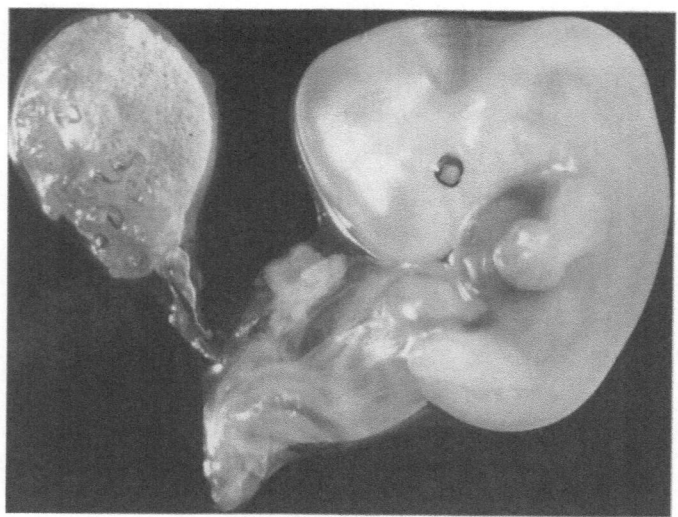

Fig. 5. Embryo and its yolk sac at 60 days of gestation. The yolk sac shows a honeycomb pattern and is covered by numerous small vessels merging at the basis of the vitelline duct

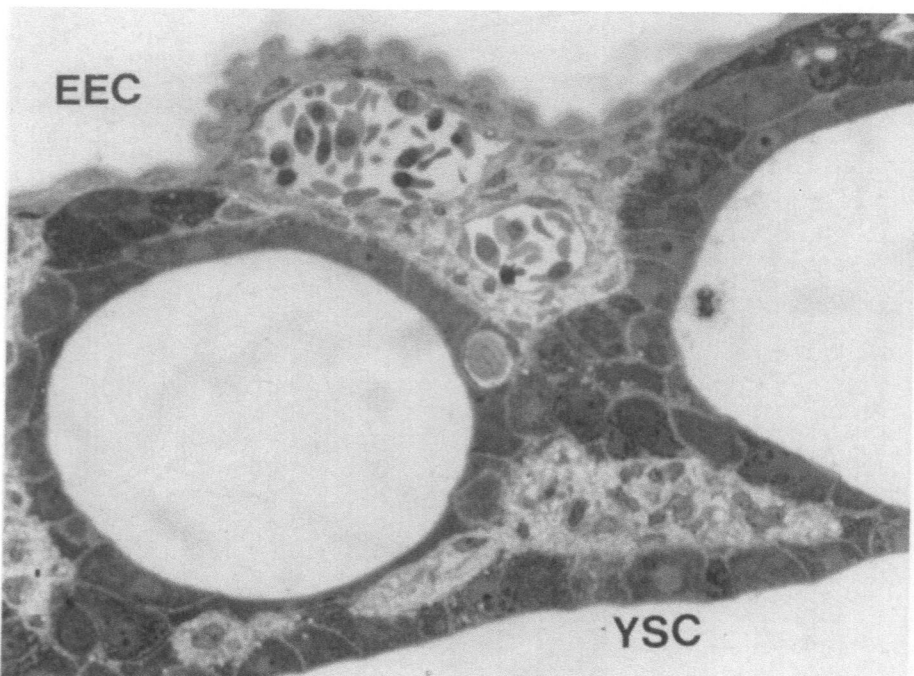

Fig. 6. Semi-thin section through the wall of the yolk sac of Fig. 5 showing the three layers. The endodermal layer faces the yolk sac cavity (*YSC*). It is made up of large cells and contains several cavities of various sizes and shapes. The mesenchymal layer contains the blood vessels. The mesothelial layer covers the yolk sac surface in contact with the extraembryonic coelom (*EEC*) and is made up of one layer of cells with a scalloped outline. Toluidine blue, ×300

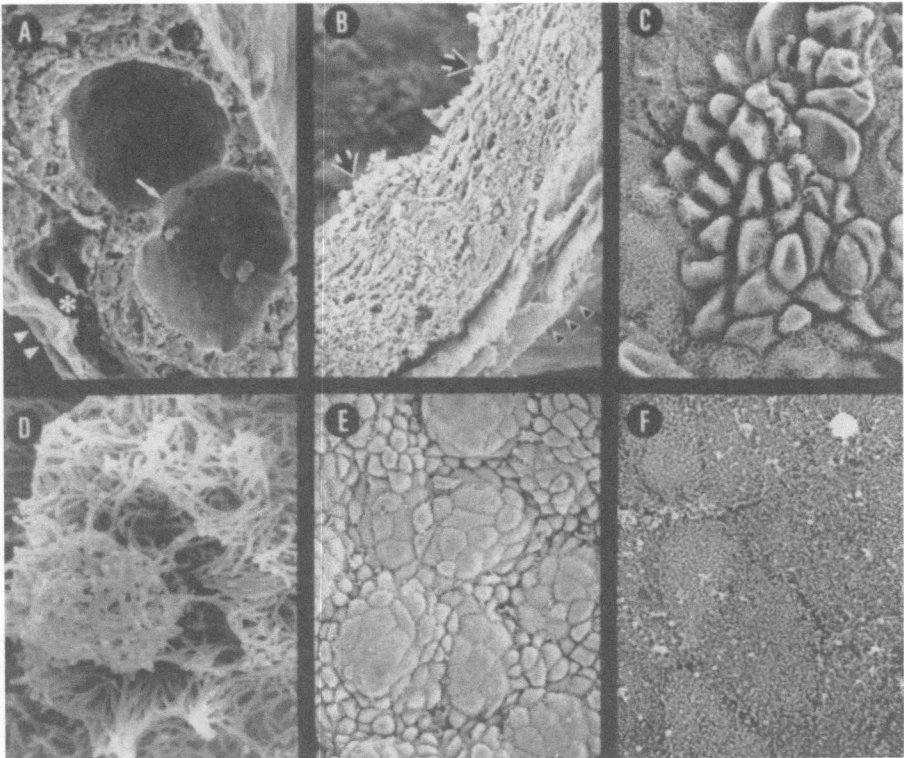

Fig. 7.A Yolk sac wall at 32 days of gestation. Note two blood-free ascinar structures closer to the endodermal inner surface, separated by a shallow ridge (*arrow*); empty spaces within the mesenchyma (*asterisk*); continuous mesothelial layer (*arrowheads*). B Yolk sac wall at 11 weeks of gestation. Note the absence of ascinar structures. The wall appears fibrous from the mesothelial (*arrowheads*) to the endodermal surfaces (*arrows*). C Mesothelial outer surface of the yolk sac at 30 days of gestation. Note the presence of cells with indistinct intercellular bounderies. D Same specimen as C. Note the long and thin microvilli of similar thickness covering the mesothelial surface. E Mesothelial surface of the yolk sac at 9 weeks. The intercellular spaces are better seen due to partial loss of microvilli. F Endothelial surface at 9 weeks. Endodermal cells have numerous short microvilli. SEM, A ×320; B ×180; C ×1300; D ×6000; E ×270; F ×900

Microscopic Features

The wall of the active secondary yolk sac is made up of an endodermal layer facing the yolk sac cavity, a vascular mesenchyme, and an external mesothelial layer (Figs. 6, 7).

The *endodermal layer* is irregular in thickness and is composed of large columnar cells united by tight junctions or occasionally by desmosomes (Fig. 8). The endodermal cells have a diameter of approximately 10–20 μm and

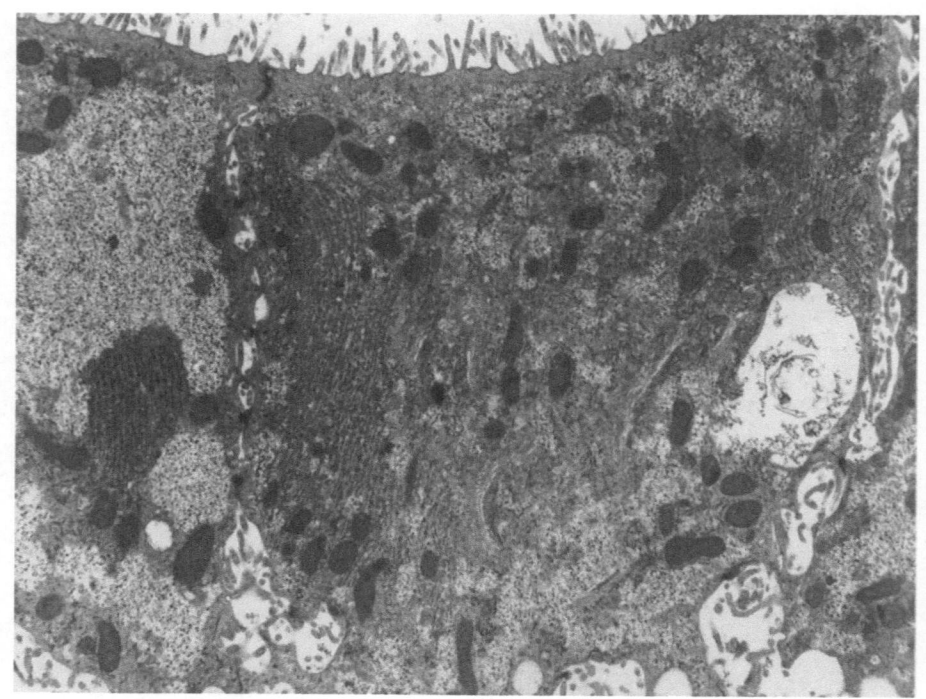

8

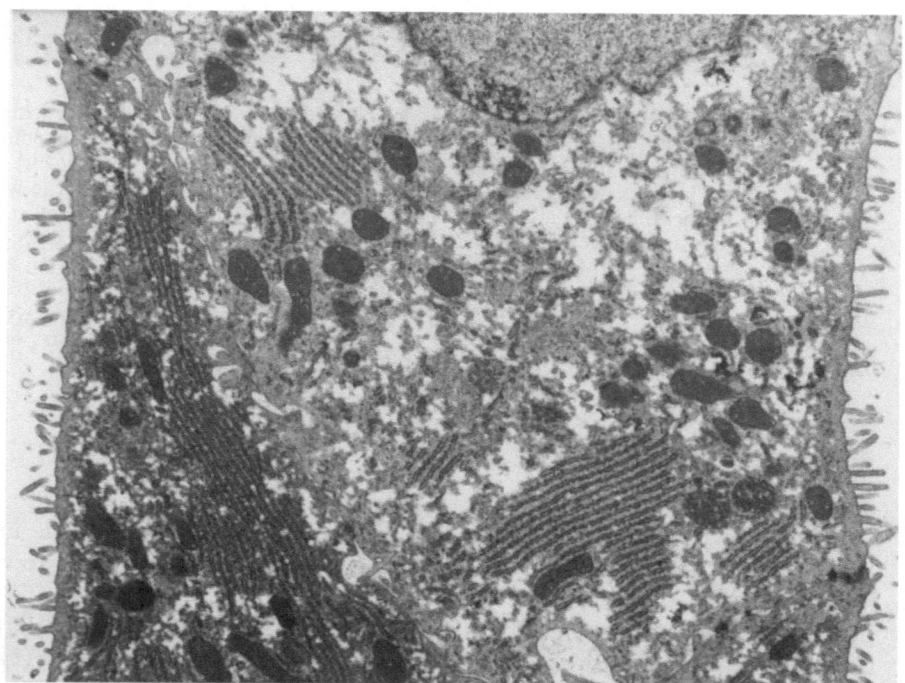

9

are covered by sparse short microvilli, 0.5–1.0 μm in length (Figs. 7–9). Microvilli are also present in intercellular channels (Fig. 8) and at the junction of the endoderm and the mesenchyme where there is a more pronounced folding of the endodermal cell membranes. The cytoplasm of endodermal cells is rich in rough endoplasmic reticulum, Golgi bodies, glycogen granules, mitochondria, and lysosomes (Figs. 8, 9). These features are consistent with a well-developed capacity for active biosynthetic processes. Hoyes [4], in his original description of the ultrastructure of the human yolk sac, reported five different types of endodermal cells, according to the proportion of intracytoplasmic organelles in each. Endodermal cells are directly in contact with mesenchymal cells as there is usually no basement membrane between the two layers. Another important feature of the endodermal layer is the presence of intracellular vacuoles of variable sizes and shapes (Figs. 7, 10), often bounded by a single unit membrane [4]. Tubular spaces presenting small microvilli in their interior and junctional structures can also be found inside the endodermal layer (Fig. 11). They resemble the bile canaliculi of the hepatic parenchyma and probably form a communicating system which opens at the surface of the endodermal layer [4]. Although, in our experience, these tubular spaces do not contain blood cells on light or transmission electron microscopy, Takashina [5] has recently described, by means of scanning electron microscopy (SEM), the presence of large orifices on the endodermal cell surface containing blood cells. He suggests that these orifices correspond to a channel system originating in the mesenchyme and are capable of transporting blood cells directly inside the yolk sac cavity.

The *mesenchymal layer* contains fusiform or stellate cells, free collagen filaments, and sinusoidal blood vessels with endothelial cells often projecting into their lumen (Figs. 6, 12). The endothelium of small vessels is usually continuous and lacks a basal lumina (Fig. 13). Intercellular spaces can be present between adjacent endothelial cells in larger vessels [4]. Masses of endodermal cells (columns) frequently invaginate within the mesenchymal layer, often reaching the mesothelial layer (Fig. 14). Isolated cells which closely resemble endodermal cells and probably correspond to erythroblast precursors can be found within the extravascular mesenchyme [4]. Occasionally, fully developed erythrocytes and macrophages can also be demonstrated outside the vessel lumen [4].

◀ **Fig. 8.** Ultrastructural appearance of the endodermal layer at 8 weeks of gestation. Endodermal cells are united by tight junctions and covered by short microvilli. Microvilli are also found in the interstitial spaces. ×8900

◀ **Fig. 9.** Endodermal cells showing well-developed parallel profiles of rough endoplasmic reticulum and numerous mitochondria. ×9900

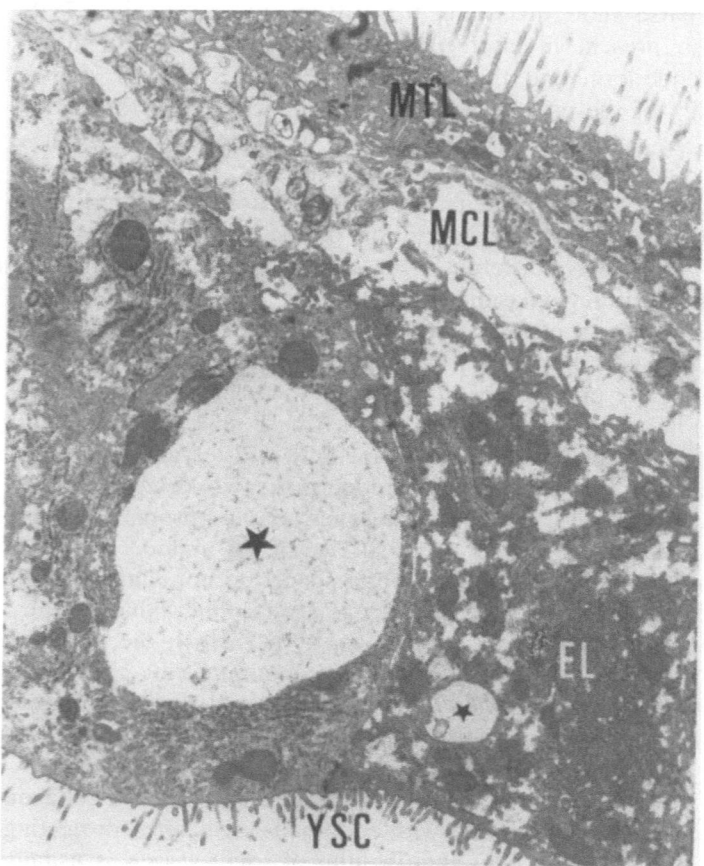

Fig. 10. Section of the wall of the yolk sac showing the endodermal (*EL*), the mesenchymal (*MCL*), and the mesothelial (*MTL*) layers. The endodermal cells face the yolk sac cavity (*YSC*) and contain empty spaces of various sizes and shapes (*stars*). ×8200

The *mesothelial layer* is fairly uniform in thickness, varying between 5 μm and 10 μm, and covers the external or coelomic surface of the yolk sac (Figs. 6, 10). The mesothelial cells are flattened cells, joined laterally by tight junctions and desmosomes, often forming complex interdigitations. They have a scalloped outline and are organized in one continuous layer (Figs. 6, 15). Their brush border is composed of numerous long microvilli, 2.0–5.0 μm in length, and their cytoplasm contains much fewer organelles than the cytoplasm of endodermal cells (Figs. 7, 15, 16). Small uncoated and coated vesicles are frequently found beneath the apical cell membrane (Fig. 17). A rather thin and often indistinct basal lamina can be identified between the mesothelial and the mesenchymal layers (Fig. 15).

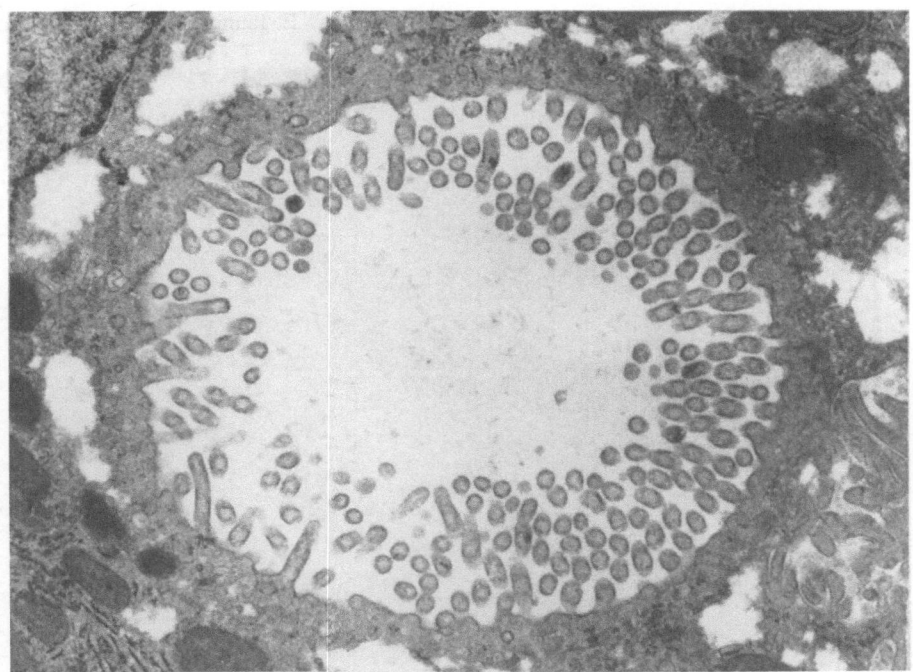

Fig. 11. Yolk sac canaliculi located inside the endodermal layer. ×20 000

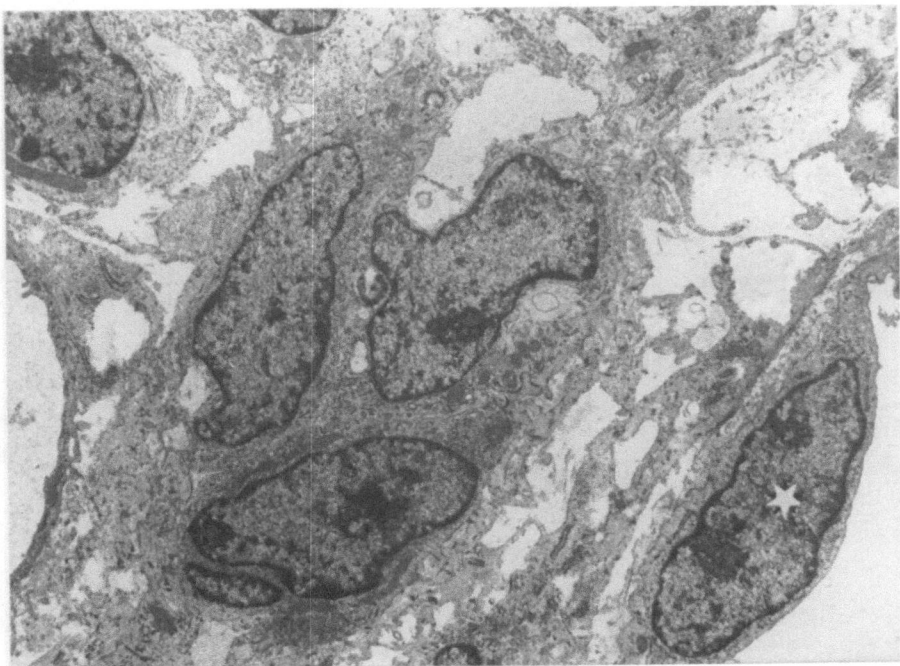

Fig. 12. Fusiform mesenchymal cells and endothelial cell (*star*) projecting into the lumen of a capillary. ×7600

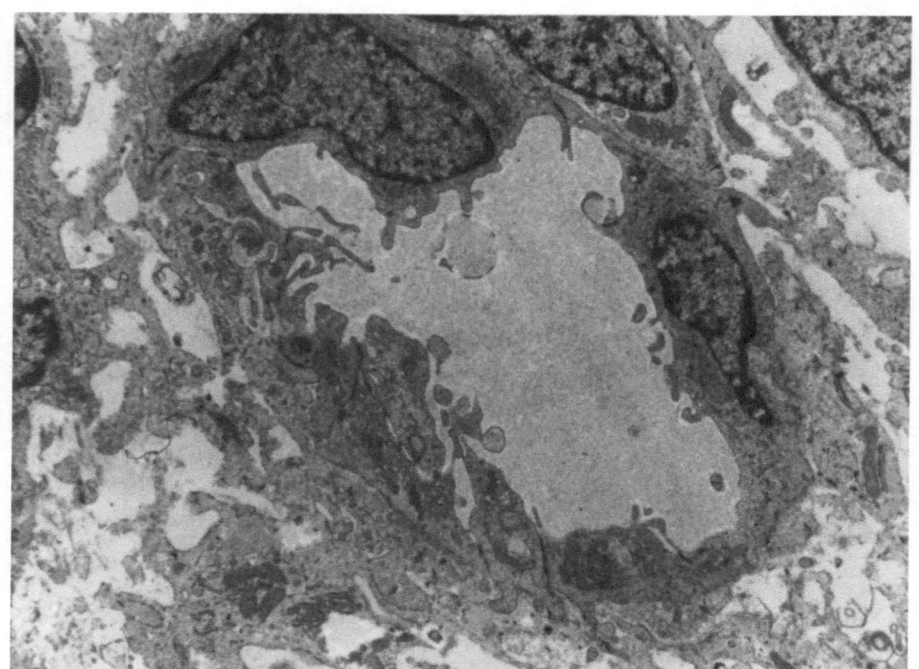

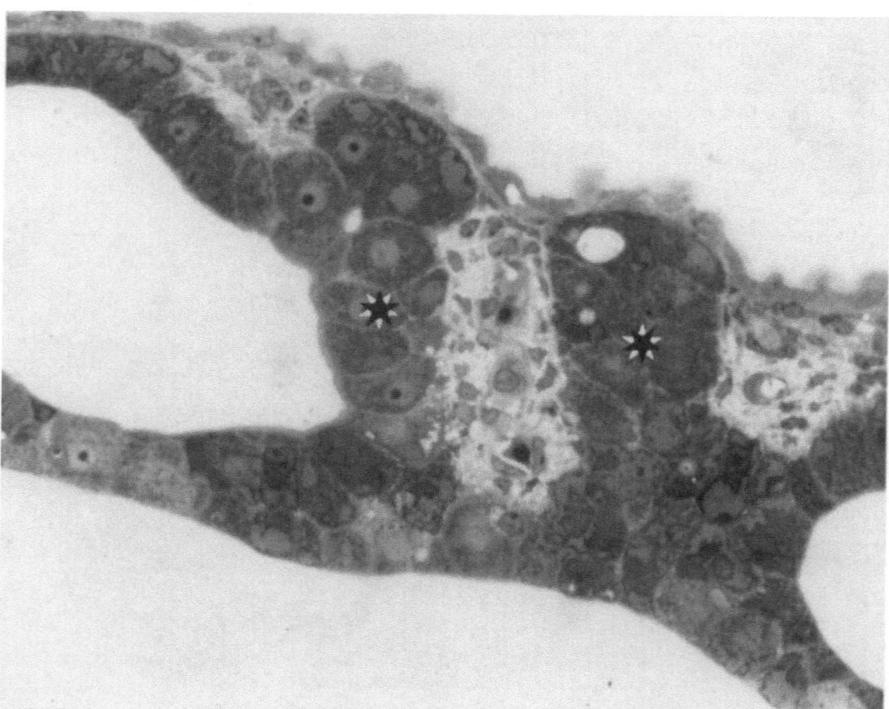

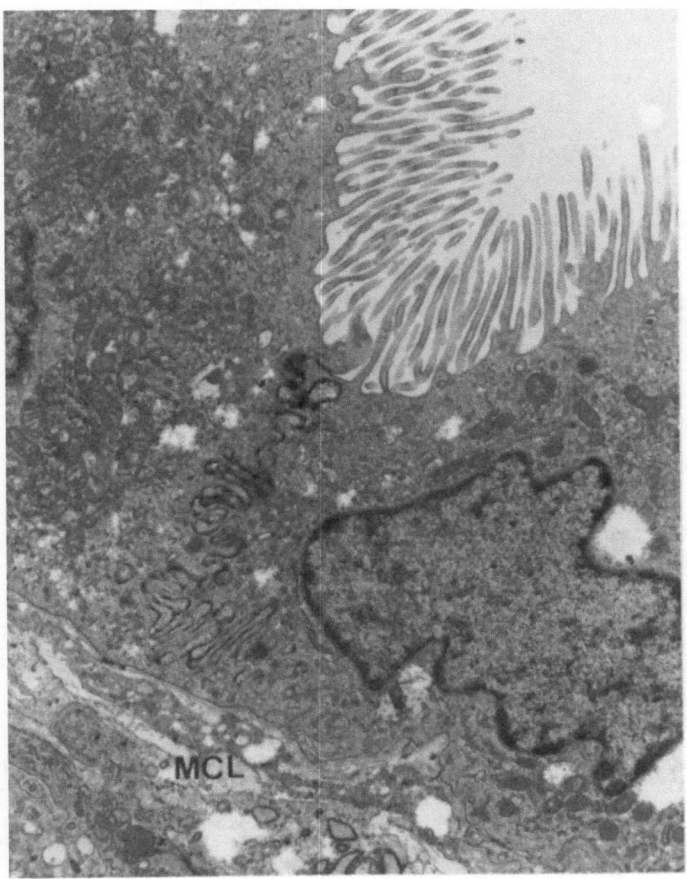

Fig. 15. Neighboring mesothelial cells bound by a sinusoidal junctional device. The basal membrane separating these cells from the mesenchymal layer (*MCL*) is almost phantomatic. ×11500

◀ Fig. 13. Small capillary inside the mesenchymal layer showing an irregular but continuous wall and no basal membrane. ×9200

◀ Fig. 14. Semi-thin section through the wall of the yolk sac at 8 weeks of gestation showing endodermal columns (*stars*) occupying the whole thickness of the wall. Toluidine blue, ×300

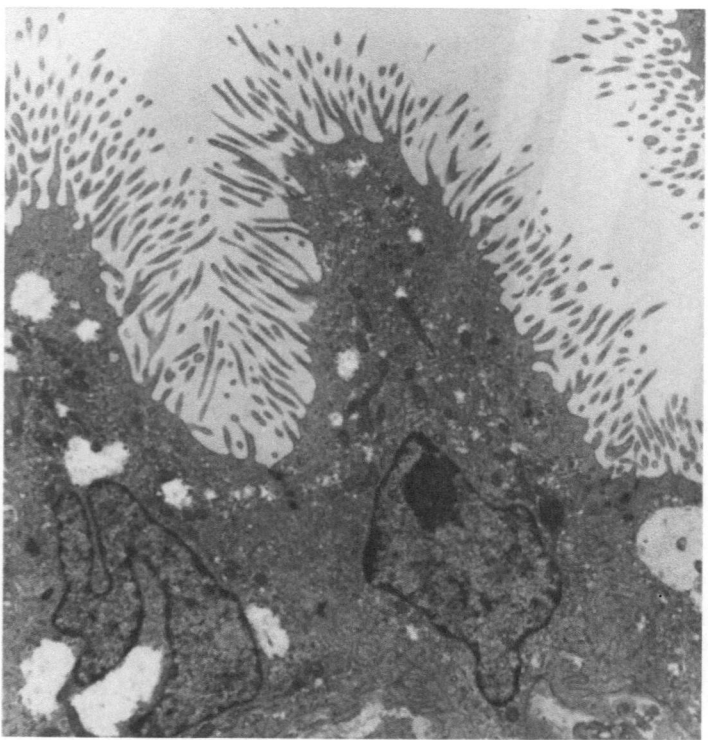

Fig. 16. Mesothelial cells with a scalloped outline showing irregular long microvilli and few organelles. ×8200

The thickness and the microscopic appearance of the different layers of the yolk sac will change according to the area of sampling and with gestational age. At the end of the first trimester, the yolk sac degenerates and major microscopic changes can be observed, mainly in the endodermal and mesothelial layers (Figs. 18, 19). Beyond 9 weeks of gestation, both layers become progressively thinner, and the microvilli of the brush border on both sides get shorter and decrease in number (Figs. 19, 20). The intercellular channels of the endodermal layer gradually disappear, the folding of the endodermal cell membrane gets less pronounced, and a basement membrane appears (Figs. 7, 19). There is also a marked decrease, as pregnancy advances, of the amount of cellular organelles in the cytoplasm of both endodermal and mesothelial cells (Figs. 18, 19). In addition, extremely large vacuoles containing a heterogeneous material can be found in the cytoplasm of these cells, sometimes occupying the whole thickness of their cytoplasm. Few changes are observed in the morphology of the mesenchymal cells, even at the end of the first trimester (Figs. 18–20). However, the amount

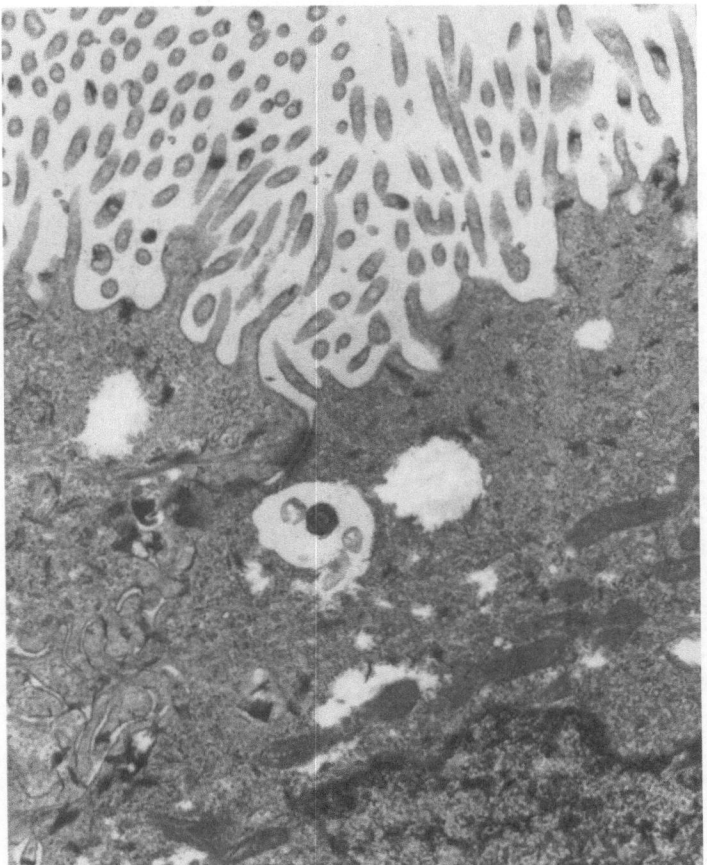

Fig. 17. Small uncoated vesicles beneath the apical mesothelial cell membrane. ×20400

of collagen increases progressively in the extravascular mesenchyme, and precursors of erythroblasts are no longer identified [4]. During the last 2 weeks of the 3rd trimester, mature (anucleated) erythrocytes become the predominant type of cell present in the vessels of the mesenchymal layer and in the circulation of the embryo [6]. All these changes reflect a rapid decline in the functional activity of the yolk sac after 9 weeks of gestation.

In Vivo Investigation of the Anatomy of the Yolk Sac

The primary yolk sac is not detectable by ultrasound, its size being well beyond the resolution limits of all available equipment. Conversely, the

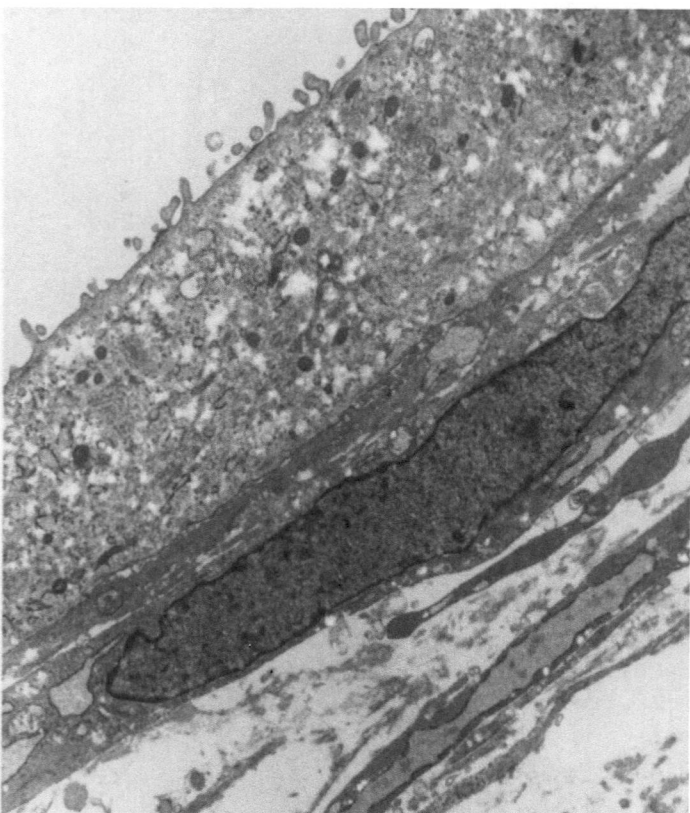

Fig. 18. Ultrastructural appearance of the yolk sac cavity surface at 11 weeks of gestation. The endodermal layer (*EL*) is much thinner than before 10 weeks (compare with Figs. 8, 9), and the microvilli and intracellular organelles have almost vanished. The mesenchymal cells (*star*) show fewer degenerative changes. ×11 400

secondary yolk sac is the first structure to be observed by ultrasound imaging within the gestational sac, before the embryo itself. First reported in vivo by Mantoni and Pedersen [7], the yolk sac appears sonographically as a round, translucent, cystic-like structure, often located near the periphery

Fig. 19. Mesothelial (*MTL*) and mesenchymal (*MCL*) layers at 11 weeks of gestation ▶ showing considerable degenerative changes of the mesothelial cells. ×6600

Fig. 20. Wall of the yolk sac at 10 weeks of gestation. The large cavities observed in early ▶ specimens (see Figs. 6, 10) have disappeared, and only small vessels containing erythrocytes are still present. SEM, ×2500

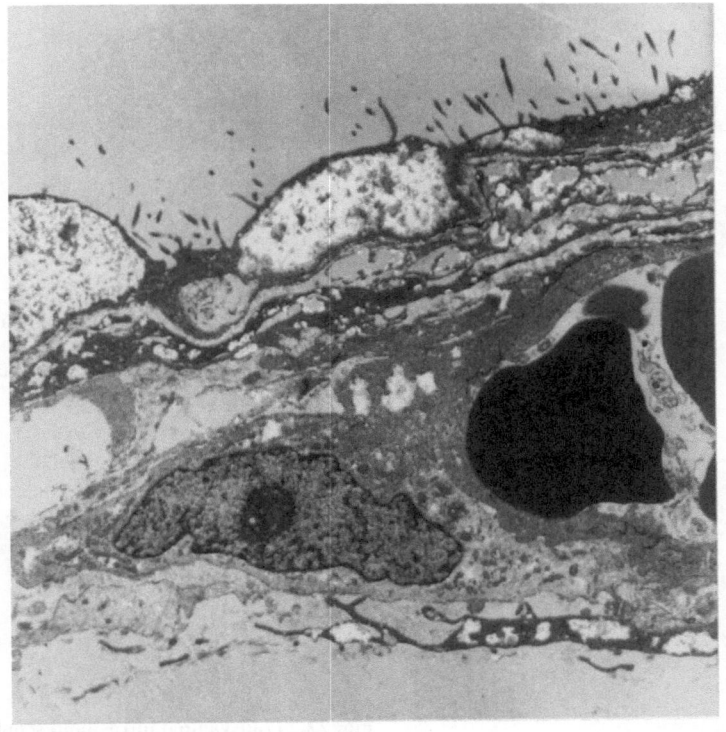

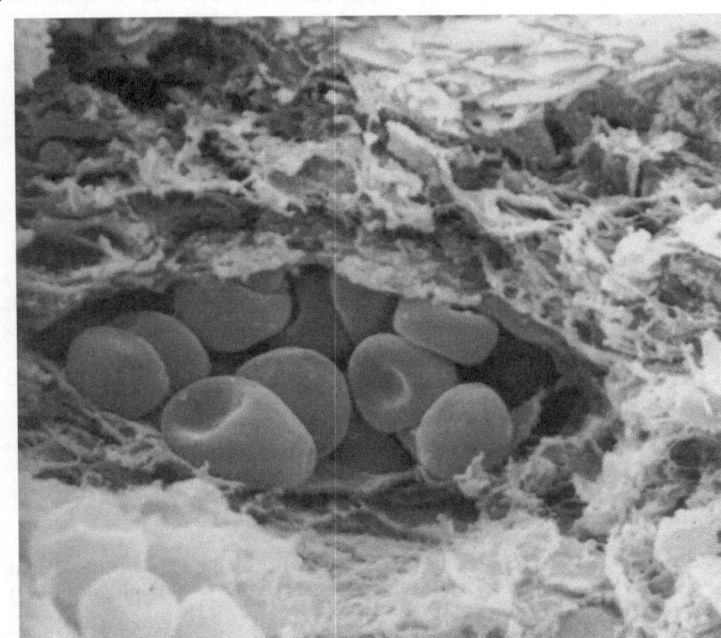

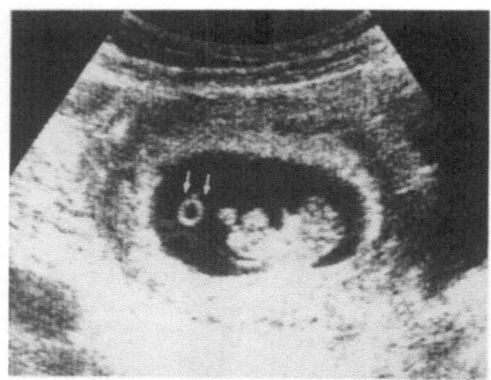

Fig. 21. Transabdominal ultrasound scan at 9 weeks of gestation showing a round and cystic-like structure (*arrows*) corresponding to the yolk sac

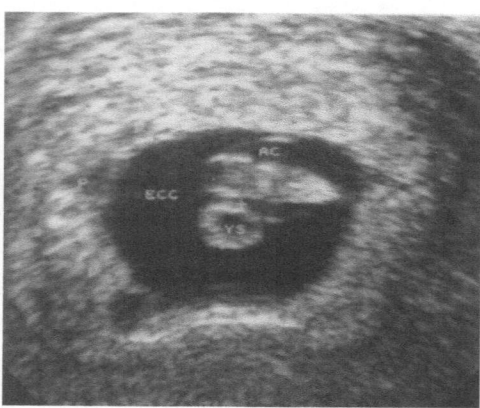

Fig. 22. Transvaginal ultrasound scan of the yolk sac (*YS*), the embryo (*#*), the amniotic cavity (*AC*), the exocoelomic cavity (*ECC*) and the placenta (*P*) at 8 weeks of gestation.

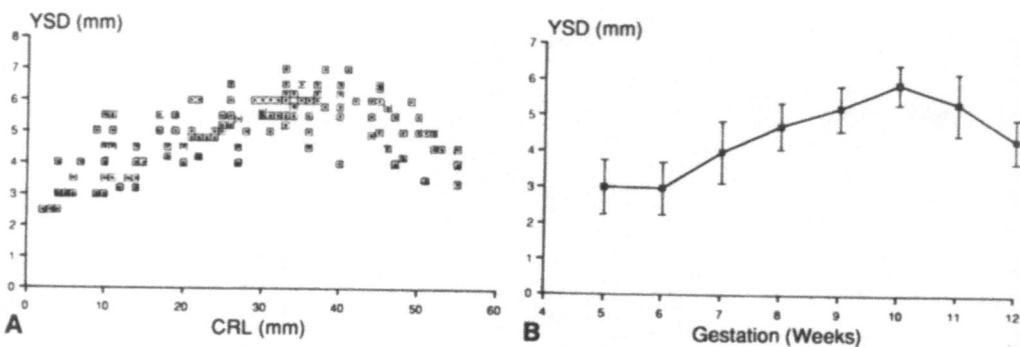

Fig. 23. A Yolk sac diameter (*YSD*) measured by means transvaginal ultrasound and related to the crown-rump length (*CRL*). **B** Mean (±2 SD) yolk sac diameter (*YSD*) related to gestational age (modified from [15])

of the gestational sac (Fig. 21). Using static ultrasound equipment, these authors were able to detect the yolk sac in 69% of normal early pregnancies [7]. Further improvement in sonographic techniques and equipment have increased our ability to visualize the yolk sac in utero [8, 9], even in the absence of a living embryo [10]. High-resolution abdominal ultrasound allows observation of the yolk sac in 80%–100% of cases depending on gestational age [9–13]. The recent use of transvaginal ultrasound [14, 15] has also aided its visualization (Fig. 22).

Demonstration of the yolk sac reliably indicates that an intrauterine fluid collection represents a true gestational sac, thus excluding the possibility of an ectopic pregnancy [13]. The yolk sac can be easily visualized by means of transvaginal ultrasound during the 5th week of gestation when the gestational sac exceeds 8 mm [14]. The discriminatory level of the serum human chorionic gonadotropin (hCG) at which the yolk sac should be systematically seen in utero by means of transvaginal ultrasound is 7200 mIU/ml [14]. Yolk sac measurements in an uneventful pregnancy demonstrate a wide scatter of dimensions when compared to the length of the embryo [8, 9, 11, 12]. However, less variability is observed when ultrasound measurements of the yolk sac are performed using the transvaginal route [15]. The yolk sac diameter increases gradually from 5 to 11 weeks of gestation and then decreases (Fig. 23). Comparison of ultrasound and morphologic findings shows that when the yolk sac reaches its maximal size, it has already started to degenerate [15]. Therefore, it is reasonable to conclude that the disappearance of the yolk sac in normal pregnancies is a spontaneous event in embryonic development rather than the result of a mechanical compression by the expanding amniotic cavity.

Sophisticated formulas have been developed to correlate ultrasound measurements of the yolk sac with gestational age or with crown-rump length [9, 11, 12]. However, the prognostic significance of the yolk sac size is not clearly established. Only a small number of abnormal early pregnancies presenting with yolk sac dimensions well outside the normal range (Fig. 24) or with increased echogenicity of the yolk sac walls have been reported [8, 9]. Most complicated pregnancies, even those with an abnormal karyotype, are associated with yolk sac measurements within normal limits [11, 12, 15]. In most of these cases, changes in yolk sac size and ultrasound appearance are probably the consequence of poor embryonic development or embryonic death rather than being the primary cause of an early pregnancy failure.

Function of the Human Yolk Sac

Although the human yolk sac contains no yolk, it has been demonstrated to have several essential roles during the early stages of embryonic development.

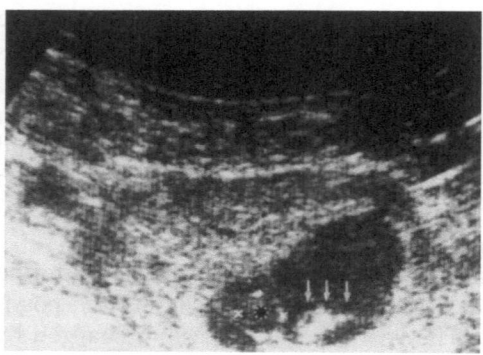

Fig. 24. Transabdominal ultrasound scan at 10 weeks of gestation showing an enlarged yolk sac (*star*) and degenerative embryonic structures (*arrows*)

Role in Embryonic Nutrition and Synthesis

Little is known about the fluid balance and nutrient exchanges between the different compartments of the gestational sac in early human gestation. However, the secondary yolk sac, because of its localization and morphologic features, must have an important role in the process of embryonic nutrition.

The secondary yolk sac floats freely in the extraembryonic coelom, and its walls are in direct continuity with the primitive digestive system throughout embryonic development [1, 2]. The presence of a well-developed microvillous border on the free surface of the mesothelial cells, numerous pinocytic vesicles within their cytoplasm, and the existence of a canalicular system at the cell interface (Figs. 7, 15–17) suggest that these cells have an intense absorptive activity and constitute an important exchange interface with the extraembryonic coelom. Therefore, the yolk sac has been considered as an expansion or satellite of the embryonic gut, especially suited for absorption of inflowing materials [16]. Indeed, the amniotic cavity does not come into contact with the chorion until the end of the first trimester, and all maternal transfers must enter the extraembryonic coelom into which the yolk sac lies. Furthermore, the fact that blood vessels first develop in the secondary yolk sac may probably also be related to the need for transport of nutrients to the embryo [1].

Because of obvious ethical considerations, the nutritional function of the yolk sac has not been directly studied in vivo in the human species. However, the role of the yolk sac in embryonic nutrition has now been well established by animal experimentation [17–19]. Exocoelomic fluid has been successfully aspirated by transvaginal puncture during the first trimester [20] and shows different biochemical characteristics than those of early amniotic fluid and maternal serum (Fig. 25). In particular, highly significant differences in total protein, creatinine and hCG levels were found between the exocoelomic and the amniotic fluids, suggesting that the thin membrane

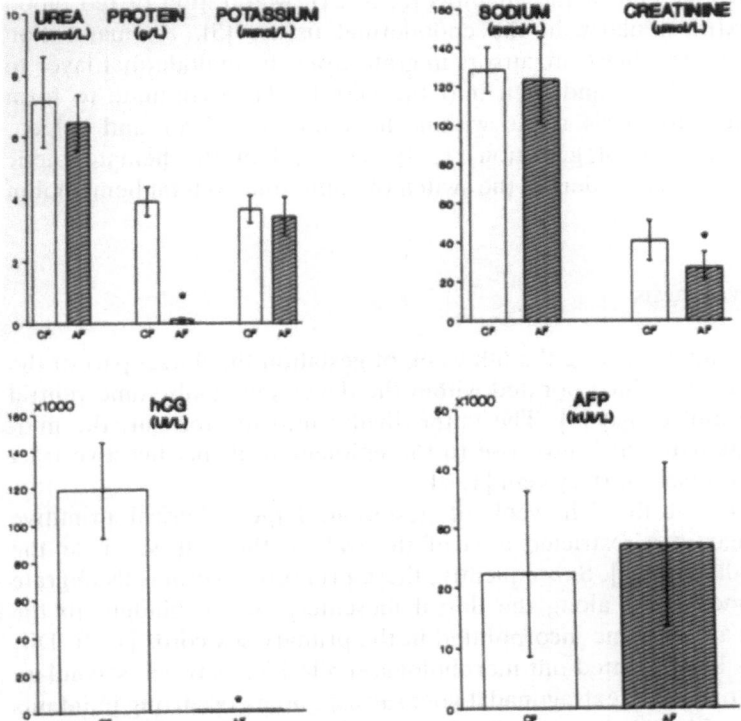

Fig. 25. Comparison of exocoelomic (*CF*) and amniotic (*AF*) fluid compositions in matched samples (*$P < .001$). (From [20])

separating these two compartments, which later becomes the amniotic epithelium, is not permeable to molecules with a high molecular weight. Therefore, most maternal or placental proteins filtered in the extraembryonic coelomic cavity are probably absorbed by the secondary yolk sac [20].

The endodermal layer of the secondary yolk sac shows morphologic similarities with the liver parenchyma (Figs. 8–12), suggestive of an intense biosynthetic activity. It has been demonstrated by Gitlin et al. [21, 22] that both organs are biologically very active and share the synthesis of numerous proteins such as α-fetoprotein (αFP), α1-antitrypsin (α1-AT), albumin, prealbumin, or transferrin. The relatively different degree of protein labeling indicates that the yolk sac is as active as the embryonic liver before 10 weeks of gestation and that its biologic activities disappear after this stage.

Role in Early Hematopoiesis

It is now well established that the first primitive blood cells develop during the 5th week of gestation from the endothelial cells inside the walls of the secondary yolk sac and of the allantois [2, 5, 21]. Maturation of red blood cells starts extravascularly in the endodermal tissues [5]. As maturation proceeds, the erythroblast precursors migrate from the endodermal layer to the mesenchymal layer and then into the vessels. They continue to form there until hematopoiesis starts within the embryonic liver and spleen, around the 8th week of gestation [1, 2]. The end of the hematopoietic activity in the yolk sac underlies the switch of embryonic to fetal hemoglobin [23].

Role in Embryogenesis

As mentioned above, during the 6th week of gestation the dorsal part of the secondary yolk sac is incorporated within the developing embryonic ventral area as the primitive gut [1]. The latter divides into the foregut, the midgut, and the hindgut which give rise to the epithelium of the digestive tract and also of the respiratory system [1, 2].

By the end of the 5th week of gestation, large spherical primitive sex cells appear in a restricted area of the wall of the yolk sac near the origin of the allantois [2]. Subsequently, these primordial germ cells migrate by ameboid movement along the dorsal mesentery of the hindgut to the gonadal ridge and become incorporated in the primary sex cords [1, 2]. Different authors have pointed out morphologic similarities between secondary yolk sac and gonadal or extragonadal yolk sac carcinoma occurring in infants and adults [16, 24, 25]. In particular, the tubules found in the endodermal layer of the normal yolk sac resemble the reticular pattern of yolk sac tumors. Furthermore, immunohistochemical studies have also shown that yolk sac tumors were able to synthesize proteins, such as aFP, which are usually produced during embryonic life by the normal yolk sac [16, 24, 25]. These findings support the hypothesis of an endomesoblastic origin for these tumors.

References

1. Hamilton WJ, Boyd JD, Mossman HW (1972) Human embryology. Heffer, Cambridge
2. Moore KL (1982) The developing human: clinically oriented embryology. Saunders, Philadelphia
3. Jauniaux E, de Munter C, Vanesse M, et al. (1989) Embryonic remnants of the umbilical cord: morphologic and clinical aspects. Hum Pathol 20:458–462
4. Hoyes AD (1969) The human foetal yolk sac: an ultrastructural study of four specimens. Z Zellforsch 99:469–490
5. Takashina T (1989) Hematopoiesis in the human yolk sac. Am J Anat 184:237–244
6. Thompson EL (1951) Time and rate of loss of nuclei by the red blood cells of human embryos. Anat Rec 111:317–325

7. Mantoni F, Pedersen JF (1979) Ultrasound visualization of the human yolk sac. JCU 7:459–461
8. Crooij MJ, Westhuis M, Schoemaker J, et al. (1982) Ultrasonographic measurements of the yolk sac. Br J Obstet Gynaecol 89:931–934
9. Green JJ, Hobbins JC (1988) Abdominal ultrasound examination of the first trimester fetus. Am J Obstet Gynecol 159:165–175
10. Reece EA, Pinter E, Green J, et al. (1987) Significance of isolated yolk sac visualized by ultrasonography. Lancet 1:269
11. Ferrazzi E, Brambati B, Lanzani A, et al. (1988) The yolk sac in early pregnancy failure. Am J Obstet Gynecol 158:137–142
12. Reece EA, Sciosca AL, Pinter E, et al. (1988) Prognostic significance of the human yolk sac assessed by ultrasonography. Am J Obstet Gynecol 159:1191–1194
13. Nyberg DA, Mack LA, Harvey D, et al. (1988) Value of the yolk sac in evaluating early pregnancies. J Ultrasound Med 7:129–135
14. Bree RL, Marn CS (1990) Transvaginal sonography in the first trimester: embryology, anatomy, and hCG correlation. Semin Ultrasound Comput Tomogr Magn Reson 11:12–21
15. Jauniaux E, Jurkovic D, Henriet Y, et al. (1991) Development of the secondary yolk sac: correlation of sonographic and anatomic features. Hum Reprod 6:1160–1166
16. Gonzalez-Crussi F, Roth LM (1976) The human yolk sac and yolk sac carcinoma: an ultrastructural study. Hum Pathol 7:675–691
17. Pinter E, Reece A, Leranth CZ, et al. (1986) Arachidonic acid prevents hyperglycemia-associated yolk sac damage and embryopathy. Am J Obstet Gynecol 155:691–702
18. Brent RL, Beckman BM, Koszalka TR (1990) Experimental yolk sac dysfunction as a model for studying nutritional disturbances in the embryo during early organogenesis. Teratology 41:405–413
19. Jollie WP (1990) Development, morphology, and function of the yolk-sac placenta of laboratory rodents. Teratology 41:361–381
20. Jauniaux E, Jurkovic D, Gulbis B, et al. (1991) Biochemical composition of exocoelomic fluid in early human pregnancy. Obstet Gynecol 78:1124–1128
21. Gitlin D, Perricelli A (1970) Synthesis of serum albumin, prealbumin, α-foetoprotein, α1-antitrypsin and transferrin by the human yolk sac. Nature 228:995–996
22. Gitlin D, Perricelli A, Gitlin GM (1972) Synthesis of α-fetoprotein by the liver, yolk sac, and gastrointestinal tract of the human conceptus. Cancer Res 32:979–982
23. Peschle C, Mavilio F, Care A, et al. (1985) Haemoglobin switching in human embryos: asynchrony of ξ → α and ε → γ-globin switches in primitive and definitive erythropoietic lineage. Nature 313:235–238
24. Nogales FF, Silverberg SG, Bloustein PA, et al. (1977) Yolk sac carcinoma (endodermal sinus tumor). Ultrastructure and histogenesis of gonadal and extragonadal tumors in comparison with normal human yolk sac. Cancer 39:1462–1474
25. Takashina T (1987) Yolk sac tumors of the ovary and the human yolk sac. Am J Obstet Gynecol 156:223–229

Assisted Conception

12 Pregnancy Rate and First-Trimester Outcome Following Ovulation Induction

G. Oelsner, D. Bider, and S. Mashiach

Ovulation induction with various drugs can achieve successful pregnancy in the anovulatory patient. The overall pregnancy rate achieved approaches that found in the normal population. Several epidemiologic reports have shown that the expectancy of establishing a clinically detectable pregnancy during one menstrual cycle among normal healthy women was approximately 25% [1, 2]. Out of these, approximately 15% would abort [3]. Pregnancies which aborted early or preclinically were not considered in the overall pregnancy rate. Leridon [4] reported a clinical pregnancy rate of 42% per cycle of exposure and a 26% clinical abortion rate in a normal population. Higher pregnancy rates in normal conceptional cycles were reported by others [5, 6]. Edmonds et al. [6] found a positive assay of urinary human chorionic gonadotropin (hCG) in 60% of exposed ovulatory cycles, of which 62% aborted with the majority of these losses occurring subclinically.

Human reproduction is astonishingly inefficient. The natural expectancy of pregnancy during any month of exposure depends on many variables such as age, parity, weight, frequency of coitus, and regularity of ovulation. Approximately 25% of patients conceive in the first month of exposure, 55% after 3 months, and 70% after 6 months [7].

The incidence of clinical, spontaneous abortions is approximately 10%–15% of all pregnancies [3, 8]. Similar to the pregnancy rate, the abortion rate depends on multiple variables, such as age, parity, and weight of the woman. The abortion rate among first pregnancies in patients under the age of 30 was reported to be in the range of 8%–11% [9]. In a recent report where pregnancy detection was based on an early β-hCG assay, a total abortion rate of 31% was calculated; 22% ended before clinical detection, and 9% aborted after the pregnancy was detected clinically [10].

Ovulation induction strives to simulate the pregnancy rates achieved by nature, or to improve them, without increasing the abortion rate. The pharmacokinetics and clinical side effects of the drugs used for the induction of ovulation are well known. In this chapter we will summarize the updated knowledge about these drugs and their mode of administration. The pregnancy rate achieved and the pregnancy outcome up to the first 12 weeks of gestation will be presented.

Table 1. Results of GnRH therapy in group 1 and 2 patients[a]

Group	Patients (n)	Route of administration	Pregnancies (n)	(%)
1	166	i.v.	82	49
1	76	s.c.	28	37
2	127	i.v.	53	42
2	28	s.c.	8	29

[a] World Health Organization classification.

Gonadotropin-Releasing Hormone

Following the observations that hypothalamic extracts could trigger the release of gonadotropins, a neurotransmitter called gonadotropin-releasing hormone (GnRH) was isolated [11]. The use of synthetic GnRH for the induction of ovulation has been introduced especially in patients who lack endogenous gonadotropins and who have pituitary glands capable of responding to GnRH. Initially, the efficacy of GnRH for ovulation induction was disappointing, however, some beneficial results were achieved using long-term intermittent infusion [12]. The administration of GnRH via an s.c. catheter appears to be less effective than the i.v. administration through a computerized minipump. The pregnancy rate achieved in group 1 infertility (according to the World Health Organization classification) treated i.v. was 49%, and in group 2 it was 42%, compared to 37% in group 1 and 29% in group 2 treated by s.c. administration [12] (Table 1). The abortion rate was 15%, which is similar to the normal abortion rate in the general population. Multiple gestation rate was 7%. The literature [12–14] and our own experience suggest that this treatment is suitable for patients who have failed to ovulate with clomiphene citrate (CC), human menopausal gonadotropins (hMG), or pure follicle-stimulating hormone (FSH). Table 2 summarizes the outcome of GnRH therapy in patients classified according to their endocrine profile. The overall ovulation rate is high (87%), with an overall pregnancy rate of 25% per cycle. In patients with polycystic ovarian disease (PCOD) or with a luteal phase defect, the treatment with GnRH is less effective. This is apparently due to the increased sensitivity and unpredictable response of the pituitary of such patients to GnRH [13]. The results of pulsatile GnRH therapy in 388 patients over 912 therapy cycles have been published worldwide [12]: 219 patients (56%) conceived; 16 (7%) had multiple pregnancies, and the abortion rate was 14.5%. It is noteworthy that the hyperstimulation rate of 1.1% in these patients is low [12].

In a retrospective multicenter study [15], 223 pregnancies were induced with pulsatile GnRH; the abortion rate in patients with PCOD and those

Table 2. GnRH therapy and various endocrine profiles

Endocrine profiles	Cycles (n)	Ovulatory rate per cycle	Pregnancy rate per cycle
Gonadotropin deficiency (isolated)	12	92	33
Hypothalamic amenorrhea	10	80	30
Anovulation	4	100	25
Luteal phase defect	9	100	11
Polycystic ovaries	7	57	14
Hyperprolactinemia	5	100	40

Table 3. Results of GnRH therapy – world literature

Reference	Route of administration	Patients (n)	Cycles (n)	Ovulatory cycles (n)	Pregnancies per ovulatory cycle (n)	Abortions (n)
Saffan and Seibel [16]	s.c.	10	16	15	4	1
Loucopoulos et al. [17]	s.c.	5	14	0	0	0
Jacobs et al. [18]	s.c.	25	83	80	25	8
Hurley et al. [19]	s.c.	14	36	30	13	2
Molloy et al. [20]	s.c.	14	50	15	2	1
Leyendecker and Wildt [21]	i.v.	33	143	143	38	9
Miller et al. [22]	i.v.	8	23	20	7	3
Berg et al. [23]	i.v.	27	40	32	11	3
Schriock and Jaffe [24]	i.v.	14	24	23	7	2
Molloy et al. [20]	i.v.	9	23	3	1	1
		159	452	361 (80%)	108 (30%)	30 (28%)

with hypothalamic amenorrhea (HA) was similar. Higher multiple pregnancy rates were observed in the HA group compared to PCOD patients and were correlated to the pulse dose, but not to the pulse interval. The first treatment cycle resulted in more multiple pregnancies than did subsequent cycles. The incidence of congenital anomalies was comparable to that with spontaneously conceived pregnancies. Tables 1, 2, and 3 summarize the general experience with GnRH therapy correlated especially to pregnancy rate and outcome [25].

The GnRH mode of therapy has undergone multiple basic and clinical investigations as a tool for induction of ovulation. GnRH therapy is a safe and effective office procedure. Disagreement exists concerning patient selection, optimal doses, rate and route of administration. There is apprehension both from the physicians and the patients concerning the cost and inconvenience of medication pumps. This drawback prevents GnRH from enjoying the widespread acceptance it deserves.

Human Menopausal Gonadotropin and Human Chorionic Gonadotropin

The first pregnancy achieved after combined urinary hMG/hCG therapy was reported by Lunenfeld et al. in 1962 [26]. Since then, thousands of patients have received this therapy with an overall pregnancy rate of 40% per patient [27].

The efficacy of gonadotropin therapy is related to the etiology of the infertility disturbance, the age of the patient, and the mode of gonadotropin administration.

For clinical purposes and for valuable comparison of results we use a simple classification according to which the patients are divided into two main groups: group 1 included women with primary or secondary amenorrhea, low levels of endogenous gonadotropins, and lack of endogenous estrogen activity, who did not bleed following medroxyprogesterone acetate withdrawal (MAP⁻). Group 2 included patients with anovulation associated with a variety of menstrual disorders (including amenorrhea), but with gonadotropin levels within the normal range, and who had evidence of endogenous estrogen activity and bled following MAP withdrawal (MAP⁺). All patients in this group had failed to conceive after CC treatment.

Human menopausal gonadotropin/human chorionic gonadotropin was administered according to the individually adjusted treatment scheme to induce follicular maturation and ovulation [28]. Vast experience was gained in our department in the administration of combined hMG/hCG therapy; 1216 anovulatory women received 5146 treatment cycles with hMG (Pergonal, Serono Laboratories, Inc., Kfar Saba, Israel) and hCG. Pregnancy was achieved in 484 (40%) patients [28]. In a previous study [29] we examined the overall results of gonadotropin administration in accordance with patient classification into groups 1 and 2. Of 510 patients, 184 conceived, with an overall pregnancy rate of 36%. There were 278 pregnancies which resulted in 206 (72%) deliveries (250 neonates) and 72 (28%) abortions.

Of the 510 patients, 192 were classified into group 1. Their pregnancy rate was 60.4%. Among the 318 patients in group 2, the pregnancy rate was 21.4%. In a subsequent report Blankstein et al. [12] found an 82% pregnancy rate in 279 group 1 patients. March [30] summarized a great deal of literature data with hMG/hCG therapy and he found an overall pregnancy rate of 60%.

In order to advise the patients regarding the achievement of pregnancy over several cycles, the reports on cumulative pregnancy rates have assisted us in our positive attitude towards patients' guidance. The cumulative pregnancy rate in group 1 patients after six treatment cycles was 91.2%. In 77 patients in this group, further treatment was given for a second pregnancy after the first gonadotropin conception. Here, the cumulative conception rate was 93.6% after eight treatment cycles. In group 2, the cumulative conception rate was 50% after 12 treatment cycles. In both groups, the results were better in patients who were younger than 35 years of age. The incidence of multiple gestation was 10% [31].

The spontaneous abortion rate in pregnancies obtained by gonadotropin therapy is higher than in spontaneous pregnancies. The abortion rate calculated by Lunenfeld et al. [32] was 21.5%. Oelsner et al. [29] reported an abortion rate of 28%, and March [30] documented an abortion rate of 25%. Ben-Rafael et al. [33] made an interesting observation: there was a significant difference in the abortion rate between the first and second pregnancy achieved by hMG/hCG administration, namely 28.5% in the first pregnancy and 12% in the second pregnancy. This is similar to the rate in spontaneous conceptions.

The relatively high rate of spontaneous abortion following induction of ovulation by hMG/hCG therapy is puzzling and is as yet unexplained. There are several possible causes for this enigma: exogenous stimulation of the ovaries may result in the ovulation of a faulty ovum. The high estrogen levels common in induced ovulations may cause abnormal tubal motility, with accelerated passage of the fertilized ovum and its implantation in a suboptimally prepared endometrium. A higher incidence of corpus luteum insufficiency may also contribute to the elevated abortion rate [9]. In contrast, a prospective study performed in nine Japanese university hospitals compared the abortion rate between pregnancies induced by hMG/hCG and spontaneous pregnancies. The abortion rates were 19.4% and 13.9%, respectively, and were not significantly different [34]. Table 4 summarizes

Table 4. Results of gonadotropin therapy – world literature

Reference	Patients (n)	Pregnant patients[a] (n)	(%)	Abortions (n)	(%)	Multiple pregnancies (n)	(%)
Thompson and Hansen [35]	1286	299	23	51	17	57	19
Butler [36]	269	85	32	23	27	19	22
Ellis and Williamson [37]	77	37	48	4	12	12	32
Tsapoulis et al. [38]	320	163	51	36	22	15	9
Spandoni et al. [39]	62	22	35	3	12	8	36
Blankstein et al. [12]	1107	424	38	106	25	114	27
Bettendorf et al. [40]	756	239	32	53	22	72	30
Brown [41]	228	167	73	36	18	44	26
Schwartz et al. [42]	232	136	59	50	36	41	31
	4337	1572	36	362	23	382	24

[a] First pregnancy only.

several large series on the pregnancy rate and outcome after hmG/hCG therapy. The total number of patients was 4337: the pregnancy and abortion rates were 36% and 23%, respectively. The multiple pregnancy rate was 24%.

A special subgroup of patients in group 2 are those with PCOD. These patients are characterized notably by low levels of FSH, with an inverted luteinizing hormone (LH): FSH ratio.

High ovulation rates can be achieved in women with PCOD by hMG/hCG administration; however, the pregnancy rate is low. The decreased FSH levels in these patients encouraged endocrinologists to prescribe pure FSH in these patients. Nevertheless, patients with PCOD treated with pure FSH have a low pregnancy rate [43]. Although preliminary data suggested that ovulation and pregnancy rates after hMG/hCG and pure FSH were comparable [44, 45], pure FSH appears to result in fewer complications, especially ovarian hyperstimulation syndrome (OHSS) [46]. Table 5 summarizes the outcome of treatment of PCOD patients according to the mode of therapy.

OHSS is the most serious complication of ovulation induction due to hMG/hCG administration. It has been associated with severe morbidity and mortality (thromboembolic phenomena and death) [46]. The overall hyperstimulation rate is about 3%–5% per cycle [12], mild and moderate hyperstimulation accounts for over 4%, and severe hyperstimulation occurs in fewer than 1% of the patients per cycle [29]. Patients with PCOD are especially susceptible to this complication with an incidence of 18%–60% per cycle [48]. Claman and Seibel [25] and Seibel et al. [45] reported the occurrence of mild OHSS in 27% of treatment cycles. In contrast, Venturoli et al. [49], in a series of ten patients, reported no OHSS among patients with PCOD. Nakamura et al. [55] reported 16 patients treated with hMG or FSH, of whom 54.5% and 60%, respectively, developed OHSS.

A high pregnancy rate was observed in cases of OHSS, and hence it was suggested that OHSS may increase the likelihood of conception. Bider et al. [46] reported a higher pregnancy rate with OHSS. The overall pregnancy rate in hMG/hCG-treated patients was 15%–20% per cycle [12], while in the selected group of patients hospitalized with OHSS it was 35% per cycle. Nevertheless, it is yet to be determined whether OHSS increases the likelihood of pregnancy, or whether pregnancy increases the incidence of OHSS due to ovarian exposure to hCG. Though there is an increased rate of conception in cycles complicated by OHSS, there is also, unfortunately, an increased rate of early pregnancy loss.

In our recent study, the abortion rate in a group of 79 patients with grades II and III OHSS who conceived after hMG/hCG therapy was 30%–50%, respectively. Shoham et al. [9] also reported a high rate of abortion in OHSS, namely 35% and 33.3% with mild and severe OHSS, respectively.

GnRH agonist (GnRH-a) is administered for pituitary suppression and ovarian down-regulation before and during gonadotropin therapy. In

Table 5. PCOD patients: results of various modes of therapy

Reference	Type of treatment	Patients (n)	Pregnancy rate (%)	Abortion rate (%)
Kemmann et al. [47]	hMG/hCG	24	58	21
Seibel et al. [48]	hMG/hCG	13	30	_a
Venturoli et al. [49]	hMG/hCG	5	40	50
Coutts et al. [50]	GnRH-a/hMG/hCG	6	90	_a
Fleming and Coutts [51]	GnRH-a/hMG/hCG	40	35	10
Fleming et al. [52]	GnRH-a/hMG/hCG	8	87	14
Flamigni et al. [53]	FSH/hCG	21	38	10
Venturoli et al. [49]	FSH/hCG	5	20	100
Seibel et al. [48]	FSH	10	10	_a
Birkhausen et al. [54]	FSH/hCG	30	56	17
Nakamura et al. [55]	Pulsatile hMG/hCG	11	36	25
Nakamura et al. [55]	Pulsatile FSH/hCG	5	80	0

[a] Not stated.

group 2 patients, a state of hypogonadal hypogonadism ("group 1-like") enables better control of follicular growth. This mode of treatment prevents the phenomena of the premature LH surge and thereby avoids early luteinization.

Patients with a suboptimal response to gonadotropin therapy, "low responders," may improve their ovarian response after GnRH-a administration. Fleming et al. [52] treated eight PCOD patients with GnRH-a and hMG/hCG; seven conceived, after having failed in previous treatment cycles with hMG/hCG alone.

Amenorrheic women with high levels of gonadotropins have a very low chance of achieving pregnancy. Some success has been achieved in the induction of ovulation by hMG/hCG therapy in these patients due to the introduction of gonadotropin suppression by GnRH-a. Check et al. [56] treated 100 women consecutively. The pregnancy rate per cycle was 5%, with 2% viable pregnancies. This mode of therapy may be attempted in these almost hopeless cases.

The incidence of ectopic gestation in women treated with hMG/hCG has been reported to be higher (2.9%) than that in spontaneous pregnancies [57, 58]. This has been solely attributed to the administration of the drug. It has been postulated that the high estrogen levels induced by gonadotropin therapy cause abnormal tubal embryo transport. In addition, the more follicles ripen and rupture, the greater is the likelihood of one egg remaining in the tube [57, 58]. Careful retrospective analysis of 5146 treatment cycles administered by us over the past 25 years [28] failed to support this observation. The overall ectopic rate was 1.44% after hMG/hCG therapy versus 0.98% in spontaneous pregnancies ($P < 0.05$). However, among 291 pregnancies in group 2 patients, an elevated ectopic gestation rate of 2.4% was found ($P < 0.05$). Among 193 pregnancies in group 1 patients, no ectopic

Table 6. Pregnancy rate after CC therapy – world literature

Reference	Patients (n)	Pregnancy rate (n)	(%)
Whitelaw et al. [62]	37	17	46
Rabau et al. [63]	101	37	37
MacGregor et al. [64]	6714	2195	33
MacLeod et al. [65]	118	37	31
Murray et al. [66]	328	82	25
Evans and Townsend [67]	145	81	56
Garcia et al. [68]	159	76	48
Gysler et al. [69]	428	184	43
	8030	2730	34

gestations occurred. This clearly demonstrates that the cause of ectopic pregnancy lies in the patient, and not in the drug. Group 2 patients include those with secondary infertility, tubal factors, and past intrauterine interventions, which are all risk factors for ectopic pregnancy. In contrast, group 1 patients are homogenous, and their only problem is usually anovulation. The previous reports of an increased incidence of ectopic gestation in hMG/hCG therapy apparently pertain to group 2 patients.

It has been postulated that ovulation induction with gonadotropins increases the incidence of chromosomal anomalies and congenital malformations in the conceptus. This has been supported by the classic study of Boué and Boué [59] who reported an 83% incidence of chromosomal anomalies in abortuses after hMG/hCG therapy in comparison to 61% in abortuses from spontaneous pregnancies. Unfortunately, comparison of maternal age, previous obstetric history, and other prognostic factors between these two groups was not addressed in this study. The incidence of chromosomal abnormalities and congenital malformations in the newborn, following pregnancy achieved by hMG/hCG administration, was examined by Hack and Lunenfeld [60]. There was no significant difference between the general population and the study group.

Clomiphene Citrate

The use of clomiphene citrate (CC) in the induction of ovulation was first reported by Greenblatt et al. in 1961 [61]. The property of CC to induce ovulation in anovulatory women was a surprising incidental finding of their research on the contraceptive potential of CC. Since then, CC has apparently helped more infertile women than any other mode of infertility therapy. CC is presumed to induce ovulation in 71% of cycles; however, the pregnancy rate is 34%, of which 17% abort (Tables 6, 7).

Table 7. Abortion rate after CC therapy – world literature

Reference	Patients who conceived (n)	Abortion rate (%)
MacGregor et al. [64]	2195	17
Karow and Payne [70]	140	15
Goldfarb et al. [71]	160	11
Ahlgren et al. [72]	159	10
Garcia et al. [68]	76	25
Kurachi et al. [34]	1034	14

The reported results of CC therapy differ markedly. Table 6 presents several large series. The reported pregnancy rate varies between 25% and 56%. Patients with anovulatory cycles or oligomenorrhea are better responders to CC therapy than those with amenorrhea [73]. The true conception rate per assumed ovulatory cycle simulates that of spontaneous cycles [9, 12, 74]. The cumulative pregnancy rate after three ovulatory cycles is 56% [65]. The day of initiation of CC therapy (the 2nd, 3rd, 4th, or 5th days of the menstrual cycle) does not affect the outcome of therapy.

The abortion rate varies between 11% and 25%, with a 17% average incidence. (Table 7). This is similar to the incidence in spontaneous pregnancies. The authors who reported higher abortion rates in CC pregnancies suggested the following reasons: (a) faulty implantation in the atrophied endometrium in the first treatment cycle; (b) ovulation and fertilization of defective ill-prepared ova; (c) chromosomal abnormalities.

CC has been "accused" of having deleterious effects on fertilization and implantation. These studies were performed in vitro. We have subsequently shown that the CC levels examined were several orders of magnitude higher than those found in vivo. We measured the levels of CC in the serum and the follicular fluid (FF) in the human. The levels ranged between 12 and 65 ng/ml in the serum, and between 8 and 50 ng/ml in FF. These levels were not detrimental to fertilization in our study [75].

The incidence of multiple pregnancy is eight to nine times higher than in spontaneous pregnancies [12]. A survey of the literature confirmed that there was no significant difference in the outcome as compared to spontaneous pregnancies. There is also no increase in the extrauterine pregnancy rate [34].

In patients treated with CC in whom follicular development is documented and yet no ovulation occurs, the exogenous administration of hCG is indicated. The hCG substitutes the midcycle gonadotropin surge which is apparently lacking in these patients. This combination is very popular, however, the pregnancy rate is not high [12, 74]; in such patients, the pregnancy rate obtained was 31.4% [76]. It is probably due to a problem of timing. Human chorionic gonadotropin induces ovulation within 36 h when administered in the presence of adequate mature follicles. However, it may

Table 8. BM therapy in idiopathic hyperprolactinemia – world literature

Reference	Patients (n)	Pregnancies (n)	(%)	Abortion rate (%)
Crosignani et al. [79]	54	26	48	_[a]
Bergh et al. [80]	22	21	95	29
Scrabanek et al. [81]	33	25	76	_[a]
Scrabanek et al. [81]	147	101	69	12
Thorner et al. [82]	54	39	72	_[a]
Turkalj et al. [83]	_[a]	_[a]	_[a]	11

[a] Not stated.

lead to atresia and inhibit ovulation when administrated earlier than the peak of follicular activity. In one report, 14 of 21 (67%) patients conceived after previously failed ovulation with CC alone [77].

The combination with hMG is widely used and, in particular today, has a greater advantage for IVF programs and in patients who are considered "clomiphene failure." The rationale of this combination is to increase the FSH level in its initial follicular phase. A pregnancy rate of 28%–49% was found with this mode of therapy.

Bromocriptine Mesylate

Elevated levels of prolactin are associated with a wide spectrum of menstrual disturbances, ranging from anovulatory cycles to amenorrhea. This is partially due to suppression of the pulsatile secretion of GnRH by the elevated levels of prolactin.

Bromocriptine mesylate (BM) is a dopamine agonist, derived from ergot alkaloids, that suppresses prolactin secretion from the pituitary. The introduction of BM into clinical use in 1969 was a major breakthrough, offering a simple and highly successful mode of therapy to hyperprolactinemic women. In about 75% of cases of galactorrhea and hyperprolactinemia associated with amenorrhea, in the absence of a demonstrable pituitary tumor, BM suppresses galactorrhea and reinitiates normal ovulatory menstrual cycles [78].

Crosignani et al. [79] reported the treatment of 54 hyperprolactinemic women with BM; 26 (48%) conceived. Bergh et al. [80] treated 22 patients who had suffered from hyperprolactinemia and secondary amenorrhoea with BM and noted 21 pregnancies (95%) with six abortions (29%). The pregnancy rate in idiopathic hypoprolactinemia ranges between 48% and 95%, with an abortion rate of 11%–29% (Table 8).

BM crosses the placenta and it has been suggested that it could have a detrimental effect on the development of the early pregnancy. However, analysis of the outcome of a large number of pregnancies exposed to BM failed to reveal an increased incidence of anomalies or of spontaneous abortions [84]. Nevertheless, it is recommended to stop BM during the luteal phase (as documented by a rise in basal body temperature).

Hyperprolactinemia has been observed in 20%–40% of PCOD patients [78]. BM decreases the circulating levels of prolactin and androgens, associated with PCOD, and successfully achieves assumed ovulation in approximately 40% of these patients [85].

The role of BM in normoprolactinemic PCOD patients is controversial. Controlled studies [78] have failed to show that BM alone had a beneficial effect in anovulatory women with PCOD. However, others have reported that some subgroups of patients with PCOD could respond to BM with an increase of up to 40% in the ovulation and pregnancy rate [78, 85].

BM has been administered to normoprolactinemic patients with unexplained infertility. Lenton et al. [86] reported a cumulative pregnancy rate of 63% after almost 1 year of BM therapy in 40 infertile, ovulatory, normoprolactinemic women unable to conceive after CC therapy. Other reports did not succeed in demonstrating the role of BM in idiopathic infertility [87, 88]. An overall pregnancy rate of 36% during 1 year of therapy was documented by McBain and Pepperell [88] using either BM or placebo in unexplained infertility. Wright et al. [87] reported nonbeneficial effects of such therapy after 6 months of BM administration.

In some infertile patients, a midluteal phase elevation of prolactin has been documented and has been blamed as the cause of infertility. The administration of BM in these patients may be beneficial. Ben-David and Schenker [89] studied prolactin levels in a group of 48 patients with unexplained infertility; 45 patients (94%) demonstrated elevation of prolactin in the midluteal phase. When BM was then administered, 18 (40%) conceived within 3 months of therapy. Tipet et al. [90] analyzed 240 ovulatory cycles after hMG/hCG therapy and demonstrated that midluteal phase prolactin elevation is rare and, if it appears, it is usually mild and has no influence on treatment outcome.

The introduction of BM as a therapeutic agent in infertility provides an additional possibility for the improvement of pregnancy rates in infertile couples. BM should be administered to all patients suffering from hyperprolactinema with or without galactorrhea. Patients who fail to achieve assumed ovulation or pregnancy should be treated as any other anovulatory patient. In addition to BM, the classic drugs (CC or hMG/hCG) should be administered for the induction of ovulation. Endocrinologists do not routinely support its use in cases of PCOD, idiopathic infertility, or normoprolactinemic luteal phase defects.

Dexamethasone

A significant amount of data on the performance and functions of gluco-corticosteroids in reproduction have been reported. Today it is accepted that glucocorticoids have central and peripheral actions, and that glucocorticoids in excessive or diminished amounts may interfere with the normal menstrual cycle.

Owing to high androgen levels and a high androgen/estrogen ratio in the FF, infertile patients may not respond adequately to the induction of ovulation by CC or hMG/hCG. The administration of glucocorticoids decreases androgen levels by suppressing their adrenal secretion. Thus, the addition of dexamethasone (DEX) to the various protocols for the induction of ovulation could increase the ovulation and pregnancy rate.

In 1956, Greenblatt et al. [91] used cortisone to treat 37 patients who had menstrual disorders. Pregnancy was achieved in ten (27%) women, with an abortion rate of 30%. Steinberger et al. [92] showed an overall pregnancy rate of 50% after prednisone therapy in 146 infertile couples with high plasma testosterone levels. Evron et al. [93] treated 27 infertile women who had PCOD and who had failed to conceive with CC/hMG/hCG. After the addition of DEX to hMG/hCG, 22 (81%) ovulated, 20 (74%) conceived, and five (25%) aborted.

In some protocols, the addition of glucocorticoids to different regimens of therapy such as hMG/hCG was reported. In infertile patients, a chronic low dose of DEX could depress plasma testosterone and induce ovulation. In PCOD patients who failed to respond to gonadotropins, the addition of DEX was effective [93]. The combination of CC with DEX has been proposed, although the mechanisms underlying the ability of DEX to synergize with CC are unknown. In patients with elevated adrenal androgens, this combination may promote ovulation. Detailed reports on such combination therapy are lacking.

To date there are no conclusive data in the literature on the true value of glucocorticoid therapy or its impact on pregnancy and abortion rate.

Conclusion

Knowledge of the integral requirements for the establishment of pregnancy has been extensively researched. During the last 30 years, remarkable progress has been made and a better understanding of the physiologic processes inherent to ovulation, fertilization, and implantation has been achieved. The introduction of ultrasensitive blood assays, 17-β-estradiol, and progesterone, as well as the application of high-frequency ultrasound scanning enable us to monitor induction of ovulation more accurately.

After careful investigation and patient selection, administration of the appropriate drug or drug combination for the induction of ovulation will

Table 9. Ovulation induction: pregnancy rate and first-trimester outcome

Treatment protocol	Ovulation rate (%)	Pregnancy rate (%)	Abortion rate (%)	Multiple pregnancy (%)	Ectopic pregnancy (%)	Chromosomal abnormalities of abortions (%)
Spontaneous[a]	95	25	15	1	1	60
GnRH	80	24	15	7	1	60
hMG/hCG	85	40	25	25	1–3	60–86
CC	77	40	17	8	0.5	60?
Bm DEX[b]	40	36	15	2.5	1	60

[a] Ovulation and pregnancy rates are reported per cycle. In the other protocols the results are per patient.
[b] See text, p. 228.

achieve pregnancy in most anovulatory infertile patients. In patients in whom assumed ovulation has successfully been achieved, and yet conception has not occurred, the reason for infertility should be redefined as unexplained. These "normally infertile" couples should then be referred to an assisted reproduction program.

Despite the great progress achieved in the treatment of the anovulatory women, the problem of corpus luteum insufficiency remains unsolved. A high incidence of luteal phase defects has been documented in stimulated cycles. This may cause a decrease in the pregnancy rate and an increased rate of early abortion. Therapy with progesterone has usually been effective in lengthening the luteal phase, however, the benefit of this therapy is yet to be proven. Nevertheless, in patients with documented corpus luteum insufficiency with low blood levels of progesterone, supplementation with progesterone or hCG is indicated.

The prospects for anovulatory sterility are promising. A large variety of treatment protocols has been developed and successfully applied to the anovulatory patient. Table 9 summarizes the results of the various protocols available for the induction of ovulation. The pregnancy rate in spontaneous cycles reaches 25% per cycle. The administration of various drugs for the induction of ovulation achieves ovulation in 40%–85% of patients, with a pregnancy rate of 24%–40%. The abortion rate in spontaneous pregnancies is 15%, and in treatment cycles varies between 15% and 25%.

References

1. Henry L (1965) French statistical research in natural fertility. In: Sheps MC, Ridley JC (eds) Public health and population change. University of Pittsburgh Press, Pittsburgh, p 86

2. Muasher SJ, Garcia JE (1986) Pregnancy and its outcome. In Vitro Fertilization, Norfolk, p 238
3. Wentz AC, Cartwright PS (1988) Recurrent and spontaneous abortion. In: Jones HW, Wendz AC, Burnett LS (eds) Novak's textbook of gynecology, 11th edn. Williams and Wilkins, London, p 328
4. Leridon H (1977) Human fertility. The basic components. University of Chicago Press, Chicago
5. Miller JF, Williamson E, Glue J (1980) Fetal loss after implantation. A prospective study. Lancet 2:554
6. Edmonds DK, Lindsay KS, Miller JF, Williamson E, Wood PJ (1982) Early embryonic mortality in women. Fertil Steril 38:447
7. Obel R (1940) Pregnancy rates and coitus rates. Hum Biol 12:545
8. Tietze C (1953) Introduction to the statistics of abortion. In: Engle ET (ed) Pregnancy wastage. Thomas, Springfield, p 135
9. Shoham Z, Zosmer A, Insler V (1991) Early miscarriage and fetal malformations after induction of ovulation (by clomiphene citrate and/or human menotropins), in vitro fertilization, and gamete intrafallopian transfer. Fertil Steril 55:1
10. Wilcox AJ, Weinberg CR, O'Connor JF, Baird DD, Schlatterer JP, Canfield RE, Armstrong EG, Nisula BC (1988) Incidence of early loss of pregnancy. N Engl J Med 319:189
11. Schally AV, Arimura A, Kartin AD (1971) GnRH: one polypeptide regulates secretion of luteinizing and follicular stimulating hormone. Science 173:1036
12. Blankstein J, Mashiach S, Lunenfeld B (1986) Ovulation induction and in vitro fertilization. Year Book Medical Publishers, Chicago, p 103
13. Liv JH, Yen SSC (1984) The use of gonadotropin-releasing hormone for the induction of ovulation. Clin Obstet Gynecol 27:925
14. Blankstein J, Lunenfeld B (1986) Hypothalamic gonadotropin releasing hormone. In: Insler V, Lunenfeld B (eds) Infertility, male and female. Livingstone, New York p 351
15. Braat DD, Aylon D, Blunt SM, Bogohelman D, Coelingh-Bennink HJ, Handelsman DJ, Heineman MJ, Lappohn RE, Lorijn RH, Rolland R (1989) Pregnancy outcome on luteinizing hormone-releasing hormone induced cycles. A multicentre study. Gynecol Endocrinol 3(1):35
16. Saffan D, Seibel MM (1986) Ovulation induction with subcutaneous pulsatile gonadotropin-releasing hormone in various ovulatory disorders. Fertil Steril 45: 475
17. Loucopoulos A, Ferin M, van der Wiele RL, et al. (1984) Pulsatile administration of GnRH for induction of ovulation. Am J Obstet Gynecol 148:895
18. Jacobs HS, Adams J, Franks S, et al. (1984) Induction of ovulation with luteinizing hormone-releasing hormone: problems, indications and contraindications. J Steroid Biochem 20:A36
19. Hurley DM, Brian R, Outch K, et al. (1984) Induction of ovulation and fertility in amenorrheic women by pulsatile low-dose gonadotropin-releasing hormone. N Engl J Med 310:1069
20. Molloy BG, Hancock KW, Glass MR (1985) Ovulation induction in clomiphene nonresponsive patients: the place of pulsatile gonadotropin-releasing hormone in clinical practice. Fertil Steril 43:26
21. Leyendecker G, Wildt L (1984) Induction of ovulation with pulsatile administration of gonadotropin-releasing hormone in hypothalamic amenorrhea. J Steroid Biochem 20:1382
22. Miller DS, Reid RL, Cetel NS, Rebar RW, Yen SSC (1983) Pulsatile administration of low-dose gonadotropin-releasing hormone: ovulation and pregnancy in women with hypothalamic amenorrhea. JAMA 250:2937
23. Berg D, Mickan H, Michael S, et al. (1983) Ovulation and pregnancy after pulsatile administration of gonadotropin-releasing hormone. Arch Gynaekol 233:205
24. Schriock ED, Jaffe RB (1986) Induction of ovulation with gonadotropin-releasing hormone. Obstet Gynecol Surv 41:414

25. Glaman P, Seibel MM (1990) Ovulation induction: GnRH. In: Seibel MM (ed) Infertility, a comprehensive text. Appleton and Lange, Norwalk, p 333
26. Lunenfeld B, Sulimovici S, Rabau E (1962) L'induction de l'ovulation dans les aménorrhées hypophysaires d'un traitement combiné de gonadotrophines urinaires menopausiques et de gonadotropines chorioniques. C R Soc Fr Gynecol 5:287
27. Lunenfeld B, Lunenfeld E (1990) Ovulation induction: hMG. In: Seibel MM (ed) Infertility, a comprehensive text. Appleton and Lange, Norwalk, p 311
28. Oelsner G, Menashe Y, Tur-Kaspa I, Ben-Rafael Z, Blankstein J, Serr DM, Mashiach S (1989) The role of gonadotropins in the etiology of ectopic pregnancy. Fertil Steril 52:514
29. Oelsner G, Serr DM, Mashiach S, Blankstein J, Snyder M, Lunenfeld B (1978) The study of induction of ovulation with menotropine. Analysis of results of 1897 treatment cycles. Fertil Steril 30:538
30. March CM (1984) The use of pergonal for the induction of ovulation. Clin Obstet Gynecol 27:965
31. Dor J, Itzkowic DJ, Mashiach S, Lunenfeld B, Serr DM (1980) Cumulative conception rates following gonadotropin therapy. Am J Obstet Gynecol 136:102
32. Lunenfeld B, Blankstein J, Ron E (1987) Short and long term survey of patients treated with hMG/hCG and follow-up offspring. In: Genazzani AR, Volpe A, Faechinettle (eds) Proceedings of the 1st International Congress on Gynecological Endocrinology. Parthenon, Lancashire, p 459
33. Ben-Rafael Z, Dor J, Mashiach S, Blankstein J, Lunenfeld B, Serr DM (1983) Abortion rate in pregnancies following ovulation induced by human menopausal gonadotropin/human chorionic gonadotropin. Fertil Steril 39:157
34. Kurachi K, Aono T, Minagawa J, Miyake A (1983) Congenital malformations of newborn infants after clomiphene-induced ovulation. Fertil Steril 40:187
35. Thompson C, Hansen LM (1970) Pergonal (menotropins): a summary of clinical experience in the induction of ovulation and pregnancy. Fertil Steril 21:844
36. Butler JK (1970) Oestrone response patterns and clinical results following various pergonal dosage schedules. In: Butler JK (ed) Development in the pharmacology and clinical uses of human gonadotropins. Searle, High Wycombe, p 42
37. Ellis JD, Williamson JC (1975) Factors influencing the pregnancy and complication rates with human menopausal gonadotrophin therapy. Br J Obstet Gynaecol 82:52
38. Tsapoulis AD, Zourlas PA, Comninos AC (1978) Observations on 320 infertile patients treated with human gonadotropins. Fertil Steril 29:492
39. Spadoni LR, Cox DW, Smith DC (1974) Use of human menopausal gonadotropin for the induction of ovulation. Am J Obstet Gynecol 120:988
40. Bettendorf G, Braendle W, Sprotte C, et al. (1981) Overall results of gonadotropin therapy. In: Insler V, Bettendorf G (eds) Advances in diagnosis and treatment of infertility. Elsevier/North-Holland, New York, p 21
41. Brown GB (1986) Gonadotropins. In: Insler V, Lunenfeld B (eds) Infertility: male and female. Churchill Livingstone, New York, p 359
42. Schwartz M, Jewelewicz R, Dyrenfurth I, Tropper P, vande Wiele RL (1980) The use of human menopausal and chorionic gonadotropins for induction of ovulation. Am J Obstet Gynecol 138:801
43. Seibel MM (1990) Ovulation induction: FSH. In: Seibel MM (ed) Infertility, a comprehensive text. Appleton and Lange, Norwalk, p 323
44. Claman P, Seibel MM (1986) Purified human follicle-stimulating hormone for ovulation induction. A critical review. Semin Reprod Endocrinol 4(3):287
45. Seibel MM, McArdle C, Smith D, Taymor MC (1985) Ovulation induction in polycystic ovary syndrome with urinary follicle-stimulating hormone or human menopausal gonadotropins. Fertil Steril 43:703
46. Bider D, Menashe Y, Oelsner G, Serr DM, Mashiach S, Ben-Rafael Z (1989) Ovarian hyperstimulation syndrome due to exogenous gonadotropin administration. Acta Obstet Gynecol Scand 68:511
47. Kemmann E, Tavakioli F, Shelden RM (1981) Induction of ovulation with menotropins in women with polycystic ovary syndrome. Am J Obstet Gynecol 141:58

48. Seibel MM, McArdle C, Smith D, Taymor ML (1985) Ovulation induction in polycystic ovary syndrome with urinary follicle-stimulating hormone or human menopausal gonadotropin. Fertil Steril 43:703
49. Venturoli S, Paradisi R, Fabbri R, Mimmi P, Franceschetti F, Bolelli G, Flamigni C (1984) Comparison between human urinary follicle-stimulating hormone and human menopausal gonadotropin treatment in polycystic ovary. Obstet Gynecol 63:6
50. Coutts JRT, Hamilton MPR, Black WP, et al. (1984) A new successful treatment for infertile women with polycystic ovarian syndrome. Excerpta Med Int Congr Ser 652:608
51. Fleming R, Coutts JRT (1989) The use of exogenous gonadotrophins and GnGH-analogues for ovulation induction in PCO syndrome. Res Clin Form 11(4):77
52. Fleming R, Haxton MJ, Hamilton MPR, McCune GS, Black WP, MacNaughton MC, Coutts JRT (1985) Successful treatment of infertile women with oligomenorrhoea using a combination of an LHRH agonist and exogenous gonadotrophins. Br J Obstet Gynecol 92:369
53. Flamigni C, Venturoli S, Paradisi R, Fabbri R, Porcu E, Magrini O (1985) Use of human urinary follicle-stimulating hormone in infertile women with polycystic ovaries. J Reprod Med 30:184
54. Birkhausen MH, Huber PR, Neuerschwander E, Natslin S (1988) Induction of follicle maturation with "pure" FSH in polycystic ovarian syndrome. Geburtshilfe Frauenheilkd 48:220
55. Nakamura Y, Yosahimura Y, Yamada H, Ubukata Y, Yoshida K, Tomaoka Y, Suzuki M (1989) Clinical experience in the induction of ovulation and pregnancy with pulsatile subcutaneous administration of human menopausal gonadotropins; a low incidence of multiple pregnancy. Fertil Steril 51(3):423
56. Check JH, Nowroozi K, Chare TS, Nazari A, Shapse D, Vaze M (1990) Ovulation induction and pregnancy in 100 consecutive women with hypergonadotropic amenorrhea. Fertil Steril 53(5):811
57. McBain JC, Evans JH, Pepperell RJ, Robinson HP, Smith MA, Brown JB (1980) An unexpectedly high rate of ectopic pregnancy following the induction of ovulation with human pituitary and chorionic gonadotropins. Br J Obstet Gynecol 87:5
58. Gemzell C, Guillame J, Fu Wang C (1982) Ectopic pregnancy following treatment with human gonadotropins. Am J Obstet Gynecol 143:761
59. Boué JY, Boué A (1973) Increased frequency of chromosomal anomalies in abortions after induced ovulation. Lancet 1:679
60. Hack M, Lunenfeld B (1979) The influence of hormone induction of ovulation on the fetus and newborn. Pediatr Adolesc Endocrinol 5:191
61. Greenblatt RB, Barfield WE, Jungck EC (1961) Induction of ovulation with MRL/41. Preliminary report. JAMA 178:101
62. Whitelaw MJ, Kalman CE, Grams LR (1970) The significance of the high ovulation rate versus the low pregnancy rate with clomid. Am J Obstet Gynecol 107:865
63. Rabau E, Serr DM, Mashiach S, et al. (1967) Current concepts in the treatment of anovulation. Br Med J [Clin Res] 4:446
64. MacGregor AH, Johnson JE, Bunde CA (1968) Further clinical experience with clomiphene citrate. Fertil Steril 19:616
65. MacLeod SC, Mitton DM, Parker AS, et al. (1970) Experience with induction of ovulation. Am J Obstet Gynecol 108:814
66. Murray M, Osmond-Clarke F (1971) Pregnancy results following treatment with clomiphene citrate. J Obstet Gynaecol Br Commonw 78:1108
67. Evans J, Townsend L (1976) The induction of ovulation. Am J Obstet Gynecol 125:321
68. Garcia J, Jones GS, Wenz AC (1977) The use of clomiphene citrate. Fertil Steril 28:707
69. Gysler M, March CM, Mishell DR Jr, Bailey EJ (1982) A decade's experience with an individualized clomiphene treatment regimen including the effect on the postcoital test. Fertil Steril 37:161

70. Karow WG, Payne SA (1968) Pregnancy after clomiphene citrate treatment. Fertil Steril 19:351
71. Goldfarb AF, Morales A, Rakoff AE, Protos P (1968) Critical review of 160 clomiphene-related pregnancies. Obstet Gynecol 31:342
72. Ahlgren M, Kallen B, Rannevik G (1976) Outcome of pregnancy after clomiphene therapy. Acta Obstet Gynecol Scand 55:371
73. Adashi EY (1990) Ovulation initiation: clomiphene citrate. In: Seibel MM (ed) Infertility, a comprehensive text. Appleton and Lange, Norwalk, p 303
74. Gorlitsky GA, Kare NG, Speroff L (1978) Ovulation and pregnancy rates with clomiphene citrate. Obstet Gynecol 51:265
75. Oelsner G, Admon D, Dor J, Ben-Rafael Z, Almog S, Dany S, Bassan M, Rudak E, Mashiach S (1989) Quantitation of CC levels measured simultaneously in serum and follicular fluid of women undergoing IVF and ET (Abstr). 6th World Congress on In Vitro Fertilization and Alternate Assisted Reproduction, Jerusalem
76. Taubert HHD, Kuhl H (1986) Steroids and steroid-like compounds. In: Insler V, Lunenfeld B (eds) Infertility, male and female. Churchill Livingstone, New York, p 413
77. O'Herlihy C, Pepperell RJ, Robinson M (1982) Ultrasound timing of HCG administration in clomiphene-stimulated cycles. Obstet Gynecol 59:40
78. Katz E, Adashi EY (1990) Treatment of infertility using bromocriptine mesylate. In: Seibel MM (ed) Infertility, a comprehensive text. Appleton and Lange, Norwalk. p 351
79. Crosignani PG, Ferrari C, Liuzzi A, Benco R, Mattei A, Rampini P, Dellabonzana D, Scardaelli C, Spela B (1982) Treatment of hyperprolactinemic states with different drugs: a study with bromocriptine, metergoline, and lisuride. Fertil Steril 37:61
80. Bergh T, Nillius JS, Wide L (1978) Bromocriptine treatment of 42 hyperprolactinemic women with secondary amenorrhea. Acta Endocrinol (Copenh) 88:435
81. Scrabanek P, McDonald D, Meagher D, et al. (1980) Clinical course and outcome of thirty-five pregnancies in infertile hyperprolactinemic women. Fertil Steril 33:391
82. Thorner MD, Edwards CRW, Charlesworth M, et al. (1979) Pregnancy in patients presenting with hyperprolactinemia. Br Med J 2:221
83. Turkalj I, Braun P, Krupp P (1982) Surveillance of bromocriptine in pregnancy. JAMA 247:1589
84. Weil C (1986) The safety of bromocriptine in long term use: a review of the literature. Curr Med Res Opin 10:25
85. Polson DW, Mason HD, Franks S (1987) Bromocriptine treatment of women with clomiphene-resistant polycystic ovary syndrome. Clin Endocrinol (Oxf) 26:197
86. Lenton EA, Sobowale S, Cooke ID (1977) Prolacin concentration in ovulatory but infertile women; treatment with bromocriptine. Br Med J 2:1179
87. Wright CS, Steale SJ, Jacobs HS (1979) Value of bromocriptine in unexplained primary infertility. A double-blind controlled study. Br Med J 1:1037
88. MacBain JC, Pepperell RJ (1982) Use of bromocriptine in unexplained infertility. Clin Reprod Fertil 1:145
89. Ben-David M, Schenker JG (1983) Transient hyperprolactinemia: a correctable cause of an idiopathic female infertility. J Clin Endocrinol Metab 57:642
90. Tippet PD, Simon JA, Rifka SM, Falk RJ (1989) Luteal phase hyperprolactinemia during ovulation induction with human menopausal gonadotropins. Incidence, recurrence and effect on pregnancy rates. Obstet Gynecol 73:613
91. Greenblatt RR, Barfield WE, Lampros CP (1956) Cortisone in the treatment of infertility. Fertil Steril 7:203
92. Steinberger E, Smith KD, Tchalakian RK, Rodrigues-Rigan LJ (1979) Testosterone levels in female partners of infertile couples. Am J Obstet Gynecol 133:9
93. Evron S, Navot D, Laufer N, Diamont Z (1983) Induction of ovulation in women with polycystic ovarian disease. Fertil Steril 40:183

13 Implantation After Embryo Transfer

M.C. Macnamee and G.M. Hartshorne

Introduction

The widespread clinical application of in vitro fertilization (IVF) and other advanced techniques of assisted conception has emphasized the relative inefficiency of human reproduction. Currently, only 10%–15% of in vitro-fertilized oocytes implant following transfer into a "synchronous" endometrium. This figure is still well below 20%, even when adverse factors associated with superovulation are negated by transfer of frozen-thawed embryos during natural or hormone replacement cycles. Making some fundamental assumptions regarding the frequency of unprotected intercourse and in vivo fertilization rates, it has been estimated that 25%–30% of naturally created embryos implant [1]. So it would seem that, at present, assisted conception techniques can only match nature when three or more embryos are transferred. A failure to achieve pregnancy in both natural and IVF cycles can be attributed to embryonic factors, endometrial factors or a combination of both.

The practice of IVF has afforded us the opportunity to examine in some detail factors which influence the potential of the human embryo [2, 3]. Factors which influence the functionality of the endometrium have proved less easily defined and are a fertile ground for future research. The incompatibility of physically sampling the endometrium and the transfer of embryos in the same cycle necessitates the use of more indirect methods, such as the assessment of endometrial thickness and luminosity on ultrasound scan, the analysis of uterine blood flow by Doppler ultrasound techniques [4], and the biopsy and immunohistochemical characterization of endometrial proteins in "mock" cycles [5].

For implantation to occur, both the embryo and the endometrium must synchronously develop the ability to implant and the ability to permit implantation, respectively. For our purposes, human implantation requires the acquisition of functionality of both embryo and endometrium, their proximation and adhesion, followed by stromal invasion and co-mingling of trophoblastic and decidual cells. What follows is a brief review of factors which may influence implantation in the human with particular emphasis on processes which are amenable to study through the practice of assisted conception.

Oocyte Quality

Follicular selection and growth are exceptionally complex processes. Nevertheless, one factor which above others relates to oocyte quality is luteinizing hormone (LH) levels during the final stages of follicle growth.

An increasing body of evidence links high basal levels of LH in the follicular phase with the failure of pregnancy, implantation and fertilization [6–8]. The degree to which reproductive success is compromised seems to be directly proportional to endogenous LH secretion in stimulation cycles for IVF and in natural cycles in which natural conception is attempted [9].

The most likely cause of this detrimental linkage between basal LH and reproductive outcome is its influence on oocyte maturation. Developing oocytes are normally held in an arrested state until the mid-cycle gonadotrophin surge re-establishes meiosis. High endogenous secretion of LH may partially preempt the midcycle surge, reducing the normal tight hold on oocyte meiotic arrest. The harvesting or induced ovulation of eggs of suboptimal maturity may compromise reproductive success by reducing embryo quality and viability. The widespread use of the gonadotrophin-releasing hormone (GnRH) agonist analogues as adjuncts to superovulation has improved success rates [7, 10], principally by reducing endogenous secretion of LH in the late follicular phase.

Less obvious are the endocrine factors and culture conditions which may affect the zona pellucida. Apart from its role in fertilization [11, 12], the function of the zona pellucida is principally protective [13]. In the mouse and other species, the constituent glycoproteins of the zona pellucida are produced by the oocyte during the whole of its growth phase, production being terminated at ovulation [14]. Unless co-cultured with reproductive tract cells, only 25% of human blastocysts derived from IVF will complete the hatching process [15], suggesting that these embryos are in some way compromised in their ability to hatch. It has also been demonstrated that active thinning of the zona pellucida occurs in a significant proportion of early human embryos [16], and implantation following embryo transfer was enhanced if their "zonae" had "thin spots" [17]. Mechanical "thinning" of embryos derived from an IVF programme has shown to enhance implantation rates significantly, up to as much as 25% per embryo transferred [18].

Fertilization

It is possible that embryonic development may be compromised by events in vitro during the oocyte's culture and fertilization. Fertilization in vivo occurs with only very small numbers of spermatozoa, perhaps 20 at most [19]. Any change from the in vivo situation may influence the resulting embryos. In vitro, the presence of large numbers of spermatozoa and possibly some debris, toxins, non-sperm cells and non-functional spermatozoa which have not been removed by passage through the female reproductive tract influence the biochemical environment of the oocyte, particularly by the release of free radicals and acrosomal enzymes. Such release occurs especially if the preparation methods used are aggressive [20] and causes oxidation, affecting proteins and reducing membrane fluidity by lipid peroxidation. Effects on the function of membrane proteins could possibly affect the sensitivity of receptors in the spermatozoa, oocyte or cumulus cells, and the membrane maturation, sperm/egg fusion and cortical granule extrusion implicit in normal fertilization.

Embryo Growth and Development

IVF may also disadvantage the resulting embryo as communication with the maternal tract is lost. While this seems to be of little importance for fertilization, certain factors in serum are necessary for normal growth of human embryos beyond the cleavage stages [21], although other factors, such as anti-trophoblastic factors in serum of women with repeated spontaneous abortion may cause abnormalities in blastocysts of mice [22]. Furthermore, it is likely that all stages of embryo growth have individual biochemical requirements [23, 24], and that frequent changes in the composition of culture media would be necessary to maintain optimal embryonic growth in vitro.

Before implantation, the embryo must become functionally competent, involving the initiation of cleavage, activation of the embryonic genome, compaction, cavitation and expression of specific factors including receptors, macromolecules and hormones. The early stages of embryo development, perhaps as far as the two- to four-cell stage in humans, seem fairly independent of the surrounding medium provided that certain basic conditions are met. However, activation of the embryonic genome (four- to eight-cell stage) [25, 2] might confer specific requirements upon the human embryo, as has been observed in the mouse in overcoming the two-cell block. Once differences between cells have been established [26], inductive influences further the differentiation of tissue types, including the inner cell mass and the trophectoderm; such induction may be uniquely sensitive to external influences, as known for many years in the newt [27]. Various growth factors induce mesoderm in amphibia [27] and in mice, the addition of epidermal growth factor (EGF) in vitro was recently shown to improve mouse blastocyst development [28] though not supported by Colver et al. [29]. It is likely that other factors assist in the normal expression of receptors on blastomeres and in generating an appropriate blastocoelic microenvironment [3].

Specific cell-cell interactions are essential for normal embryonic growth and implantation, and communication and adhesion molecules are expressed at various stages of development, from the oocyte-granulosa cell cooperation, through sperm-egg fusion and among the cells of the developing embryo where they may confer polarity [30].

Many embryos arrest and degenerate in vitro and others which appear normal are in fact not [25]. The possible rescue of embryos without developmental potential is futile, since these are naturally lost in vivo, either before or soon after implantation [31–33]. Our objective is therefore to obtain and support the development of those with the greatest potential to succeed. This means both improving the environment for growth so that normal embryos are not hampered in their development, and improving our selection procedures so that embryos without the potential for development can be detected and discarded.

Our ability to reliably grow embryos in vitro which are capable of normal development and implantation depends upon our efforts to determine embryonic requirements for nutrients and the embryo/maternal communication during the preimplantation period. We may then be able to devise adequate culture conditions with some grounding in physiology. Such studies are exceptionally difficult in the human, but recent methods of embryonic assessment [34], co-culture of rat blastocysts and polarized endometrial cells [35], and even organ culture of the uterus [36] give hope that such aims are attainable.

Adhesion

Evidence from many animal species and the higher than normal rate of ectopic pregnancy following assisted conception suggest that the healthy embryo will implant almost anywhere given the chance and a copious blood supply. The endometrium is a site in which implantation can be said to occur rarely. The permissive disposition of the endometrium is a prerequisite to adhesion of the hatched blastocyst. The initial apposition of trophoblast and endometrial epithelium may result from the active resorption of luminal fluid. The so-called attachment reaction, the progressive interdigitation of epithelial microvilli and the increasingly close association of the blastocyst membranes and luminal epithelium are closely regulated by oestrogen and progesterone [37], as is the generalized endometrial oedema and increased membrane turnover seen at the time of apposition. The close proximity of the trophoblastic and uterine cells (10–20 nm) allows intercellular communication and the process of adhesion to begin.

Among the factors controlling cell-cell adhesion is galactosyl transferase. This glycoprotein is involved in mammalian fertilization and is also frequently found to be the only plasma membrane-associated glycosyl transferase in embryonic cells [38]. Galactosyl transferase mediates attachment between blastomeres of mouse embryos, and its disruption using specific monoclonal antibodies will even cause dissociation of morulae. Galactosyl transferase activity has not been examined in human embryos; however, the high degree of cell-cell adhesion in human cleavage-stage embryos has been associated with pregnancy resulting from frozen-thawed human embryos [39]. This might reflect the later potential for implantation and is a feature of embryos which could be scored during routine assessment.

Proteolytic activity produced by the embryo may also be involved in implantation by degrading the extracellular matrix and enabling invasion of the trophoblast. Mouse embryos with a mutation which reduces levels of plasminogen activator in conditioned medium were also less competent to invade the endometrium when replaced in vivo [40]; however, the exact role of plasminogen activator in implantation is uncertain [41]. Such non-invasive methods of analysis may be applied in human embryos, and assays of plasminogen activator and other enzymes and secreted products might become routine practice if their predictive value were great enough and accurate enough for clinical use.

Heparin and heparan sulphate proteoglycans are synthesized by peri-implantation mouse embryos and mediate attachment and outgrowth, particularly on substrates of laminin and fibronectin. Outgrowth can be disrupted by the use of soluble heparin or RGD (asg-gly-asp) containing proteins [42]. In mice, sulphated proteoglycans may be present on the embryo surface from the two-cell stage [43].

No non-invasive method of determining the levels of sulphated proteoglycans on human embryos has yet been determined, but the following may also be a potentially useful test for embryo implantational ability. Adhesion may be provoked by changes in the expression of complementary glycoproteins on both embryo and endometrium. While the exact nature of the changes in the uterine glycocalyx that result in adhesiveness remain controversial [37], more recent work has demonstrated the probable importance of extracellular matrix proteins in cell-cell adhesion in general and implantation specifically. Amongst the components of intercellular matrices are the glycoproteins fibronectin and laminin. Fibronectin has two subunits of MW 220 000 and 20 000, linked by a disulphide bond. Of most importance in the present context is the occurrence of RGD sequences which bind to receptors of the integrin family present in cell membranes. The relationship between membrane-bound receptors and cytoskeletal components of the cell has been demonstrated, and this phenomenon has been linked with many processes involving cell migration, including implantation [44]. Recent evidence has demonstrated that binding of fibronectin to its receptor can regulate differentiated gene expression [45]. The possibility exists that stromal cell fibronectin could attach to membrane-bound integrin receptors and initiate adhesion.

Laminin is a large glycoprotein similar to, but chemically and structurally distinct from fibronectin and was first isolated from a mouse tumour producing basement membranes [46].

Recently, it has been demonstrated [47] that human trophoblast binds to laminin in vitro and that decidual cells in vivo and in vitro are surrounded by this glycoprotein. This has led Loke and colleagues [47] to suggest that trophoblast-laminin interaction showing high affinity and some specificity may provide the initial epithelial anchorage and facilitate trophoblast migration through the decidua. Auto-antibodies to laminin have been implicated in the degradation of trophoblastic adhesion to the placental basement membrane in pre-eclampsia [48].

Using immunohistochemistry in our own laboratories, laminin has been localized to the surface of the luminal epithelium in endometrial biopsies taken following oocyte collection in superovulation cycles. The possibility of manipulating the expression of laminin and fibronectin by altering the steroid environment is an exciting area for future experimentation.

Clinical Aspects of Embryo Transfer

In contrast to the situation in animals, clinical and ethical constraints apply to work in humans. Much information has already been gained from the treatment of infertile patients by IVF, but rigorously controlled or invasive studies are frequently impossible. Nevertheless, certain factors of importance may be discerned from the limited data from humans.

Selection of Viable Embryos

Embryonic implantation combines the need (a) for adequate trophectoderm function for hatching, attachment to the luminal epithelium, trophectoderm invasion and placental functioning; and (b) for the totipotency of the inner cell mass, capable of normal fetal development. Some means of selecting embryos with an adequate potential for development is necessary.

Growth in culture, since it currently provides suboptimal conditions, may act to select out embryos capable of withstanding such adversities. Growth to the blastocyst in vitro may be too extreme a selection since the implantation rates of such blastocysts seem to be poor [49]. Nevertheless, results from Mills [50] have shown that culture for 1 extra day (i.e., 3 days instead of 2) may change the decision of which embryos are chosen for replacement based on subjective appearances, leading to significantly higher pregnancy rates.

Trial by freezing and thawing has also been suggested to select the most viable embryos; however, cleaving human embryos with at least half of their cells intact after freezing appear to have an equivalent chance of implantation to those with all cells intact [51]. Also, frozen-thawed human embryos produce similar pregnancy rates per embryo transferred to those transferred fresh [52], suggesting that selection has not occurred, despite some embryos having been lost during freezing and thawing. Similar results were obtained in mice when one or more cells were removed by micromanipulation, although a clearly reducing trend was evident, with none to four cells removed [53].

Other means of selecting those embryos which are most likely to implant include light-microscopic observation during routine clinical treatment; however, the results are highly variable with large numbers of normal-looking, rapidly dividing embryos failing to implant, and those with abnormal features or which are damaged during freezing often implanting. Routine embryology dictates that any method employed must be rapid and reliable, and the results must be available before the embryos are replaced. For these reasons, the most promising method, observing the nutrient uptake of embryos growing in vitro [34], is unlikely to be used clinically.

Embryo Number

As soon as numbers of patients treated by IVF were sufficient, the first attempts to mathematically describe implantation were performed [54]. Clearly, the factor most influencing the likelihood of implantation following IVF is the number of healthy embryos transferred. The second most important factor is "maternal" age, with those of 40 years and over having a greatly reduced pregnancy rate. The simplest mathematical model describing the likelihood of implantation following IVF is a binomial expansion with sample sizes reflecting the numbers of embryos replaced, and each embryo being assigned the same potential for implantation. A modification of this model was made to allow for the patient's ability to allow implantation so that an endometrial factor was entered into the equation. Both of these models shared a considerable degree of homogeneity with recorded data from several centres, but the more simple model failed to explain the higher than expected rate of multiple implantation. The most recent elaboration of these theories is the beta-binomial model exposed by Walters [55].

The conclusion of the beta-binomial model and that proposed by Bouckaert et al. [56] for repeated IVF attempts is that patients do not share a common and constant probability of uterine receptivity.

What is clear is that the likelihood of pregnancy is optimized by the replacement of three or more embryos, and that the replacement of more than three, or exceptionally four, cannot be justified ethically as the proportion of higher-order multiple births rapidly increases.

Luteal Phase Support

The process of implantation and the elaboration of human chorionic gonadotrophin (hCG) in the maternal circulation by the implanting embryo rescues the waning corpus luteum, and the termination of the reproductive cycle is prevented, and endometrial integrity is maintained. The conversion of a monovular species to a polyovular one by the use of superovulation drugs suggests that the normal steriodogenic balance of the multiple corpora lutea may be compromised both physically, by removal of cells during follicular aspiration and by drug-induced biochemical effects [7, 57]. Work from our own and other centres suggests that while absolute levels of oestradiol in the late follicular and early luteal phase are supraphysiological, the most likely parameter influencing implantation is the ratio of progesterone to oestradiol [8]. However, in many attempts to regulate or "normalize" the drug-stimulated luteal phase, the rates of implantation have not been significantly enhanced although some of these are demonstrably successful in terms of plasma concentrations of steroids [8, 58–60]. Very complex strategies of luteal phase support are currently used but with little demonstrable effect.

Conclusion

The usual close synchrony between the developing oocyte and the endometrium throughout the follicular phase may not be maintained following superovulation. However, with IVF we can at least be sure that cleaving embryos have been transferred into a steroid-primed uterus, thus achieving the close apposition of an apparently functional embryo and the endometrium. It is likely that in the mid to late luteal phase, luminal resorption brings the hatched blastocyst into close contact with the luminal epithelium. Once the expanding blastocyst and the luminal epithelial cells are brought into close contact, adhesion may begin. The role of complementary surface glycoproteins and other cell surface indicators remains controversial.

Much current research interest has been shown in the role in implantation of the glycoproteins which make up the extracellular matrix. Most notably, fibronectin and laminin have been considered as possible candidates for primary adhesion molecules.

The ability of the endometrium to react to the implanting embryo is dependent on its prior exposure to oestrogen and progesterone in a species-dependent manner.

No non-destructive model exists to test the implantational ability of embryos. Therefore all embryonic assessments are indirect, relying on individual features or functions of the embryo. The identification of embryonic features which are important in implantation has been complicated by our inability to identify a receptive endometrium, and hence we are dealing at present with a multitude of variables. In the short history of assisted conception, the greatest advances have come from improvements in superovulation strategies and oocyte quality. Non-invasive methods of assessing embryo "fitness" are greatly needed and will surely develop in parallel with endometrial assessments which can predict the fertility of a particular cycle.

The characteristic accumulation of glycogen by the differentiating decidual cells formed from cells in the endometrial stroma and the localized changes in vessel permeability are the earliest responses to deciduogenic stimuli. In some species, including humans, these stimuli may be generated prior to blastocyst hatching.

References

1. Roberts CJ (1975) Where have all the conceptions gone? Lancet i:495–499
2. Braude P, Bolton V, Moore S (1988) Human gene expression first occurs between the four and eight-cell stages of preimplantation development. Nature 332:459–462
3. Hartshorne GM, Edwards RG (1991) Role of embryonic factors in implantation: recent developments. Baillieres Clin Obstet Gynaecol 5(1):133–158
4. Steer CV, Campbell S, Pampiglione JS, Kingsland CR, Mason BA, Collins WP (1990) Transvaginal colour flow imaging of the uterine arteries during the ovarian and menstrual cycles. Hum Reprod 5:391–395
5. Manners CV (1990) Endometrial assessment in a group of infertile women on stimulated cycles for IVF: immunohistochemical findings. Hum Reprod 5:128–132
6. Stanger JD, Yovich JL (1985) Reduced in-vitro fertilization of human oocytes from patients with raised basal luteinizing hormone levels during the follicular phase. Br J Obstet Gynaecol 92:385–393
7. Macnamee MC, Howles CM, Edwards RG, Taylor PJ, Elder KT (1989) Short-term luteinising hormone-releasing hormone agonist treatment, prospective trial of a novel ovarian stimulation regimen for in-vitro fertilization. Fertil Steril 52:264–269
8. Howles CM, Macnamee MC, Edwards RG (1987) Follicular development and early luteal function of conception and non-conceptual cycles after human in-vitro fertilization. Hum Reprod 2:17
9. Regan L, Owen EJ, Jacobs HS (1990) Hypersecretion of luteinizing hormone, infertility and miscarriage. Lancet 336:1141–1144
10. Rutherford AS, Suback-Sharpe RJ, Dawson KJ, Margova RA, Franks S, Winston RML (1988) Improvement of in-vitro fertilization after treatment with Buserelin, an agonist of luteinizing hormone releasing hormone. Br Med J 296:1765
11. Bleil JD, Wasserman PM (1980) Structure and function of the zona pellucida: identification and characterization of the proteins of the mouse oocyte zona pellucida. Dev Biol 76:185–202
12. Yu SF, Wolf DP (1981) Polyspermic mouse eggs can dispose of supernumerary sperm. Dev Biol 82:203–210
13. Bronson RA, McLaren A (1970) Transfer to mouse oviduct of eggs with and without zona pellucida. J Reprod Fertil 22:129–136

14. Wasserman PM (1990) Biochemistry and functions of mouse zona pellucida glycoproteins. In: Edwards RG (ed) Establishing a successful human pregnancy. Raven, New York (Serono symposia publications, vol 66)
15. Lindenberg C, Hyttel P, Sjogren A, Greve T (1989) A comparative study of attachment of human, bovine and mouse blastocysts to uterine epithelial monolayer. Hum Reprod 4:446–456
16. Wright G, Wiker S, Elsner C, Kort H, Massey J, Mitchell D, Toledo A, Cohen J (1990) Observations on the morphology of human zygotes and implications for cryopreservation. Hum Reprod 5:109–115
17. Cohen J, Wiemer KE, Wright G (1988) Prognostic value of morphological characteristics of cryopreserved embryos: a study using videocinematography. Fertil Steril 49:827–834
18. Cohen J, Elsner C, Kort H, Malter H, Massey J, Mayer MP, Wiemer K (1990) Impairment of the hatching process following IVF in the human and improvement of implantation by assisting hatching using micromanipulation. Hum Reprod 5:7–13
19. Corselli J, Talbot P (1987) In vitro penetration of hamster-oocyte cumulus complexes using physiological numbers of sperm. Dev Biol 122:227–242
20. Aitken RJ, Clarkson JS (1988) Significance of reactive oxygen species and antioxidants in defining the efficacy of sperm preparation techniques. J Androl 9:367–376
21. Ashwood-Smith MJ, Hollands P, Edwards RG (1989) The use of Albuminar 5 (TM) as a medium supplement in clinical IVF. Hum Reprod 4:702–705
22. Chavez DJ, McIntyre JA (1984) Sera from women with histories of repeated pregnancy losses cause abnormalities in mouse peri-implantation blastocysts. J Reprod Immunol 6:273–281
23. Gott AL, Hardy K, Winston RML, Leese HJ (1990) Non-invasive measurement of pyruvate and glucose uptake and lactate production by single human preimplantation embryos. Hum Reprod 5:104–108
24. Schini SA, Bavister BD (1988) Two-cell block to development of cultured hamster embryos is caused by phosphate and glucose. Biol Reprod 39:1183–1192
25. Tesarik J, Kopecny V, Plachot M, Mandelbaum J (1988) Early morphological signs of embryonic genome expression in human preimplantation development as revealed by quantitative electron microscopy. Dev Biol 128:15–20
26. Johnson MH, Pickering SJ, Dhiman A, Radcliffe GS, Maro B (1988) Cytocortical organisation during natural and prolonged mitosis of mouse 8-cell blastomeres. Development 102:143–158
27. Gurdon JB, Mohun TJ, Sharpe CR, Taylor MV (1990) Induction, gene activation and embryonic differentiation. In: Edwards RG (ed) Establishing a successful human pregnancy. Raven, New York, pp 155–172 (Serono symposia publications, vol 66)
28. Paria SC, Day SK (1990) Preimplantation embryo development in vitro: cooperative interactions among embryos and role of growth factors. Proc Natl Acad Sci USA 87:4756–4760
29. Colver RM, Howe AM, McDonough PG, Boldt J (1991) Influence of growth factors in defined culture medium on in vitro development of mouse embryos. Fertil Steril 55:194–199
30. Vestweber D, Grossler A, Boller K, Kessler R (1987) Expression and distribution of cell adhesion molecule Uvomorulin in mouse preimplantation embryos. Dev Biol 124:442–447
31. Plachot M, DeGrouchy J, Junca AM, et al. (1987) From oocyte to embryo: a model deduced from in vitro fertilization for natural selection against chromosome abnormalities. Ann Genet (Paris) 30:22–32
32. Buster JE, Bustillo M, Rodi IA, Cohen SW, Hamilton M, Simon JA, Thorneycroft IH, Marshall JR (1985) Biologic and morphologic development of donated human ova recovered by non-surgical uterine lavange. Am J Obstet Gynecol 153:211–217
33. Formigli L, Roccio C, Belotti G, Stangalini A, Coglitone MT, Formigli G (1990) Non-surgical flushing of the uterus for pre-embryo recovery: possible clinical applications. Hum Reprod 5:329–335

34. Gardner DK, Leese HJ (1986) Non-invasive measurement of nutrient uptake by single cultured pre-implantation mouse embryos. Hum Reprod 1:25–27
35. Glasser SR, Julian JA, Decker GL, Tang JY, Carson DD (1988) Development of morphological and functional polarity in primary cultures of immature rat uterine epithelial cells. J Cell Biol 107:2409–2423
36. Bulletti C, Jasunni VM, Tabarelli S, et al. (1988) Early human pregnancy in vitro utilizing an artificially profused uterus. Fertil Steril 49:991–996
37. Weitlauf, HM (1988) Biology of Implantation In: Knobil E, Neill J, et al. (eds) The physiology of reproduction. Raven, New York
38. Hathaway HJ, Sur BD (1988) Novel cell surface receptors during mammalian fertilization and development. Biol Essays 9:153–158
39. Cohen J, Inge K, Suzman M, Wicker SR, Wright G (1989) Videocinematography of fresh and cryopreserved embryos: a retrospective analysis of embryonic morphology and implantation. Fertil Steril 51:820–827
40. Axelrod HR (1985) Altered trophoblast function in implantation-defective mouse embryos. Dev Biol 108:185–190
41. Glass RH, Aggeler J, Spindle A, Pedersen RA, Werb Z (1983) Degradation of extracellular matrix by mouse trophoblast outgrowths. A model for implantation. J Cell Biol 96:1108–1116
42. Farach MC, Tang JP, Decker GL, Carson DP (1987) Heparin/heparan sulphate is involved in attachment and spreadity of mouse embryos in vitro. Dev Biol 123:401–410
43. Dziadek M, Fugiwara S, Paulsson M, Timpl R (1985) Immunological characterization of basement membrane types of heparan sulphate proteoglycan. EMBO J 4:905–912
44. Ruoslahti E, Pierschbacher MD (1987) New perspectives in cell adhesion: RGD and integrins. Science 238:491–497
45. Adams JC, Watt FM (1989) Fibronectin inhibits the terminal differentiation of human keratinocytes. Nature 340:307–309
46. Timpl R, Rohde H, et al. (1979) Laminin – a glycoprotein from basement membranes. J Biol Chem 9933–9937
47. Loke YW, Gardner L, Burland, King A (1989) Laminin in human trophoblast – decidua interaction. Hum Reprod 4:457–463
48. Foidart JM, Hunt J, Lapiere CM, Nusgens B, De Rycker C, Bruwier M, Lambotte R, Bernard A, Mahieu P (1986) Antibodies to laminin in preeclampsia. Kidney Int 29:1050–1057
49. Bolton VN, Hawes SM, Taylor CT, Parsons JH (1989) Development of spare human preimplantation embryos in vitro: an analysis of the correlations among gross morphology, cleavage rates and development to the blastocyst. J In Vitro Fert Embryo Trans 6:30–35
50. Mills C (1991) Factors affecting embryological parameters and embryo selection for IVF/ET. In: Brinsden PR, Rainsbury PA, Yovich J (eds) The Bourn Hall textbook of in vitro fertilization and assisted reproduction. Parthenon Park Ridge (In press)
51. Hartshorne GM, Wick K, Elder K, Dyson H (1990) Effect of cell number at freezing upon survival and viability of cleaving embryos generated from stimulated IVF cycles. Hum Reprod 5:857–861
52. Bhattacharya J, Elder K, Macnamee M, Sathanandan M, Wick KL, Brinsden P (1990) A comparison of implantation rates and multiple pregnancy rates between the transfer of fresh embryos in super-ovulation cycles and frozen-thawed embryos in natural or artificial cycles. Proceedings of the 7th World Congress on Human Reproduction (abstr 157)
53. Van Steirteghem A (1990) IVF and preimplantation diagnosis. Proceedings of the 1st International Symposium on Preimplantation Genetics, Chicago, 14–19 Sept 1990 (abstr)
54. Edwards RG (1985) In vitro fertilization and embryo replacement. NY Acad Sci 442:1–22
55. Walters DE (1989) On the utility of statistical modelling in reproductive studies. Hum Reprod 4:341–345

56. Bouckaert A, Baeten S, de Cooman S, Thomas K, Loumaye E (1989) In-vitro fertilization: how many attempts before success? Hum Reprod 4:261–264
57. Hutchinson-Williams KA, DeCherney AH (1990) Clinical aspects of implantation. In: Edwards RG (ed) Establishing a successful human pregnancy. Raven, New York (Serono symposia publications, vol 66)
58. Leeton J, Trounson A, Jessup D Jr (1989) Support of the luteal phase in in vitro fertilization programmes: results of a controlled trial with intramuscular proluten. J In Vitro Fertil Embryo Transfer 2:166–169
59. Trounson A, Howlett D, Rogers P, Hoppen HO (1986) The effect of progesterone supplementation around the time of oocyte recovery in patients superovulated for in vitro fertilization. Fertil Steril 45:532–535
60. Kuperminc MJ, Lessing JB, Amit A, Yovel I, David MP, Peyser MR (1990) A prospective randomized trial of human chorionic gonadotrophin or dydrogesterone support following in-vitro fertilization and embryo transfer. Hum Reprod 5:271–273

14 Artificial Endometrial Preparation for Implantation

J.S. Younis and N. Laufer

Introduction

The birth of the first baby conceived by in vitro fertilization and embryo transfer (IVF-ET) in 1978 [1] paved the way for the development of a new technology using donated oocytes for the treatment of disorders previously defined as incurable. Indeed, since the first successful attempts to establish and maintain pregnancies in ovarian failure patients [2, 3], oocyte donation (OD) has become an established treatment practiced by many assisted reproduction units around the world. The main indication for OD remains premature ovarian failure (POF) characterized by amenorrhea, high gonadotropins, and low estradiol levels before the age of 35–40 years. This condition is relatively common; about 4% of women suffer from POF at 30 years of age, and as many as 10% of them develop it by the age of 40 [4, 5]. OD is also indicated in patients with certain transmissible genetic diseases or women with a low response to ovarian stimulation. In addition, women with abnormal oocytes that do not fertilize by the husband or donor sperms are candidates for treatment by OD.

Basically, treatment with OD is analogous to that of sperm donation. However, the relative inaccessibility of the oocytes, the need for appropriate endometrial preparation in the recepient, and the difficulty of synchronization between the transferred embryos' age and endometrial maturation in the recipient make this procedure different and more complex.

Several issues are to be considered in the context of OD: donors' selection and anonymity, recipients' evaluation, endometrial preparation, embryonic-endometrial synchronization, mode of oocyte retrieval, route of gamete/embryo transfer, pregnancy maintenance, and ethical and legal considerations. It is the purpose of this chapter to summarize current concepts involved in artificial endometrial preparation and embryonic-endometrial synchronization. The possible reasons for the superiority of an artificial endometrial environment to achieve implantation will be reviewed, and the factors affecting the clinical outcome in such an environment will be discussed.

Endometrial Preparation

The natural ovarian-menstrual cycle is 28 days long, divided about equally into a follicular and a luteal phase. In this cycle, the ovary has two functional roles that are distinct but intertwined. The first is concerned with gametogenesis and the second with steroidogenesis. The dominant follicle provides an oocyte capable of ovulation and fertilization, and concomitantly produces hormones which stimulate the endometrium in a synchronized manner to assure successful implantation. Endometrial stimulation is performed sequentially by estradiol (E_2) in the follicular phase, and by E_2 and progesterone (P_4) in the luteal phase; these are secreted by the dominant follicule and the corpus luteum, respectively. A precise coordination of these events is required to maintain effective reproduction.

The basic assumption in the initial clinical experiments of OD was that, in order to assure optimal endometrial development, normal implantation, and sustained embryonic growth, the artificial preparation of the recipient's endometrium must imitate, as closely as possible, the hormonal milieu of the natural cycle. The first OD attempts employed exogenous estrogen (E) and P_4 in a variable dosage to prepare the endometrium to simulate the natural cycle [2, 3]. Estrogen was administered incrementally in the artificial follicular phase (AFP) with a late follicular peak, a post-ovulatory fall, and a second midluteal rise. Progesterone was initiated on day 15, increased on day 17, and reduced again on day 27 of the recipient's cycle. However, during the cycle in which ET took place, E and P_4 were maintained at midluteal phase levels to avoid the late luteal phase decrease in steroid levels observed in nonconceptual cycles. Figure 1 illustrates our protocol of treatment in the first patients treated by OD.

Recently, Serhal and Craft [6] introduced a simplified approach to artificial endometrial preparation using a fixed dose of E and P_4 replacement therapy. Interestingly enough, this protocol resulted in a high pregnancy rate comparable to that achieved with the classic protocol of endometrial preparation. These clinical results were later confirmed by other workers [7–9]. Moreover, morphologic studies of endometrial specimens prepared by a fixed dose of E and P_4 were shown to be adequately developed [10, 11]. Since the metabolism of such drugs differs from one patient to the other, E_2 and P_4 levels varied tremendously between cycles, and in most cases supraphysiologic levels were achieved. However, this fact did not seem to affect the success rate of the OD program [8, 9]. These morphologic and clinical results clearly demonstrate that endometrial developmental capacity and receptivity are preserved even if endometrial preparation does not mimic the exact hormonal levels of the natural cycle. Because of its simplicity, the fixed protocol may be a preferable choice of treatment for POF patients undergoing OD. In the last 2 years we have used this mode of endometrial preparation successfully in most of the OD cycles [9].

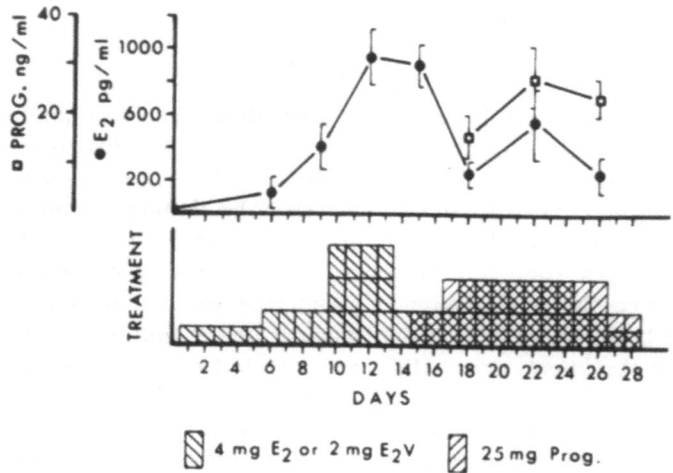

Fig. 1. Protocol of endometrial preparation by exogenous estrogen and progesterone mimicking the hormonal profile of the natural cycle in POF patients. E_2, estradiol; E_2V, estradiol valerate; *Prog.*, *PROG.*, progesterone

Several regimens of E and P_4 protocols have been employed in OD programs. Estrogen has been delivered by the oral route using E_2 valerate, micronized E_2, micronized E_2 and estriol in a 2:1 ratio, or conjugated estrogens. Alternatively, the vaginal route was utilized using polysiloxane-impregnated E_2 rings, or the transdermal route was used employing patches of E_2. Progesterone has been administered orally or vaginally, but most frequently by the intramuscular route using P_4 in oil. Nearly all possible combinations of the above-mentioned therapies have been utilized to induce endometrial maturation in POF patients. No controlled studies have been performed to examine the effectiveness of the various protocols except for one work which demonstrated the efficiency of transdermal E_2 therapy [12]. It seems that all of the regimens employed are capable of producing appropriate serum ovarian steroid levels, adequate endometrial morphologic development, and successful implantation and pregnancy.

Preparatory Cycles

In order to evaluate the effect of various hormonal replacement therapies on endometrial development, preparatory ("mock") cycles are undetaken. In these cycles, endometrial biopsies are performed following endometrial preparation, usually on day 21, the day of assumed implantation. Endometrial dating is carried out in accordance with the criteria of Noyes et al. [13]. The desired goal in the mock cycle is to achieve secretory changes which compare favorably with the expected day of a natural cycle. Our group [3] was the first to show that, in paticnts with POF, an adequately developed endometrium could be produced by administering exogenous E and P_4. Endometrial sampling performed on days 18 and 22 of the cycle was found to correspond to days 17 and 22, respectively (Fig. 2). Electron microscopy confirmed the adequacy of the endometrial response and demonstrated the presence of the nucleolar channel system, a prominent feature of the early secretory endometrium (Fig. 3).

Some investigators recommend the performance of two endometrial biopsies, a midluteal one and a late luteal one, in two preparatory cycles in order to prevent any influence of the first on the second biopsy [14]. Nearly 80 years ago Loeb [15, 16] discovered that decidualization of the endometrium could be induced by different stimuli to uteri of animals that were suitably prepared with ovarian steroid hormones. However, while a traumatic stimulus could cause a decidual reaction in various species of animals, this reaction was not documented in the human. We did not observe any effect of the first biopsy on the morphology of the second in terms of glandular or stromal enhancement of maturation or decidualization [11]. Our routine is to perform two biopsies in the same preparatory cycle with the aid of a disposable curette (Pipelle de Cornier, Prodimed, Neuilly-en-Thelle, France).

Two phenomena specific for a midluteal biopsy following artificial preparation of the endometrium are described: a lag in glandular dating and a disparity between glandular and stromal development [14, 17]. The lag in glandular maturation was also found by our group [11] and could be explained by the lack of an early P_4 rise which starts in the normal cycle on day 14. In artificially prepared cycles the endometrium is exposed to P_4 1 day later, a fact that should be taken into acount when normalizing the dating to natural cycles. Alternatively, it could be that supraphysiologic levels of E_2 in the artificial follicular phase may interfere with the normal process of P_4 receptor development [18], which in turn may require a longer exposure to P_4 to achieving a normal developmental morphologic appearance. The fact that biopsies obtained in the late luteal phase were adequetaly developed [14, 17] may serve as indirect evidence that this later mechanism may indeed be the possible explanation for the midluteal glandular lag in development.

Glandular stromal disparity in the midluteal biopsy was reported by some workers [14, 17], but not confirmed by our work or that of others [3, 10, 11, 19, 20]. It is possible that these discrepancies may stem from different interpretations of similar histologic findings by different pathologists since the criteria for assessing the stromal elements seem to be less stringent and clear than those of the glandular component.

The lag in endometrial development in artificially prepared cycles is relatively small (1–3 days) and does not appear to interfere with the relatively high conception rate of OD programs. It seems that the implantation

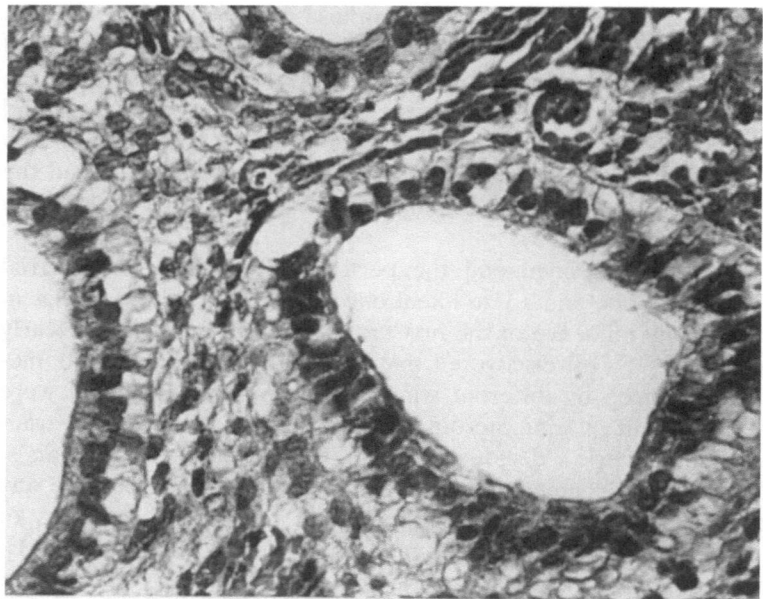

Fig. 2. Tubular glands with subnuclear vacuolization corresponding to day 17 of the cycle. Hematoxylin and eosin; ×375

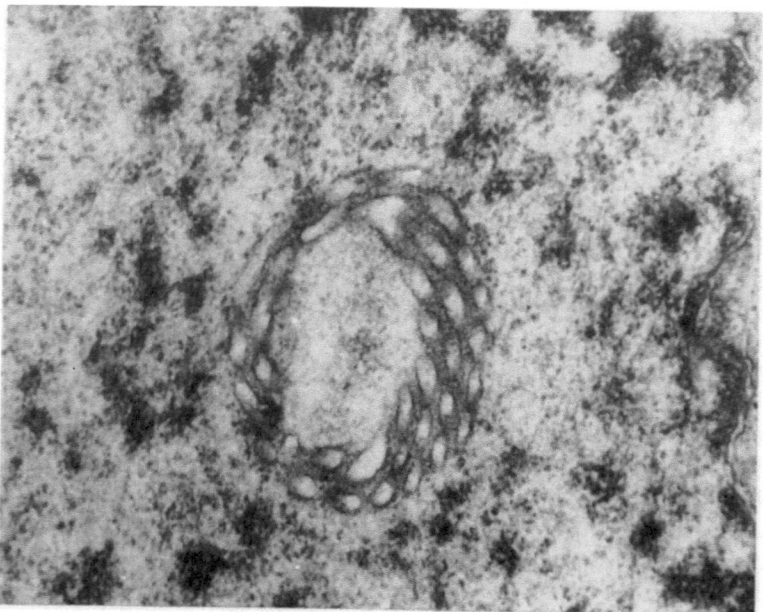

Fig. 3. Electron microscopy showing a prominent nucleolar channel system, a prominent feature of early secretory endometrium

process is tolerant to some lag in endometrial development and reflects another facet of the relatively wide temporal window of endometrial receptivity observed in OD programs. Clinical data are gathering to suggest that the window of implantation is wider than has been initially suggested, and pregnancies have occurred when embryos were transferred on days 16–20 of the cycle [21].

Embryonic-Endometrial Synchronization

In addition to an appropriately prepared endometrium, an adequate synchronization between the age of the embryos transferred and the age of the endometrium is crucial for success in OD. In order to achieve such a synchronization, several methods of treatment had been developed. Lutjen et al. [2] established their first pregnancy by synchronization of cycles between the recipient and the donor, and transfer of embryos to the uterus on day 16. The donor was chosen from a group of patients undergoing routine IVF-ET. Our group used a similar approach employing, however, anonymous volunteers dedicated solely for OD, thus facilitating the act of synchronization between recipient and donor [3]. Exogenous hormonal treatment of the recipients was begun 2–5 days before the expected menses of the prospective donors, so that the embryos could be transferred to the recipients on cycle days 16–20 (Fig. 4). For better matching of donors and recipients, some donors were given oral contraceptives during the cycles preceding treatment. This approach was possible since the effect of the pills on the donors' endometria did not interfere with the procedure.

Currently most programs, including ours, use "spare" oocytes donated by patients undergoing routine IVF-ET. Therefore the precise time of donation in most instances cannot be accurately predicted, a fact that imposed severe logistic limitations in planning the treatment. Consequently, a need to develop other approaches to embryonic-endometrial synchronization was created. One possible solution to this problem, which eliminated completely the temporal interdependence between donor and recipient, is the use of frozen embryos for later thawing and transfer [22, 23]. Another possibility is the simplified method of endometrial preparation proposed by Serhal and Craft [6]. This treatment employs a fixed dose of exogenous hormonal steroids and allows for the manipulation of the duration of the follicular phase in accordance with oocyte availability. These workers reported that their recipients were maintained on a fixed E replacement regimen for 2–4 weeks before the expected donation, thus allowing more flexibility in the timing of transfer. Several investigators, including us, have applied the latter approach successfully in OD programs, eliminating the need for synchronization between donor and recipient or the use of cryopreserved embryos [7–9, 24].

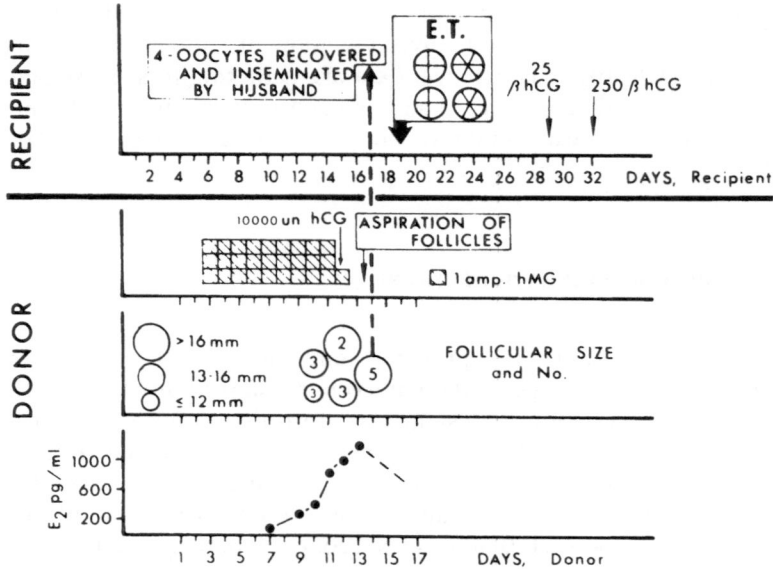

Fig. 4. Synchronization between an ovariectomized recipient and a matched oocyte donor. Singleton pregnancy was established after the transfer of four embryos in the four-to-six cell stage on cycle day 19. Cycle day 1 of the donor corresponds to cycle day 4 of the recipient. *ET*, embryo transfer; *hCG*, human chorionic gonadotropin; *hMG*, human menopausal gonadotropin; E_2, estradiol

The use of gonadotropin-releasing hormone agonist (GnRH-a) to synchronize between the recipient and donor has been recently suggested [25, 26]. This treatment may be given to either a dedicated donor [25] or a recipient with an ovarian function [26] in order to achieve an agonadal state facilitatig hormonal manipulation and synchronization. Our results demonstrate that in oligomenorrheic or normal recipients with normal cycles, treated by the simplified approach, the use of GnRH-a is not always obligatory [9].

The Limits of Follicular Phase Duration

The main theoretical advantage of the simplified approach of embryonic-endometrial synchronization is the ability to stretch or shorten the length of an AFP beyond its normal duration. In the natural ovulatory cycle, a short or a prolonged follicular phase duration is not a common occurrence; only 7% and 5% of these cycles are ≤11 and ≥20 days, respectively [27]. The limits of normal follicular phase duration are not clear. Moreover, it is not known if the duration of the follicular phase could affect endometrial developmental capacity and whether it could affect implantation and pregnancy.

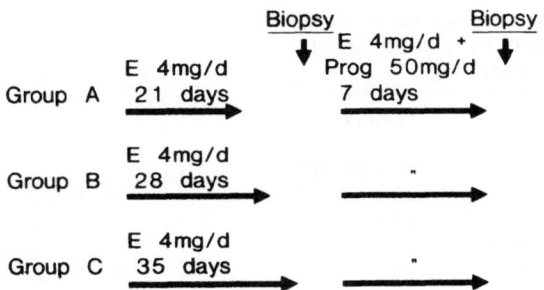

Fig. 5. Protocol of the study – prolonged AFP durations. *E*, oral estradiol and estriol in a 2:1 ratio; *Prog*, intramuscular injection of progesterone in oil

Table 1. Mean serum E_2 and P_4 levels ($\pm$SE) in the follicular and luteal phases in the various groups (groups A, B, and C correspond to AFP durations of 21, 28, and 35 days, respectively)

	Follicular	Luteal	
	E_2		P_4 (ng/ml)
	(pg/ml)		
Group A	1091 ± 147^a	683 ± 174^a	15.9 ± 5.2^b
Group B	762 ± 109^a	794 ± 205^a	14.9 ± 3.6^b
Group C	776 ± 111^a	761 ± 111^a	16.5 ± 5.0^b

[a,b] Values with the same letters are not significantly different.

We studied the effect of a prolonged follicular phase duration on endometrial development in 18 women suffering from POF with a mean age of 34.9 ± 5.5 years. An AFP was initially created in all women by oral micronized estradiol and estriol (in a ratio of 2:1) (Estrofem, Novo Industri A/S, Copenhagen, Denmark), two tablets per day. The patients were randomly and prospectively divided into three treatment groups of different AFP durations lasting 21–35 days, as shown in Fig. 5. At the end of this prolonged AFP, intramuscular injections of P_4 in oil 50 mg/day (Gestone, Paines & Byrne Limited, Greenford, England) were started in all patients, combined with the same dosage of E for another 7 days. Two endometrial biopsies were performed in each cycle: a late follicular one and a midluteal one.

None of the women in this study had any breakthrough bleeding, including those treated by unopposed E for 35 days. Mean serum E_2 and P_4 levels were higher than those achieved in natural cycles and similar in all three groups (Table 1). Women in all three groups demonstrated a normal proliferative endometrium with no signs of glandular cystic hyperplasia. The results of the second biopsy are shown in Table 2. Group B seemed to have a more appropriately dated endometrium than either group A or C. However, using Fisher's exact test, the number of women who had a >2

Table 2. Dating (mean ± SD) of the midluteal biopsy in the various groups (groups A, B, and C correspond to AFP durations of 21, 28, and 35 days, respectively)

	Glands	Stroma
Group A	18.6 ± 1.8^a	$18.7 \pm 1.8^{a'}$
Group B	21.8 ± 0.8^b	$22.2 \pm 0.4^{b'}$
Group C	18.6 ± 1.5^c	$19.4 \pm 2.5^{c'}$

[a] versus [b], [b] versus [c]: $P < 0.005$.
[a] versus [c]: not significant.
[a] versus [a'], [b] versus [b'], [c] versus [c']: not significant.

Table 3. Indications for oocyte donation

	Patients (n)	Cycles (n)	Pregnancy (n)
Turner's mosaic	4	5 (4)	2
Swyer's syndrome	1	2	0
1° IPOF	6	11 (10)	4
2° IPOF	14	23 (22)	7
Surgical castration	2	2	0
Genetic disease	2	5	1
Failed IVF[a]	2	6	1
	31	54 (51)	15

[a] Owing to abnormal oocytes as indicated by previous recurrent fertilization failures by both husband and donor sperm.
IPOF, idiopathic premature ovarian failure. Figures in parentheses are numbers of ET cycles.

days difference did not differ among the groups. These results demonstrate that an artificially prolonged follicular phase of up to 35 days results in an adequately prepared secretory endometrium. Moreover, it has been shown that prolonged E stimulation had no adverse effects on endometrial developmental capacity and that the length of E priming did not affect endometrial response to P_4.

Our findings corroborate a similar preliminary observation by Navot et al. [17], who found that prolonged E stimulation (up to 35 days) did not adversely affect endometrial maturation. Conversely, the same group found that a very short exposure (as short as 6 days) to E will also allow normal development of the endometrium with the addition of P_4.

In order to examine the effect of the duration of an artificial E endometrial preparation on the pregnancy rate in an OD program, we recently evaluated 31 women that were treated in our OD program using the simplified approach over 54 cycles. The age at admittance ranged from 20 to 41 years (mean 33.5 ± 4.7 years). The indications for OD are listed in Table 3;

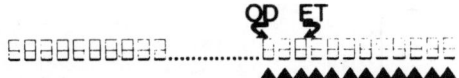

Fig. 6. The simplified approach of oocyte donation in the Hadassah IVF program. *OD*, oocyte donation; *ET*, embryo transfer; *squares*, micronized estradiol 2 mg; *triangles*, progesterone 50 mg

Table 4. Recipients and donors characteristics

Variable	Range	Mean ± SD
Recipients		
Age (years)	20–41	33.5 ± 5.0
Oocytes (n)	2–5	3.2 ± 0.9
Embryos (n)	1–4	2.3 ± 0.9
AFP duration (days)	4–29	15.7 ± 6.2
E_2 FP (pg/ml)	186–1954	748 ± 357
E_2 LP (pg/ml)	309–1314	692 ± 227
P_4 LP (ng/ml)	10–136	38.7 ± 18.7
ET (day)	2–3	2.4 ± 0.5
Donors		
Age (years)	23–40	30.1 ± 3.9
Oocytes (n)	10–31	20.5 ± 4.3
E_2 on hCG day (pg/ml)	873–3894	2092 ± 803

FP, follicular phase; LP, luteal phase.

22 women were amenorrheic, five oligomenorrheic, and four had normal cycles. Our protocol is shown in Fig. 6. The fixed E replacement therapy was continued from 5 and up to 35 days according to oocyte availability. Simultaneously, on the day of donation, an intramuscular injection of P_4 in oil 50–100 mg/day was added to the regimen. The donated oocytes were fertilized with the husband's sperm and the embryos transferred to the uterus 48 or 72 h after the donation.

Oocytes were donated solely by young patients undergoing routine IVF-ET in our unit in an anonymous manner. Only women with "spare" oocytes (usually above 15) were asked to donate; 46 women with a mean age of 30.7 ± 4.0 years donated all the oocytes to the 31 women in the study. A superovulation protocol in all donors employed the midluteal GnRH-a with high-dose human menopausal gonadotropin.

Among 54 cycles studied, three were exluded owing to lack of fertilization of the donated oocytes. The number of oocytes donated for each cycle ranged from two to five (mean 3.2 ± 0.9), and the number of embryos transferred per cycle ranged from one to four (mean 2.3 ± 0.9) (Table 4).

Fifteen clinical and four chemical pregnancies were achieved. Only clinical pregnancies were included in the statistical evaluation. The clinical pregnancy rate per transfer, and the delivery and ongoing pregnancy rate

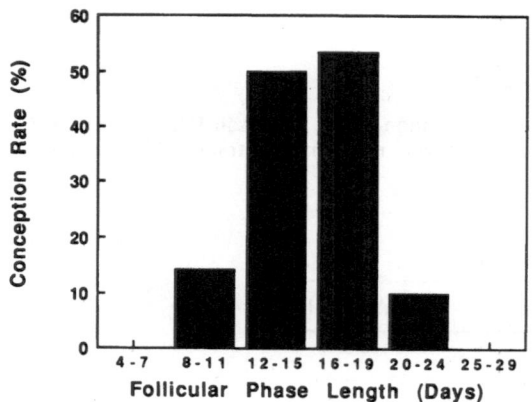

Fig. 7. The influence of the artificial follicular phase duration on
the pregnancy rate in the oocyte
donation program

were 29.4% and 25.5%, respectively. About 45% of all the women conceived, and 42% of them delivered or are in the final stages of pregnancy.

Among the 51 OD cycles, 13 ETs were performed after a short AFP
duration of 4–11 days, 24 transfers following an intermediate duration of
12–19 days, and another 13 after a prolonged duration of 20–29 days.
Thirteen out of the 15 pregnancies were achieved after an AFP duration of
12–19 days, corresponding to a success rate of 54.2%. On the other hand,
only one pregnancy was achieved in each of the short and prolonged durations corresponding to a success rate of 7.7% (Fig. 7).

In order to examine if these differences were statistically significant
since other clinical factors could have affected the results, a logistic regression analysis was performed. Using this model, the success rate in our
OD program was found to be closely linked to the AFP duration. The
success rate was over 40-fold and 20-fold higher for transfers performed
after an intermediate AFP duration than after a short or a prolonged
duration, respectively.

Our results demonstrate that manipulating the duration of the follicular
phase to allow for greater logistic flexibility in an OD program does not
adversely affect the clinical pregnancy rate (29.4%). However, it seems that,
for a better clinical outcome, the AFP length should not be extremely
manipulated to very short or prolonged durations and should be kept at
12–19 days.

This finding that the conception rate was affected by AFP duration is in
sharp contrast to the previous morphologic studies demonstrating adequate
endometrial development following extreme preparatory manipulations.
Recently, preliminary data were reported to suggest that routine histologic
endometrial assessments do not always seem to be reliable. In a group of
infertile women undergoing ovarian stimulation for IVF, 33% of endometrial biopsies were found to be histologically in-phase, but at the same time
to have an atypical pattern of protein secretion [28]. It was suggested that in

these cases a functional endometrial deficiency may exist in spite of a normal morphologic appearance. The discrepancy found in our work between the histologic data in the preparatory mock cycles and the actual clinical results may be explained by a similar biochemical dysfunction not evident morphologically. Since it seems that histologic evaluation of the endometrium might be too crude a measure for assessing actual endometrial receptivity, it may be that more refined biochemical markers of endometrial adequacy should be developed and applied.

Factors Affecting Success

Pregnancy rate after OD (25%–50%) is significantly higher than after IVF-ET performed owing to a purely mechnical factor (10%–25%). Generally, factors affecting embryo implantation and pregnancy could be grouped into two categories. The first is concerned with embryo quality and the second with endometrial receptivity. It is assumed that young donors, that usually respond favorably to superovulation protocols, will eventually lead to the production of oocytes and embryos of "good quality," which will in turn affect the higher conception rate in OD cycles. However, since most OD programs use only young donors (usually below 35 years of age), this factor has not been investigated in controlled studies. On the other hand, evidence exists to suggest that superovulation in itself may lead to a less receptive endometrial environment and in turn adversely affect IVF results. Paulson et al. [29] found, after controlling for embryo quality and matching between patients' characteristics, that implantation rates per individual embryo were significantly higher in OD than in standard IVF (35% versus 10.7%). In addition, clinical and ongoing pregnancy rates per cycle were likewise higher in the donor group than in the IVF group (67% versus 39%, and 61% versus 30%, respectively). This study suggests that since endometrial maturation in OD programs is simple and clear of "background noise," it is better controlled and results in improved endometrial receptivity.

Cryopreservation of human embryos significantly compromises their quality and reduces their capacity to implant in OD programs [21, 30, 31]. It is therefore recommended to reserve the use of frozen-thawed embryos only to those cases in whom fresh transfers are not possible. The number of embryos transferred is another factor that seems to affect the outcome of OD [21, 30]; however, it seems that for optimal clinical results the number of embryos needed is not as high as in IVF-ET. Our impression is that transferring more than two fresh embryos does not increase the success rate of OD.

The effect of other factors on the success of OD remains controversial. These include the recipient's age and the route of embryo-gamete transfer. Abdalla et al. [30], studying 152 cycles of OD, found that the pregnancy rate was 37.5% in the 20–29 age group and steadily dropped to 25.8% in the 40–49 age group. On the other hand, Sauer et al. [32] were able to induce a normal-appearing endometrium in seven women aged 40–44 years and succeeded in producing six pregnancies in nine cycles of OD performed in these women. Moreover, when compared to younger women with POF, these results were no different. Although preliminary data have shown that gamete intrafallopian transfer is a more efficient way to achieve a pregnancy in OD [21, 30, 33], controlled prospective studies have not yet been performed to resolve this issue.

Conclusion

Previously incurable disorders can now be treated successfully by OD. This new technology provides a unique tool to investigate and answer basic questions related to the physiology of reproduction. In order to induce an artificial environment for implantation, the hormonal milieu of the natural cycle does not have to be closely imitated. Moreover, it seems that the duration of an AFP may be manipulated without impairing endometrial developmental capacity. However, contrary to endometrial morphology which seems to be tolerant to extreme estrogen priming durations, functional receptivity is less permissive and seems to be adversely affected by such manipulations. An optimal priming duration is probably 10–20 days. The temporal window of endometrial receptivity and implantation seems to be relatively wide (between days 16 and 20) and tolerant to some endometrial inadequacy.

Because clinical results are affected by embryo quality as well as endometrial development, it is tempting to speculate that at times a favorable endometrium may compensate for embryos of poorer quality. The model of OD in POF patients may be an ideal instrument to test this hypothesis.

References

1. Steptoe PC, Edwards RG (1978) Birth after the reimplantation of a human embryo. Lancet 2:336
2. Lutjen P, Trounson A, Leeton J, Findlay J, Wood C, Renou P (1984) The establishment and maintenance of pregnancy using in vitro fertilization and embryo donation in a patient with primary ovarian failure. Nature 307:174–175
3. Navot D, Laufer N, Kopolovic J, Rabinowitz R, Birkenfeld A, Lewin A, Granat M, Margalioth E, Schenker JG (1986) Artificially induced cycles and establishment of pregnancies in the absence of ovaries. N Engl J Med 314:806–811
4. Leridon H (1977) Human fertility: the basic components. University of Chicago Press, Chicago
5. Tulandi T, Kinch RAH (1981) Premature ovarian failure. Obstet Gynecol Surv 36:521–527

6. Serhal P, Craft I (1987) Simplified treatment for ovum donation. Lancet 1:687–688
7. Cameron IT, Rogers PA, Caro C, Harman J, Healy DL, Leeton JF (1989) Oocyte donation: a review. Br J Obstet Gynecol 96:893–899
8. Ben-Nun I, Ghetler Y, Gruber A, Jaffe R, Fejgin M (1989) Egg donation in an in vitro fertilization program: an alternative approach to cycle synchronization and timing of embryo transfer. Fertil Steril 52:683–687
9. Younis JS, Mordel N, Lewin A, Simon A, Shenker JG, Laufer N (1991) The effect of "follicular phase" duration on clinical outcome in an oocyte donation program: a logistic regression analysis. (submitted for publication)
10. Hung TT, Ribas D, Tsuiki A, Preyer J, Slackman R, Davidson OW (1989) Artificially induced menstrual cycle with natural estradiol and progesterone. Fertil Steril 51:968–971
11. Younis JS, Mordel N, Ligovetzky G, Lewin A, Schenker JG, Laufer N (1991) The effect of a prolonged artificial follicular phase on endometrial development in an oocyte donation program. J In Vitro Fert Embryo Transfer 8:84–88
12. Droesch K, Navot D, Scott R, Kreiner D, Liu HC, Rosenwaks Z (1988) Transdermal estrogen replacement in ovarian failure for ovum donation. Fertil Steril 50:931–934
13. Noyes RW, Hertig AT, Rock J (1950) Dating the endometrial biopsy. Fertil Steril 1:3–25
14. Rosenwaks Z, Navot D, Veeck L, Liu HC, Steingold K, Kreiner D, Droesch K, Stumpf P, Muasher S (1988) Oocyte donation-the Norfolk program. Ann NY Acad Sci 541:728–741
15. Loeb L (1907) Wounds of the pregnant uterus. Proc Soc Exp Biol Med 4:93–96
16. Loeb L (1908) The production of deciduomata and the relation between the ovaries and formation of the decidua. JAMA 50:1897–1901
17. Navot D, Anderson TL, Droesch K, Scott RT, Kreiner D, Rosenwaks Z (1989) Hormonal manipulation of endometrial maturation. J Clin Endocrinol Metab 68:801–807
18. Molina R, Castilla JA, Vergara F, Perez M, Garrido F, Herruzo AJ (1989) Luteal cytoplasmic estradiol and progesterone receptors in human endometrium: in vitro fertilization and normal cycles. Fertil Steril 51:976–979
19. Lutjen PJ, Findlay JK, Trounson AO, Leeton JF, Chan LK (1986) Effect on plasma gonadotropins of cyclic steroid replacement in women with premature ovarian failure. J Clin Endocrinol Metab 62:419–423
20. Rogers P, Murphy C, Cameron I, Leeton J, Hosie M, Beaton L, Macpherson A (1989) Uterine receptivity in women receiving steroid replacement therapy for premature ovarian failure: ultrastructural and endocrinological parameters. Hum Reprod 4:349–354
21. Scott RT, Rosenwaks Z (1990) Oocyte donation – state of the art 1989. In: Mashiach S et al. (eds) Advances in Assisted Reproductive Technologies. Plenum Press, New York, p 633
22. Devroey P, Braeckmans P, Camus M, Khan I, Smitz J, Staessen C, van den Abbeel E, van Waesberghe L, Wisanto A, van Steirteghem AC (1986) Pregnancies after replacement of fresh and frozen-thawed embryos in a donation programs. In: Feichtinger W, Kemeter P (eds) Future aspects in human in vitro fertilization. Springer, Berlin Heidelberg New York, p 133
23. Salat-Baroux J, Cornet D, Anlvarez S, Antoine JM, Tibi C, Mandelbaum J, Plachot M (1988) Pregnancies after replacement of frozen-thawed embryos in a donation program. Fertil Steril 49:817–821
24. Kennard EA, Collins RL, Blankstein J, Schoyer LR, Kanoti G, Reiss J, Quigley MM (1989) A program for matched, anonymous oocyte donation. Fertil Steril 50:655–660
25. Devroey P, Smitz J, Camus M, Wisanto A, Deschart J, van Waesberghe, van Steirteghem AC (1989) Synchronization of donor's and recepient's cycles with GnRH analogues in an oocyte donation programme. Hum Reprod 4:270–274
26. Meldrum DR, Wisot A, Hamilton F, Gutlay-Yeo AL, Marr B, Huynh D (1989) Artificial agonadism and hormone replacement for oocyte donation. Fertil Steril 52:509–511

27. World Health Organization (1983) Task force on methods for the determination of the fertile period. Special programme of research, development and research training in human reproduction: a prospective multicentre trial of the ovulation method of natural family planning. III. Characteristics of the menstrual cycle and of the fertile phase. Fertil Steril 40:773–778
28. Manners CV (1990) Endometrial assessment in a group of infertile women on stimulated cycles for IVF: immunohistochemical findings. Hum Reprod 5:128–132
29. Paulson RJ, Sauer MV, Lobo RA (1990) Embryo implantation after human in vitro fertilization: importance of endometrial receptivity. Fertil Steril 53:870–874
30. Abdalla H, Kirkland A, Barber R, Leonard T, Burton G, Power M, Studd JW (1990) Oocyte donation, factors affecting outcome in 152 cycles (Abstr 92). 2nd Joint Meeting of the European Society of Human Reproduction and Embryology, Aug 29–Sept 1, Milan
31. Levran D, Dor J, Rudak E, Nebel L, Ben-Shlomo I, Ben-Rafael Z, Maschiach S (1990) Pregnancy potential of human oocytes – the effect of cryopreservation. N Engl J Med 323:1153–1156
32. Sauer MV, Paulson RJ, Lobo RJ (1990) A preliminary report on oocyte donation extending reproductive potential to women over 40. N Engl J Med 323:1157–1160
33. Asch RH, Balmaceda JP, Ord T, Borrero C, Cefalu E, Gastaldi C, Rojas R (1988) Oocyte donation and gamete intrafallopian transfer in premature ovarian failure. Fertil Steril 49:263–267

Failures of Placentation and Embryonic Development

15 Epidemiology and Etiology of Early Pregnancy Disorders

E.R. Barnea

Magnitude of the Problem

Pregnancy loss can be defined as an abortion which happens before 20 weeks of gestation. We will call those abortions occurring within the first 12 weeks *early pregnancy loss* (EPL). In 1956 Hertig et al. [1] described the morphologic features of abnormal eggs for the first time. In their study, they addressed the problem of pregnancy loss at the previllous stage and considered that it amounted to more or less 32% before the establishment of gestational amenorrhea. It is supposed that roughly 50% of all conceptions abort [2]. Moreover, the frequency of recurrent abortion (repeated pregnancy loss) increases after each incident: it has been evaluated at 12% after one loss, 25% after two, and 39% after three. The rate may increase even further with subsequent abortions [3]. By definition, habitual (recurrent) abortion needs three consecutive losses. Some authors consider that two consecutive incidents might be sufficient since the likelihood of pregnancy loss has increased twofold. Conversely, after a normal pregnancy, with a healthy child, the likelihood of subsequent abortion decreases.

The reported frequency of pregnancy loss varies markedly. This may be due to the different methodologic approaches used in the early detection of pregnancy [4, 5]. Clinical assessment is the least sensitive, ultrasound can be credited as being of intermediate value, while biochemical assays are extremely accurate and valuable (see Chaps. 20, 21). Recent human chorionic gonadotropin (hCG) measurements allow detection of pregnancy within a very short time after implantation. That is why it is now possible to suggest that there could be a 50% pregnancy loss. Modern hCG assays are completely free of cross-reactivity with luteinizing hormone (LH). More crucial still was the introduction of new technologies in assisted reproduction. Each embryo transfer (ET) must be monitored accurately (see Chaps. 12, 13) and its implantation characterized. It ensues that the latest papers dealing with in vitro fertilization (IVF) and ET give an order of 20–30% EPL [7].

Causes for Increased Rate of EPL in Modern Society

It might be that the rate of spontaneous abortion has increased in the recent years. Several factors could be involved:

1. More pregnancies are obtained in a subfertile or infertile population after correction of various disorders. Obviously these pregnancies are more fragile and are at high risk for early loss.
2. All kinds of technologies for assisted reproduction have a recognized high rate of early placental loss.
3. In Western countries, there is a clear trend towards delayed child-bearing. Older women, i.e., over 35, have an overall loss rate of 31%, while under 34 it is only 17%.
4. In the recent years, evironmental problems have become prominent and pollution has been added as a cause of abortion. However, circumstantial evidence is not easy to demonstrate in all instances; but it is well known that heavy smoking may lead to increased early second trimester abortions. Accidental irradiation (Three Mile Island nuclear plant, Chernobyl) is also associated with an increased abortion rate. Similarly, the rate of pregnancy loss increases with the effect of drug abuse.
5. Other environmental problems are linked to the different jobs in which a woman may be employed nowadays, for example, exposure to various chemicals and toxins in laboratories or factories, or to volatile anesthetics in operating rooms.
6. Human reproduction is adversely affected by stress. We live in a society in which emotional disturbances are prominent and affect all facets of daily life like (education, job, health, etc.).
7. Serious illnesses are better treated, and modern treatments allow such patients to become pregnant with, however, a high risk of demise.

Table 1. Risk factors for EPL: reproductive aspects

Previous EPL or ectopic pregnancy
Previous infertility
Irregular cycles
Galactorrhea
Acne/hirsutism
Endometriosis
Uterine fibroids, malformations
Cervical incompetence
Diethylstilbestrol exposure
Pregnancy occurring less than 3 months after stopping oral
   contraception
Pregnancy occurring less than 2 months after removing IUD
Poor sperm quality

Table 2. Risk factors for EPL: general aspects

Advanced maternal age, advanced paternal age
Debilitating diseases
Autoimmune and collagen disorders
Infectious diseases, i.e., rubella, toxoplasmosis
Severe dietary problems
Severe exertion
Heavy stress exposure
Obesity
Endocrine disorders
Surgical conditions
Drug use and substance abuse
Cigarette smoking
Adverse environment (female/male)

---

8. In our society, childbearing is usually a well-planned event with a strong emotional surrounding. Early abortion is usually felt as a frustrating event. The pursuit of the goal is further enhanced. This does not necessarily help to solve the problem; on the contrary, it may worsen it. The necessity of psychologic support for infertile patients or habitual aborters is obvious.

This chapter deals with etiologies of early pregnancy dysfunction and loss. We will emphasize those factors which are preventable and those which can be treated.

Etiology of EPL: A Systematic Approach

It must be clearly stated that the etiologies of EPL are not evident in the majority of cases. Many possible factors exist, however. They are detailed in Tables 1 and 2. Nature fails to achieve a successful pregnancy in a very large proportion of cases. It has been estimated that only one-third of all natural conceptions eventually reach term. This could be related to the classic concept of the survival of the fittest. Very early abortion can be classified as:

1. Biochemical pregnancy: this is documented by transient circulating hCG elevations and is associated with a brief menstrual delay.
2. Blighted ovum where the embryo fails to develop from very early stage, an empty sac is visible by ultrasound.
3. Clinical spontaneous or missed abortion where significant hCG levels are usually present, and embryonal parts may be identifiable by ultrasound or by direct examination of products of conception.

The introduction of new techniques in human reproduction in the last 2 decades has helped us to gain a much better understanding of the various etiologies involved in abortion. The pathogenesis and placental lesions are described in Chap. 16. In many cases, several etiologies may be simultaneously involved, which makes specific identification of a causal relationship quite difficult and in a large percentage of cases even impossible.

Preimplantation Failure

IVF has taught us that fertilization is a complex event that requires both competent sperm and egg. Dysfunction of one or both of these may lead to failure of fertilization. It became clear in the last decade that fertilization can easily be achieved in laboratory conditions. However, the information regarding events that follow from that point onwards is rather fragmentary. Thus, once the zygote starts dividing and travels into the fallopian tube and the uterus either naturally or through IVF or gamete intrafallopian transfer (GIFT), the information is very limited. This is also the period when most failures actually take place.

The number of experiments carried out using cultured human embryos is growing. Current data indicate that embryos can be grown in vitro until the blastocyst stage with quite a high success rate. This again indicates that preimplantation embryonal development can proceed under in vitro conditions without requiring a complex environment as has been repeatedly suggested. In vivo, the situation is somewhat different and perhaps less favorable; the spermatozoon must first travel and meet the oocyte in the ampullary portion of the tube. It ensues that any damage to the tube will be harmful, leading either to failure of fertilization or to ectopic implantation (see Chap. 17) when hatching of the embryo occurs in the tube.

Failure of Implantation

The "black box" in human reproduction remains the understanding of the process of implantation. For successful implantation, a unique interaction between the embryo (an allograft) and the maternal immune system (tolerance) has to take place at a favorable site (endometrium). Implantation is understandably very difficult to study in vivo in the human. Several animal models, including primates, are useful in the study of such events; none, however, mimics precisely the processes in the human (see Chap. 27). In vitro, human models do provide important information in this regard, and a number of them are being currently developed.

Embryonic Failure

It has been shown that after several days of culture the human embryo can also attach to the culture dish [6]. This attachment is associated with a major increase in the rate of hCG secretion; this is one of the signals (with, for example, platelet-activating factor, PAF) which is needed for implantation to take place. Hatching means dissolution of the membrana pellucida, which is necessary for implantation. Once the pellucida has dissolved, the embryonic cells are no longer contained in a limited volume and rapid cellular growth can follow. Hatching must be initiated by a signal intrinsic to the blastocyst; external influences may also be involved, however (Chap. 1). One may speculate on the regulation.

Embryonal attachment may depend on specific local signals as evidenced by the changes present in the adjacent endometrium (see Chap. 2). These changes also depend on the embryonal ability to penetrate through the endometrial surface epithelium. The lack of a putative preimplantation signal may lead to failure at this stage. Some evidence for the presence of such an amplifying signal was suggested by the finding that transfer of more than one embryo into the uterus following IVF increases pregnancy rate. However, one may argue that multiple embryo transfer increases the likelihood of transferring a "good" and viable embryo [8]. Failure of the blastocyst to attach to the endometrial wall within 3–4 days after arriving in the uterine cavity will inevitably lead to demise of the embryo. It has also been suggested [9] that the four-cell embryo stage may require genome activation in order to produce hCG. Variations of the carbohydrate side chains have a strong implication in the biologic function and thereby might prevent the corpus luteum rescue.

Failure of the Maternal Immune Response

The mechanisms involved in successful implantation necessitate the immune-tolerant recognition of the embryo by the mother. Lack of development of immunologic tolerance can induce failure of implantation, early pregnancy rejection, and severe growth retardation in late gestation, depending on the type and gravity of the disorder (see Chap. 26). It is widely accepted that the uterus is an immunologically privileged site (see Chap. 9) where decidual cells, lymphoid cells, and epithelial cells interact to protect the growing conceptus (see Chap. 1).

Early embryonic signals are also involved. In particular, early pregnancy factor (EPF), one of the earliest signals, is an immunomodulator involved in the protection of the conceptus as an allograft in several species [10]. Before implantation, EPF is possibly maternal in origin. Nahas and Barnea [10] have demonstrated that preimplantation blastocyst does not produce EPF, while postimplantation embryo and trophoblast are fully capable of doing so. Chemical pregnancies after ET are often associated with reduced or absent EPF levels in maternal serum. It has recently been suggested [11] that progesterone may also be an immunomodulator. It remains to be proven, however, that an immunologic factor is also involved in abortions associated with corpus luteum deficiency.

Uterine Factor

The implantation window is 3–4 days (see Chap. 2) in midluteal phase when the endometrium is primed by optimal progesterone levels. Prior to and after these times the endometrial receptivity for the embryo is poor or nonexistent. An altered milieu can be caused by ovulatory or postovulatory dysfunctions (e.g., corpus luteum deficiency, see below). In IVF technologies, it is believed that several pregnancies fail to implant correctly because of a poor endometrial receptivity at that time.

As regards the minimal hormonal support required for implantation, a lot of information has been gained from patients enrolled in an embryo donation program because of premature ovarian failure. Acceptable success rates with combined regimens of progesterone and estrogen point to the importance of these steroids in the implantation process (see Chap. 2). Altered or deficient production by the endometrium of various proteins (prolactin, placental protein 14 α_2PEG, etc.; Chap. 2) must still be documented, but might be of importance. However, there is a low implantation rate following in vitro fertilization. This may be partly due to the fact that the endometrium is rarely in phase. To overcome this problem, it has been suggested that the embryo should be frozen and transfer delayed to the next cycle. This procedure has, however, proved to be only moderately successful. Most probably other factors are involved, and the freezing techniques are not the true answer.

Implantation in an Abnormal Location

Extrauterine pregnancy is proof that the intimate presence of a true decidua is not critical for pregnancy development. Ectopic implantations are characterized by an unfavorable environment as the result of an inadequate blood supply and poor nutritional support. This leads to tissue hypoxia which is reflected by the isolated increase in serum α-hCG levels as compared to those of the native hormone [12]. Detailed discussion on this topic appears in Chap. 20. The likelihood of a pregnancy developing and proceeding outside the uterine corpus is very low. Cervical or interstitial pregnancy will inevitably fail.

Adverse Maternal Environment

Corpus Luteum Deficiency

Corpus luteum failure is a classic cause of implantation failure. Insufficient luteal function leading to low progesterone secretion may be due to several endocrine and nonendocrine disorders (Table 3). Corpus luteum deficiency is a complex problem, and, though a number of criteria have been put

Table 3. Etiology of corpus luteum defect

Central disorders (reduced FSH, premature LH/FSH peak)
Disrupted corpus luteum
Elevated prolactin levels
Polycystic ovary (increased LH)
Hyperandrogenism
Perimenopause
Caloric restriction
Strenuous exercise
Hormonal treatments
Endometriosis
Defective follicular development
Drug use/abuse

forward for diagnosis, no unanimous view has been reached so far [13, 14]. According to progesterone levels, there may be aluteal phase, inadequate luteal phase, and short luteal phase [15]. Generally, the clinical picture involves shortened cycles, i.e., with a luteal phase 3 or more days shorter than the usual. There may be spotting and premenstrual symptoms. An out-of-phase endometrial biopsy and low progesterone levels are considered diagnostic. The cut-off point for diagnostic low progesterone levels is not well defined. Under normal conditions this hormone is produced in pulses, and values of 8 ng/ml can be recorded. Soules et al. [16] have recently proposed that only the decrease in pulse amplitude could be diagnostic.

If implantation has occurred during a deficient luteal phase, there will usually be a transient increase in hCG levels. Most probably correct implantation is defective as the endometrium has not been adequately prepared. Thus, the most severe forms of corpus luteum defect are associated with failed implantation and pregnancy loss which is unnoticed during menstruation. It might be that milder forms will allow pregnancy to continue, but with a loss before 6–7 weeks. The classic problem of late luteal failure (rupture of lutein cyst, "accidental" lutectomy) involves "late" subsequent abortion before 8 weeks. Afterwards, progesterone production is essentially derived from the placenta and the corpus luteum only has an accessory function. It must remembered that progesterone could have an immunosuppressor role. Low steroid levels could prevent adequate maternal recognition of pregnancy and thus facilitate rejection of the implanted embryo [11].

The corpus luteum also secretes other factors such as oxytocin, vasopressin, and relaxin [17]. Their altered secretion could also influence the maintenance of pregnancy. We have recently found that the two neuropeptides have a modulatory role on placental hCG secretion in vitro. This is suggestive of an ovarian/placental link in early pregnancy. It has also been shown that defective follicular development or experimental manipulation of the follicular phase may also cause corpus luteum defect. In athletes with or without caloric restrictions and in certain emotional states there is a corpus luteum defect. It is believed to be produced by activation of the stress axis [18] and could be related to the adverse effect of endogenous opioids and peptides, among others (see below).

Polycystic Ovaries

Polycystic ovaries are associated with hyperandrogenism, menstrual irregularity, and altered follicle-stimulating hormone (FSH)/LH ratios together with ovarian morphologic changes. The syndrome may encompass amenorrhea, anovulation, premature follicular luteinization, and corpus luteum defect. Patients with polycystic ovaries who have elevated circulating LH levels are also likely to be habitual aborters. Fertilization is achieved, but the subsequent formation of a defective blastocyst seems likely [19].

Hyperprolactinemia

Increased pituitary prolactin production is often associated with menstrual abnormalities. Among others, there may be corpus luteum defect. This is due to central suppression of FSH and LH secretion, and to a direct ovarian effect of prolactin which decreases gonadotropin binding. A transient rise in prolactin either at midcycle or also in the late luteal phase was shown to cause infertility via corpus luteum deficiency and presumably failure of implantation [20].

Different forms of prolactin have been identified: there are mono- and dimer- and isoforms (glycosylated) which differ in their affinity for the receptor. Differences in the biologic activity might ensue [21] and might separate pituitary from decidual prolactin. Decidual prolactin is not inhibited by dopamine in contrast to pituitary prolactin. Decidual prolactin could act as a paracrine factor on the adjacent placenta. We have demonstrated [22] that prolactin added in superfusion experiments inhibits the secretion of hCG by placental explants. This could be evidence for a detrimental effect of elevated prolactin levels on pregnancy. It has been claimed that hyperprolactinemia is one cause of habitual abortion. This has not been confirmed. Jouppilla and Ylikorkala [23] could not find any differences in serum prolactin levels between normal and pathologic early pregnancies, while Crosignani et al. [24] demonstrated that 16 patients out of 64 with hyperprolactinemia experienced spontaneous abortion.

Endometriosis

Endometriosis is a complex disorder that may lead to infertility and is associated with an increased rates of spontaneous abortion [25]. The endometrial implants present in extrauterine locations may lead to inflammation, bleeding, and subsequent scar formation. Infertility is caused by mechanical factors affecting the fallopian tubes, ovary, and peritoneum. Simultaneously there is a release of several humoral and immunologic mediators (prostaglandins and bradikinins) and activation of macrophages which release various cytokines. These factors may prevent or affect fertilization and subsequently implantation. There is also a direct adverse effect on the corpus luteum and endometrium due to the formation of autoantibodies. In the more severe cases of endometriosis, the infertility rate is elevated, while in mild cases fertility does not appear to be impaired.

Thyroid Gland

Changes in the thyroid function may be associated with spontaneous abortion; this is true for hyper- and hypothyroidism. Hyperthyroid patients often fail to conceive and have a high frequency of pregnancy loss. In uncorrected hypothyroidism there is a low progesterone and LH secretion suggestive of corpus luteum defect.

Adrenal Gland

Hypercorticism of any etiology, including Cushing's syndrome or adrenal hydroxylase deficiency, are associated with hyperandrogenism and disturbed menstrual function resembling a polycystic ovary syndrome. The condition should be diagnosed and treated according to the specific etiology before any attempt at pregnancy: high rates of abortion are recorded. In 21 hydroxylase deficiency, there is an increased risk for female embryo virilization.

Endocrine Pancreas

Diabetic patients are prone to menstrual disorders, with a frequent polycystic ovary-like syndrome. It has been well recognized that such patients, especially when the disease is not amenable to strict control, experience a high frequency of spontaneous abortion and congenital anomalies of the offspring [26]. Hyper- or hypoglycemia induce large variations in insulin levels. Moreover, the adverse effects of ketone molecules have been demonstrated in various animal models [27].

Evidence for a direct effect of these compounds on the placenta and neural tissue of the human embryo was recently provided in vitro ([28] Barnea et al. submitted, 1991). In contrast, D-glucose appeared to have no effect in these cultures. In these studies elevated ketone molecules and insulin decreased hCG, free β-hCG, and progesterone secretion of placental explants from early pregnancies (less than 9 weeks). Inhibition of free β-hCG secretion may be one of the main mechanisms for the reduced hCG secretion and the increased rate of abortion in diabetes. In contrast, after 11 weeks, the effect of ketone molecules and insulin was mildly stimulatory on hCG secretion.

Uterine Anomalies

The immediate surrounding of the conceptus may be harmful for pregnancy development. Mechanical disorders can be caused by either congenital or acquired defects. It is estimated that 15% of pregnancy losses can be attributed to mechanical factors. Presence on the endometrial surface of fibroids, polyps, or adhesions prevent implantation or, rarely, trigger abortion. This latter effect is significant when implantation has taken place on the pathologic area or in a nearby location. Subsequently, restricted maternal blood flow or uterine irritability will lead to pregnancy loss. The most severe form is caused by Asherman's syndrome in which the uterine walls adhere, practically obliterating the uterine cavity.

It has been well documented that diethylstilbestrol affects 69% of exposed women and causes increased abortion rates. This is caused by various uterine anomalies such as incompetent cervix, T-shaped uterus, a wide lower uterine segment, midfundal constriction, irregular uterine margins, and an increased smooth muscle/collagen ratio. In addition, it has been shown that the fallopian tubes in these women have a bizarre shape, leading to infertility and a higher incidence of ectopic pregnancy.

Defects of the fusion of müllerian ducts are congenital; they are present in 2%–3% of women. When the defect is significant (large septum or unicornuate uterus), patients undergo severe reproductive problems, with a 50% miscarriage rate. It has been estimated that müllerian defects could be involved in 10% of all abortions. Evidently, milder defects may just be incidental findings.

The incompetent cervix often leads to spontaneous interruption of pregnancy. This defect is congenital or acquired. It occurs in about 1% of all pregnancies and usually leads to midtrimester abortion. However, it has been suggested that cerclage could also reduce the rate of habitual abortion (see Chap. 24).

Infectious Agents

In the past, great emphasis was placed on infection as a major cause of reproductive disorders. It has been shown that tubal damage was a frequent consequence. This is especially true for chlamydia and gonorrhea, which lead to an increased rate infertility and ectopic gestations. Whether endometritis impairs implantation or causes abortion has not been settled. The endometrium is replaced on a monthly basis and therefore the formation of chronic endometritis is a rare event during the reproductive age, except following dilation and curettage or shortly after delivery. Tuberculous endometritis is almost always associated with pelvic, i.e., tubal lesions, leading to irreversible sterility.

Systemic Disorders

Any kidney or liver disease seriously modifying the equilibrium of the internal milieu can cause abortion. Moreover, serious diseases may require the use of medications which are potentially harmful to the embryo. Auto-immune diseases may or may not be associated with clinical syndromes (e.g., systemic lupus erythematosus). The rate of pregnancy loss in collagen disorders such as lupus erythematosus is 20%–30% in mild cases and over 70% in severe cases, especially if kidney function is impaired. In these cases, the placenta and decidua are affected by the early formation of immune deposits and fibrinoid necrotic areas. In some women, there is a circulating lupus coagulant factor which decreases the number of blood platelets, increases coagulation time, and induces thrombosis in the uterine and placental blood vessels. In patients with active rheumatoid arthritis there is a 25% rate of spontaneous abortion. More recently, a reproductive auto-immune failure syndrome was described [29] which may be associated with infertility, endometriosis, and habitual abortion. Such women have poly-clonal B cell activation, antiphospholipid antibodies and autoantibodies against histone and nucleotides (see Chap. 26).

Fertility Drugs

Gonadotropins, clomiphene citrate, gonadotropin-releasing hormone (GnRH) and its analogues are commonly used fertility drugs. These medications are intended to increase pregnancy rates, but they also rise abortion rates. Obviously, the overall effect is beneficial, but calculation of the doses may be critical. An excess or underdosage of clomiphene citrate may induce a corpus luteum defect [30]. This may be linked to a prolonged half-life of the drug in the follicular fluid and the circulation [31]. Induction of ovulation with human menopausal gonadotropin (hMG)/hCG is associated with an increased rate of pregnancy loss. These is an increased estrogen/progesterone ratio which appears to be detrimental for implantation and early pregnancy development (see Chap. 12). GnRH analogues have a luteolytic effect [32]. The data on inadvertent use of GnRH analogues during early pregnancy is very limited, and in some cases there appears to be reversible interference with corpus luteum function. The use of pulsatile GnRH for the treatment of a variety of ovulatory disorders is associated with a high (27%) abortion rate [33]. An increased abortion rate within a short time after stopping danazol administration (for treatment of endometriosis) has also been documented. It does seem to be due to a corpus luteum defect.

Drug Use and Abuse

Several conditions require the use of medications during pregnancy. The first trimester is the most vulnerable period for such drug exposure. The use of several compounds is obviously contraindicated during morphogenesis: they include neuroleptics, chemotherapeutic agents, oral anticoagulants, antibiotics such as sulfonamides and tetracyclins, retinoids, and antiepileptics among others. The interested reader is referred to recent texts on these drugs and their consequences [34]. All are potentially very harmful to the embryo, leading either to abortion or to different effects on the different steps of organogenesis and thus to malformation. Moreover, some substances (e.g., medroxyprogesterone acetate) can affect the luteal function adversely. Drug abuse causes medical, social, and political concern. Tranquilizers, hallucinogens, opiates, and tobacco are currently used. Some substances can induce specific malformations or early abortion (see Chap. 7 for biochemical evidence). Acute alcohol exposure is also harmful since, in healthy women, it can decrease the hCG-induced progesterone peak and increase estradiol and prolactin levels [35].

Exposure to a large variety of toxins present in the working or general environment has also been questioned. Such substances can be inhaled (fumes, anesthetics) [36] or digested. Others penetrate through the skin. They are either teratogens or can induce early abortion. Jelousek [37] has recently published a detailed list of such substances among which trichloroethylene, benzene, and formaldehyde are of note.

Stress

We recently addressed the subject of stress and reproductive failure [18]. There is evidence that stress can interfere with various aspects of reproduction. This effect encompasses amenorrhea, anovulation and probable failure of fertilization, corpus luteum defect, probable failure of implantation, and probably too EPL by a direct effect on the placenta. It was previously suggested that lowering the level of stress could solve infertility problems and might lead to successful completion of pregnancy. It is not an easy task to document pregnancy loss related to stress. There are many additional factors which may interfere. Stress is associated with activation of the hypothalamo/pituitary/adrenal and hypothalamo/pituitary/ovarian axes. The blood levels of various hormones increase such as opiates, oxytocin and vasopressin, adrenocorticotropic hormone (ACTH), growth hormone (GH), prolactin and enkephalins, corticosteroids and androgens. The adrenal medulla also produces increased amounts of catecholamines.

We have recently examined [22] the effect of isolated stress related factors on hCG secretion in static (long-term) and dynamic (short-term) cultures of placental explants. Our studies suggest that prolactin, vasopressin, and oxytocin significantly affect hCG secretion. The oxytocin effect is probably receptor dependent since the stimulatory effect on hCG secretion was inhibited by co-administration of a specific receptor antagonist [22]. We have also found that the opiate tested, namely β-endorphin, inhibits, while dynorphin stimulates hCG secretion in culture; naloxone, a specific opioid antagonist, blocks these effects [22a] (see also Chap. 8). These findings suggest that stress-related factors of maternal origin affect placental hormone secretion.

Postimplantation Disorders

Genetic Disorders and the Embryo

The reasons for failed early development are multifactorial; embryonal factors may play a prominent role. A defective embryo implants poorly and also fails early. The embryo may develop abnormally due to its altered genetic make-up as demonstrated by a chromosomal aberration. On the other hand, it may be affected by its inability to develop because of the adverse environment causing delayed or improper cell division and/or harming the blastomer function and specialization, i.e., gene expression.

It is estimated that 50% of early losses are chromosomal. Except for balanced translocation, nearly all cases are caused by spontaneous mutation. The type of the chromosomal disorder depends on the moment of its formation during the reproductive cycle: trisomy, monosomy, or triploidy will appear during oocyte meiosis; triploidy may be produced during fertilization division; tetraploidy may occur following the first cleavage division; mosaicism is a consequence of abnormal mitotic division after fertilization.

Pregnancies observed before 8 weeks are associated with chromosomal anomalies in more than 15% of cases. Between 8 and 12 weeks, the frequency drops to 7%, and to 1% between 12 and 17 weeks. Advanced maternal age is significantly associated with such pathologies, especially with autosomal trisomies [38]. In spontaneous abortion cases, a good yield (90%) of chromosomal analysis can be obtained; chromosomal aberrations are detected in more than 75% of cases (with 70% demonstrating an aneuploid pattern) [39]. Aneuploid conceptuses are usually diagnosed as monosomy X, polyploidy (triploidy or tetraploidy), and autosomal trisomies.

Most autosomal trisomy cases are quickly arrested in their development, except when the chromosomal pair is No. 13, 18, 21, 22 (evolution to term is usually possible then). Sex chromosome anomalies (monosomy, trisomy) are not necessarily associated with inevitable abortion. Lastly, triploidy cases may develop in rare cases up to late in the second trimester. Aging of the gametes and irradiation may induce genetic anomalies: this can be demonstrated by the relation between frequency of Down's syndrome and maternal age. It cannot be absolutely excluded that maternal age is also influenced by environmental factors [40]. Balanced translocations have been associated with habitual abortion in fewer than 0.3% of studied cases. However, Boué et al. [5] have demonstrated in a large study that one parent had a structural chromosomal anomaly in 7.2% of abortion cases. There does not seem to be any difference in respect of chromosomal constitution between habitual and sporadic abortion. In fact, in one study of 69 habitual aborters, the rate was 17%, even lower than in sporadic cases.

Nongenetic Disorders – Defective Embryonal Development

Why the abnormal embryo fails in a substantial proportion of cases is not clear at present. Why some of the chromosomally defective embryos escape rejection and survive until delivery also remains unanswered. Unfortunately, even if the embryo is abnormal, the local development procedures are effective and therefore abnormal embryo development proceeds.

Apparent nongenetic failure of the embryo may be induced by absent expression of genes. These could act at the critical time (before 5–6 weeks) when embryogenesis takes place. The result is polygenic malformation. It is probable that the abnormal embryo is not able to send signals of viability to the whole embryoplacental unit and to the maternal organism. Usually too, the early circulatory system fails and the placenta stops developing. It is also highly probable that the early placenta also participates in the polygenic anomaly with reduced trophoblastic proliferation, leading to villous hypoplasia and to insufficient decidual colonization. This hypothesis introduces failure or insufficiency of placentation as the true etiologic factor for spontaneous abortion (see Chap. 16) [41].

Defective Placenta and Altered Embryoplacental Relationship

Abortion occurring after a menstrual delay indicates that the corpus luteum functioned at least for a while. In contrast, there are pregnancies with implantation, but where menses comes at the expected time; a corpus luteum defect must be suspected. These pregnancies have a delayed hCG doubling time compared to those in which the loss occurs after a menstrual delay [42]. A defective placenta may be the underlying cause for such failures (see Chap. 16). Failure to rescue the corpus luteum in due time may also occur [43]. This could be produced either by a local enzymatic defect or by a lack of stimulus to produce progesterone or estradiol. We speculate that this stimulus could be embryo derived (see Chap. 8).

We have also found that the pattern of pulsatile hCG secretion in superfusion experiments shows major changes following heavy maternal cigarette smoking or following in vivo exposure to neuroleptic drugs; both are now accepted etiologies for pregnancy failure. Furthermore, we have recently provided evidence that the embryo-placental relationship may be altered in cases of adverse environment in vitro (see Chap. 7). Thus, study of the defective placenta should give an insight of its interaction with the embryo. The fact that clinical signs of spontaneous abortion are frequently evident only 2–3 weeks after embryonal demise is also suggestive of a local trophic role of the embryo on the placenta. A recent study dealing with ultrasound performed at 8 and 12 weeks also points to this possibility [44]. Among 220 women who had a viable pregnancy at 8 weeks, only seven

(3.2%) experienced a fetal loss thereafter. This indicates that most clinically recognized spontaneous abortions occurring after 8 weeks are actually pregnancies in which fetal demise occurred before 8 weeks.

Habitual Abortion

Habitual abortion is a complex and frustrating condition. A specific etiology cannot be determined in all cases in spite of major investigative efforts. It is classically defined as three or more consecutive spontaneous abortions. Habitual abortion is either primary or secondary. The former includes patients who have not previously had a live birth: they can be treated more easily. The prognosis is less favorable for secondary aborters because therapeutic measures are less effective and often not specific. It has been postulated that habitual abortion and fetal growth retardation could represent various degrees of rejection of the fetal allograft by the mother (see Chap. 26).

The apparent incidence of habitual abortion is estimated to be 0.4%–1% of the general population. However, its true incidence is not known, since many women do not report to medical care, and therefore certain categories of patients may be overrepresented. Habitual aborters are usually older and attempt pregnancy later in life.

Diagnostic Work-up

Immunologic etiology is discussed in Chap. 26. For all other patients, an extensive work-up is mandatory. However, a definitive diagnosis will be reached in only 20%–30% of cases. The prognosis for all cases is not too bad with 60% of patients achieving a successful pregnancy. Interestingly enough, the birth weight is significantly lower than in the general population [45]. It is also of note that fetal loss remains higher than in control population irrespective of any form of treatment. It is probably more important for the patient to feel that she is being helped, regardless of the modalities.

References

1. Hertig AJ, Rock J, Adams CD (1956) A description of thirty four human ova within the first seventeen days of development. Am J Anat 98:439–493
2. Reganl (1988) A prospective study of spontaneous abortion. In: Beard RW, Sharp F (eds) Early pregnancy loss: mechanisms and treatment. Springer, Berlin Heidelberg New York, pp 23–28
3. Hassold TJ (1980) A cytogenetic study of repeated spontaneous abortion. Am J Hum Genet 32:723–730
4. Warburton D, Strobino B (1987) Recurrent spontaneous abortion. In: Bennet CJ, Edmonds DK (eds) Spontaneous and recurrent abortion. Blackwell, Oxford, pp 193–213

5. Boué A, Boue J, Gropp A (1985) Cytogenetics of pregnancy wastage. Adv Hum Genet 14:1–57
6. Edwards RG (1983) Current status of human conceptions in vitro. Proc R Soc Lond 223:417–448
7. Edwards RG (1986) Causes of early embryonic loss in human pregnancy. Hum Reprod 1:185–198
8. Osborn JC, Moor RM (1988) An assessment of the factors causing embryonic loss after fertilization in vitro. J Reprod Fertil Suppl 36:59–72
9. Cooke ID (1988) Failure of implantation and its relevance to subfertility. J Reprod Fertil 35:155–159
10. Nahas F, Barnea ER (1990) Early pregnancy factor origin before and after implantation. Am J Reprod Immunol 22:105–108
11. Van Vlasselaer P, Vandeputte M (1985) Local immune suppression by progesterone. In: Ellendorff F, Koch E (eds) Early pregnancy factors. Perinatology Press, Ithaca
12. Barnea ER, Oelsner G, Benveniste R, Romero R, Decherney AH (1986) Progesterone, estradiol, and alpha HCG secretion in patients with ectopic pregnancy. J Clin Endocrinol Metab 65:529–531
13. Balasch J, Vanrell JA (1986) Luteal phase deficiency: an inadequate endometrial response to normal hormone stimulation. Int J Fertil 31:368–371
14. Vanrell JA, Balasch J (1986) Luteal phase defects in repeated abortion. Int J Obstet Gynecol 24:111–115
15. Gennazani AR, D'Ambrogio G, Fachinetti F, Alessandrini G, Petraglia F, Volpe A (1985) Luteal phase defect. In: Labrie F, Prouix L (eds) Endocrinology. Elsevier, New York
16. Soules MR, Clifton DK, Cohen NL, Bremner WJ, Steiner RA (1989) Luteal phase deficiency: abnormal gonadotropin and progesterone secretion patterns. J Clin Endocrinol Metab 69:813–820
17. Csapo AL, Pulkkinen MO, Wiest WG (1973) Effects of luteectomy and progesterone replacement in early pregnant patients. Am J Obstet Gynecol 115:759–763
18. Barnea ER, Tal Y (1991) Stress related reproductive failure. J In Vitro Fertil 8:17–28
19. Homburg R, Armar NA, Eshel A, Adams J, Jacobs HS (1988) The influence of serum luteinizing hormone concentrations on ovulation, conception, and early pregnancy loss in patients with polycystic ovarian syndrome. Br Med J 297:1024–1026
20. Ben David M, Shenker JG (1984) Transient hyperprolactinemia: a correctible cause of idiopathic female infertility. J Clin Endocrinol Metab 57:442–445
21. Ben David M, Chrambach A (1980) Method for isolation by gel electro-focussing of isohormones B and C of human prolactin from amniotic fluid. J Endocrinol 84:125
22. Tal Y, Sharf M, Barnea ER (1991) Effect of stress related factors upon placental hCG secretion in vitro. Human Reprod (in press)
22a. Barnea ER, Ashkenazy R, Sarne Y (1991) The effect of dynorphin on placental pulsatile human chorionic gonado tropin secretion in vitro. J Clin Endocrinol Metab 73:1093–1098
23. Jouppilla P, Ylikorkala O (1984) Role of maternal prolactin in early pregnancy failure. Obstet Gynecol 64:373–375
24. Crosignani PG, Mattei AM, Scarduelli C, Cavioni V, Boracchi P (1989) Is pregnancy the best treatment for hyperprolactinemia. Hum Reprod 4:910–912
25. Galle PC (1989) Clinical presentation and diagnosis of endometriosis. Obstet Gynecol Clin North Am 16:29–42
26. Sutherland HW, Pritchard CV (1986) Increased incidence of spontaneous abortion in pregnancies complicated by maternal diabetes mellitus. Am J Obstet Gynecol 155:135–138
27. Frienkel N, Dooley SL, Metzger BE (1985) Care of the pregnant woman with insulin dependent diabetes mellitus. N Engl J Med 313:96–103
28. Garcia-Segura LM, Barnea ER, Biggers W, Naftolin F, Sanyal MK (1986) Insulin modulates neuronal plasma membrane development in human fetal spinal cord neurons in culture. Neurosci Lett 65:283–286

29. Glicher N, Elroey A (1988) The reproductive autoimmune failure. Am J Obstet Gynecol 159:223–230
30. Hammond MG (1984) Monitoring techniques for improved pregnancy rates during clomiphene ovulation induction. Fertil Steril 42:499–509
31. Oelsner G, Barnea ER, Admon D, Mikkelson TJ, de Cherney AH (1987) Levels of clomiphene citrate isomers in follicular and peripheral plasma of patients undergoing IVF-ET. N Eng J Med 317:317–318
32. Barnea ER, Maheux R, Cladwell BV, de Fazio J, de Cherney AH, Naftolin F (1985) LRF D_6a has a dose dependent stimulatory or inhibitory effect on the ovary in normal luteal phase women. J Endocrinol Invest 8:297–302
33. Homburg R, Eshel A, Armar NA, Tucker M, Mason PW, Adams J, Kilborn J, Sutherland IA, Jacobs SH (1989) One hundred pregnancies after treatment with pulsatile luteinizing hormone releasing hormone to induce ovulation. Br Med J 298:809–812
34. Friedman JM, Little BB, Brent RL, Cordero JF, Hanson JW, Shepard TH (1991) Potential human teratogenicity of frequently prescribed drugs. Obstet Gynecol 75:554–599
35. Teoh SK, Mendelson JH, Mello NK, Skupny A, Ellingboe J (1990) Alcohol effects on hCG-stimulated gonadal hormones in women. J Pharmacol Exp Ther 254:407–411
36. Pharoah POD, Alberman E, Doyle P, Chamberlain G (1977) Outcome of pregnancy among women in anesthetic practice. Lancet i:34–36
37. Jelousek FR (1989) Prediction of risk of human developmental toxicity: how important are animal studies for hazard identification. Obstet Gynecol 74:4–13
38. Creasy MR (1988) The cytogenetics of spontaneous abortion. In: Beard RW, Sharp F (eds) Early pregnancy loss: mechanisms and treatment. Springer, Berlin Heidelberg New York, pp 293–304
39. Guarneri S, Bettio D, Simoni G, Brambati B, Lanzani A, Fraccaro M (1987) Prevalence and distribution of chromosome abnormalities in a sample of first trimester internal abortions. Hum Reprod 2:735–739
40. Angell RR, Hillier SG, West JD, Glasier AF, Rodger MW, Baird DT (1988) Chromosome anomalies in early human embryos. J Reprod Fertil Suppl 36:73–81
41. Khong TY, Liddel HS, Robertson WB (1987) Defective haemochorial placentation as a cause of miscarriage: a preliminary study. Br J Obstet Gynaecol 94:649–655
42. Lenton AE, Woodward A (1988) The endocrinology of conception cycles and implantation in women. J Reprod Fertil Suppl 36:1–15
43. Aspilaga MO, Whittaker PG, Grey CE, Lind T (1983) Endocrinologic events in early pregnancy failure. Am J Obstet Gynecol 147:903–908
44. Simpson JL, Mills JL, Homes LB, Ober CL, Aarons J, Jovanovic L, Knopp RH (1987) Low fetal loss rates after ultrasound-proved viability in early pregnancy. JAMA 258:2555–2557
45. Beard RW (1988) Clinical association of recurrent miscarriage. In: Beard RW, Sharp F (eds) Early pregnancy loss: mechanisms and treatment. Springer, Berlin Heidelberg New York, pp 3–8

16 Morphology and Mechanisms of Abortion

J. Hustin and E. Jauniaux

Introduction

Chorion villous sampling in the first trimester of pregnancy is associated with a low (7%) rate of losses [1, 2], probably not different from that occurring spontaneously in the general population after 9 weeks of gestation. That such an invasive procedure has so few consequences is surprising (see Chaps. 23, 25), but quite in keeping with the old concept that normal pregnancies are securely tight up to term.

The mechanisms by which a pregnancy interrupts and is eventually expelled are not only poorly understood, but have also never attracted much interest. It is generally held that either the placenta is retained in utero for significant periods of time or could be partly expelled through retroplacental hemorrhage and abruptio. Khong et al. [3] have recently suggested that some miscarriages could be due to insufficiency of implantation which could subsequently lead to circulatory disturbances.

Early abortions, i.e., of less than 5 weeks' age, are usually expelled in toto. They are frequently surrounded by a blood clot which also occupies the intervillous space. The accompanying decidua is more or less totally necrotic, and trophoblastic infiltration is therefore difficult to ascertain. However, a reduced or even hypoplastic extravillous trophoblast is a frequent finding when villous development is scant or truly hypoplastic.

Numerous studies have been devoted to the expression at the placental level of various histologic features suggestive of anomalies of the conceptus [4–9]. Though some criteria have been met, the predictive value of histology is still the subject of much controversy [6, 7] (see also Chap. 28). Some authors even suggest that correct diagnosis is no more frequently met than that obtained by pure chance [6]. All investigators have accepted a fairly high percentage of errors in the diagnosis of chromosomal anomalies, but there has been no analysis of the causes of discrepancies between cytogenetics and histology. In other terms, all cases which were not correctly diagnosed have not been submitted to revision.

Two main directions will be followed in this chapter. In the first part, some features of spontaneous abortion during the 1st trimester will be discussed. In particular, we will try to explain why the level of wrong

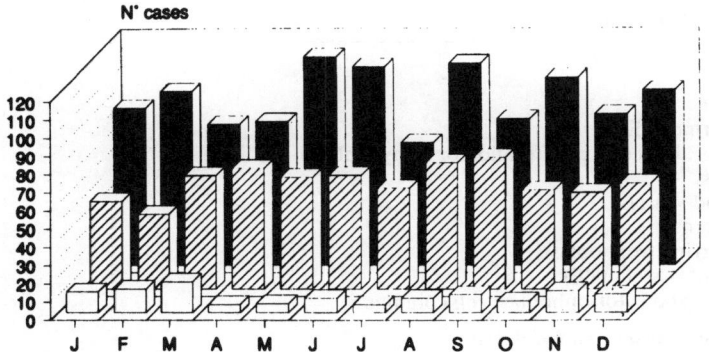

Fig. 1. The rate of spontaneous abortion during the years 1989–1990; *x-axis*, the different months. *Open columns*, abruptio placentae; hatched columns, histologically defined anomalies of the placenta; *stippled columns*, embryonic death and retention. There is no significant seasonal or monthly variation

diagnosis is so high and whether there are some clues to proper evaluation. In the second part, the real mechanisms which are involved in pregnancy arrest and abortion will be investigated. By this pathogenetic approach we would like to propose a functional classification which, in a substantial number of cases, correlates with placental anomalies, but which might also explain why a number of specimens are declared histologically either normal or with only features of retention while their karyotype is abnormal.

Pathology of Spontaneous Abortion

Our institution has two main commitments: one is devoted to the histopathologic study of various specimens, the other is directed towards cytogenetics. Both orientations are complementary in the field of pathology of reproduction. That is why we could consider some epidemiologic data. Over 2 consecutive years we reported the frequencies of the different types of spontaneous abortion specimens observed (Fig. 1). There was no statistical difference between the respective percentages of expulsion "en bloc" of ovular anomalies or of retention not otherwise specified during the 24 months under study. There was also no indication of a possible, seasonal variation.

A systematic approach was conducted as detailed in Table 1. All data concerning placental tissue were noted in order to separate as completely as possible retention features from anomalies (possibly chromosomal) of the conceptus.

In Table 2, we detail the different histologic diagnoses obtained in a continuous series of 4479 cases of the first and second trimesters. Of par-

Table 1. Histologic examination

Placental Tissue

Villous development
Percentage of abnormal villi
Villous trophoblast – degree of proliferation or of hypoplasia
Villous core – Cells
                – Blood vessels
                – Cisterns
                – Incurring trophoblasts
Villous arborization (← stem villi)
Trophoblastic shell – and cytotrophoblast cell columns

Chorionic plate (blood vessels) and yolk sac

Maternoembryonic Interface

Importance of interstitial trophoblast (depth of penetration)
Intravascular trophoblast (plugs – depth of penetration)
Physiologic changes of spiral arteries
Necrosis of the decidua – extent
Inflammatory cells
Blood in the intervillous space
Decidual hemorrhage
Status of decidual veins

---

ticular interest is the low (2.5%) percentage of first-trimester infections as opposed to a high percentage (31.5%) after 14 weeks. Anomalies of the conceptus had been suspected in 28.6% of first-trimester cases against 5.8% in the second trimester. From these findings we can conclude that the pathology of abortion is markedly different before and after 12 weeks. The mechanisms of spontaneous abortion of first trimester pregnancies can be divided in four categories: abnormal karyotypes, endocrine abortions, maternal diseases, and tubal (ectopic) pregnancy.

Abnormal Karyotypes

The incidence of chromosomal disorders in early pregnancies, complicated by spontaneous abortion, varies between 30% and 50% [7–9]. Correlation between cytogenetic and pathologic findings has been reported by several authors with a wide range of results [4–9]. There is, however, a rather important discrepancy between the criteria used by the different authors. In order to understand the possible relationship which existed between karyotype and villous development, we conducted a retrospective study on 600 karyotypic analyses for spontaneous abortion. Cytogenetic results were obtained in 434 cases, including 122 (28%) cases presenting with chromosomal abnormalities. The overall correlation between cytogenetics

Table 2. Spontaneous abortion: histologic findings

	First trimester (n)	(%)	Second trimester (n)	(%)
Normal viable placenta	386	10.4	151	19.7
Anomaly of the conceptus	1062	28.6	45	5.8
Evolution arrest NOS	1988	53.5	219	28.5
Abruptio placentae	184	5.0	112	14.5
Inflammatory changes	94	2.5	242	31.5
	3714	100	765	100

NOS, not otherwise specified.

Table 3. Discrepancies between histology and cytogenetics – results of a "second look" survey

Karyotype	Cases (n)	Cases correctly identified by second look histology
Normal male	13	3
Trisomy 21	9	1
Trisomy 16	3	0
Various trisomies and deletions	21	8
Monosomy X	19	6
Triploidy	10	3
Tetraploidy	5	2
	80	23

and standard histology was 45% for this series. In fact, 95 cases with a complete discrepancy were observed. All cases where a female normal karyotype ($n = 15$) was obtained were eliminated since it is always extremely difficult to exclude the possibility of culturing maternal cells.

The final study group included 13 cases of normal male karyotypes with histologic features of placental anomaly and 67 specimens related to an abnormal karyotype where the possibility of chromosomal anomaly was completely unsuspected at histology. The details of these cases are reported in Table 3. As we wanted to ascertain that the histologic diagnosis was clear-cut, we looked at all these 80 specimens again and compared the results obtained by two independant observers with the first diagnosis (Table 3).

It is particularly interesting to note that almost all cases of trisomy 21 and monosomy X, and several tetraploidies had been previously mis-diagnosed by histology. Interestingly enough, 12 of 13 cases with normal male karyotype were not recognized as normal. Either the histologic aspects linked with retention blurred the picture, or the conceptus, though chromosomally normal, was grossly malformed with obvious consequences at the placental level. However, this hypothesis still remains unproven.

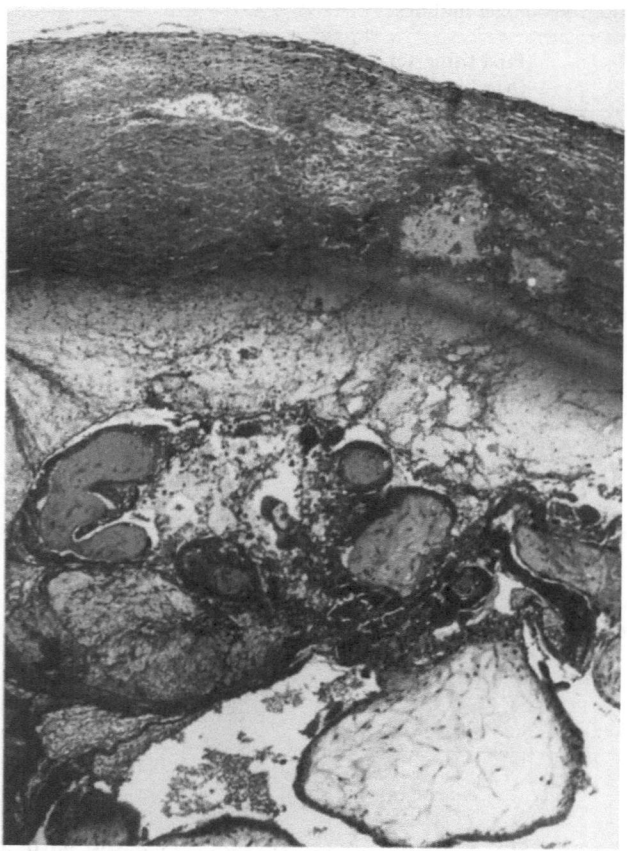

Fig. 2. Voluntary interruption of pregnancy at 8 weeks. The chorion laeve (covered with a largely necrotic operculum) countains a few villi which appear edematous or fibrotic and avascular. Hemotoxylin and eosin (H&E); ×25

At the second look, eight cases out of nine trisomies 21 were once again lost to correct diagnosis just as the majority of monosomy X cases. It is well known that trisomy 21 and monosomy X are not usually associated with prominent alterations of villous development even late in gestation [10]. It is therefore not surprising that all these cases were missed both at intra- and interobserver assays. Recently Röckelein et al. [8] concurred that monosomy X escaped systematic recognition whatever assay was used. Tetraploidy is not necessarily associated with morphologic anomalies [7]. Curiously, there were ten cases of unsuspected triploidy, of which seven remained undiagnosed at second look. Triploidy is known to be associated with caricatural villous anomalies which are sometimes confused with true molar changes [8, 9]. It has been shown, also in our own series, that 70% of triploid cases could be diagnosed by histology only. All cases which escape detection display, after review, a placental appearance of simple evolutionary arrest. Minguillon et al. [7] had a 22.5% incidence of abnormal placental findings (observed by two pathologists) in cases with normal karyotypes and two cases out of eight triploidies were also judged as normal.

The frequent finding of villous anomalies in cases with normal karyotypes deserves more comment. From the moment of differentiation between chorion frondosum and chorion laeve, the latter area will go through important regressive changes. The number of villi is reduced, and they are either hydropic and avascular or fibrotic. The trophoblast layer is markedly hypoplastic (Fig. 2). All these changes are, of course, found in true trisomic cases. Therefore, it is mandatory to record the percentage of normal and abnormal placental villi in the histologic survey of spontaneous abortion. It might be wise to discard as probably of no significance the anomalies present in fewer than 20% of all villi. Similarly, a mixture (probably fraught with artifact) of strictly normal villi with some others which are distended and hypoplastic must not be considered as pointing to possible malformations of the conceptus.

There are thus 20 definitively abnormal cases, which were added to the overall series of concordant histologic-cytogenetic observations. However, this figure represents only 29% of the 67 specimens with defective karyotype and nonsignificant histologic findings. In the grand total of 122 cases, there is thus a rescue of 16% abnormal cases and thus, at the end, 75 cases (61%) of chromosomal anomaly were suggested by histology alone.

All these comments lead to the important notion that histologic examination of spontaneous abortion products is not sufficient to give a proper etiologic diagnosis, especially in the concept of chromosomal anomaly. However, in most countries standard histology is more frequently used than systematic karyotyping because of the cost of the latter procedure and the rather low yield of culture growths. In our experience, 684 specimens (46%) failed to grow from a total of 1486 specimens cultivated between 1984 and 1990 (Jauniaux et al., submitted). There was, moreover, a significant number of cases where only maternal cells were cultivated. Recent studies [11] suggest, however, that chorion villus sampling at the time of diagnosis of embryonic death might be associated with a high success rate.

What are the main features pointing to an anomaly of the karyotype in placental specimens? Classically [12], it has been suggested that the presence of numerous intravillous recurring trophoblasts together with a reduced or absent capillary network is strongly suggestive of a chromosomal anomaly. How does this occur? The budding of secondary and tertiary villi is reduced or nonexistent [9] (Fig. 3). There are no lateral cytotrophoblastic proliferative nests and, if the villous core pushes outward for some time, it does so without "cellular logic" and in contiguous locations (Fig. 4). Eventually this leads to incorporation of trophoblastic lining which in turn degenerates

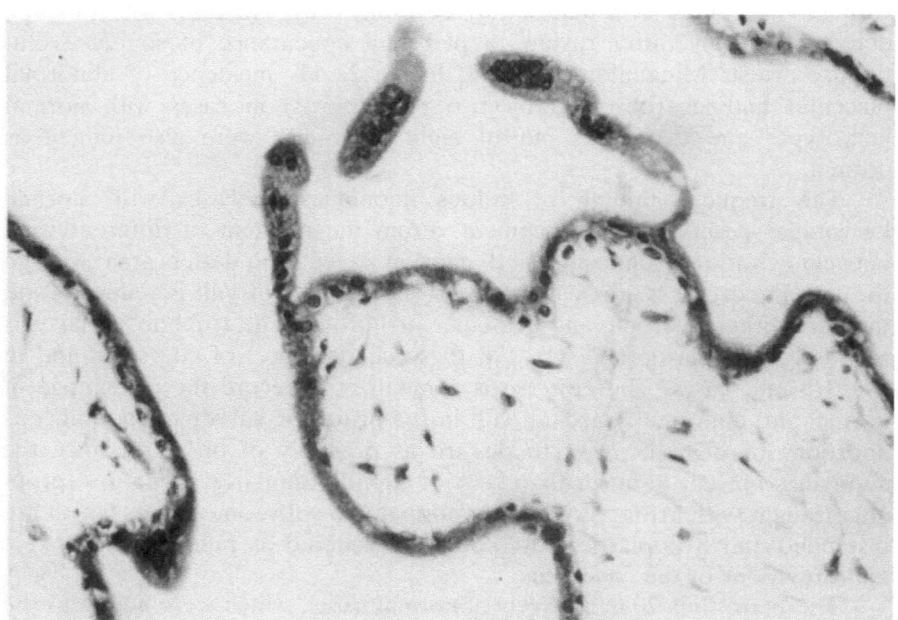

Fig. 3. Trisomy 2. Trophoblastic hypoplasia with defective villous arborization. H&E; ×63

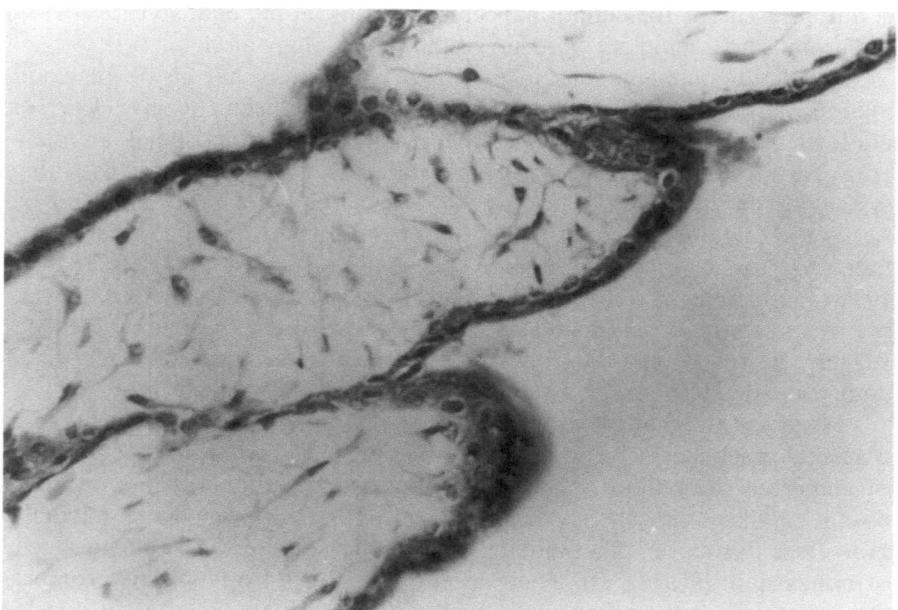

Fig. 4. Trisomy 2. Note the abortive villous arborization leading to trophoblastic inclusion and the total absence of vascularization. H&E; ×63

leaving only isolated cells in the core. This points, in fact, to a highly defective arborization of the villous tree. Incurring trophoblasts share the immunohistochemical properties of syncytiotrophoblast (*human chorionic gonadotropin*, hCG; *Schwangerschafts protein* 1, SP1; and, in early cases, prolactin, PRL; Hustin et al., unpublished results). This, of course, constitutes an argument for their derivation from villous lining.

We have observed (see below) that trophoblastic hypoplasia was often a diffuse phenomenon, identified both at the villous and extravillous laeve. It might be that, in characteristic cases, trophoblast hypoplasia is a very early phenomenon with defective cell columns and reduced villous formation. It can be suggested that this failure of trophoblastic development occurs as early as the differentiation of trophoectoderm. Another feature which must also be looked for is the general or localized absence of blood vessels [13], pointing either to an absent embryo and/or to defective yolk sac function (see Chap. 11). A common finding is that the most edematous villi are usually devoid of any capillaries. This suggests that:

1. Transfer or diffusion of water and electrolytes occurs whatever the villous state.
2. Water accumulates in loco since no capillary resorption is possible. This edema culminates in cistern formation which is often found in triploid specimens.

A more detailed histologic study of such pathologic cases might shed some light on the enigmatic rationale of secondary and tertiary villous budding. It is well accepted that, in normally evoluting pregnancies, villous arborization proceeds continuously according to a well-equilibrated scheme [14]. However, the primary mechanism remains to be discovered. There are some cytotrophoblastic "nests" on the lateral sides of stem villi. They could indicate the site of future development and act as a trigger for mesenchymal and capillary growth.

Endocrine Abortions

A number of early gestations are interrupted before amenorrhea is present. This is the most logical scheme for so-called endocrine abortions (except, of course, for the exceptional case of accidental lutectomy during the first weeks). Endocrine abortion must be defined as a failure of pregnancy to continue owing to adverse effects of hormone imbalance. It is quite difficult to prove this etiologic factor since there are no characteristics of the expelled conceptus. Most probably, abnormal hormonal patterns may have a detrimental effect on pregnancies at the crucial time of implantation when the endometrium must be in a critical state (see Chap. 2).

Luteal insufficiency may be accompanied by a number of pregnancy losses. These are very early and usually expelled with menstrual endo-

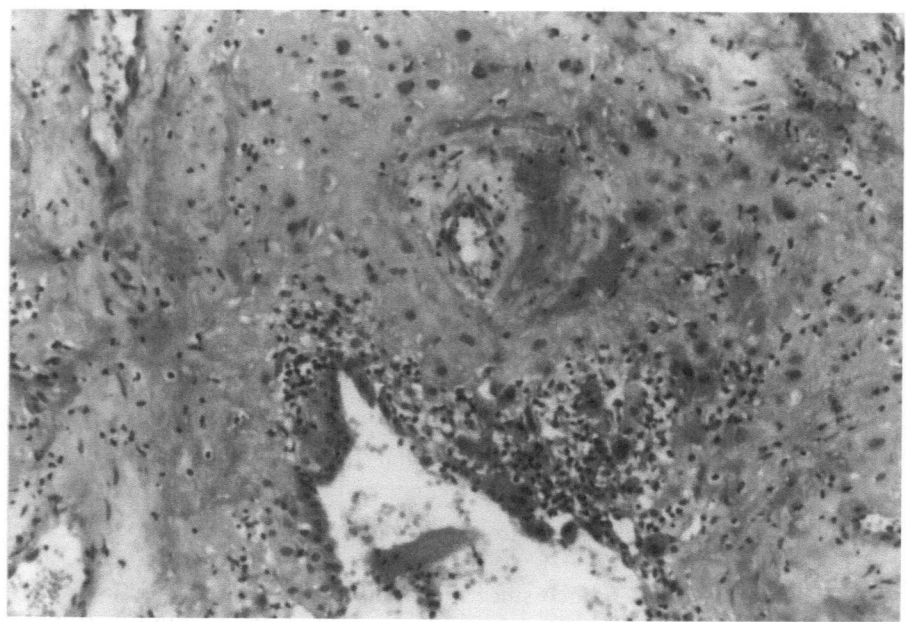

Fig. 5. Uteroplacental artery demonstrating fibro-occlusive changes in a case of spontaneous abortion of unknown etiology. H&E; ×63

metrium. They are rarely demonstrated histologically unless specifically searched for. Logically one has to postulate that shedding of the mucosa linked to corpus luteum involution occurs without any influence of early trophoblast. This means, in particular, that early trophoblast productions (i.e., hCG, prostaglandins), though possessing a true luteotropic effect [15, 16], do not directly influence the endometrium. It has recently been demonstrated [17] (see also Chap. 20) that with the use of an ultrasensitive and specific β-hCG assay, premenstrual pregnancy loss could account for as much as 22% of all conceptions. Thus, in the case of early (i.e., preamenorrhea) abortion, two hypotheses may be put forward:

1. One is that the corpus luteum is unsufficient and the decidua does not develop; corpus luteum is defective though embryonic signals (platelet-activating factor, PAF; early pregnancy factor, EPF; hCG; prostaglandins) are clearly sent and interpreted.
2. The other is that of a defective embryo – with defective embryonic signals and thus an absence of interpretation by the maternal organism of an ongoing pregnancy. These cases are those where disorganized eggs are found either without an embryo or without any regular trophoblastic development [18].

Maternal Diseases

There are arguments for some analogy in pathogenesis between late (i.e., end of first-trimester and second-trimester) abortions and so-called pre-eclamptic changes in the placental bed (Fig. 5) [19]. It is not clear whether these changes are a real clue to etiology or whether they merely reflect circulatory disturbances due to a lack of sufficient trophoblastic effect on the vessels. It may be that the primary anomaly is not associated with an insufficiency of placentation, but that it precludes normal evolution and is possibly a signal for poor obstetric outcome. In this respect, it must be remembered that repeated fetal loss, i.e., habitual abortion cases, are significantly associated with lupus anticoagulant and anticardiolipin antibodies [20]. The presence of these factors, which are possibly involved in the pathology of coagulation, is highly suggestive of a vascular – possibly thrombotic, hence ischemic – disorder as the underlying cause of abortion. Moreover, Woodhams et al. [21] have recently, suggested that there is an increased activation of coagulation during normal pregnancy, counterbalanced by a simultaneous rise in fibrinolytic activity. In spontaneous abortion, Woodhams et al. [21] demonstrated that only the activation of coagulation increased.

The Problem of Tubal (Ectopic) Pregnancy

Fox has suggested (see Chap. 17) that conceptuses could implant almost everywhere and particularly well in tubes. Our findings support this hypothesis and we have discussed (see Chap. 2) the possibility that the endometrium is a less convenient site for implantation unless specific embryonic requirements (hormonal, secretory, etc.) are met. What differs markedly is the rapidity with which such pregnancies are expelled by an abruptio phenomenon long before tubal distension has occurred. Three possibilities could be proposed:
1. The mean blood pressure at the tubal level may be higher than in endometrium.
2. Physiologic changes are less prominent, even though there are intravascular trophoblastic plugs which, however, do not extend deeply. The tortuosity of tubal arteries is markedly less than that of endometrial spiral arteries.
3. There is no contact between trophoblast and decidua. It ensues that strong anchoring is probably reduced (see Chap. 2).

The Mechanisms of Spontaneous Abortion

It is highly probable that a number of abortion cases are still in search of a valid reason. Khong et al. [3] have recently suggested that the phenomenon

Fig. 6. Spontaneous abortion. Placental bed. Note the reduced pattern of interstitial trophoblast infiltration, the absence of intravascular plugs, and the fragmented appearance of the hypoplastic trophoblastic shell. H&E; ×40

of placentation itself could be defective and could explain some cases of miscarriage. Considering the delicate and highly precise problems needed by correct implantation (see Chaps. 2, 6), we thought that a closer look at the pathogenesis of abortion could be fruitful. It is, in fact, curious that all studies and all textbooks which are devoted to the subject are usually completely silent about the mechanisms which lead to expulsion.

In a previous study, we suggested that during the first trimester the normal conceptus is probably completely protected from the maternal environment and that no true blood flow is initiated (see Chap. 5). Ultrasound studies have now demonstrated that, in pregnancies with a dead embryo, pulsatile echoes could be demonstrated in the intervillous space and appeared synchronous to the maternal pulse [22]. Therefore, we put forward the hypothesis that pathogenesis of abortion could be linked with vascular disturbances at the basal plate level or even within the intervillous space. We have thus undertaken a thorough survey of spontaneous abortion products which were submitted for histology during a 6-year period (a total of 3714 specimens among which 184 were obtained as complete undisturbed eggs). In this retrospective study, only these 184 cases were included in order to differentiate confidently between real pathology and the almost

Table 4. Maternotrophoblastic interface

	Normal pregnancies (%)	Embryonic death (%)	Abnormal villous development (%)
Extra villous trophoblastic hypoplasia	0	56	82
Physiologic changes of the spiral arteries	100	64	77

unavoidable artifacts linked to instrumental obtainment of the specimens [23].

The control series consisted of 222 products from voluntary terminations of pregnancies performed during the first 12 weeks (usually for psychosocial reasons). In order to obtain reproducible results, we adhered to the same strict scheme of investigation as summarized in Table 1. However, more interest was devoted to the maternoembryonic interface and to the status of maternal blood vessels. The 184 cases under study were not exactly representative of the whole series in that 79 (43%) displayed anomalies of villous development, with trophoblastic and vascular hypoplasia pointing to a possible abnormal karyotype as opposed to 28.6% in the complete series (see Table 2) [24]. No infectious etiology could be elicited in any case.

The normal appearance of the early placenta has already been described (see Chap. 3). In contrast, spontaneous abortion specimens displayed features of reduced or absent placentation. There was a considerable reduction of thickness of the trophoblast shell. Cytotrophoblastic cell columns were also scant, if at all present (Fig. 6). Curiously, in most abnormal cases, the remaining columns were not oriented towards the trophoblastic shell, but rather in various directions. This also points to a defective arborization scheme of the villous tree (see above). Though cases of ovular death could also have features of reduced placentation, these were not as prominent as in the abnormal group (Table 4).

At the maternotrophoblastic interface, pathologic groups were characterized by a marked decrease in extravillous trophoblast infiltration both extra- and intravascularly. It ensued that physiologic changes of spiral arteries were lacking in more than 60% of spontaneous abortion cases. Lastly, abundant blood and clots were present in the intervillous space in the vast majority of ovular anomalies or ovular deaths (Figs. 7, 8).

There is a gradient of trophoblastic infiltration within the decidua at the placental site. The deeper the penetration, both interstitially and intravascularly, the more normal the outcome of pregnancy is supposed to be. On the other hand, highly pathologic conceptuses were usually associated with a reduced or absent infiltration (Fig. 9). This paradoxically culminates in cases of hydatidiform moles where a more or less thick, compact layer of atypical trophoblast is apposed to the decidua. This is not, however, equiv-

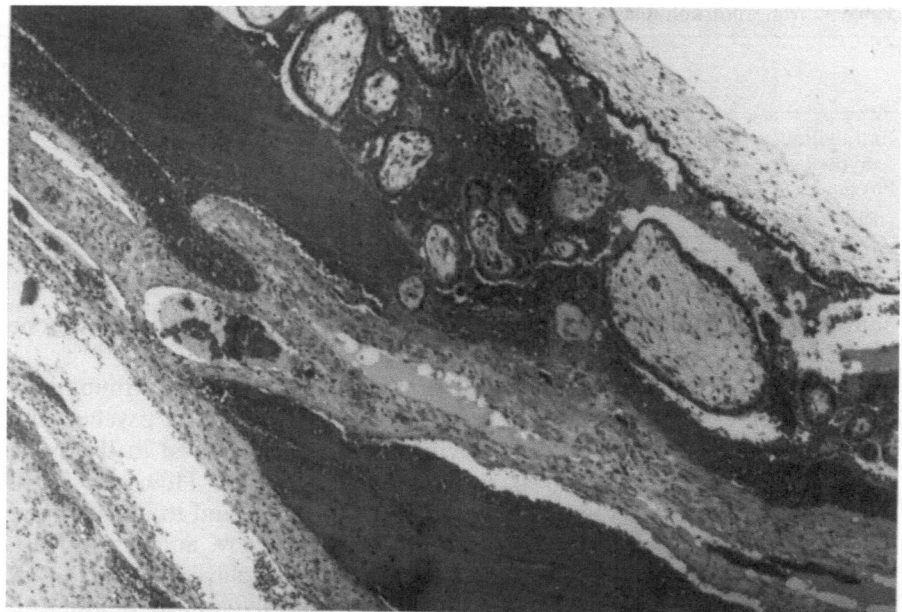

Fig. 7. Spontaneous abortion. Irruption of maternal blood in the intervillous space. The number of villi is reduced. Decidua is also infiltrated by maternal blood. H&E; ×25

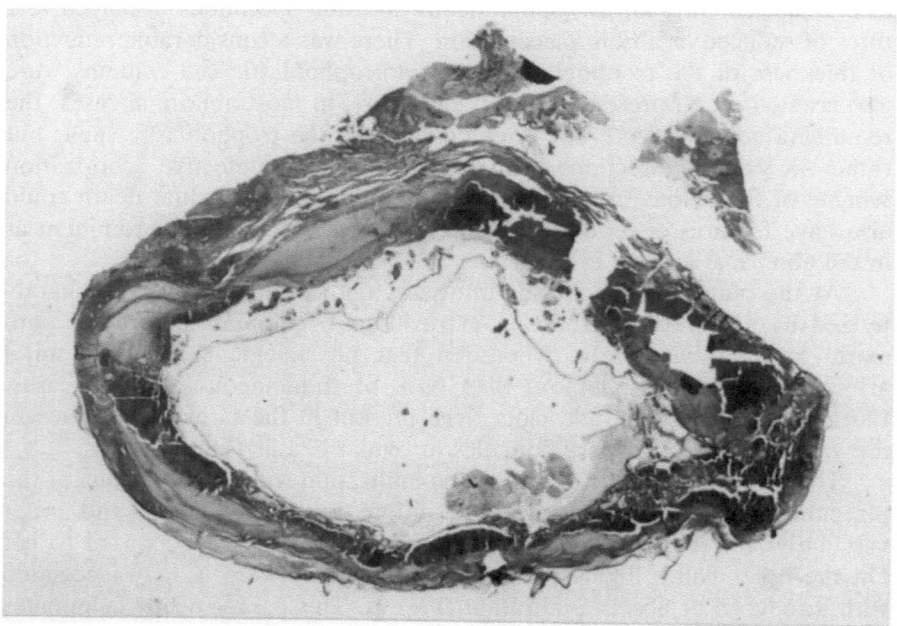

Fig. 8. Spontaneous abortion. Complete egg. Macerated embryo is conspicuous. Note the extreme placental hypoplasia and the presence of clots in the IVS and decidua. H&E; ×4

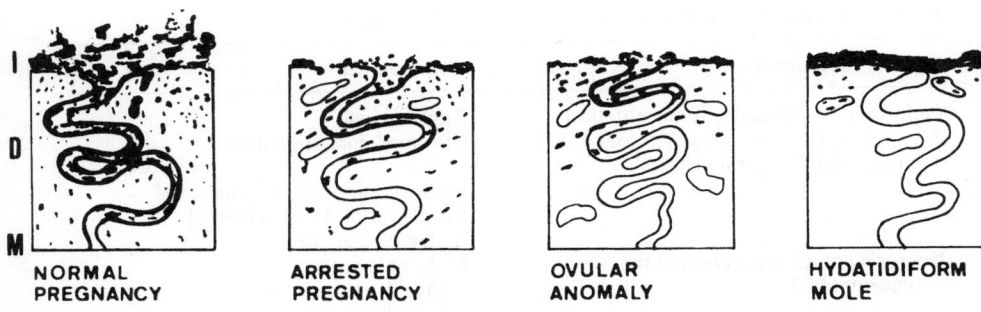

NORMAL PREGNANCY   ARRESTED PREGNANCY   OVULAR ANOMALY   HYDATIDIFORM MOLE

Fig. 9. Extravillous trophoblast infiltration of the decidua in various pathologic states; I = trophoblast-decidual interface; D = decidua; M = myometrium

alent to the trophoblastic shell of normal pregnancies. Only thus is intra-vascular trophoblastic invasion possible. In triploid cases, the appearance of the placental bed is suggestive of a limited trophoblastic infiltration only around tapped spiral arteries and not diffusely throughout the decidua. There is thus a distinct possibility that pregnancy failures during the first trimester could be separated into two categories according to the complete-ness of placentation.

In the first group, we will include all those cases where trophoblastic infiltration outside the trophoblastic shell is distinctly abnormal (i.e., essen-tially reduced). These are cases of ovular pathology, chromosomal or highly malformative or blighted ova. If extravillous trophoblast is quantitatively reduced (it remains to be defined whether qualitative alterations also occur), the trophoblastic shell and intravascular plugs will oppose a markedly reduced cohesion and resistance to the physical maternal environment and, in particular, to the sluggish blood flow in the uteroplacental arteries. If maternal blood is allowed to flow freely in the intervillous space, several conditions must be met. First, the embryonic heart must be completely formed with a vigorous and regular beat. Secondly, embryonic arterial diameter must be wide enough for a significant blood pressure and blood flow to the capillaries. Thirdly, allantochorial arteries must be present in the chorionic plate and stem villi in order to obtain a sufficient intravillous flow. Fourthly, the blood pressure in the interventricular septum (IVS) must be inferior to that in the villous capillaries.

If any of these conditions is not present, embryonic blood flow will quickly cease at the placental level, eventually leading to death, and mater-nal blood will enter the IVS and simultaneously infiltrate the decidua via the interface between trophoblastic shell and endometrium. A more or less complete cleavage line will be obtained and placental separation (complete or incomplete) will happen.

Other cases are probably normal conceptuses with a normal implanta-tion. In these cases pregnancy failure might be maternal in origin. Either inflammatory processes (infections are, however, rare during the first 12

Table 5. Spontaneous abortion: etiology versus pathogenesis

Primary mechanism	Common causes
Failure or insufficiency of placentation	Blighted ova
Absence or reduction of physiologic changes of the spiral arteries	Chromosomal anomalies
	Malformations of the conceptus
	Trophoblast insufficiency (?)
	Maternal diabetes
Necrosis and/or separation at the placental bed	Maternal causes
	– Menstruation
	– So-called endocrine causes:
	• Nontransmission or non-reception of embryonic signals
	• Early lutectomy
	– Infections
	– Maternal causes (hypertension, diabetes)
	– Uterine and cervical malformations
Physiologic changes of the spiral arteries may be reduced	
Normal placentation:	Normal pregnancy
– Normal Extravillous trophoblast	
– Physiologic changes of the utero-placental arteries are present	
Untimely initiation of blood flow in the IVS with subsequent abruptio	

weeks) are prominent and give rise to abcess formation with extensive necrosis (and sloughing) of the decidua, or the necrotic foci (which are a part of the physiologic processes) are so extensive that physical support to the conceptus no longer exists. This is particularly true when shedding of the uterine mucosa occurs during either normal or delayed menstruation. This is probably the process by which so-called endocrine abortions are expelled.

There remains a last group, the importance of which cannot, of course, be fully estimated. It comprises specimens where the conceptus is apparently normal, with a sufficient trophoblastic infiltration of the placental bed. No etiologic factors are present except for the untimely occurrence of blood in the intervillous space. The hypothesis is that spontaneous initiation of a physiologic blood flow in the IVS occurs too early. This may be due to an elevation of the pressure within the spiral artery and/or to a premature loosening of the intravascular plugs and trophoblastic shell disruption.

The study of the mechanisms leading to spontaneous abortion enables us to propose a classification based on pathogenesis. It is summarized in Table 5. It ensues that histologic examination of the embryo and its placenta is a prerequisite for proper diagnosis. The etiology of abortion cannot be approached in a significant number of cases. However, care must be exercised in the search of so-called normal specimens: abortion could be purely fortuitous due to a lack of synchronism in the different physiologic events. Moreover, even if histology is a poor predictor of embryonic anomalies, it is fairly inexpensive and widely available. Infections, though infrequent during the 1st trimester, can also be readily detected. Uterine malformations leading to cervical incompetence are of little importance since the bulging of the gestational sac in the uterine cavity and its possible sliding through the endocervical isthmus occurs only at the end of the first 12 weeks. Numerous studies are now needed to replace the context of habitual aborters in this new approach. It is, in our opinion, mandatory to restrict this term to those women who have experienced three or more consecutive miscarriages with a complete histologic study of the placenta and its surroundings, leading to the conclusion that all accidents must be included in the subgroup "failure or insufficiency of placentation". The presence of other stigmata pointing to a generalized trophoblastic hypoplasia would be recorded as additional evidence. Such reports could conceivably add to the knowledge on embryonic loss and focus the data on the different factors involved in the etiology [25].

References

1. Brambati B, Oldrini A, Ferrazzi E, Lanzani A (1985) Chorionic villi sampling: general methodological and clinical approach. In: Fracarro M, Simoni G, Brambati B (eds) First trimester fetal diagnosis. Springer, Berlin Heidelberg New York, pp 7–18
2. Rhoads GG, Jackson LG, Schlesselman SE, et al. (1989) The safety and efficacy of chorionic villus sampling of early prenatal diagnosis of cytogenetic abnormalities. N Engl J Med 320:609–617
3. Khong TY, Liddel HS, Robertson WB (1987) Defective haemochorial placentation as a cause of miscarriage: a preliminary study. Br J Obstet Gynaecol 94:649–655
4. Berry CL (1980) The examination of embryonic and fetal material in diagnostic histopathology laboratories. J Clin Pathol 33:317–326
5. Göcke H, Muradow I, Cremer H (1982) Morphologische und Zytogenetische Befunde bei Frühaborten. Verh Dtsch Ges Pathol 66:141–146
6. Novak R, Agamanolis D, Dasu S, Igel H, Platt M, Robinson H, Shekata B (1988) Histologic analysis of placental tissue in first trimester abortions. Pediatr Pathol 8:477–482
7. Minguillon C, Eiben B, Bähr-Porsch S, Vogel M, Hansmann I (1989) The predictive value of chorionic villus histology for identifying chromosomally normal and abnormal spontaneous abortions. Hum Genet 82:373–376
8. Röckelein G, Ulmer R, Schröder J (1990) Karyotype and placental structure of first trimester spontaneous abortions: a morphological study. Eur J Obstet Gynecol Reprod Biol 38:25–32
9. Röckelein G, Ulmer R, Schwille R (1990) Surface and branching of placental villi in early abortion: relationship to karyotype. Virchows Arch [A] 417:151–158
10. Oberweiss D, Gillerot Y, Koulisher L, Hustin J, Philippe E (1983) Le placenta des trisomies dans le dernier trimestre de la festation. J Gynecol Obstet Biol Reprod (Paris) 12:345–349
11. Johnson HP, Drugan A, Koppitch FC, Uhlman WR, Evans MI (1990) Post mortem chorionic villus sampling is a better method for cytogenetic evaluation of early fetal loss than culture of abortus material. Am J Obstet Gynecol 163:1505–1510

12. Philippe E (1986) Pathologie foeto-placentaire. Masson, Paris
13. Meedges BHLM, Ingenhoes R, Peeters LLH, Exalto N (1988) Early pregnancy wastage: relationship between chorionic vascularization and embryonic development. Fertil Steril 49:216–220
14. Castellucci M, Scheper M, Scheffen I, Celona A, Kaufmann P (1990) The development of the human placenta villous tree. Anat Embryol (Berl) 181:117–128
15. Hahlin M, Dennefors B, Johanson C, Hamberger L (1988) Luteotropic effects of prostaglandin E_2 on the human corpus luteum of the menstrual cycle and early pregnancy. J Clin Endocrinol Metab 66:909–914
16. Bennegird B, Hahlin M, Hamberger L (1990) Luteotropic effects of prostaglandins I_2 and D_2 on isolated human corpora lutea. Fertil Steril 54:459–464
17. Wilcox AJ, Weinberg CR, O'Connor JF, Baird DD, Schlatterer JP, Canfield RE, Armstrong EG, Misula BC (1988) Incidence of early loss of pregnancy. N Engl J Med 319:189–194
18. Philippe E (1980) Les nidations précocément abortives. J Gynecol Obstet Biol Reprod (Paris) 9:513–521
19. Rushton DI (1988) Placental phatology in spontaneous miscarriage. In: Baird RW, Sharp F (eds) Early pregnancy loss. Springer, Berlin Heidelberg New York, pp 149–159
20. Creagle MD, Malia RG, Cooper SM, Smith AR, Duncan SLB, Greaves M (1991) Screening for lupus anticoagulant and anticardiolipin antibodies in women with fetal less. J Clin Pathol 44:45–47
21. Woodhams BJ, Candotti G, Shaw R, Kernoff PB (1989) Changes in coagulation and fibrinolysis during pregnancy: evidence of activation of coagulation preceding abortion. Thromb Res 55(1):99–107
22. Schaaps JP, Hustin J (1988) Dynamic imaging of the utero-placental border in the first trimester of pregnancy. Troph Res 3:37–45
23. Hustin J, Jauniaux E, Schaaps JP (1990) Histological study of the materno-embryonic interface in spontaneous abortion. Placenta 11:477–486
24. Houwert-de Jong MH, Bruinse HW, Eskes TKAB, Mantingh A, Termijtelen A, Kooyman CD (1990) Early recurrent miscarriage: histology of conception products. Br J Obstet Gynaecol 97:533–535
25. Edwards RG (1986) Causes of early embryonic loss in human pregnancy. Hum Reprod 1:185–198

17 Ectopic Sites of Placentation

H. Fox

Introduction

An ectopic pregnancy is one in which the fertilized ovum implants in a site other than the uterine cavity. In practice 95%–98% of ectopic gestations are in the fallopian tube, other sites being, in probable descending order of frequency, the ovary, cervix, abdominal cavity, liver, spleen and vagina.

Almost all that is known about ectopic pregnancy, in terms of epidemiology, aetiology and pathology, refers to tubal pregnancies and, indeed, many accounts of ectopic pregnancy make no reference to sites other than the tube, tending to regard "ectopic pregnancy" and "tubal gestation" as being synonymous. This account will therefore, of necessity, be devoted largely to tubal pregnancies.

Tubal Pregnancy

Incidence

Many difficulties are encountered in attempting to define the true incidence of tubal pregnancies. Optimally, their incidence should be expressed in terms of the community at risk, i.e. the number of tubal gestations per 100 000 women in the population aged between 14 and 44 years [1]. Community-based data of this type have rarely been reported, but nevertheless it is clear that there has been, over the last 20 years, an approximate doubling, at least, of the incidence of tubal pregnancies [2, 3]. The reasons for this increase are debatable, but possible factors are considered later.

Epidemiology

Tubal pregnancies are more common in urban than in rural populations, are encountered with greatest frequency in women of low socio-economic status and occur much more often in black than in white women. These factors account for some of the geographical variations in the incidence of tubal

pregnancy, e.g., high incidence in Jamaica, Africa and certain urban populations in the United States, but do not adequately explain the high incidence
in Finland and the low incidence in Malaysia. Furthermore, there has been
no real attempt to establish whether these ethnic and social factors are
independent or not of the incidence of pelvic inflammatory disease.

The incidence of tubal pregnancy rises with increasing maternal age [4],
and the highest incidence is in women with two or three previous pregnancies. Despite the fact that most patients with tubal gestations are parous,
there is a high incidence of antecedent one child infertility, primary infertility, spontaneous abortion and ectopic pregnancy, whilst many patients
have a history of pelvic inflammatory disease [5].

Aetiology and Pathogenesis

Tubal Factors

Tubal abnormalities clearly play an important aetiological role, and it is
commonly thought that pelvic inflammatory disease, with presumed salpingitis, is a key factor in tubal nidation. It is, however, difficult to asess the
true quantitative importance of tubal infection in this respect for there is a
striking lack of correlation between a clinical history of pelvic inflammatory
disease and pathological evidence of tubal inflammation. It is therefore not
surprising that there is a poor epidemiological association between clinically defined pelvic inflammatory disease and tubal pregnancy [4]. Equally
unsurprisingly, but somewhat disconcertingly, the reported incidence of
clinically diagnosed pelvic inflammatory disease in patients with tubal pregnancies has varied from 0.5% to 82%, whilst that of histologically proven
salpingitis has ranged from 18% to 55% [5]. Despite these discrepancies, it
would, however, be a reasonable assumption that 30%–40% of tubes in
which a conceptus is lodged show evidence of present or past salpingitis.
This, in itself, does not prove that tubal infection predisposes to tubal
pregnancy unless the incidence, in the same population, of salpingitis in
women with intrauterine pregnancies is known. In the only case control
study of this relationship, Westrom [6] calculated that pelvic inflammatory
disease increased the incidence of tubal pregnancies seven- to tenfold.

In Western countries chlamydial infection is probably the most important cause of salpingitis predisposing to tubal pregnancy [7, 8], but in many
parts of the world tubal tuberculosis and schistosomiasis are still important
factors.

Salpingitis isthmica nodosa is a probable, though not fully proven,
aetiological factor in tubal pregnancy [9], but the conceptus does not
actually implant in a diverticulum. It is possible, indeed probable, that the
hypertrophied muscle which is a feature of this condition undergoes spasm
and obstructs the tubal lumen.

Tubal endometrioisis is probably not an important cause of tubal implantation, but scanning electron microscopy has shown that there is often a marked deficiency of cilia in the tubes of women who have a tubal gestation [10]. The significance of this finding merits further study.

A significant proportion of pregnancies which occur after a failed tubal sterilization, probably 10%–15%, nidate in the tube, whilst there is also a high incidence of tubal implantation following reconstructive tubal surgery.

Congenital abnormalities of the tube, such as accessory ostia, segmental deficiency or diverticula, predispose to tubal pregnancy, whilst tubal neoplasms are an extremely rare cause of tubal implantation.

Ovulatory Dysfunction

Iffy [11] noted that the fetus in a tubal gestation often appeared to be 3–4 weeks older than would be expected from the period of amenorrhoea and suggested that conception had actually occurred before the bleeding episode which the patient considered as her last menstrual period. Iffy therefore suggested that a conception leading to a tubal gestation occurred during a cycle in which there was delayed ovulation and a short, inadequate luteal phase. Hence at the time of corpus luteum decay the conceptus would not have reached the stage when it was producing sufficient human chorionic gonadotropin (hCG) to prevent luteolysis and menstruation. During the subsequent menstrual bleeding the fertilized ovum could either be arrested in its transit through the tube by a reflux of menstrual blood or, if it had already reached the uterine cavity, be flushed back into the tube by menstrual regurgitation. This theory is supported by the fact that the corpus luteum of pregnancy is not uncommonly on the opposite side to a tube containing an ectopic gestation.

The presence of a corpus luteum of pregnancy in the gonad contralateral to the gravid tube could also be due to transuterine migration of the fertilized ovum, whilst transperitoneal migration, the ovum trekking across the pelvis to enter the contralateral tube, is also a possibility.

Hormonal and Contraceptive Factors

If pregnancy occurs in a woman using an intrauterine device (IUD), it will be ectopically sited in an unusually high proportion of cases. This does not, however, mean that IUDs cause ectopic pregnancy, but rather that they inhibit intrauterine implantation much more successfully than they do ectopic nidation; and, in fact, an IUD user has no higher risk of an ectopic pregnancy than does a non-user [12]. Neither combined oral steroid contraception nor condom usage is associated with any excess of tubal gestations, but a high incidence of tubal pregnancies has been noted in women using low-dosage progesterone-only contraception, possibly because of the effects of progesterone on tubal muscular contractility.

Low progesterone levels have also been invoked as a possible cause of tubal gestation in cases of presumed luteal insufficiency, whilst it has been suggested that reduced oestrogen levels, due to dieting or excessive exercise, may also interfere with tubal contractility and predispose to tubal implantation [13]. Low oestrogen levels may also be a factor in the suggested link between cigarette smoking and tubal pregnancies [14].

There have been a number of reports of tubal pregnancies following the use of ovulatory agents, both gonadotrophins and clomiphene, being implicated in this respect. These reports are, however, largely anecdotal, and it remains an open question as to whether there is a truly increased incidence of tubal implantation following administration of ovulatory agents [5].

Abnormal Conceptus

It has intermittently been suggested that the incidence of fetal chromosomal abnormalities is much higher in tubal pregnancies than in intrauterine gestations [15]. This hypothesis appeared to have been refuted by cytogenetic studies [16, 17], but has recently been revived by a flow cytometric investigation of tissue from tubal conceptuses in which a very high incidence of fetal aneuploidy was noted [18]. This question is therefore still not fully resolved.

Miscellaneous Factors

The incidence of ectopic nidations in pregnancies following in vitro fertization is higher than in those occurring spontaneously [3]. The reasons for this are far from clear. The incidence of tubal pregnancies is also increased in women who have been exposed pre-natally to diethylstilboestrol (DES).

Causes of Increased Incidence of Tubal Pregnancy

A consideration of the aetiological factors detailed above provides some reasons for the increasing incidence of tubal pregnancies. Clearly, the worldwide increase in the incidence of sexually transmitted diseases is one factor. It is, however, probable that it is medical treatment of salpingitis, as much as the disease itself, which contributes to the high incidence of tubal implantation, for this results in a scarred and damaged, rather than a blocked, tube and hence predisposes to ectopic nidation rather than to infertility. Similarly, untreated tubal tuberculosis causes infertility, whilst treatment of this disease results in residual tubal damage and hence a propensity for tubal nidation.

Other factors include the increased usage of IUDs, the more widespread resort to tubal sterilization, the greater frequency of tubal restorative surgery and the readier availability of in vitro fertilization (IVF).

The increased incidence of tubal pregnancy could be, to some extent, a cohort phenomenon [19], whilst the more widespread indulgence in cigarette smoking, exercise and dieting by women may also be of some importance. Nevertheless, this increased incidence appears to be due, at least in part, to medical conversion of women who are actually or potentially infertile into women at high risk of a tubal implantation and to increasing attempts to control fertility.

Implantation and Placentation in the Tube

The early stages of implantation in the tube have rarely been observed, but there are thought to be four forms: fimbrial, plical, mural and muroplical. Fimbrial implantation occurs in about 7% of tubal gestations [5], and in many such cases it is difficult to know if nidation has, in fact, occurred in the tube or in the ovary, as it is not uncommon to have to resort to the somewhat unsatisfactory diagnosis of "tubo-ovarian pregnancy". In plical implantation the conceptus implants on, or near, the tip of a mucosal fold and does not establish any contact with the tubal wall [20].

In a mural implantation the conceptus is in direct contact with the luminal aspect of the wall of the tube, between adjacent plicae, whilst in a muroplical implantation the fertilized ovum embeds between plical folds with trophoblast invading both the adjacent plicae and the tubal wall.

Definitive placentation takes place only in mural and muroplical implantations, plical implantation offering too small a volume of tissue to allow for adequate nidation and implantation. In a mural implantation, the process of placentation is identical to that which occurs in an intrauterine gestation [21]. It is not known if, after implantation in the tubal wall, the trophoblast becomes orientated at one pole of the conceptus to form a definitive placenta. In some tubal pregnancies the trophoblast does appear to invade a localized area of the wall to form an approximately discoid placenta, but in most cases there is a circumferential placental growth to produce a structure akin to a placenta membranacea in an intrauterine pregnancy.

The early development of the placenta in a tubal gestation exactly replicates that seen in an intrauterine site with the formation of primary villous stems, primary stem villi and arborization into secondary and tertiary stem villi. The primary stem villi are radially orientated, and their distal portions are formed by aggregates of cytotrophoblast which are anchored to the tubal tissues (Fig. 1). A distinct trophoblastic shell is apparent in most cases, but Nitabuch's layer tends to be poorly developed. Cytotrophoblastic cells from the tips of the primary villous stems eventually break through the prophoblastic shell to infiltrate and extensively colonize the subjacent tubal wall. These invading interstitial extravillous cytotrophoblastic cells extend into the muscular wall of the tube, not uncommonly penetrating

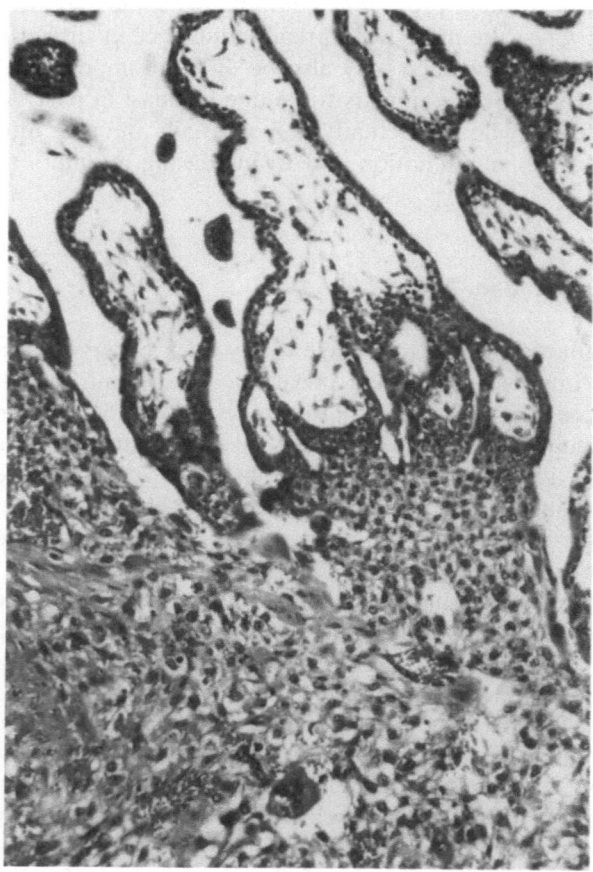

Fig. 1. Mural implantation in a tubal pregnancy. Primary villous stems, cytotrophoblastic cell columns and a trophoblastic shell are all apparent. Haematoxylin and eosin; ×240

its full thickness to reach the subserosa. In many cases, these infiltrating cells spread out laterally in the longitudinal axes of the mucosal and muscular layers of the tube. This degree of invasion of the muscle coat by cytotrophoblastic cells is more exuberant than is the similar invasion of the myometrium in intrauterine pregnancy and may be indicative of an absence of regulating factors in this site [22].

Intravascular invasion by extravillous cytotrophoblastic cells is also seen. These cells appear to grow directly into the lumens of the vessels of the tubal wall from the proliferating cytotrophoblast of the anchoring columns (Fig. 2). This penomenon of intravascular invasion is apparent in most, but not all, mural and muroplical implantations, but it is not present in chronic tubal abortions. The intravascular cytotrophoblast forms intraluminal cel-

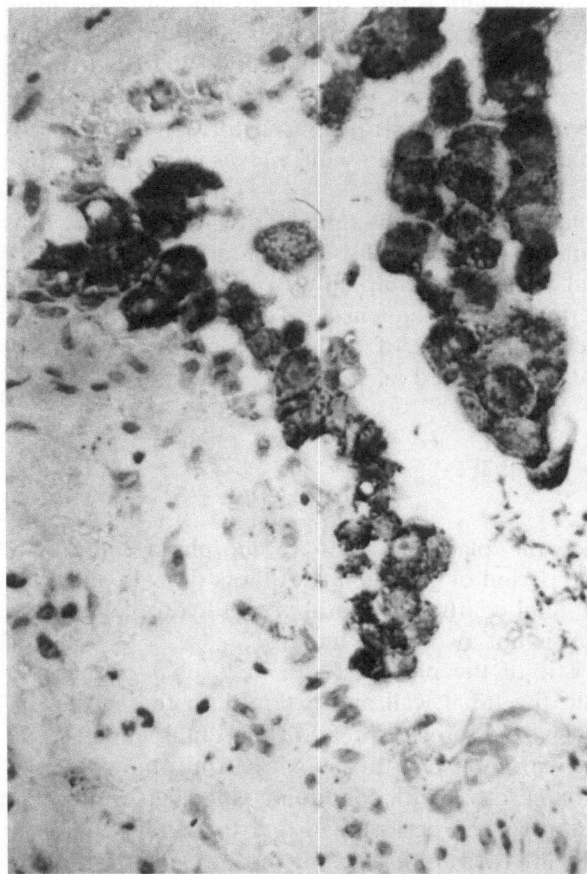

Fig. 2. Intravascular extravillous cytotrophoblastic cells within the lumen of a tubal vessel in an ectopic pregnancy. The cells have been stained for hPL. ×350

lular plugs which often appear almost to fill the vessel. These cells destroy and replace the endothelial cells of the invaded vessels, infiltrate the vessel wall and destroy the medial elastic and muscular tissues, these being replaced by fibrinoid material. In the vast majority of cases, the vessels invaded by extravillous trophoblast are clearly arteries, but in some instances veins are also involved, this being in contrast to the absence of venous involvement in intrauterine gestations.

Changes within the tubal vessels extend in most tubes throughout the full length of their course through the muscular layer of the wall and often extend into vessels in the subserosa. Exceptionally, involved vessels are seen outside the serosa in the peritubal loose mesenchymal tissues. Several points about placentation in the tube merit special comment.

1. It is clear that if a fertilized ovum is able to find a sufficiently adequate
    area of tubal tissue into which it can implant, then nidation and placen-
    tation occur in a manner which is virtually identical to that occurring in
    a normal intrauterine gestation.
2. Invasion of the placental bed by extravillous cytotrophoblastic cells
    occurs in a tubal pregnancy in exactly the same way as in an intrauterine
    gestation.
3. Invasion of maternal blood vessels in the tube by cytotrophoblastic cells
    occurs in the same manner as in the uterus, even though the structure of
    the vessels in the tubal wall differs markedly from that of the uterine
    spiral arteries. Furthermore, the sequence of changes which is seen
    during the physiological conversion of the spiral arteries into utero-
    placental vessels is exactly replicated in the tube.

Morphology of the Placenta in Tubal Pregnancies

In many tubal pregnancies the placental villi are morphologically fully
normal for the length of the period of gestation. In others the villi show the
usual changes which follow fetal death. The features of a hydropic abortion
are sometimes seen, but this is not a common finding.

A particular characteristic of the placental villi in a tubal gestation is
their propensity to infiltrate the tubal wall and extend into the subserosa.
There may indeed be extensive subserosal spread of placental villous tissue
along the length of the tube within the subserosa and, not uncommonly, villi
penetrate through the overlying serosa with continuing villous growth on the
serosal surface [21, 23]. It must again be assumed that this undue invasiveness of placental villi is indicative of an absence of those regulatory factors
which exist in the uterus.

Immunological and Immunohistochemical Features of Trophoblast
in Tubal Pregnancies

The distribution of major histocompatibility antigens in the villous and
extravillous trophoblast in tubal gestations is identical to that noted in the
trophoblast populations of an intrauterine pregnancy [24]. Similarly, the
distribution of trophoblastic membrane antigens, low molecular weight epithelial cytokeratins, epithelial membrane antigen, hCG, human placental
lactogen (hPL) and *Schwangerschafts protein* 1 (SP1) in a tubal pregnancy
is similar to that described in normal early intrauterine placentas [25].
Immunohistochemical studies have also shown that the distribution of
proteinases and proteinase inhibitors in trophoblast is identical in tubal and
intrauterine pregnancies [22].

Tubal Response to Tubal Implantation

The tube can show a decidual response in a tubal pregnancy. Whether this is, in fact, a direct response to the tubal nidation is a moot point, for decidual change is not uncommonly seen in tubes that are removed immediately post partum following an intrauterine pregnancy, and it has not been clearly shown that the degree and extent of decidual change in a tube harbouring a pregnancy exceeds that normally associated with an intrauterine gestation.

The reported incidence of tubal decidual change in a tubal pregnancy varies considerably [9, 21], but it is rarely a conspicuous feature, usually being patchy and often involving only scattered isolated cells. Even in those few tubes in which there is a widespread decidual change, the "decidua" is, of course, much thinner than that found in the uterus.

Outcome of Tubal Pregnancies

A very high proportion of tubal pregnancies abort during the early stages of gestation. This is invariably the rule in cases of fimbrial and plical implantation, simply because these sites offer insufficient tissue for placentation, but it is also common in mural placentations. Abortion in a mural implantation is usually associated with intraluminal and intramural haemorrhage, and it is thought that this bleeding is the direct cause of fetal death. The haemorrhage is largely a consequence of trophoblastic invasion of the tubal wall and vasculature. The tube lacks the haemostatic mechanisms which are operative in the uterus, and minor degrees of bleeding which would be easily controlled in the uterus may, in a tube, assume a magnitude sufficient to endanger fetal viability.

Tubal rupture occurs in about 50% of tubal gestations and appears to be due partly to the limited distensibility of the tube, which is unable to dilate adequately to accommodate the developing conceptus, and partly to transmural trophoblastic invasion with eventual penetration of the serosa. Tubal rupture is usually accompanied by fetal demise, but occasionally the trophoblast grows out through the rupture site and forms a secondary placental site in the abdomen or on the broad ligament.

It will be apparent that most tubal gestations terminate at an early stage. This is not, however, an indication of any failure of placentation, but rather of the mechanical and haemostatic inadequacies of the tube as a site of nidation.

Despite this very high incidence of early fetal loss in tubal pregnancies, there are many reported instances (well over 100) of tubal pregnancies which have persisted until the 3rd trimester, or even to term, in an unruptured tube [5, 26]. The reasons why some tubal gestations are capable of progressing to term without rupturing the tube are obscure and nothing is known about the vascular supply of the placenta and fetus in such cases.

Ovarian Pregnancy

An ovarian pregnancy may be primary or secondary. A primary ovarian gestation occurs if the ovum is fertilized whilst still within the follicle, or if an ovum, fertilized outside the ovary, primarily implants on the ovary. A secondary ovarian pregnancy occurs when the ovum is fertilized within the tube, implants primarily in the tube and is then regurgitated to implant secondarily on the ovary, usually as a result of tubal rupture.

The aetiology of primary ovarian pregnancy is largely unknown, but it has been suggested that ovulatory dysfunction may be a factor in primary intrafollicular pregnancy [27]. The incidence of ovarian pregnancies is relatively high in women using an IUD, but, as already noted, this is simply a reflection of the inability of the IUD to prevent ovarian implantation.

Nothing is known about placentation in the ovary, but ovarian pregnancies usually terminate during the first trimester. There have, however, been a number of reports of ovarian gestations which have proceeded on term [28].

Cervical Pregnancy

The incidence of ectopic cervical implantation is very difficult to determine, largely because of the use of differing criteria for distinguishing between a primary cervical pregnancy, a placenta praevia and the cervical stage of an intrauterine abortion [29].

The aetiology of cervical implantation is poorly understood. Studdiford [30] suggested that unusually rapid transport of the fertilized ovum resulted in its reaching the endocervical canal before having attained a stage when it is capable of implantation. Schneider and Dreizin [31] proposed, as an alternative view, that ovum transport was normal, but maturation of the conceptus was delayed, thus allowing it to enter the endocervical canal before reaching a stage of development compatible with nidation. Others have maintained that cervical implantation is secondary to endometrial insufficiency [32], a view possibly supported by the association between cervical pregnancy and previous sharp curettage [33].

Cervical pregnancies usually abort at an early stage, and information as to the mode of placentation in this site is completely lacking. A small number of cervical pregnancies have proceeded to the third trimester and have resulted in the birth of live infants [29].

Abdominal Pregnancy

Abdominal implantation may be either primary or secondary. Most are of the secondary type and are a consequence of either tubal or uterine rupture, whilst even those which are apparently primary are probably a result of regurgitation of a fertilized ovum from the tube.

Most abdominal pregnancies are in the cul-de-sac and involve either the posterior surface of the uterus or the anterior aspect of the rectosigmoid colon. Other reported sites of implantation include the broad ligament, colonic mesentery, lumbar gutters, mesentery of the small bowel, omentum, ileum, lesser peritoneal sac, liver and spleen [5].

A gestation situated within the abdominal cavity is not subject to the same restraints of space as is a tubal pregnancy, and the principal factor determining whether abortion occurs or not is the ability of the placenta to establish an adequate blood supply. This being the case, it is perhaps surprising that as many as 25% of abdominal gestations do achieve adequate placentation and progress to an advanced stage [34]. In advanced abdominal pregnancies, the placenta usually has a broad base of attachment, but it has to be admitted that the mechanisms of placentation within the abdominal cavity remain totally unexplored.

General Comments and Conclusions

A consideration of ectopic pregnancies leads inexorably to the view that implantation and placentation are purely a function of the trophoblast, the maternal tissues playing a purely passive role. Thus, once a fertilized ovum reaches a specific stage of maturation it will implant, irrespective of where it happens to be situated at that time and no matter how apparently inhospitable the maternal tissues may appear. Hence the lack of factors often thought to be of critical importance in early pregnancy, such as endometrial "receptivity", the microenvironment of the endometrium and endometrial-trophoblastic "messages", do not deter successful implantation and placentation.

The process of placentation in an ectopic site appears to follow the same course as in an intrauterine gestation and is equally successful. Early pregnancy loss is the rule in an ectopic site, but this is largely because of mechanical and haemostatic reasons, and it is clear that a fully adequate maternal blood supply to the placenta can be established even in such unlikely sites as the abdominal cavity or the uterine cervix.

Ectopic pregnancies serve as a model for early pregnancy in which the uterine factor is removed, and the apparent success with which implantation and placentation are achieved casts considerable doubt on many of the presumed verities of early gestation.

References

1. Barnes AB, Wennberg CN, Barnes BA (1983) Ectopic pregnancy: incidence and review of determinant factors. Obstet Gynecol Surv 38:345–356
2. Drife JO (1990) Tubal pregnancy. Br Med J 301:1057–1058
3. Stabile I, Grudzinskas JG (1990) Ectopic pregnancy: a review of incidence, etiology and diagnostic aspects. Obstet Gynecol Surv 45:335–347
4. Beral V (1975) An epidemiological study of recent trends in ectopic pregnancy. Br J Obstet Gynaecol 62:775–782
5. Fox H, Buckley CH, Randall S (1987) Ectopic pregnancy. In: Fox H (ed) Haines and Taylor: obstetrical and gynaecological pathology, 3rd edn. Livingstone, Edinburgh, p 818
6. Westrom L (1975) Effect of pelvic inflammatory disease on fertility. Am J Obstet Gynecol 121:707–713
7. Brunham RC, Binns B, McDowell J, Paraskevas M (1986) *Chlamydia trachomatis* infection in women with ectopic pregnancy. Obstet Gynecol 67:722–726
8. Kihlstrom E, Limdgren R, Ryden G (1990) Antibodies to Chlamydia trachomatus in women with infertility, pelvic inflammatory disease and ectopic pregnancy. Eur J Obstet Gynecol Reprod Biol 35:199–204
9. Green LK, Kott ML (1989) Histopathologic findings in ectopic tubal pregnancy. Int J Gynecol Pathol 8:255–262
10. Vasquez G, Winston RML, Brosens IA (1983) Tubal mucosa and ectopic pregnancy. Br J Obstet Gynaecol 90:468–474
11. Iffy L (1963) The role of premenstrual, post-mid cycle conception in the aetiology of ectopic gestation. J Obstet Gynaecol Br Commonw 70:996–1000
12. Ory HW (1981) Ectopic pregnancy and intrauterine contraceptive devices: new perspectives. Obstet Gynecol 57:137–143
13. James WH (1989) A hypothesis on the increasing rates of ectopic pregnancy. Paediatr Perinatol Epidemiol 3:189–193
14. Handler A, Davis F, Ferre C, Yeko T (1989) The relationship of smoking and ectopic pregnancy. Am J Public Health 79:1239–1242
15. Poland BJ, Dill FJ, Styblo C (1976) Embryonic development in ectopic human pregnancy. Teratology 14:315–321
16. Elias L, LeBeau M, Simpson JL, Martin AD (1981) Chromosome analysis of ectopic human conceptuses. Am J Obstet Gynecol 141:698–703
17. Schafer D, Pfuhl JP, Baum R, Rossler M, Baumann R (1989) Cytogenetische, endokrinologische sowie immunologische Studien an kultiviertem Gewebe ektoper Schwangerschaften. Zentralbl Gynakol 111:1476–1485
18. Aine R, Karikoski-Leo R, Heinonen PK (1990) Flow cytometric DNA in ectopic pregnancy. J Clin Pathol 43:963
19. Makinen JI (1989) Increase of ectopic pregnancy in Finland – combination of time and cohort effects. Obstet Gynecol 73:21–24
20. Falk HC, Hassid R, Dazo EP (1975) Tubal pregnancy: a report of a very early luminal form of embedding. Obstet Gynecol 45:215–215
21. Randall S, Buckley CH, Fox H (1987) Placentation in the fallopian tube. Int J Gynecol Pathol 6:132–139
22. Earl U, Morrison L, Gay C, Bulmer JN (1989) Proteinase and proteinase inhibitor localization in the human placenta. Int J Gynecol Pathol 8:114–124
23. Budowick M, Johnson TRB, Genadry R, Parmley TH, Woodruff JD (1980) The histopathology of the developing ectopic pregnancy. Fertil Steril 34:1169–1171
24. Earl U, Wells M, Bulmer JN (1985) The expression of major histocompatibility antigens by trophoblast in ectopic tubal pregnancy. J Reprod Immunol 8:13–24
25. Earl U, Wells M, Bulmer JN (1985) Immunohistochemical characterization of trophoblast antigens and secretory products in ectopic tubal pregnancy. Int J Gynecol Pathol 5:132–142

26. Obolet W, van der Merwe JV, van Assche FA (1988) Advanced extrauterine pregnancy: description of 38 cases with literature survey. Obstet Gynecol Surv 43:386–397
27. Boronow RC, McElin TW, West RH, Buckingham JC (1965) Ovarian pregnancy. Am J Obstet Gynecol 91:1095–1106
28. Williams PC, Malvar TC, Kraft JR (1982) Term ovarian pregnancy with delivery of a live female infant. Am J Obstet Gynecol 142:589–591
29. Yankowitz J, Leake J, Huggins G, Gazaway P, Gates E (1990) Cervical ectopic pregnancy; review of the literature and report of a case treated by single-dose methtrexate therapy. Obstet Gynecol Surv 45:405–414
30. Studdiford WE (1945) Cervical pregnancy. Am J Obstet Gynecol 49:169–174
31. Schneider P, Dreizin DH (1957) Cervical pregnancy. Am J Surg 93:27
32. Thomsen M, Johansen F (1961) Two cases of cervical pregnancy. Acta Obstet Gynecol Scand 40:99–107
33. Parente JT, Ou C, Levy J, Legatt E (1983) Cervical pregnancy analysis: a review and report of five cases. Obstet Gynecol 62:79–82
34. Golan A, Sandback O, Adronikou A, Rubin A (1975) Advanced extrauterine pregnancy. Acta Obstet Gynecol Scand 64:21–25

18 Hydatidiform Mole and Related Disorders

T. Maruo and M. Mochizuki

Introduction

"Hydatidiform mole" basically means a fluid-filled cystic mass, it is derived from the Greek word *hydatis* meaning a drop of water, and the Latin word *moles* meaning a mass. A classic hydatidiform mole is an abnormal conceptus without an embryo, with gross hydropic swelling of the placental villi and usually pronounced trophoblastic proliferation of both cytotrophoblastic and syncytial elements. The villous swelling leads to central cistern formation with a concomitant compression of the connective tissue that has lost its vascularity. Microscopically there is complete villous hydrops, absence of chorionic blood vessels, and a variable degree of trophoblastic hyperplasia. The hydrops might be caused by hyperplasia or dysplasia of the trophoblast, leading to oversecretion of fluid into the villous stroma. This may explain the mechanism in molar change. When there is a failure of the villous vessels to link up with those of the embryo, partial hydatidiform mole may develop in the presence of an embryo.

However, hydropic change is a well-recognized and common finding in spontaneous abortion. The importance of making the correct diagnosis in cases of plcental hydrops stems from the fact that, like the complete mole, a partial mole carries a risk of persistent trophoblastic disease, while a simple hydropic abortion does not. Thus, the distinction between simple hydropic abortion and partial mole should be made on the appearance of the trophoblast. Only if trophoblastic hyperplasia is seen should a diagnosis of partial mole be made. It is therefore important to distinguish three separate clinicopathologic entities: complete hydatidiform mole, partial hydatidiform mole, and hydropic abortion.

Furthermore, when we discuss hydatidiform mole and related disorders, attention must be paid to gestational trophoblastic disease. The generic term "gestational trophoblastic disease" indicates certain specific clinical and morphologic patterns of disease which have complex pathogenetic interrelationships. These disease forms are designated: hydatidiform mole, invasive hydatidiform mole, and choriocarcinoma. The greatest importance of hydatidiform mole is the risk of subsequent choriocarcinoma. Although in the majority of cases of hydatidiform mole there are no neoplastic sequelae,

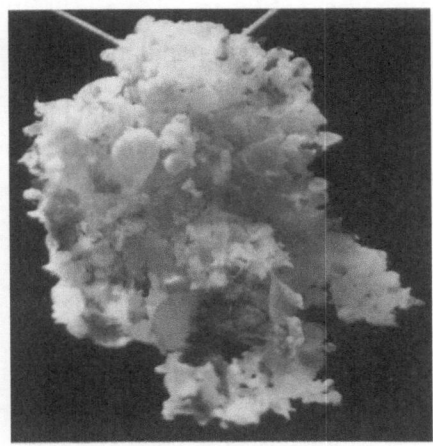

Fig. 1. Vesicles of hydatidiform mole

the important point is that in at least 50% of cases choriocarcinoma is preceded by a hydatidiform mole. The inability to grow hydatidiform mole in immunosuppressed animals is consistent with other evidence that hydatidiform moles are not in themselves malignant, in contrast to choriocarcinoma.

With the recent advent of accurate methods for estimating human chorionic gonadotropin (hCG) and successful use of chemotherapeutic agents, an increasing number of patients are treated on the basis of elevated hCG levels in the serum or urine, and operative surgery plays a much smaller role in the management of the disease. As a result, the precise histologic diagnosis of trophoblastic neoplasia has become less frequently available. This does not invalidate a histologic subdivision which is very important for determining prognosis.

Complete Hydatidiform Mole

A complete hydatidiform mole is an abnormal product of gestation; it is characterized by complete hydropic change with a variable degree of trophoblastic proliferation and is usually associated with the absence of a fetus. Hydatidiform swelling results from progressive accumulation of fluid within the villous stroma (Fig. 1).

Epidemiology

Many hospital-based studies have suggested that the incidence of hydatidiform mole is considerably more elevated in Asia, parts of Africa, and in Latin America than in North America or Europe. Rates have been shown to vary between ten cases of hydatidiform mole per 1000 pregnancies in Indonesia and 0.7 cases per 1000 pregnancies in the United States. In population-based studies, geographic variations are less evident, but incidence is still higher in Asia; 1.96 hydatidiform moles per 1000 pregnancies are recorded in Japan and 0.6 per 1000 pregnancies in Sweden. Thus, hydatidiform mole appears to be at least three to four times more frequent in the Third World countries than in the western hemisphere [1, 2].

A number of attempts have been made to relate low socioeconomic status with a relatively high frequency of hydatidiform mole. The available data indicate that not all areas of high incidence are to be regarded as economically deprived, nor has substantial economic development altered the high incidence rate. Japan serves as an excellent example of a country which has experienced substantial economic development and still retains a relatively high incidence of hydatidiform mole. Nevertheless, it is of great interest to note that the incidence of choriocarcinoma per 10000 pregnancies in Japan has decreased remarkably in recent years [3, 4].

Hydatidiform mole occurs at any time during the reproductive period, but there is an increased risk for women over the age of 40 and at the younger end of the child-bearing range. There appears to be no direct relationship with gravidity.

Genetics

Complete hydatidiform moles appear to result from fertilization of an egg from which the nucleus has been lost or inactivated. A number of studies have demonstrated that the majority of complete hydatidiform moles are of normal female 46 XX chromosomal constitution and arise from a single sperm bearing a 23X set of chromosomes which duplicates [5, 6]. The androgenetic origin of the nucleus of complete hydatidiform moles has been confirmed by polymorphic banding, HLA gene loci, and enzyme marker studies [7–9]. However, Jacobs et al. [10] found two different hydatidiform moles out of the 24 they studied. One was 46 XY but its origin could not be determined; and the other, although of 46 XX constitution, appeared to be the product of a normal conception containing both a paternal and maternal genome. Clearly, although most complete moles are of androgenetic derivation, in a small minority other mechanisms may operate.

Histopathology

The macroscopic appearances of a complete hydatidiform mole are characteristic. A hydatidiform mole consists of a mass of grapelike structures or strings of vesicles mixed with a variable quantity of clotted blood. The vesicles range in diameter from approximately 1 to 30 mm. It is rare for a gestational sac to be identified.

The chorionic villi become greatly distended by the accumulation of stromal fluid, resulting in variable degeneration of the connective tissue, from mild edema to central liquefaction. The second feature is the virtual absence of fetal stromal blood vessels. In addition, there is always some degree of proliferation of both cytotrophoblast and syncytiotrophoblast. It has been advocated that the variation in degree of trophoblastic hyperplasia can be used to identify the potential malignancy at the time of molar evacuation and to predict which hydatidiform mole subsequently progresses to choriocarcinoma. However, it is now generally accepted that it is not possible to predict potential malignant behavior from the histologic appearances of a hydatidiform mole [5].

Diagnosis

The most common signs and symptoms in women with molar pregnancy are vaginal bleeding, anemia, excessive uterine enlargement, toxemia-like syndrome, and hyperemesis gravidarum. The mean time for the onset of vaginal bleeding is the 8th week of gestation. Intermittent bright red vaginal bleeding results from the separation of molar tissue from the uterine wall, thereby exposing maternal vascular channels. In addition to vaginal bleeding, many women pass molar tissue mixed with blood clots through the vagina. When molar vesicles are identified histologically, the diagnosis of hydatidiform mole can be made conclusively.

Excessive uterine enlargement for gestational age is seen in approximately half the patients with hydatidiform mole. Although the uterus may fill the abdominal cavity, fetal heart sounds and fetal signs are absent. The uterus may feel doughy on palpation because the myometrium is distended with hydatid vesicles. The excessive uterine enlargement is generally associated with markedly elevated hCG titers.

Toxemia-like syndrome (hypertension, edema, and proteinuria) and hyperemesis gravidarum occur almost exclusively in patients with excessive uterine enlargement and markedly elevated hCG titers. Although the relationship between hydatidiform mole and toxemia-like syndrome remains unexplained, the diagnosis of molar gestation must be considered when a patient presents with toxemia early in pregnancy [11].

Hyperthyroidism in patients with hydatidiform mole is uncommon, although increased thyroid function may be detected biochemically. It may be due to markedly elevated hCG levels, as hCG has been demonstrated to have an intrinsic thyroid-stimulating activity. Recent studies have shown that hCG binds to specific hCG/luteinizing hormone (LH) receptors in the thyroid and induces thyroid stimulation. Patients with markedly elevated hCG levels may also present with thyroid storm: this may be precipitated by infection, surgery, anesthesia, or toxemia in a patient with uncontrolled hyperthyroidism. Thus, attention must be paid to the danger of precipitat-

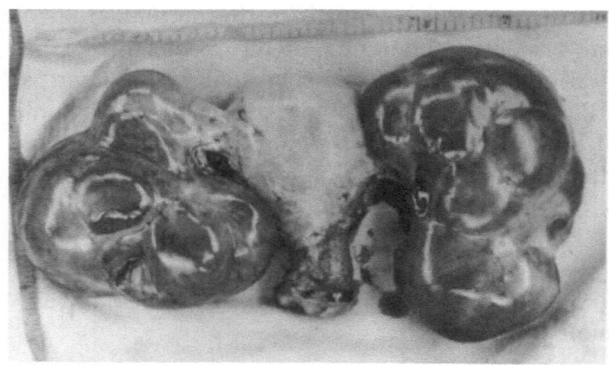

Fig. 2. Theca lutein cysts associated with molar pregnancy

ing thyroid storm when molar evacuation is performed in patients with markedly elevated hCG levels. However, increased thyroid function and, in some cases, thyrotoxicosis with crisis has been reported despite low concentrations of hCG in patients with trophoblastic neoplasms. In accordance with such clinical experience, human chorionic thyrotropin (hCT), which is a glycoprotein different from hCG, has been reported to be isolated from trophoblastic tissues. Hence, both hCG and hCT may act as a thyroid stimulator in patients with hydatidiform mole [12, 13].

A few percent of patients with hydatidiform moles may develop trophoblastic pulmonary embolization and present with chest pain, tachypnea, and cyanosis.

The presence of theca lutein cysts (Fig. 2) is almost exclusively limited to patients with markedly elevated hCG titers and subsequent ovarian hyperstimulation. Patients with theca lutein cysts may present with symptoms of pelvic pressure or acute abdominal pain due to spontaneous cystic rupture or torsion. The theca lutein cysts may be difficult to palpate because of the frequently excessive uterine enlargement. However, ultrasonography can accurately document the presence of lutein cysts.

Diagnostic Examinations

Ultrasonography. Ultrasonography is a sensitive and reliable technique to distinguish molar pregnancies from normal intrauterine gestations. Hydatidiform moles produce a characteristic pattern of echoes on ultrasound examination that appear like a "snowstorm." If a normal pregnancy is present, a gestational sac should be visualized at least from the 6th week of gestation.

Measurement of hCG. When molar gestation contains abundant trophoblastic growth, the hCG titer generally exceeds the levels observed in normal single pregnancy of the same gestational age. Hence, marked elevation of hCG levels in the serum or urine relative to the normal range of the same gestational age has been regarded as a valuable diagnostic marker of hydatidiform mole. When a patient presents with hCG levels exceeding 100 $\times$ 10^4 mIU/ml in serum or urine, the clinician must consider the high possibility of the presence of molar gestation. This level of hCG is rarely seen in a normal pregnancy even during the peak period between 8 and 10 weeks of gestation, except in patients with multiple gestations. On the other hand, in case of molar gestation where trophoblastic proliferation is not active, the hCG titers may actually be lower than the levels seen in normal pregnancies of the same gestational age. Therefore, the hCG titer alone cannot be used to support conclusively the diagnosis of hydatidiform mole.

Prognosis

In the great majority of patients who abort a hydatidiform mole there are no sequelae, and postmolar persistent trophoblastic disease is diagnosed in only approximately 10%. Nevertheless, the risk is sufficiently great to warrant careful follow-up for all patients with hydatidiform mole, since there is no convincing evidence in favor of histologic grading of the malignant potentiality of hydatidiform mole.

The availability of registration of patients with hydatidiform mole and a monitoring system for hCG titers following molar evacuation can contribute to an improvement in prognosis.

Follow-Up

After molar evacuation all patients must be followed with quantitative hCG determinations to facilitate the early detection of proliferative sequelae. Ideally, the measurement of hCG should be obtained on serum using the enzyme immunoassay (EIA) for β-hCG-carboxyl terminal peptide (β-hCG-CTP), which has the capability of measuring hCG specifically and quantitatively from high levels down to extinction. It is also advisable to obtain a chest X-ray 1 month following evacuation to rule out pulmonary lesions.

Following molar evacuation, hCG determinations should be obtained weekly until normal for 3 consecutive weeks; thereafter assays should be performed monthly for 6 consecutive months from the first normal titer. After 6 consecutive months of normal (i.e., absent) hCG titers, a new pregnancy may be authorized.

A question commonly asked concerns the length of time for which it is safe to follow a slowly resolving hCG titer. The mean time for complete regression of the hCG titer after molar evacuation is 15 weeks (Fig. 3) [4,

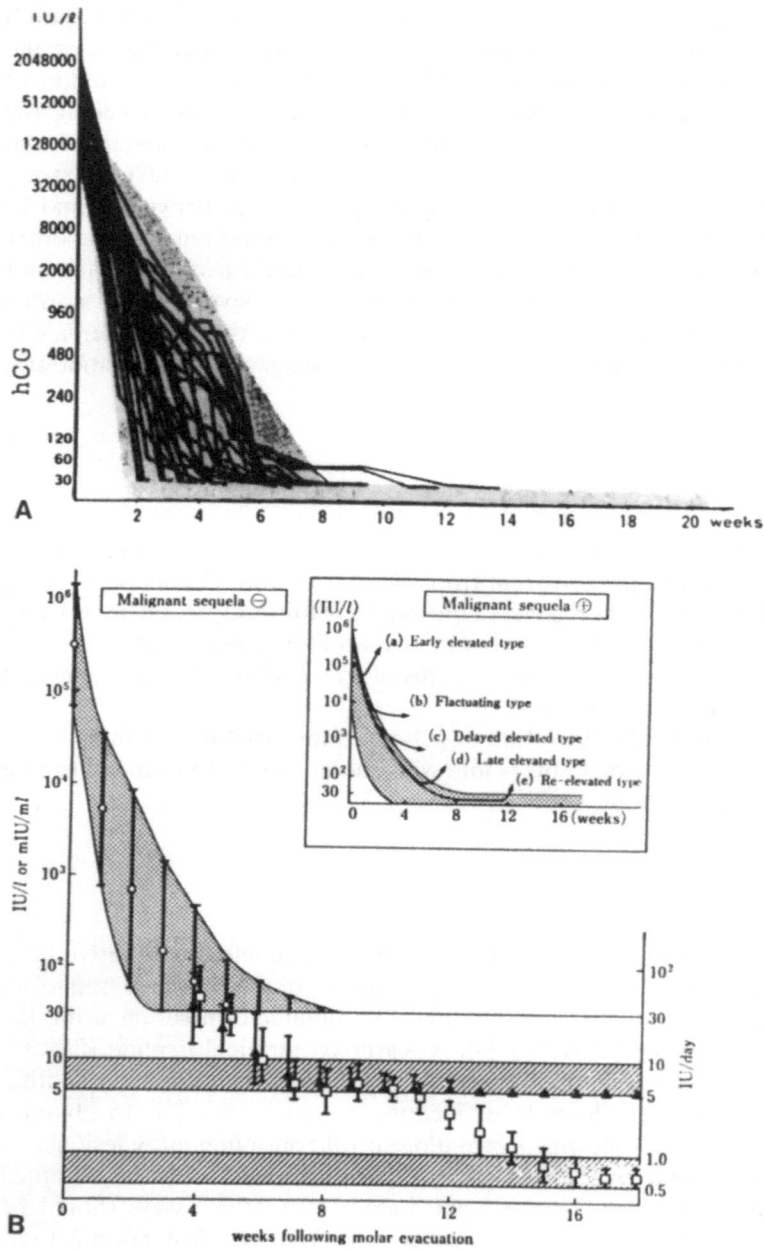

Fig. 3A,B. Normal regression curve of hCG following evacuation of molar pregnancy. Follow-up result (**A**) and the schematic illustration (**B**). *Circles*, urinary hCG by Gonavislide or Hi-Gonavis, serum hCG by UCG Titration Set; *triangles*, serum hCG by β-hCG RIA or EIA; *squares*, urinary hCG by β-hCG CTP EIA

14]. The regression time appears to be independent of morphology, but dependent on a number of other factors, such as renal clearance, the completeness of the initial evacuation, and the preevacuation hCG titer. For this reason, we use the concept of a plateau or reelevation of the hCG level rather than a set period of time for complete regression.

After molar evacuation, patients who show any of the following changes are readmitted for further evaluation and treatment: (a) the hCG titer plateaus for 3 or more consecutive weeks or increases again; (b) metastases appear subsequent to uterine evacuation; or (c) the final pathologic specimens from the molar tissues demonstrate myometrial invasion or frank choriocarcinoma.

Partial Hydatidiform Mole

Partial hydatidiform mole is an abnormal conceptus with an embryo which tends to die early, with a placenta with focal villous swelling leading to vesicle formation, and with focal trophoblastic hyperplasia. Existence of an embryo can be ascertained by the presence of fetal red blood cells within villous capillaries. The unaffected villi appear normal, and vascularity of the villi disappears following fetal death.

Genetics

Most partial moles have a triploid karyotype with an additional set of chromosomes derived from the father [15]. Although early chromosomal studies of partial hydropic abortuses were difficult to interpret because of a lack of specificity of the morphologic appearance, it has been demonstrated that the majority of cases of partial hydrops with definite trophoblastic hyperplasia are triploid.

Histopathology

Grossly, a product of conception specimen may appear to be an ordinary abortus, but microscopic examination reveals hydropic villi with an unusual degree of trophoblastic hyperplasia. The feature which histologically distinguishes a partial mole from a simple hydropic abortion is the presence of definite trophoblastic hyperplasia. This is usually focal, and some villi show little or no hyperplasia.

Prognosis

Choriocarcinoma has not been recorded after partial hydatidiform mole, but chemotherapy has been required in some patients after partial hydatidiform mole because of concern about elevated hCG titers in serum or urine. Although it has been suggested that partial hydatidiform moles are not followed by persistent trophoblastic disease, this is not quite true, and there is considerable evidence that patients with partial mole require follow-up in the same way as complete those with hydatidiform moles.

Hydropic Abortion

Hydropic change is an extremely common finding in spontaneous abortion. When a partial mole is separated from a hydropic abortion on the basis of trophoblastic hyperplasia, hydropic abortion is known to be associated with a variety of chromosomal abnormalities. These include triploidy, tetraploidy, and various trisomies.

The placenta in hydropic abortion may appear normal macroscopically, but hydropic abortion is microscopically characterized by atrophic and attenuated trophoblasts. Fetal parts may be seen. Since there is no evidence that the presence of simple hydropic change predisposes to the development of trophoblastic disease, follow-up is not required after the hydropic abortion.

Invasive Hydatidiform Mole

Invasive mole is a hydatidiform mole that has extended into the myometrium. Since trophoblast has the ability to invade vascular channels, molar villi may extend into vascular channels of the myometrium. Invasive mole is characterized by persistence of villous structures with trophoblastic hyperplasia. The presence of chorionic villi is a clear histologic marker for a lesion which carries a more favorable prognosis than choriocarcinoma.

The invasive hydatidiform mole commonly results from a complete hydatidiform mole and may do so from partial hydatidiform mole. It does not often progress to choriocarcinoma. Transportation nodules may appear anywhere, but there is no real progression like a true cancer; and spontaneous regression may occur. Prior to the use of chemotherapy, deaths in patients with invasive mole were rarely due to metastatic or malignant disease but rather to sepsis, hemorrhage, embolic phenomenon, uterine perforation, or complications from surgery. With the advent of effective chemotherapy, histologic verification of invasive mole is seldom performed.

Histopathology

The macroscopic appearance depends on the extent of invasion. Less commonly the uterus is perforated or the mole penetrates into adjacent struc-

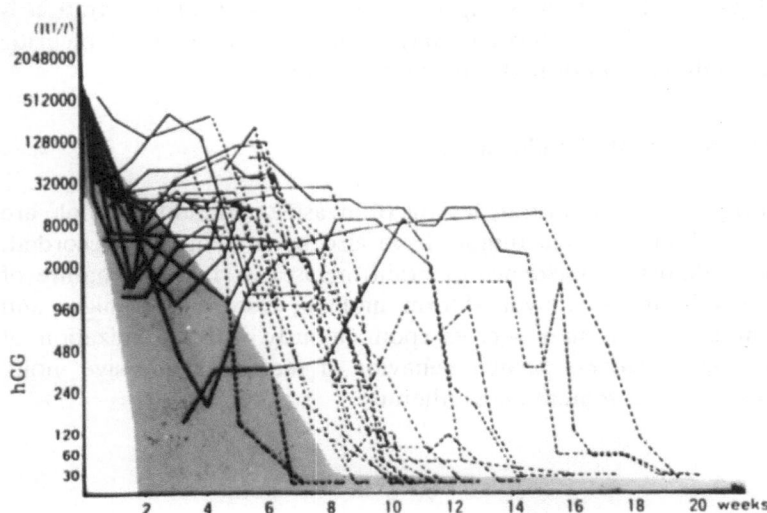

Fig. 4. Follow-up of hCG levels in patients who were diagnosed as having invasive mole following evacuation of molar pregnancy. *Solid lines*, before hysterectomy; *broken lines*, after hysterectomy

tures such as the broad ligaments. In some cases the original hydatidiform mole is still retained in the uterine cavity.

The characteristic microscopic feature is the presence of hydropic villi with trophoblastic proliferation within the myometrial vessels. Any uterus containing proliferating trophoblastic islands requires a careful search for the presence of hydropic villi to avoid an erroneous diagnosis of choriocarcinoma.

Diagnosis

The diagnosis of invasive hydatidiform mole should be made unequivocally only from histologic examination of the whole uterus. However, if curettings contain fragments of myometrium where molar trophoblast is directly conspicuous without any intervening endometrium, a diagnosis of focally invasive mole may be made; it is suggestive, but not proof, that there may be invasive mole elsewhere in the uterus.

Invasive mole usually occurs after a hydatidiform mole. Following evacuation of the original hydatidiform mole the patient may remain symptom free for several weeks or even months before reelevation of hCG values (Fig. 4) [16], although in a small proportion of cases the diagnosis of invasion is made at hysterectomy with the original mole in situ. Amenorrhea may occur, but irregular uterine bleeding is the usual presenting symptom.

Abdominal pain and tenderness suggest that perforation or penetration is imminent. More rarely, hemoptysis may occur, with or without pleuritic pain, due to embolization of molar tissue in the lung.

Metastases in Invasive Hydatidiform Mole

In the majority of cases metastatic lesions of invasive hydatidiform mole are detected in the lung. Vaginal tumors have also been frequently recorded, while there is almost no incidence of brain metastasis [17]. The nature of the metastases in invasive hydatidiform mole is essentially benign, and metastatic molar lesions may regress spontaneously with hyalinization of villi. Despite the relatively benign behavior of metastatic invasive mole, many of these patients require chemotherapy.

Prognosis

There is certainly little evidence to suggest that a patient with invasive mole is more likely to develop choriocarcinoma than one with an ordinary non-invasive mole. It has been suggested that the trophoblast of invasive mole, which is capable of forming villi, is biologically different from that of the more aggressive malignant trophoblast of choriocarcinoma. The presence of chorionic villi in an invasive or metastatic trophoblastic lesion may, there-fore, be an indication of good differentiation and limited malignancy, while their absence may be a marker of poorer differentiation and greater malig-nancy. The mortality of invasive hydatidiform mole was essentially discussed in the prechemotherapy era. Nowadays, most cases of invasive hydatidiform mole are naturally included as part of persistent trophoblastic disease, and accurate statistics are not readily available. Nevertheless, no deaths have occurred in cases of histologically proven invasive hydatidiform mole during the past decade in our hospital [17]. Most deaths in the past were due to local events such as hemorrhage or uterine perforation rather than the development of metastatic disease. It can be concluded that the current mortality of invasive mole is extremely low [1, 18].

Choriocarcinoma

Choriocarcinoma is a malignant tumor arising from the trophoblastic epi-thelium that includes both cytotrophoblastic and syncytiotrophoblastic elements. The abnormal trophoblastic elements are highly anaplastic and disposed in sheets. Choriocarcinoma is distinguished morphologically from invasive mole by its lack of villous structures. It is essentially an intravascular growth without any supporting connective tissue, with frequent areas of extensive hemorrhage, necrosis, and marked invasiveness of adjacent tissues.

Choriocarcinoma may arise from conceptions that give rise to a live birth, a stillbirth, an abortion at any stage, an ectopic pregnancy, or a hydatidiform mole [19]. The percentage of cases of choriocarcinoma preceded by hydatidiform mole is approximately 50% [20]. Because of a closer follow-up, choriocarcinoma developing after a molar pregnancy is usually detected at a less advanced stage than choriocarcinoma developing after a full-term delivery or nonmolar abortion.

Epidemiology

Incidence rates vary: in population-based studies, there are 0.53 cases of choriocarcinoma per 10000 pregnancies in Japan and 0.2 per 10000 pregnancies in Sweden; in hospital-based studies 15.3 cases of choriocarcinoma per 10000 pregnancies are detected in Indonesia and 0.3 per 10000 pregnancies in the United States [21].

An extraordinarily increased incidence of postmolar choriocarcinoma is noted over 45 years of age. Furthermore, it is reported that choriocarcinoma occurred in 50% of patients who had had heterozygous hydatidiform moles [6]. This incidence is much higher than that of homozygous hydatidiform mole, suggesting that the heterozygosity observed in molar cells may be a risk factor for postmolar choriocarcinoma.

Histopathology

There is usually a typical trophoblastic nodule which is composed of a central area of hemorrhage and necrosis, with a variable peripheral rim of tumor tissue. This is a unique feature of choriocarcinoma which lacks an intrinsic stromal support and vasculature. As a morphologic point of distinction from invasive hydatidiform mole, formed chorionic villi are absent in choriocarcinoma.

Choriocarcinoma rarely invades muscle directly, but rather via venous sinuses. Since choriocarcinoma lacks an intrinsic vessel supply and relies entirely on invasion of maternal blood vessels for nutrition, there is always some permeation of the dilated venous sinuses at the periphery of the tumor, and this is the first step in the progression of myometrial invasion towards the uterine venous plexus.

Apart from local spread to the vagina and vulva, lung metastases occur most commonly with a reported frequency of 60%. Cerebral metastases are the next most common, followed by liver, kidney, and intestine metastases. Metastasis of choriocarcinoma occurs exclusively by hematogenous spread.

Prognosis

Although choriocarcinoma is highly malignant and fatal, dramatic improvement in survival has been obtained due to the use of chemotherapeutic agents. Sustained remission can now be achieved in over 80% of patients if treatment is initiated early in the course of the disease. Particularly, the sustained remission rate for localized choriocarcinoma is almost 100% [20], although the remission rate for choriocarcinoma with metastases to multiple distant organs still remains low.

Molecular Endocrine Aspects of Hydatidiform Mole and Related Disorders

Quantitative Aspects

Trophoblast occupies a unique position as the site of production of both protein and steroid hormones. It is clinically evident that in patients with hydatidiform mole hCG is present in serum in higher concentrations than those of normal gestation, whereas human placental lactogen (hPL) is present in lower serum concentrations as compared to those in normal gestation (Fig. 5). This is one of the characteristic endocrine features of hydatidiform mole [14].

The discrepancy between hCG and hPL levels observed in patients with hydatidiform mole may be explained by the different gene expression of α-hCG and β-hCG and hPL during trophoblast differentiation. Namely, in situ hybridization studies with ^{3}H-labeled cDNA probes complementary to α-hCG, β-hCG and hPL mRNA have revealed that α-hCG mRNA can be first synthesized in less differentiated trophoblast (cytotrophoblast). Synthesis of β-hCG mRNA starts later during trophoblast differentiation, whereas hPL mRNA may be synthesized only in fully differentiated trophoblast of normal placenta [22]. By contrast, in the case of choriocarcinoma, few signals with an hPL probe have been found despite the presence of abundant messages for α-hCG and β-hCG in the in situ hybridization studies [23]. Since choriocarcinoma consists of undifferentiated trophoblasts, it is assumed that the differentiation of trophoblast is blocked before the expression of hPL gene (Fig. 6). Accordingly, hydatidiform molar trophoblast is thought to occupy the middle position between normal trophoblast and choriocarcinoma trophoblast from a viewpoint of gene expression of α-hCG and β-hCG and hPL during trophoblast differentiation [24].

Increasing concentrations of progesterone in normal placental tissues selectively suppress the production of α-hCG and β-hCG through the inhibition of cellular levels of α-hCG and β-hCG mRNAs without affecting

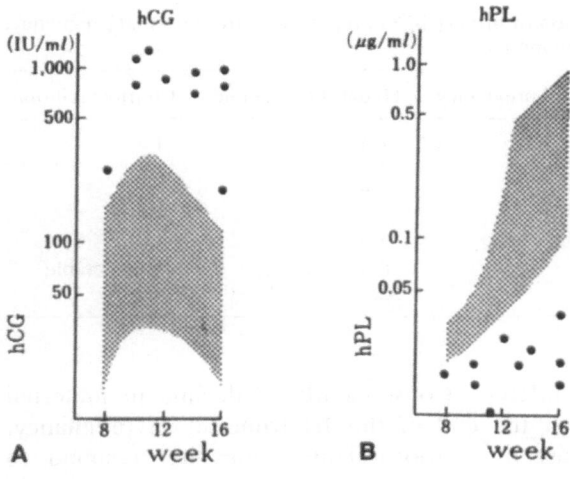

Fig. 5A,B. Serum levels of hCG (**A**) and hPL (**B**) in patients with complete hydatidiform mole relative to those in normal single pregnancy. The *shaded area* shows the normal range of hCG and hPL at the same gestational age of normal single pregnancy

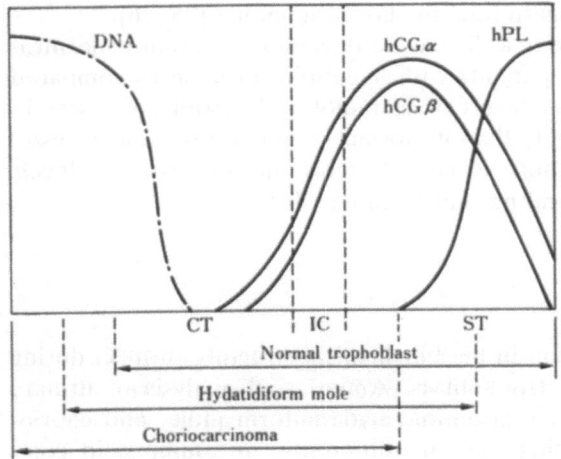

Fig. 6. Gene expression of α-hCG, β-hCG and hPL during differentiation and dedifferentiation of trophoblast. *CT*, cytotrophoblast; *IC*, intermediate cell; *ST*, syncytiotrophoblast. *Broken line*, DNA synthesis as labeling index; *solid lines*, mRNA transcription as mean grain count per nucleus

Fig. 7A–C. Serum levels of estradiol (**A**), estriol (**B**), and progesterone (**C**) in patients with complete hydatidiform mole relative to those in normal single pregnancy ▼

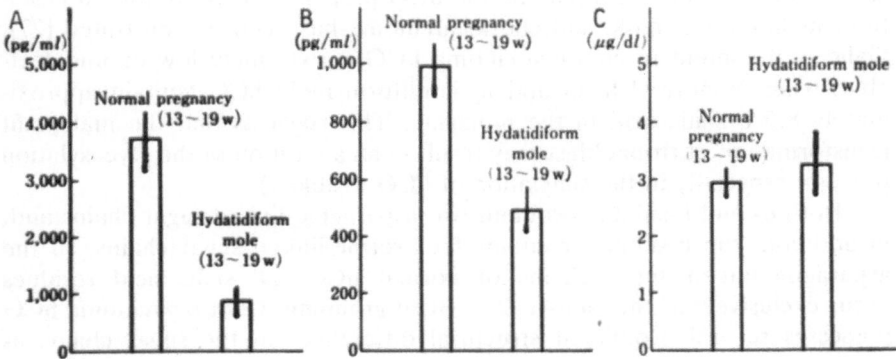

Table 1. Carbohydrate composition of urinary hCG preparations from normal pregnancy, hydatidiform mole, and choriocarcinoma

Carbohydrate	Normal pregnancy	Hydatidiform mole	Choriocarcinoma
Fucose	1.5	1.4	1.4
Galactose	5.3	9.0	7.6
Mannose	5.3	4.8	5.7
N-Acetyl galactosamine	3.6	3.6	3.9
N-Acetyl glucosamine	8.5	4.9	5.0
Sialic acid	8.6	8.5	Undetectable
Total carbohydrate	32.8	32.2	23.9

the cellular levels of hPL mRNA. Consequently, a decline in maternal hCG concentration occurs at the end of the 1st trimester of pregnancy. By contrast, the trophoblast of choriocarcinoma does not respond to progesterone owing to the low levels of receptor protein for progesterone in the tumor. Thus, unlike the normal placenta, there is no inhibitory regulation of hCG synthesis by progesterone in choriocarcinoma [25, 26].

On the other hand, there is a distinctive deviation in serum concentrations of steroid hormones in patients with hydatidiform mole as compared with normal gestation. Serum levels of estradiol and estriol are lower in patients with hydatidiform mole than in normal pregnancy of similar gestational age (Fig. 7), while there is no difference in progesterone levels between hydatidiform mole and normal gestation [14].

Qualitative Aspects

There is a qualitative alteration in the biochemical structure of hCG during malignant transformation of trophoblast. Amino acid analysis of urinary hCG preparations from normal gestation, hydatidiform mole, and choriocarcinoma has shown that there are no differences in amino acid composition of these three hCG preparations. However, a great difference in carbohydrate composition among the hCG preparations from normal gestation, hydatidiform mole and choriocarcinoma has been demonstrated [27]. Sialic acid content in choriocarcinoma hCG is extremely low or undetectable, whereas normal hCG and hydatidiform mole hCG contain approximately 8.5% sialic acid in the molecule. This suggests that the malignant transformation of trophoblast may result in an alteration of the glycosylation process, especially in the sialylation of hCG (Table 1).

Both α- and β-subunits contain two asparagine-linked sugar chains and, in addition, the β-subunit contains four serine-linked sugar chains. In the asparagine-linked sugar chains of normal hCG, all sialic acid residues occur exclusively as the NeuAcα2 → 3Gal grouping. Choriocarcinoma hCG possesses several prominent structural differences in the sugar chains as

Table 2. Structural alterations in sugar chains of choriocarcinoma hCG

1. Appearance of sugar chains free from sialic acid
2. Occurrence of triantennary sugar chains with Galβ1 → 4GlcNAcβ1 → 4Manα1 → group
3. Increase in sum total of fucosylated sugar chains

Table 3. Detection of desialylated hCG in patients with hydatidiform mole, invasive mole, persistent trophoblastic disease, and choriocarcinoma using radioimmunoassay for desialylated β-hCG-COOH-terminal peptide

Diagnosis	Cases (n)	Asialo-hCG detected cases (n)
Hydatidiform mole	11	0
Invasive hydatidiform mole	5	0
Persistent trophoblastic disease	7	0
Choriocarcinoma	6	2
	29	2

compared with normal hCG. More than 97% of the sugar chains of chorio-carcinoma hCG are free from sialic acid, whereas the sugar chains of normal hCG are mostly sialylated. Choriocarcinoma hCG contains unusual biantennary complex-type sugar chains in addition to regular tri-, bi-, and monoantennary sugar chains. Furthermore, the occurrence of the Galβ1 − 4GlcNAcβ1 → 4Manα1 → group is a characteristic feature of the sugar chains of choriocarcinoma hCG, as normal hCG does not contain any triantennary sugar chains. The total of fucosylated sugar chains of chorio-carcinoma hCG is twice as much as normal hCG, indicating that fucosyla-tion is also modified in choriocarcinoma (Table 2). In contrast, urinary hCG purified from patients with hydatidiform mole is found to have almost the same profile of sugar chains as that of normal pregnancy hCG [28]. It can therefore be concluded that the structural changes in the sugar chains of choriocarcinoma hCG observed are specific for malignant trophoblast.

Recently two methods have been developed in order to detect the structural change of sugar chains of choriocarcinoma hCG. One is based on the capability of the lectin from *Arachis hypogala* peanut (PNA) to bind to asialylated sugar chains and the other is an immunologic assay method using an antiserum (R141) generated against desialylated β-hCG-COOH-terminal peptide. In PNA-sepharose affinity chromatography, the entire portion of choriocarcinoma hCG containing no sialic acid is absorbed, whereas the major portion of urinary hCG from normally pregnant women and patients with hydatidiform mole is not absorbed on the column. When the PNA binding of urinary hCG from six patients with choriocarcinoma was studied, two of the six cases (33%) were highly bound to the PNA-sepharose column, suggesting that PNA-sepharose affinity chromatography is an effective means to detect the incompletely sialylated hCG in patients with choriocarcinoma.

On the other hand, with the radioimmunoassay using the antiserum (R141) generated against desialylated β-hCG-COOH-terminal peptide, no asialo-hCG immunoreactivity was found in patients with hydatidiform mole and invasive mole, whereas urinary hCG from one-third (33%) of patients with choriocarcinoma demonstrated asialo-hCG immunoreactivity (Table 3). Thus, this immunologic method to detect desialylated hCG also seems to be a useful means for the biochemical diagnosis of choriocarcinoma [29].

However, desialylation of sugar chains of hCG does not always appear in patients with choriocarcinoma. Development of a specific method to detect the prominent structure of sugar chains of Galβ1 → 4G1cNAcβ1 → 4Manα1 → group, characteristic of choriocarcinoma hCG, will be the next step in the biochemical diagnosis of choriocarcinoma.

References

1. Elston CW (1978) Trophoblastic tumours of the placenta. In: Fox HF (ed) Pathology of the placenta. Saunders, London, p 368
2. Rustin GJS, Bagshawe KD, Hammond CB (1985) Gestational trophoblastic tumors. CRC Crit Rev Oncol Hematol 3:103
3. Takeuchi S (1982) The incidence and management of gestational trophoblastic neoplasms. 10th World Congress on Obstetrics and Gynacology, San Francisco
4. Mochizuki M, Ashitaka Y, Maruo T, Masuko K, Harada A, Chough SY (1984) Sixteen years experience in treatment and folow-up of patients with trophoblastic diseases. Asia Oceania J Obstet Gynaecol 10:15
5. Elston CW (1983) Development and structure of trophoblastic neoplasms. In: Loke YW, Whyte A (eds) Biology of trophoblast. Elsevier, Amsterdam, p 187
6. Ichinoe K (1986) Mechanisms of origin of hydatidiform mole and its propensity to malignancy. In: Ichinoe K (ed) Trophoblastic diseases. Igaku-Shoin, Tokyo, p 3
7. Kajii T, Ohama K (1977) Androgenetic origin of hydatidiform mole. Nature 268:633
8. Wake N, Takagi N, Sasaki M (1978) Androgenesis as a cause of hydatidiform mole. JNCI 60:51
9. Hoshina M, Boothby WR, Hussa RO, Pattilo RA, Camel HM, Boime I (1984) Segregation patterns of polymorphic restriction sites of the hCGα gene in trophoblastic disease. Proc Natl Acad Sci USA 81:2504
10. Jacobs PA, Wilson CM, Sprenkle JA, Rosenshein NB, Migeon B (1980) Mechanism of origin of complete hydatidiform moles. Nature 286:714
11. Goldstein DP, Berkowitz RS (1982) The diagnosis and management of molar pregnancy. In: Goldstein DP, Berkowitz RS (eds) Gestational trophoblastic neoplasms. Saunders, Philadelphia, p 143
12. Saida K (1977) Thyroid function on trophoblastic neoplasias. Folia Endocrinol 53:1023
13. Tojo S, Mochizuki M, Kanazawa S (1974) Comparative assay of hCG, hCT and hCS in molar pregnancy. Acta Obstet Gynecol Scand 53:369
14. Mochizuki M, Maruo T (1987) Endocrine aspects of trophoblastic disease. In: Takamizawa H (ed) Trophoblastic diseases. Kanahara, Tokyo, p 61

15. Szulman AE, Surti U (1978) The syndromes of hydatidiform mole. II. Morphologic evolution of the complete and partial mole. Am J Obstet Gynecol 132:20
16. Mochizuki M, Maruo T (1982) New approach to prospective management of trophoblastic diseases. Clin Obstet Gynecol 38:294
17. Saida A, Maruo T, Ashitaka Y, Tojo S (1982) Clinical features of invasive hydatidiform mole and its treatment. Prog Obstet Gynecol 34:27
18. Tomoda Y, Kaseki S, Goto S, Ishizuka T, Mano H (1986) Approach for attaining complete cure of trophoblastic disease. In: Ichinoe K (ed) Trophoblastic diseases. Igaku-Shoin, Tokyo, p 111
19. Hertz R (1978) Choriocarcinoma and related gestational trophoblastic tumors in women. Raven, New York
20. Maruo T, Tojo S, Harada A, Ashitaka Y (1981) Clinical features of choriocarcinoma and its treatment. J Jpn Soc Cancer Ther 16:704
21. Yen S, McMahon B (1968) Epidermiologic features of trophoblastic disease. Am J Obstet Gynecol 101:126
22. Hoshina M, Boothby M, Boime I (1982) Cytological localization of chorionic gonadotropin and placental lactogen mRNAs during development of the human placenta. J Cell Biol 93:190
23. Hoshina M, Hussa R, Pattillo R, Boime I (1983) Cytological distribution, of chorionic gonadotropin subunit and placental lactogen mRNAs in neoplasma derived from human placenta. J Cell Biol 97:1200
24. Mochizuki M, Maruo T, Hoshina M, Nishimura R (1986) Choriocarcinoma and placental protein hormones. In: Ichinoe K (ed) Trophoblastic diseases. Igaku-Shoin, Tokyo, p 93
25. Maruo T, Matsuo H, Otani T, Hoshina M, Mochizuki M (1986) Differential modulation of chorionic gonadotropin (CG) subunit messenger ribonucleic acid levels and CG secretion by progesterone in normal placenta and choriocarcinoma cultured in vitro. Endocrinology 119:855
26. Mochizuki M (1988) Biology of trophoblast and placental protein hormones. In: Mochizuki M (ed) Placental protein hormones. Excerpta Medica, Amsterdam, p 3
27. Nishimura R, Endo Y, Tanabe K, Ashitaka Y, Tojo S (1981) The biochemical properties of urinary human chorionic gonadotropin from the patients with trophoblastic diseases. J Endocrinol Invest 4:349
28. Mizuochi T, Nishimura R, Derapee C, Taniguchi T, Hamamoto T, Mochizuki M, Kobata A (1983) Structure of the asparagine-linked sugar chains of human chorionic gonadotropin produced in choriocarcinoma. J Biol Chem 258:14126
29. Takeuchi Y, Matsuura S, Nishimura R, Mochizuki M (1984) Clinical application of an antibody produced against asialylated hCG in the biochemical diagnosis of choriocarcinoma. Folia Endocrinol 60:1046

19 Pathogenesis of Human Malformations

Y. Gillerot, E. Jauniaux, L. van Maldergem, and C. Fourneau

Introduction

Around 3% of children present at birth with either "minor" or major malformations, which can be isolated or included in a more complex set of malformations. In our experience, 15% of all congenital malformations are incompatible with life. This figure has been shown to be remarkably constant with time, as evidenced by the literature [1] and by data collected by regional, national, or international organizations which routinely investigate congenital malformations [2, 3]. One of the major roles of these organizations is to detect and report an abnormally high incidence of malformations in the general population in order to prevent accidents such as those observed with thalidomide in the early 1960s. Some centers [4], such as ours, also propose a teratology information service which can be consulted by medical professionals. Free exchange of information between regional and national centers and the creation of international information banks are fundamental steps if we want to investigate the pathogeny of human malformation on a broader basis.

The term "pathogenesis' is currently used to describe the mechanisms by which a system in deviated from its usual course, resulting in a morbid process or a disease. Pathogenesis must be distinguished from "ontogenesis," which refers to the study of the origin and development of an individual from fertilization to maturation. The science and study of the causes of malformations or diseases and their mode of action is usually called "etiology." There is no doubt that these concepts are often interdependent. The aim of the present chapter is to demonstrate how the understanding of these concepts constitutes the cornerstone of a system of prevention on which genetic counseling and early prenatal diagnosis are based.

Methodological Approach

Many different techniques have been proposed for the examination of an abnormal conceptus. Investigation of live-born malformed infants has been standardized for more than 20 years [5] and should include complete autopsy with X-ray examination, dysmorphologic examination and chromosomal analysis. Most of our present knowledge on the pathogenesis of human malformations has been gained from examination of abnormal viable fetuses. With the advent of antenatal diagnostic techniques, the pediatric pathologist is faced with more nonviable fetuses and embryos which often require a different methodological approach. Therefore, from a practical point of view, the methodology used should be adapted to the gestational age of each specimen.

Examination of a first-trimester conceptus is particularly difficult, not only because of the smallness of the material, but also because the specimen is often fragmented or lysed. The examination procedure should always start with localization of the material according to the general classification scheme proposed by Fujikura et al. [6]. Estimation of gestational age is also an important step and can be based on the following criteria:

- The last normal menstrual period of the mother if she knows it precisely. This supposes that she was not using an oral contraceptive at the time of conception and that her cycles were regular before conception. In general, the validity of gestational age evaluation based on menstrual dating is poor [7].
- Embryologic dating according to the tables of classic textbooks [8–10]. These tables allow assessment of the developmental stage of the embryo. Small parts of the embryo may be sufficient for correct dating; in particular, measurement of foot length provides an accurate evaluation of gestational age [8]. If embryonic structures are totally missing, the villous morphology can be used [11], but it is certainly a much less precise criterion.
- Ultrasound dating has become a widely used technique and is now considered as the "modern gold standard" of gestational age evaluation in routine practice. Furthermore, the recent advent of high-resolution imaging combined with the use of a transvaginal probe has significantly improved the observation of normal and abnormal early pregnancy (see Chaps. 4, 21). These new techniques have facilitated the correlation between sonograms and the anatomic descriptions of embryonic or fetal development [12].

Very early specimens and small fetal parts can be examined by means of a stereodissecting microscope or even a scanning electron microscope (Fig. 1). Such an approach is often time consuming and should only be used for well-preserved specimens. In routine practice, we must often limit our investigation to simple macroscopic and classic histologic examinations. However, pathogenic markers for specific chromosomal disorders such as generalized hydrops, cystic hygroma, or placental molar changes may already be apparent macroscopically in early pregnancy stages and may therefore orient the diagnosis. Furthermore, every examination of an early conception product should be completed by histologic examination of the placental

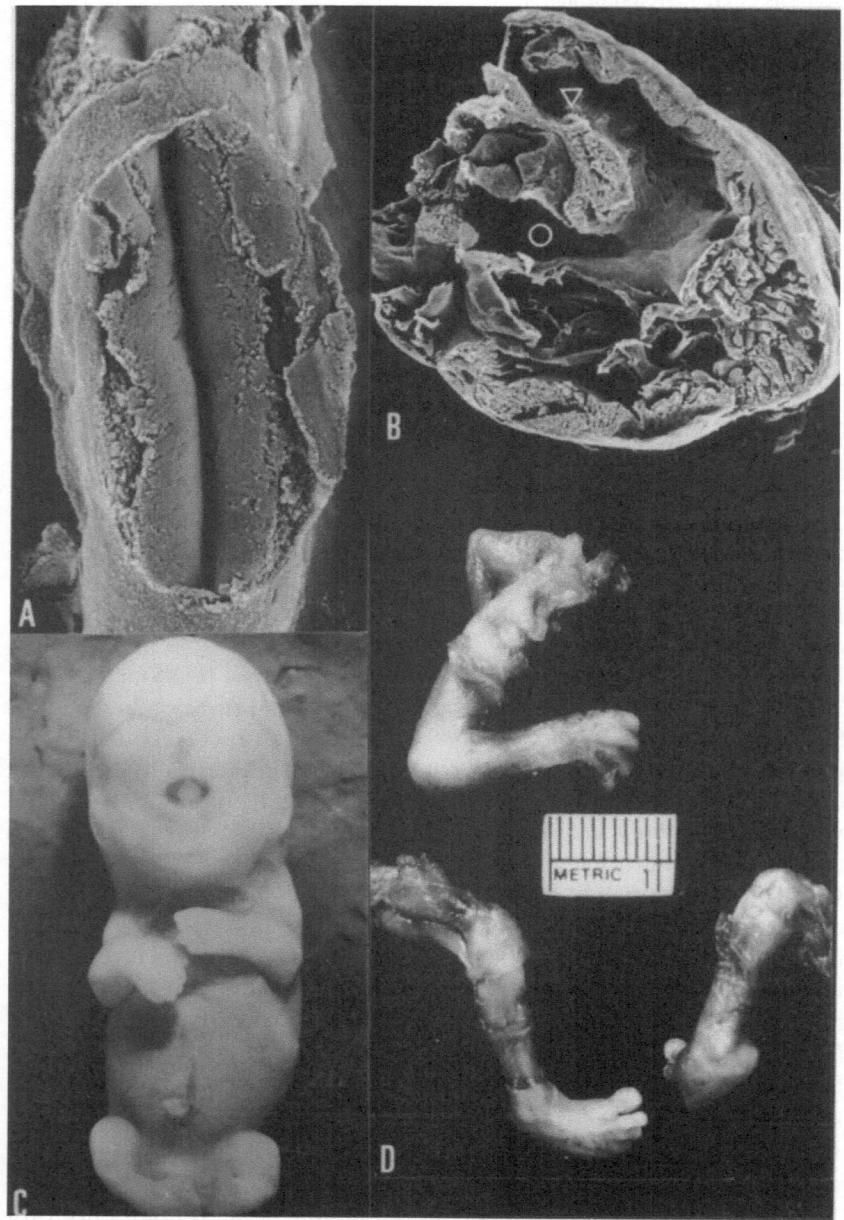

Fig. 1. A Dorsal view of a human embryo at 30 days post natural fertilization (Streeter's stage 12). Note the extensive exophytic growth of neural tissue. The defect has delineated an oval shape on the embryonic epiblast. **B** Large ventricular septal defect (*circle*) in a fetal heart at 11 weeks of gestation from a case of trisomy 18. Note the dysplastic pulmonary valve (*triangle*). The cusps are grossly obliterating the lumen of the pulmonary artery. **C** Cyclops anomaly with holoprocencephaly and fusion of the upper segments of both optic cups in a fetus with trisomy 13. **D** Arthrogryposis in a fetus at 14 weeks of gestation. (Courtesy of Dr. G. Moscoso; **B** from [41])

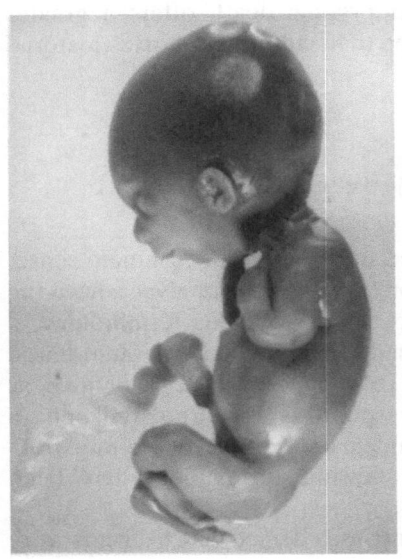

Fig. 2. An 18-week-old fetus with osteogenesis imperfecta type II. Note the deformities of the extremities due to multiple fractures

tissue and cytogenetic analysis. The placenta is often the only material available, and particular attention should be paid to the villous morphology which may be helpful in the recognition of chromosomal abnormalities, infectious complications, or metabolic disorders (see Chap. 3).

Examination of specimens obtained after therapeutic abortion is of paramount importance in order to confirm or invalidate prenatal findings. Correlation of ultrasound features with pathologic findings at all stages of gestation may help the members of a perinatal team to improve the management of subsequent pregancies presenting similar problems. In this context, single case reports of a particular syndrome [13] may be as interesting as series of several cases of the same syndrome [14]. As an example, Fig. 2 illustrates the case of an 18-week-old fetus aborted because of an ultrasound diagnosis of major hydrocephaly, for which postabortion investigations demonstrated osteogenesis imperfecta type II.

The importance of correct processing of the specimens cannot be overemphasized. The delay before fixation and/or immersion in the culture medium may considerably influence the quality of pathologic and cytogenetic results. In the case of late spontaneous abortion, investigations are often more difficult because the product of conception may have been retained for some time in utero after embryonic or fetal death. In these cases, cytogenetic and even histologic analysis may sometimes not be carried out. However, the placental tissue continues to grow long after the embryo has died, and it has recently been demonstrated that chorionic villous sampling performed at the time of diagnosis of embryonic death offers a greater chance of successfully obtaining a karyotype than culture of abortus material [15].

Pathogenesis of Congenital Malformations

The term "malformation" is generally used in a wide, indeterminate sense, thus the words "malformed infant, fetus, or embryo" do not always have the same meaning. In order to clarify and to standardize this terminology, a committee of international experts proposed, 10 years ago, a system based on the development field (DF) concept [16]. This concept is defined as "regions or parts of the embryo which respond as a coordinated unit to embryonic interactions and result in complex or multiple anatomic structures." According to this view, the following terminology has been proposed.

A *malformation* is the result of an intrinsic defect which occurs very early during the initiation and the organization of one DF. Several mechanisms may be implicated, such as a lack of development resulting, for example, in amelia, a deficiency in cellular regression as observed in cases of syndactyly or an incomplete closing as found in labiopalatal clefts. Several and sometimes most of the anatomic structures which normally develop from one of the primitive embryonic layers may be involved. Figure 3 illustrates a severe condition known as cloacal exstrophy, which results from incomplete migration of the infraumbilical mesoderm between the 6th and 7th weeks of gestation. This phenomenon leads to incomplete morphogenesis of all structures which normally develop from this portion of the mesoderm. In this particular case, the malformation complex has been reported as an OEIS complex [17] including an omphalocele, bladder exstrophy, imperforate anus, and a sacral meningocele. Figure 4 represents another example of malformation and, in particular, a polytopic field defect.

A *disruption* results from a secondary change in an otherwise normal DF. The amniotic band syndrome, also called early amnium rupture sequence, is a good example of this concept. This syndrome has been recognized for at least 300 years and consists essentially of a myriad of fetal abnormalities (Fig. 5A) due to the destruction of normally developed structures [18, 19]. The extent of the disruption process depends on the precocity of the primary event during embryonic or fetal life. The incidence of fetal disruption is close to one per 10 000 in liveborns and is much higher in stillbirth [19]. An important characteristic of the disruption syndrome is that fetal malformations are distributed asymmetrically, and it is, therefore, extremely rare to find two individuals affected in the same way. Figure 5b and c illustrates the extreme morphologic aspects that can be observed in cases of amniotic bands.

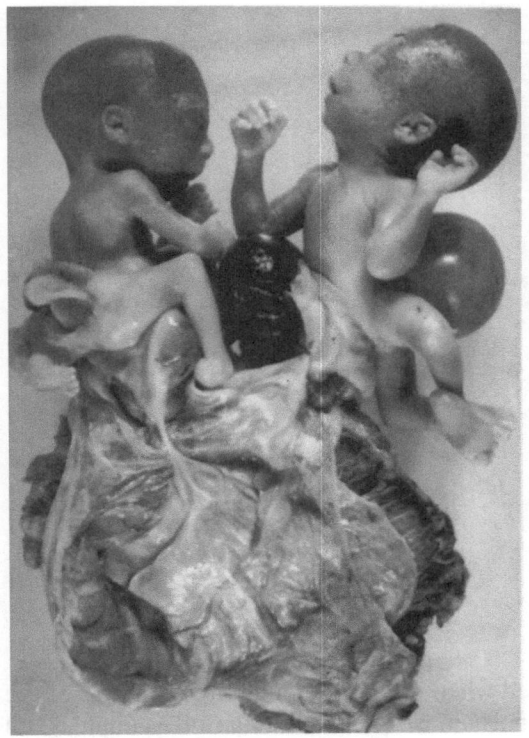

Fig. 3. OEIS complex in monozygotic twins

The results of animal experiments and clinical studies suggest that disruption results from a multifactorial process in which vascular compromise and, more specifically, hemorrhage are the principal pathogenic stimuli [18, 19]. In the amniotic band syndrome, the development of fibrous "amniotic" bands should be considered as a late, reparative event of little or no pathogenic significance [18]. This theory is supported by the fact that fetal disruption occurs predominantly in monozygotic twin gestations where placental vascular connections are extremely frequent [11, 18, 19]. Several cases of monozygotic twin pregnancies complicated by the death in utero of one the fetuses have been associated with a disruption syndrome of the surviving fetus [11, 19], probably resulting from embolization of necrotized tissue through placental anastomoses [11]. Changes in the vascular wall caused by infectious agents such as the rubella virus can also lead to fetal disruption. A similar mechanism has been suggested for the first and second branchial arch syndrome, also known as Goldenhar's syndrome or hemifacial microsomia. This fetal abnormality may be due to developmental alterations of the stapedial artery which irrigates the first and second branchial arches during developmet between days 33–40 of pregnancy [20].

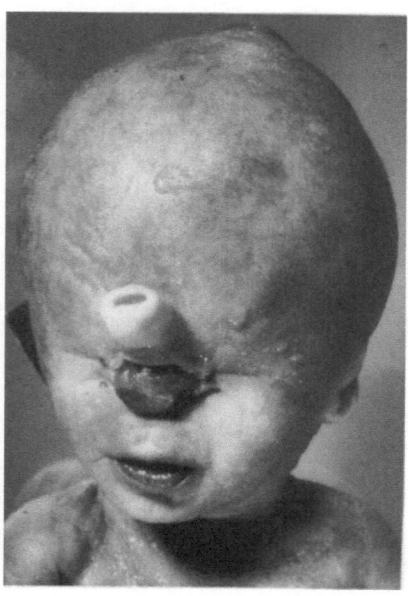

Fig. 4. Case of cyclopia, example of a polytopic field defect

The term *deformation* is used when the normal anatomy of a part of the body is deformed by external mechanical forces such as, for example, fetal head deformities secondary to uterine malformations. This mechanism may also play a role in the development of certain types of craniostenosis [21].

A *sequence* consists of a cascade of secondary malformations which are the result of a focal primary defect. The primary defect may be relatively insignificant with respect to the secondary malformations observed later in pregnancy. The early obstruction of the urethra is the typical example of a sequence. The baby in Fig. 6 died several minutes after birth, and pathologic investigations revealed a complete stenosis of the urethra. This small malformation resulted in massive distension of the bladder with a bilateral reflux and secondary renal dysplasia, gastroschisis, limb and facial deformities secondary to oligohydramnios, and pulmonary hypoplasia leading to fetal death.

A *syndrome* results from the intrinsic alterations of several DFS by one etiologic agent, such as those found in chromosomal disorders. The phenotypic expression of chromosomal disorders can already be observed in fetuses at the end of the first trimester (Fig. 7).

A *dysplasia* is an abnormal cellular organization within tissue(s) and its morphologic result(s). Since all anatomic sites containing the affected tissues are involved, dysplasia is in general not confined to a single organ. Figure 2 is an example of dysplasia: it shows an 18-week-old fetus with a lethal form of osteogenesis imperfecta (type II). Visible multiple abnormalities

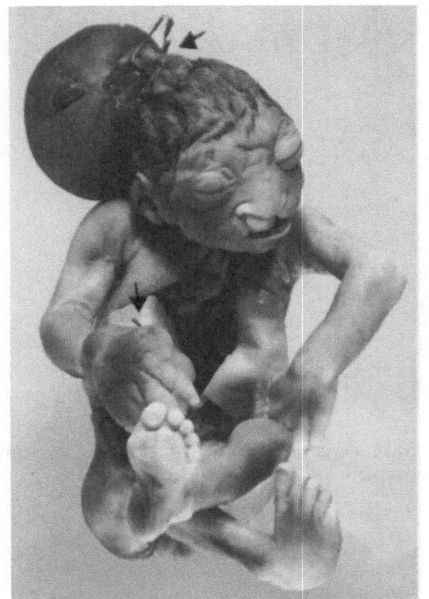

A

B

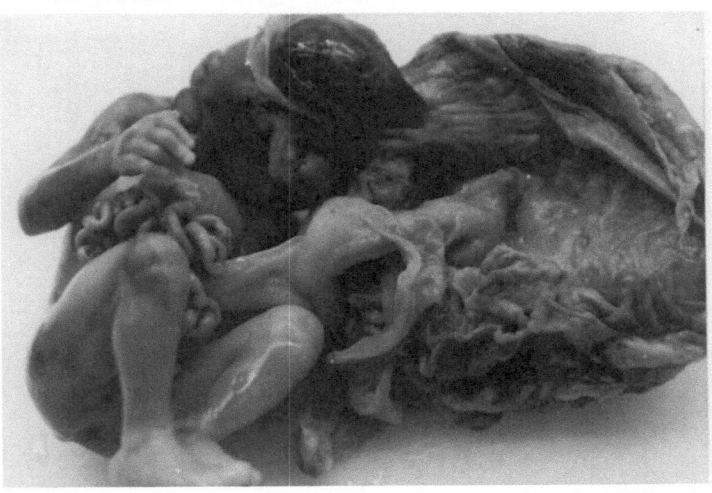

C

Fig. 5A–C. Early amnion rupture sequence in a 19-week-old fetus (**A**). Amniotic bands are clearly visible (*arrows*). **B** Very mild form of the sequence in a newborn infant. **C** Severe form in a 24-week-old fetus. The head is attached to the placenta

are related to a defect of connective tissue and subsequent widespread involvement.

An *association* refers to the nonrandom occurrence of several malformations consistently observed together and which are not the result of a

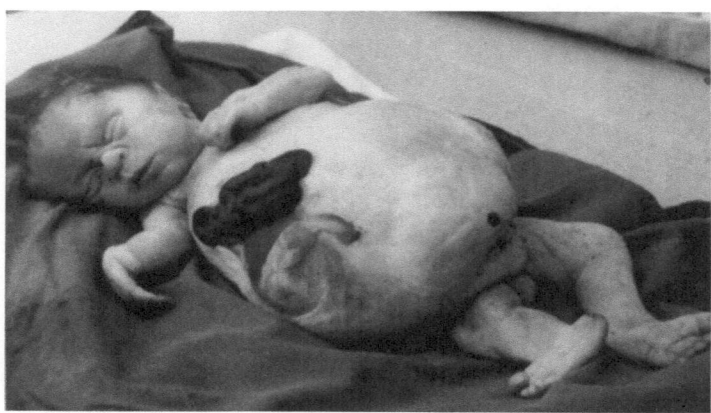

Fig. 6. Newborn with an early urethral obstruction sequence. Note typical facies, extreme distension of the bladder with secondary gastroschizis and deformities of legs

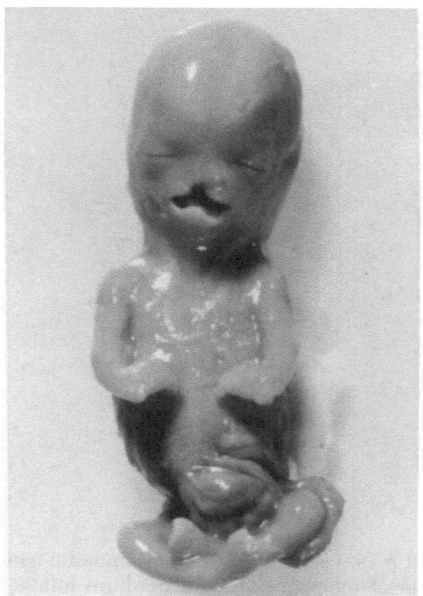

Fig. 7. A 12-week-old fetus with a trisomy 18 syndrome

known etiologic agent. Associations should not be confused with syndromes or sequences because associations occur sporadically with a very low risk of recurrence. The recognition of one characteristic malformation (i.e., coloboma) should alert the clinician and help him or her to exclude other abnormalities, such as heart defect in the case of CHARGE association, for example.

Etiology of Congenital Malformations

Genetic Causes

Chromosomal abnormalities represent the most frequent causes of human malformations during the embryonic and fetal periods of life. Products of conception that have reached the 2nd week of embryonic development are estimated to have a 78% rate of abnormal chromosomal constitution [22]. This rate decreases drastically during the following weeks of pregnancy, and the rate of chromosomal defects in liveborn infants is only around six per 1000 [23]. The incidence and rate of chromosomal defects vary with gestational age. For example, more than 95% of all monosomy X embryos spontaneously abort early in gestation, and this chromosomal defect is, therefore, rarely found during the 2nd and 3rd trimesters or after delivery. Despite the fact that these abnormalities are of genetic origin, most of them occur accidentally, and it has been estimated that nearly 2% of all spontaneous abortions with a chromosomal anomaly are due to parental chromosomal rearrangements [24].

Monogenic abnormalities correspond to inherited characteristics controlled by alleles at a single gene locus. They can be dominant, recessive, or X-linked, and are individually relatively rare as they represent only 7.5% of all congenital malformations found at birth [25]. An updated list of these abnormalities is regularly published by McKusick [26]. Monogenic malformations are, in general, less lethal than malformations secondary to chromosomal abnormalities, and, unless they are diagnosed during pregnancy, most fetuses survive until delivery. However, the combination of several recessive "lethal genes" can induce spontaneous abortion of products of conception [27]. The incidence of monogenic abnormalities is higher in fetuses presenting at birth with lethal malformations, and their incidence in this category of fetuses varies around 12.5% of the cases [28].

Multifactorial abnormalities correspond to the association of hereditary and nongenetic factors and represent the second cause of human malformations found at birth after those due to unknown factors. Neural tube defects, for example, and certain cardiopathies belong to this group [29].

Environmental Factors

Environmental factors causing malformations are relatively rare and represent only about 7% of all malformations [29]. Teratogenic factors such as certain drugs (for example, thalidomide or antiepileptics) or chemicals directly absorbed by the mother, primitive maternal disorders such as phenylketonuria or diabetes, and pregnancy complications such as viral diseases may induce severe fetal malformations. This is particularly true if these agents operate during the first 3 months of gestation. The development of new industrial technologies may also influence the incidence of pregnancy complications. The recent Chernobyl accident (April 26, 1986) has been much discussed by the mass media and the literature. A Norwegian team has recently published the results of a 3-year obstetric survey in order to assess the possible effects of the radioactive fallout from Chernobyl [30]. Compared with the 3-year period before the accident, they found a significant increase in the spontaneous abortion rate in their county during the first year after the accident. This peak was followed by a slight decrease during the second and third years, but incidences were still higher than before the accident. This report, although suggestive, does not represent absolute scientific proof, but highlights the importance of such epidemiologic studies on early reproductive failure in this context.

Unknown Factors

Many single malformations and polymalformation syndromes cannot be classified into one of the above-mentioned categories. In fact, they probably represent the largest category of human malformation. This is the main reason why all fetuses, neonates, or infants presenting with malformations should be investigated in detail and reported in a general system of malformation surveillance as some of these malformations may also be an early epidemiologic indicator [31].

Pathogenesis; A Clue to the Etiology of Malformations

Malformations and *syndromes*, as defined previously, are often genetically determined, while disruptions are usually acquired and sporadic. In some circumstances, the distinction between polymalformation syndromes and disruption syndromes may not be easy. International data base systems and computerized literature search are now available as new diagnostic tools for genetic counseling. The characteristic features of a malformation complex or a syndrome are introduced in the computer and immediately analyzed. However, parental anamnesis can already provide many important clues to the diagnosis. The following examples illustrate the importance of the complete investigations of each case:

1. Figure 8 shows a 16-week-old fetus with encephalocele and bilateral renal agenesis. This polymalformation syndrome could possibly be of genetic origin or result from a fetal disruption. In fact, medical and, especially,

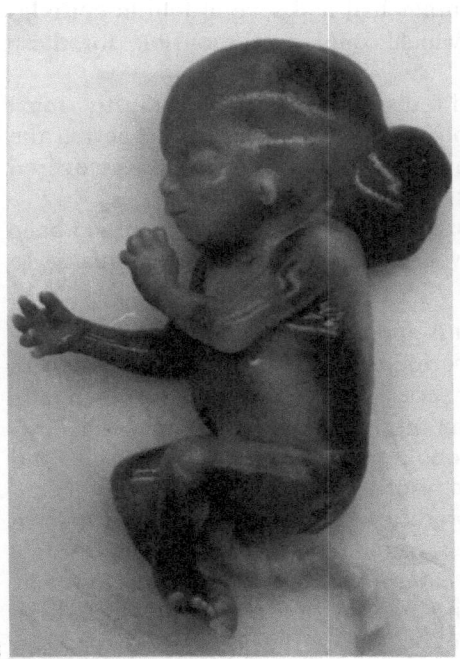 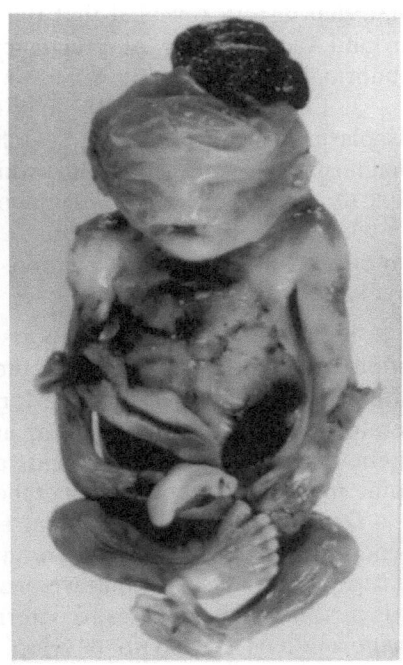

8   9

Fig. 8. Combination of a posterior encephalocele and bilateral renal agenesis, probably due to maternal hyperthermia (16-week-old fetus)

Fig. 9. Meckel's syndrome in an 18-week-old fetus. Note the hexadactly

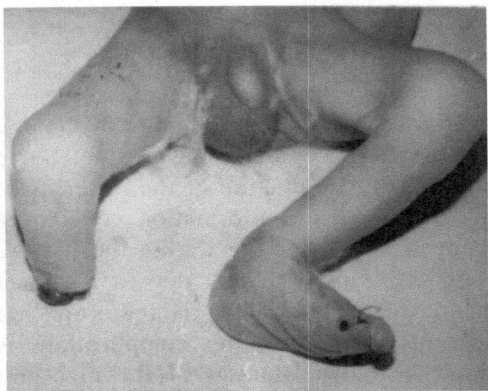

Fig. 10. Amniotic band syndrome, with amputation of the right leg

maternal history revealed that the mother had had several febrile episodes around the 25th day of pregnancy which could be responsible for these abnormalities [32].

2. Figure 9 shows an 18-week-old fetus with postaxial hexadactly, renal dysplasia, and exencephaly. Search on the data base system suggested the diagnosis of Meckel's syndrome which was confirmed when it was discovered that the parents were consanguineous,

3. Figure 10 shows a neonate born at 32 weeks with an amniotic band syndrome. Maternal interrogation revealed an interstitial pneumonia at 16 weeks of gestation.

As shown by these case reports, it is apparent that anamnesis alone may often be adequate to determine the period during which the conceptus was affected. This is especially true for a series of agents which are potentially teratogenic, including radiation, aminopterin, thalidomide, warfarin, synthetic progestin, isoretinoic acid, alcohol, specific antiepileptic drugs, and some maternal infections or metabolic disorders.

Deformations are in general sporadic, whereas *sequences* are heterogeneous. *Dysplasias* most often have a genetic origin, but there are a few exceptions. As discussed above, *associations* are always sporadic. However, it is possible in some cases to determine the time of gestation at which these malformations started to develop. For example, the VATER association results from a defect in the mesoderm development occurring before day 35 of gestation.

Regional and International Epidemiologic Data

The Particular Case of 1st- and 2nd-Trimester Spontaneous Abortions

There is no doubt that examination of spontaneously aborted human embryos and placentas is of great clinical and epidemiologic value. Some authors have also shown the importance of examination of the products of conception after legal abortion [33]. In general, information concerning direct observation of abnormal products of conception during the first 12 weeks of gestation are still very limited. This is mainly due to technical difficulties bound to the approach of such material. It is therefore not surprising that the etiologic evaluation of first-trimester complications is principally based on cytogenetic and histologic analysis of the placental tissue (see Chap. 16).

Specific epidemiologic data on the second trimester are also rare and are usually included in large series together with data on the third trimester. Shepard et al. [34] have recently estimated the defect rates during the 2nd trimester to be around 30%. Between 1986 and 1989, 380 2nd-trimester fetuses were examined in our center, and we found an incidence of defects of 32.6%. This preliminary study confirms the findings of Shepard et al.

Incidence of Malformations in Various Surveys of Perinatal Deaths

The indidence of congenital malformations varies in the different surveys of perinatal mortality published in the literature. It was, in general, lower in the earlier studies [35], where the perinatal mortality was still relatively high, than in more recent studies [36]. This trend is due to the improvement of neonatal intensive care, which allows the survival of more normal babies, and subsequent the relative increase of the overall proportion of congenital malformations among neonatal deaths. In our experience, which concerns 1200 consecutive neonatal autopsies performed during the last 10 years, the rate of congenital malformations is 35.2%. Furthermore, the decrease in family sizes has strongly increased the emotional aspect of each pregnancy, and the "right" to a normal infant is probably more pronounced than formerly. In such circumstances, systems of surveillance of congenital malformations are of paramount importance. However, should we content ourselves with this type of prevention which is passive and which, at best, only records the damage? It may be better to try to reduce the frequency of congenital malformations which still represent the principal causes of perinatal death and infantile mortality and morbidity [3]. The answer to this question is complex because there are, in fact, more than 100 causes of malformations in the human.

Modern Approach and Future Prospects

The widespread use of prenatal diagnostic techniques and the improvement in these techniques may solve, in part, the management of infants with severe abnormalities. On the other hand, it is evident that the blind use, for economic reasons, of antenatal diagnostic techniques by unexperienced clinicians may not be profitable to the patient at all. If this situation extends, it will be necessary to better circumscribed at-risk pregnancies and to promote better collaboration between private or general practitioners and multidisciplinary medical teams specialized in antenatal diagnosis.

The population of at-risk pregnancies usually includes pregnant patients who are over 35 years old, patients presenting with chronic diseases such as diabetes, and patients who have already had one abnormal conceptus. The latter group of patients presents the highest risk, and, if the origin of the abnormality has been adequately diagnosed, it is evident that the medical team carrying out antenatal diagnosis during the following pregnancy will be better armed. In this context we know that the vast majority of congenital malformations are due to a dysmorphogenetic event which takes place during the first 12 weeks of gestation. As for viable fetuses, correlation between prenatal and postnatal findings in 1st-trimester complicated pregnancies is the ideal management of these cases. Because of the difficulties in obtaining intact embryos or small fetuses, an absolute diagnosis should be made in utero. Prenatal investigation of early embryos, in particular by means of ultra sonography, requires a good knowledge of human embryology. The sonographers must be aware of the physiologic changes occurring during normal embryonic development [37]. Once again, we would like to insist on the importance of a close collaboration between obstetricians, pathologists, and geneticists for an appropriate management of these cases. Using this methodology, several authors have recently been able to document the early ultrasound features of abnormalities like tetraploidy, triploidy, trisomy 16, and monosomy X [38, 39]. More studies like these are needed. Careful ultrasound examination together with appropriate pathologic examination may help us to understand the pathogenic mechanisms of various abnormalities and shed new light on the reasons for a given chromosomal abnormality. Nature spares some fetuses and eliminates others [40]. More knowledge on this phenomenon, also called "terathanasia," may have important consequences and direct practical applications.

References

1. Warkany J (1980) Teratology in perspective. Explanations and recommendations for prevention. In: Porter IH, Hook EB (eds) Human embryonic and fetal death. Academic New York, pp 355–362
2. De Wals P, Dolk H, Bertrand F, et al. (1988) La surveillance épidémiologique des anomalies congénitales par le régistre Eurocat. Rev Epidemiol Sante Publique 36:273–282
3. De Wals P, Bertrand F, Vanderlinden M, et al. (1989) Perinatal mortality in Belgium. Biol Neonate 55:10–18
4. Garbis JM, Robert E, Peters WJP (1990) Experience of two teratology information services in Europe. Teratology 42:629–634
5. Langley IA (1971) The perinatal postmortem examination. J Clin Pathol 24:159–169
6. Fujikura T, Froehlich LA, Driscoll SG (1966) A simplified anatomic classification of abortions. Am J Obstet Gynecol 95:902–905
7. Kramer MS, McLean FH, Boyd ME, et al. (1988) The validity of gestational age estimation by menstrual dating in term, preterm, and postterm gestations. JAMA 260:3306–3308
8. Streeter GL (1951) Developmental horizons in human embryo. Carnegie Institute of Embryology, Washington
9. Hamilton JW, Mossman HW (1972) Human embryology. Heffer, Cambridge
10. Moore KL (1988) The developing human. Clinically oriented Embryology, 4th edn. Saunders, Philadelphia
11. Benirschke K, Kaufmann P (1990) The pathology of the human placenta. Springer, Belin Heidelberg New-York
12. Neiman HL (1990) Transvaginal ultrasound embryography. Semin Ultrasound Comput Tomogr Magu Reson 11:22–23
13. Jauniaux E, Vyas S, Finlayson C, et al. (1990) Early sonographic diagnosis of body stalk anomaly. Prenat Diagn 10:127–132

14. Jauniaux E, Demunter C, Pardou A, et al. (1989) Evaluation echographique du syndrome de l'artère ombilicale unique: une série de 80 cas. J Gynecol Obstet Biol Reprod (Paris) 18:341–347
15. Johnson MP, Drugan A, Koppitch FC, et al. (1990) Postmortem chorionic villi sampling is a better method for cytogenetic evaluation of early fetal loss than culture of abortus material. Am J Obstet Gynecol 163:1505–1510
16. Spranger JW, Opitz JM, Schmith W, et al. (1982) Errors of morphogenesis: concepts and terms. Recommendations of an international working group. J Pediatr 100:160–165
17. Carey JC, Greenbaum B, Hall BD (1978) The OEIS complex. Birth Defects 14(6B):253–263
18. Lockwood C, Ghidini A, Romero R, et al. (1989) Amniotic band syndrome: reevaluation of its pathogenesis. Am J Obstet Gynecol 160:1030–1033
19. Luebke HJ, Reiser CA, Pauli RM (1990) Fetal disruptions: assessment of frequency, heterogeneity and embryologic mechanism in a population referred to a community-based stillbirth assessment program. Am J Med Genet 36:56–72
20. Poswillo D (1973) The pathogenesis of the first and second brancheal arch syndrome. Oral Surg 35:312–328
21. Higginbottom MC, Jones KL, James HE (1980) Intrauterine constraint and craniosynostosis. Neurosurgery 6:1980–1986
22. Boué J, Boué A (1973) Anomalies chromosomiques dans les avortements spontanés. In: Boué A, Thibault CD (eds) Les accidents chromosomiques de la réproduction. Inserm, Paris, pp 29–55
23. Jacobs PA (1977) Epidemiology of chromosomal abnormalities in man. Am J Epidemiol 105:180–188
24. Lipmann-Hand A (1980) Genetic counseling and human reproductive loss. In: Porter IH, Hook EB (eds) Human embryonic and fetal death. Academic, New York, pp 299–314
25. Kalter H, Warkany J (1983) Congenital malformation: etiologic factors and their role in prevention. N Engl J Med 308:424–497
26. McKusick VA (1988) Mendelian inheritance in man, 8th edn. Johns Hopkins University Press, Baltimore
27. MacCluer JW (1980) Inbreeding and human fetal death. In: Porter IH, Hook EB (eds) Human embryonic and fetal death. Academic, New York, pp 241–260
28. Gillerot Y, Koulischer L (1985) La fréquence des malformations létales: expérience portant sur 600 autopsies. J Genet Hum 33:289–293
29. Brent RL (1986) The complexities of solving the problem of human malformations. Clin Perinatol 13:491–503
30. Ulstein M, Skeie-Jensen T, Irgens LM, et al. (1990) Outcome of pregnancy in one Norwegian county 3 years to and 3 years subsequent to the Chernobyl accident. Acta Obstet Gynecol Scand 69:277–280
31. Källen B (1989) Population surveillance of congenital malformations. Possibilities and limitations. Acta Paediatr Scand 78:657–663
32. Jones KL (1988) Smith's recognizable patterns of human malformations, 4th edn. Saunders, Philadelphia
33. Millar DR, Fothergill DJ (1990) The value of a histological examination of the products of conception after legal abortion. J Obstet Gynaecol 10:179–180
34. Shepard TH, Fantel AG, Fitzsimmons J (1989) Congenital defect rates among spontaneous abortuses: twenty years of monitoring. Teratology 39:325–331
35. Naeye RL (1977) Causes of perinatal mortality in the US: collaborative perinatal project. JAMA 238:228–229
36. Mutch LMM, Brown NJ, Speidel BD, et al. (1981) Perinatal mortality and neonatal survival in Avon: 1976–1979. Br Med J 282:119–122
37. Tarantal AF, Hendrickx AG, O'Rahilly R (1990) First trimester conceptus. J Ultrasound Med 9:614–615

38. Boué J, Bessis R (1990) Aspects morphologiques des erreurs chromosomiques létales. Reprod Nutr Dev [Suppl] 1:895–945
39. Ruchelli ED, Shen-Schwartz S, Martin J, et al. (1990) Correlation between pathologic and ultrasound findings in first trimester spontaneous abortions. Pediatr Pathol 10:743–756
40. Warkany J (1978) Terathanasia. Teratology 17:187–192
41. Moscoso G (1989) Biol Neonate 56:147–150

Diagnostic Methods

20 Diagnosis of Abnormalities of Early Pregnancy by Measurement of Fetoplacental Products in Biological Fluids

J.G. Grudzinskas and T. Chard

The human fetus and placenta, and the surrounding maternal endometrium, secrete a wide variety of products which can be described as "specific" in the sense that they are present in much higher concentrations than in the non-pregnant state. The products (Table 1) appear in maternal blood and urine, and amniotic fluid. Their measurement in these sites provides valuable biological and clinical information on the progress of a pregnancy [5, 7, 14]. This chapter discusses the clinical biochemistry of these compounds in the first 12 weeks of gestation.

Diagnosis of Pregnancy

Early diagnosis of pregnancy commonly depends on the detection of human chorionic gonadotrophin (hCG) in maternal urine or 1–2 days earlier in blood. The test becomes positive shortly after the time of implantation. Claims for earlier detection of human pregnancy by measurement of an "early pregnancy factor" have not been substantiated [5]. Highly specific and sensitive immunometric assays have shown the ubiquitous nature of hCG at low concentrations (<15 IU/l). In clinical practice, a single estimation of hCG should only be considered definitive if it is greater than 25 IU/l or if a lower level of hCG is seen to increase twofold at an interval of 3 days [16]. When hCG has been given therapeutically, estimations should be delayed until clearance of the exogenous hCG has occurred, a delay of up to 14 days. Under these circumstances, assays for other placental proteins may be appropriate: for example, *Schwangerschaftsprotein* 1 (SP1) [13]. Other pregnancy-associated proteins, such as human placental lactogen (hPL), pregnancy-associated plasma protein A (PAPP-A) and placental protein 5 (PP5), are not contenders as diagnostic tests since the rise in maternal blood cannot be clearly identified until after 6 weeks of amenorrhoea. Finally, positive hCG values found immediately after implantation should be interpreted with care; at this stage many pregnancies may be diagnosed which subsequently abort [4].

Table 1. Major fetal, trophoblast and decidual proteins identified in early pregnancy

Fetal
   Alpha-fetoprotein (aFP)
   Fetal antigen 1 (FA-1)
   Fetal antigen 2 (FA-2)
Trophoblast
   Human chorionic gonadotrophin (hCG)
   Schwangerschaftsprotein 1 (SP1)
   Human placental lactogen (hPL)
   Pregnancy-associated plasma protein A (PAPP-A)
   Placental protein 5 (PP5)
Decidual or endometrial
   Insulin-like growth factor-binding protein (IGF-BP: α_1-PEG, PP12, α_1-PAMG)
   Progestogen-dependent endometrial protein (PEP: α_2-PEG, PP14, α_2-CMG,
     α_2-PAMG, AUP)

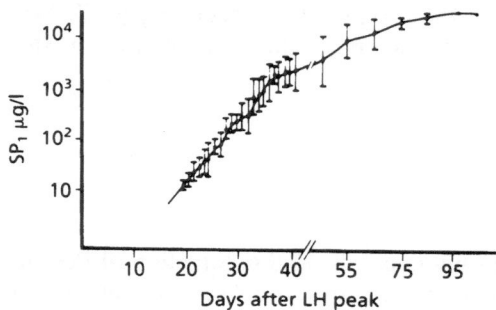

Fig. 1. Mean serum SP1 levels in nine women in the 1st trimester of normal pregnancy. *LH*, luteinising hormone, (From [27])

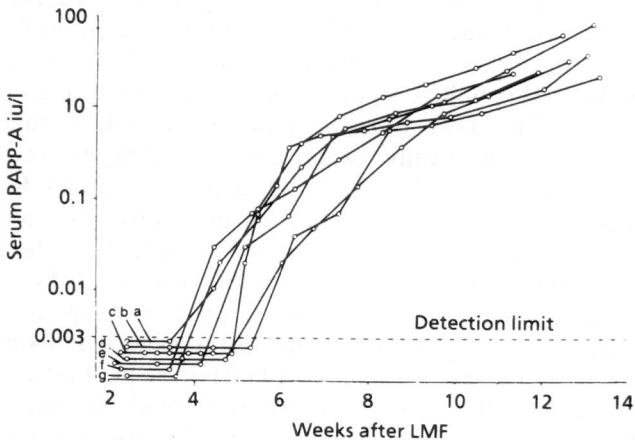

Fig. 2. Serum PAPP-A levels in seven women in the 1st trimester of normal pregnancy. (From [11])

Normal Early Pregnancy

Chorionic gonadotrophin and other placental products might be synthesised by the blastocyst prior to implantation. However, significant amounts are secreted into the mother only after the trophoblast makes direct contact with maternal blood in the course of the implantation process. There is no obvious trigger for the production of hCG, or for that of compounds such as SP1, PAPP-A, hPL and PP5. Production of these compounds appears to be an inherent, free-running function of the trophoblast with no obvious control or feedback mechanisms of the type which are normally encountered in adult physiology.

The trends in blood concentration of SP1 seem to parallel the growth rate curve of functioning trophoblast [13] (Fig. 1). The synthesis of SP1, at least in the earlier weeks of pregnancy, is independent of the presence of an embryo (it is found in women with hydatidiform mole) and of the site of implantation (it is found in women with ectopic gestation). In normal pregnancy the doubling times for hCG and SP1 are similar: the concentrations of both double in 2–3 days in the first 6 weeks of pregnancy. The disappearance rate of SP1 and hCG after removal of the placenta are also equivalent (40–60 h). Blood levels of both hPL and PP5 increase as pregnancy advances and have a relationship to functioning trophoblast mass. PAPP-A can be detected in maternal blood 28 days after conception, which is too late for use as a primary diagnostic test for pregnancy [11] (Fig. 2). PAPP-A levels increase throughout gestation with a doubling time of 4.9 days during the first trimester; the disappearance rate after removal of the placenta is several days [29].

In the first 6 weeks of gestation, the bulk of the steroids (oestrogen and progesterone) required for the maintenance of the pregnancy is provided by the corpus luteum. Thereafter most steroid production is derived from the fetoplacental unit. The synthesis of alpha-fetoprotein (aFP) by embryonal endodermal tissues (principally the yolk sac) is reflected by an increase in circulating levels during the first trimester; increasing concentrations of aFP are detected in maternal blood after 10 weeks of gestation [24]. Secretion of endometrial proteins parallel the morphological changes in this tissue [1]. PP14, which is derived from the glandular epithelium of the endometrium, has a pattern of blood levels in early pregnancy which bears a striking resemblance to hCG. By contrast, insulin-like growth factor-binding protein (IGF-BP1; PP12), which is derived from stromal cells of the gestational endometrium, reaches peak levels in the second trimester [39].

Control of Synthesis and Secretion

The mechanisms which control the synthesis of trophoblast proteins are poorly understood; uteroplacental perfusion plays a major role, influenc ng synthesis and secretion by short-short feedback or mass action. Nevertheless, there are possible effects of the ovary, the embryo and endometrium. Up to 6 weeks of pregnancy, the ovary provides steroid support for the pregnancy. It also secretes a molecule (possibly relaxin) which is responsible for endometrial secretion of PP14. Thus, in donor pregnancies in women without ovaries (e.g., in Turner's syndrome), the dramatic increase in PP14 levels seen in a normal pregnancy is almost totally absent [12].

An embryo is not essential, as evidenced by placental protein secretion in blighted ovum and hydatidiform mole. The exact site of implantation and the interaction between the trophoblast and gestational endometrium are also not definitive factors for secretion of the substances considered here, at least in the earliest days of pregnancy. Circulating placental proteins are consistently present in ectopic gestation, but often at levels somewhat lower than those observed in a normal pregnancy of equivalent gestation. The difference is particularly striking in the case of PAPP-A (reviewed in [31]). There are significant changes in placental protein synthesis in aneuploid pregnancies (see below), but the precise mechanism of these changes is uncertain [6]. In rare pregnancies, there is total absence of a particular placental protein due to gene deletion. This has been described for hPL, SP1, placental sulphatase and PAPP-A. In the latter case, the deficiency has been associated with Cornelia de Lange's syndrome in the child [41]. In the case of hPL and SP1 deficiencies, there is no obvious abnormality in either the mother or the child.

The control of endometrial protein synthesis is a combination of increased bulk of the tissue or origin (endometrial glands in the case of PP14, endometrial stroma in the case of IGF-BP1) and other factors. At one time it was believed that progesterone was the main factor controlling PP14 synthesis. However, this now seems unlikely [15], although an ovarian factor is of undoubted importance [12].

Early Pregnancy Failure

If a woman wishes to become pregnant, her chances of producing a viable offspring in any one ovarian cycle is approximately 25%. There have been many estimates of the frequency of early pregnancy failure, several based on studies on the transient appearance of hCG in association with a potentially fertile cycle. Estimates of the incidence of this phenomenon vary from 8% to 55% (for review see [4]).

Threatened and Spontaneous Miscarriage

Circulating levels of hormones and proteins of fetal, trophoblastic and maternal origin have been used to predict the outcome in women with vaginal bleeding in early pregnancy [23, 27]. In general, levels of placental products are reduced in association with threatened miscarriage in which the outcome is fetal loss, but normal in cases in which the pregnancy proceeds. As a clinical predictive tool, measurement of these compounds is probably unnecessary if fetal life can be demonstrated by ultrasound [30]. Nevertheless, a proportion of patients in whom fetal heart action has been demonstrated will spontaneously miscarry. We have described blood levels of hPL, SP1, PAPP-A, progesterone, oestradiol, aFP and pregnancy-zone protein (PZP) in this condition in 108 patients in whom the history and clinical findings were diagnostic of threatened miscarriage [42]. Eleven patients showed a fetal heart action until miscarriage occurred. In this group, the predictive value of an abnormal level in the sample obtained at presentation was greatest for PAPP-A. All 11 women who had evidence of a live fetus, but who subsequently miscarried had depressed PAPP-A levels. The levels of the other substances generally remained within the normal range.

As a result of these and other studies, it is now possible to conclude: (a) if fetal heart action is not evident ultrasonically after 7 weeks of gestation (gestational sac volumes >3 ml) then depressed levels of fetoplacental products will not provide any additional information; (b) spontaneous miscarriage after the observation of fetal heart action is extremely uncommon; in this small group, the levels of hormones and proteins will be in the normal range, with the exception of low levels of PAPP-A.

Studies of women with apparently normal pregnancies at 6–10 weeks of gestation (awaiting chorionic villus sampling for chromosomal abnormalities) have revealed depressed levels of serum aFP, PAPP-A and, to a lesser degree, hCG within 2–3 weeks of spontaneous miscarriage (B. Brambati, J.G. Grudzinskas and T. Chard, personal observations). Many of these fetuses may have had an abnormal karyotype: measurement of hormones and proteins in the first trimester may identify this particular group, as do similar measurements in the second trimester [38].

Anembryonic Pregnancy

In some cases of abortion an embryo cannot be found. This condition can be diagnosed ultrasonically if no fetus is apparent in a gestational sac of greater than 3 ml (7 weeks of gestation) [34]. Levels of placental proteins and hormones are low or normal. Serum aFP levels are within the normal range in many women with this condition, suggesting that an embryo has been present at some stage and that the term "anembryonic pregnancy" is a misnomer [34].

Mean levels of circulating hCG have also been shown to be depressed at 4 weeks of gestation in a group of women studied prospectively at a subfertility clinic who were subsequently shown to have "blighted ovum"; mean levels of maternal PAPP-A, oestradiol and progesterone fell some 3 weeks later [44]. Regardless of the pathogenesis of this condition, these findings confirm that a deviation from the normal rise in blood levels of hCG is highly suggestive of failed pregnancy at a time before ultrasound can provide useful information (i.e., in the absence of fetal heart action).

Ectopic Pregnancy

The major value of biochemical tests in this condition, especially hCG, is to alert the clinician to the possibility of a pregnancy-related disorder and to exclude such a disorder if a negative result (i.e., <25 IU/l) is obtained [28]. Stabile et al. [30] have reported a sensitivity of 100% for estimations of hCG used in conjunction with ultrasound to make the diagnosis. Quantitative estimations of hCG may also be of some value since depressed levels are commonly seen in ectopic gestation. In conjunction with ultrasound examination, hGC can often distinguish between normal or failed intrauterine pregnancy and ectopic gestation [23]. The use of a discriminatory zone for hCG in conjunction with ultrasound can also be helpful. If levels of hCG are greater than 6500 IU/l, ultrasound examination should reveal the presence of a live embryo in utero; if this is not the case, failed pregnancy, in particular ectopic gestation, should be considered [17, 26].

Similar findings have been reported with SP1 and PAPP-A. In the latter case, secretion is more severely compromised [10, 29]. In one study, PAPP-A levels were depressed in all 17 women with ectopic gestation, with PAPP-A being detectable in only two women (27). Stabile et al. [31] examined 60 women with proven ectopic pregnancy: circulating PAPP-A was detected in

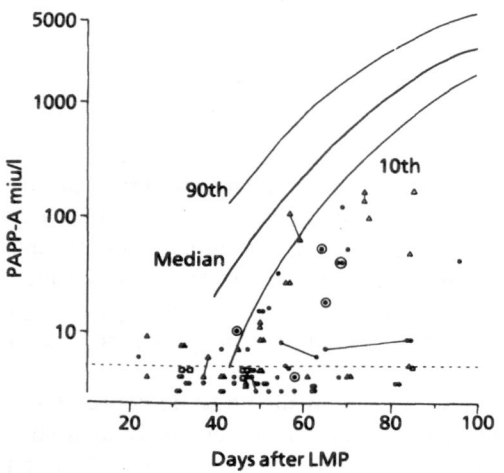

Fig. 3. Circulating PAPP-A levels in 60 women with ectopic gestation in relation to the normal range. The *encircled symbols* indicate that fetal tissue was identified. *LMP*, last menstrual period. (From [32])

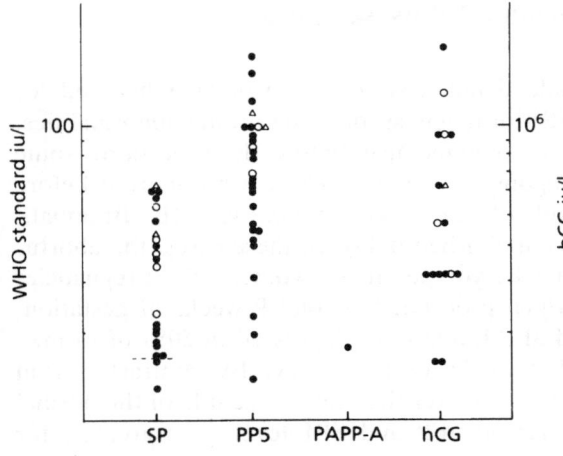

Fig. 4. Serum SP1, PP5, PAPP-A and hCG levels in 31 women with gestational trophoblastic disease before treatment: hydatidiform mole; choriocarcinoma, *open circles*, lung metastases; *triangles*, --- 16.5 IU/l WHO standard 78/ 610. Levels of SP1, PP5, PAPP-A <IU/l or hCG <100 000 IU/l not shown. (From [37])

only 30 and levels were low in 24 women (Fig. 3). Levels of IGF-BP1 were normal in this group [35].

Westergaard et al. [43] noted depressed levels of oestradiol, progesterone and PP14 in the same group of women. They concluded that it was not possible to distinguish between failed intrauterine pregnancy and ectopic gestation by the measurement of serum hormones and proteins, but that it was possible to exclude ectopic pregnancy by these measurements. At the time of clinical presentation, the presence of hCG would alert the clinician to the possibility of a pregnancy-related disorder. The finding of normal levels of progesterone or PAPP-A or PP14 would indicate a normal intrauterine pregnancy [32].

Trophoblastic Disease

Trophoblast tissue in hydatidiform mole retains its ability to synthesise hormones and proteins; the capacity is reduced for hPL and steroids, but greater for hCG.

The risk of choriocarcinoma is higher in women with very high levels of hCG in gestational trophoblast tumours. In untreated hydatidiform mole we have noted reduced levels of PAPP-A prior to treatment and elevated levels of SP1 and PP5 [37] (Fig. 4). High SP1 levels were found in a proportion of women who developed subsequent malignant disease. Than et al. [36] observed high serum concentrations of PP14 in women prior to treatment for hydatidiform mole, falling rapidly after evacuation. Circulating PP14 was not seen in women with choriocarcinoma. In patients with choriocarcinoma, serum SP1 levels are usually lower than in benign disease, and circulating PP5 cannot be detected [19, 20].

Pre-natal Diagnosis: Chorionic Villus Sampling

Brambati et al. [3] have made detailed studies on women scheduled for chorionic villus sampling (CVS) for diagnosis of chromosome abnormalities. Some of these women aborted spontaneously before the procedure, some after, while others had an ongoing pregnancy. Those who aborted before CVS had generally lower levels of aFP, PAPP-A and hCG (B. Brambati, T. Chard, J.G. Grudzinskas, unpublished data). In these cases, the abortus generally showed an abnormal karyotype. In 25 women with pregnancies complicated by aneuploidy which proceeded beyond 9 weeks of gestation, circulating SP1, PAPP-A and aFP levels were depressed in 50% of women when tested between 6 and 9 weeks of pregnancy. By contrast, serum progesterone and hCG levels were greater than the 5th centile of the normal range. The sensitivity of depressed aFP and SP1 levels was greatest for Down's syndrome [3].

It is well known that maternal serum aFP levels can show a dramatic rise following invasive procedures in early pregnancy, notably amniocentesis [8] and surgical [22] or medical [9] termination. A similar phenomenon is seen after CVS [18, 40]. Ward et al. [40] examined serum levels of aFP, hCG, SP1 and PAPP-A before and 6 h after villus sampling in 20 women. In most cases, aFP levels rose after sampling. Depressed or elevated pre-operative levels of aFP, hCG and PAPP-A were associated with fetal anomalies and pregnancy loss in seven of the 11 pregnancies which failed to progress. The two highest levels of hCG were seen in pregnancies in which the fetus was homozygous for α- or β-thalassaemia. Similar observations have been made for hPL and SP1 in pregnancies complicated by hydrops fetalis in late pregnancy [2]. There is no significant change in proteins and hormones of trophoblastic origin following CVS [40].

Conclusion

Estimation of hCG is still the mainstay of the diagnosis of early pregnancy. When used in conjunction with ultrasound it can provide much valuable clinical information in the first 12 weeks of gestation.

After detection of the fetal heart at 6–7 weeks of gestation, spontaneous miscarriage is uncommon. In some women with apparently normal pregnancies, but who will miscarry within a few weeks, depressed levels of aFP, hCG and PAPP-A have been observed. In women presenting with vaginal bleeding after 7 weeks of gestation, if the fetal heart is seen, only PAPP-A measurements may predict the small group who will subsequently miscarry.

In women with suspected ectopic gestation, detection of hCG confirms a pregnancy-related disorder. High levels of progesterone or PP14 may exclude this condition, while depressed levels of PAPP-A increase the likelihood of the diagnosis.

The identification of secretory proteins of the endometrium and decidua which can be measured in maternal blood suggests that IGF-BP1 and PP14 are active rather than passive participants in the reproductive axis. However, the fact that pregnancy may proceed normally in the absence of any rise in PP14 levels suggests that this molecule does not fulfil an irreplaceable role in the maintenance of an early pregnancy [12]. Since the endometrium produces hormones and proteins which act both locally and systemically, it is possible to consider this tissue as an endocrine organ. The ability to measure these substances, in particular IGF-BP1 and PP14, may provide both a non-invasive test of endometrial function in the assessment of fertility and an index of endometrial and decidual function in the earliest days of pregnancy.

Many fetuses in pregnancies which will fail in the first trimester have an abnormal karyotype. It is possible that the hormone and protein measurements at this time may identify these particular patients as is the case in the second trimester.

References

1. Bell SC (1988) Synthesis and secretion of proteins by the endometrium and decidua. In: Chapman MG Grudzinskas JG, Chard T (eds) Implantation. Springer, Berlin Heidelberg New York, pp 95–18
2. Bellman O, Tebbe J, Lang N, Baur MP (1980) Determination of SP1 and hPL for predicting perinatal asphyxia. In: Klopper A, Genazzani A, Crosignani PG (eds) The human placenta: proteins and hormones. Academic, London, pp 99–108
3. Brambati B, Lanzani A, Tului L (1991) Ultrasound and biochemical assessment of first trimester pregnancy. In: Chapman M, Grudzinskas JG, Chard T (eds) The embryo. Springer Berlin Heidelberg New York, pp 181–194
4. Chard T (1991) Frequency of implantation and early pregnancy loss in natural cycles. Baillieres Clin Obstet Gynaecol 5:179–189
5. Chard T, Grudzinskas JG (1987) Early pregnancy factor. Biol Res Pregnancy Perinatal 8:53–56
6. Chard T, Grudzinskas JG (1991) The endocrinology of the fetoplacental unit in the second trimester of pregnancy. In: Chapman M, Grudzinskas JG, Chard T (eds) The embryo. Springer, Berlin Heidelberg New York, pp 209–226
7. Chard T, Klopper A (1982) Placental function tests. Springer, Berlin Heidelberg New York
8. Chard T, Kitau MJ, Ledward R, Coltart T, Embury S, Seller MJ (1976) Elevated levels of maternal plasma alpha-fetoprotein after amniocentesis. Br J Obstet Gynaecol 83:33–34
9. Chard T, Olajide F, Kitau M (1990) Changes in circulating alphafetoprotein following adminstration of mifepristone in first trimester pregnancy. Br J Obstet Gynaecol 97:1030–1032
10. Chemnitz J, Tornehave D, Teisner B, Poulsen HK, Westergaard JG (1984) The localisation of pregnancy proteins (hPL, SP1 and PAPP-A) in intra- and extrauterine pregnancies. Placenta 5:489–494

11. Chemnitz J, Folkersen J, Teisner B (1986) Comparison of different antibody pre-
    parations against pregnancy-associated plasma protein A (PAPP-A) for use in
    localisation and immunoassay studies. Br J Obstet Gynaecol 93:111–118
12. Critchley HOD, Chard T, Lieberman BA, Buckley CH, Anderson DC (1990) Serum
    PP14 levels in a patient with Turner's syndrome pregnancy after frozen embryo
    transfer. Hum Reprod 5:250–254
13. Grudzinskas JG, Gordon YB, Jeffrey D, Chard T (1977) Specific and sensitive
    determination of pregnancy specific beta-1 glycoprotein by radioimmunoassay. Lancet
    1:33–35
14. Grudzinskas JG, Stabile I, Campbell S (1988) Early pregnancy failure: biochemical
    and biophysical assessment. In: Beard RW, Sharp F (eds) Early pregnancy loss.
    Mechanisms and treatment. RCOG, London, pp 183–190
15. Howell RJS, Olajide F, Teisner B, Grudzinskas JG, Chard T (1989) Circulating levels
    of placental protein 14 and progesterone following mifepristone (RU38486) and
    gemeprost for termination of first trimester pregnancy. Fertil Steril 52:66–68
16. Jones HW, Acosta AA, Andrews MC, et al. (1983) What is pregnancy? A question
    for in vitro fertilisation. Fertil Steril 40:728–733
17. Kadar N (1983) Ectopic pregnancy. In: Studd J (ed) Progress in obstetrics and
    gynaecology, vol 3. Livingstone, Edinburgh, pp 305–323
18. Knott PD, Chan B, Ward RHT, Chard T, Grudzinskas JG, Petrou M, Modell B
    (1988) Changes in circulating alphafetoprotein and human chorionic gonadotrophin
    following chorionic villus sampling. Eur J Obstet Gynecol Reprod Biol 27:277–281
19. Lee JN, Salem HT, Al-Ani ATM, Chard T (1981) Circulating concentrations of
    specific placental proteins (human chorionic gonadotrophin, pregnancy-specific beta-1
    glycoprotein and placental protein 5) in untreated gestation trophoblastic tumours.
    Am J Obstet Gynecol 39:702–704
20. Lee JN, Salem HT, Chard T, Huang SC, Ouyang PC (1982) Circulating placental
    proteins (hCG, SP1 and PP5) in trophoblastic disease. Br J Obstet Gynaecol 89:69–
    72
21. Lenton EA, Grudzinskas JG, Gordon YB, Chard T, Cooke ID (1981) Pregnancy
    specific β_1 glycoprotein and chorionic gonadotrophin in early pregnancy. Acta Obstet
    Gynecol Scand 60:489–492
22. Naik K, Kitau M, Setchell ME, Chard T (1988) The incidence of fetomaternal
    haemorrhage following elective termination of first trimester pregnancy. Eur J Obstet
    Gynecol Reprod Biol 27:355–357
23. Niven PAR, Landon J, Chard T (1972) Placental lactogen levels as a guide to
    outcome of threatened abortion. Br Med J 3:799–801
24. Olajide F, Kitau MJ, Chard T (1989) Maternal serum AFP levels in the first trimester
    of pregnancy. Eur J Obstet Gynecol Reprod Biol 30:123–128
25. Pittaway DE, Wentz AC, Maxon WS, Herbert C, Daniell HJ, Fleischer AC (1985)
    The efficacy of early pregnancy monitoring with serial chorionic gonadotrophin
    determinations and realtime ultrasonography in an infertile population. Fertil Steril
    44:190–194
26. Rottem S, Timor-Tritsch IE (1988) In: Timor-Tritsch IE, Rottem S (eds) Transvaginal
    ultrasonography. Heinemann, London, pp 125–142
27. Salem HT, Ghaneimah SA, Shaaban MM, Chard T (1984) Prognostic value of
    biochemical tests in the assessment of fetal outcome in threatened abortion. Br J
    Obstet Gynaecol 91382–385
28. Seppala M, Rantaa T, Tontti K, Stenman UH, Chard T (1980) Use of a rapid
    hCG-beta-subunit radioimmunoassay in acute gynaecological emergencies. Lancet
    1:165–166
29. Sinosich MJ (1985) Biological role of pregnancy-associated plasma protein A in
    human reproduction. In: Bischof P, Klopper A (eds) Proteins of the placenta.
    Karger, Basel, pp 158–184
30. Stabile I, Grudzinskas JG, Campbell S (1987) Ultrasonic assessment of complications
    during first trimester of pregnancy. Lancet 2:1237–1240

31. Stabile I, Campbell S, Grudzinskas JG (1988) Can ultrasound reliably diagnose ectopic pregnancy? Br J Obstet Gynaecol 95:1247–1252
32. Stabile I, Westergaard JG, Grudzinskas JG (1988) Ectopic pregnancy: diagnostic aspects. In: Chapman MG, Grudzinskas JG, Chard T (eds) Implantation: biological and clinical aspects. Springer, Berlin Heidelberg New York, pp 229–238
33. Stabile I, Grudzinskas JG, Chard T (1988) Clinical applications of pregnancy protein estimations with particular reference to pregnancy-associated plasma protein A (PAPP-A). Obstet Gynecol Surv 43:73–82
34. Stabile I, Olajide F, Grudzinskas JG (1989) Maternal serum alphafetoprotein levels in anembryonic pregnancy. Hum Reprod 4:204–205
35. Stabile I, Teisner B, Chard T, Grudzinskas JG (1990) Circulating levels of placental protein 12 in ectopic pregnancy. Arch Gynecol Obstet 247:149–153
36. Than GN, Tatra G, Szabo DG, Csaba K, Bohutt E (1988) Beta lactoglobulin homologue placental protein 14 (PP14) in serum of patients with trophoblastic disease and non-trophoblastic gynaecological pregnancy. Arch Gynaecol 243:131–137
37. Tsakok FTM, Koh M, Ratnam SS (1983) Pregnancy associated proteins in trophoblastic disease. Br J Obstet Gynaecol 90:483–486
38. Wald MJ, Cuckle HS, Densem W Nanchahal K, Royston P, Chard T, Haddow JE, Knight GJ, Palomaki GE, Canick JA (1988) Maternal screening for Down's syndrome in early pregnancy. Br Med J 2:883–887
39. Wang HS, Perry LA, Kanislus J, Iles RK, Holly JMP, Chard T (1991) Purification and assay of insulin-like growth factor-binding protein-1: measurement of circulating levels throughout pregnancy. J Endocrinol 128:161–168
40. Ward RHT, Grudzinskas JG, Bolton AE, et al. (1985) Fetoplacental products as a prognostic guide following chorionic villus sampling. In: Fraccaro M, Simoni G, Brambati B (eds) First trimester diagnosis. Springer, Berlin Heidelberg New York, pp 73–76
41. Westergaard JG, Chemnitz J, Teisner B, et al. (1983) Pregnancy-associated plasma protein A – a possible marker in the classification and diagnosis of Cornelia de Lange syndrome. Prenat Diagn 3:225–232
42. Westergaard JG, Teisner B, Sinosich MJ, Masden LT, Grudzinskas JG (1985) Does ultrasound examination render biochemical tests obsolete in the prediction of early pregnancy failure. Br J Obstet Gynaecol 92:77–83
43. Westergaard JG, Teisner B, Stabile I, Grudzinskas JG (1988) Ectopic pregnancy: diagnostic aspects. In: Tomoda S, Mizutani S, Narita O, Klopper A (eds) Endometrial and placental proteins: basic concepts and clinical applications. International Science Publishers, pp 615–622
44. Yovich JL, McColin JC, Willcox DL, Grudzinskas JG, Bolton AE (1986) The prognostic value of beta hCG, PAPP-A, oestradiol and progesterone in early human pregnancies and the effect of medroxy progesterone acetate. Aust N Z J Obstet Gynaecol 26:59–64

21 Ultrasonography of Abnormal Early Pregnancy

Z. Blumenfeld, J.M. Brandes, and M. Bronshtein

Introduction

Recent technologic advances in ultrasound have benefited many areas of obstetric practice, including first-trimester evaluation, fetal dating, and detection of abnormalities and malformations [1]. The recent improvement in ultrasound technology offered by the use of 5–7.5-MHz transvaginal probes has allowed more detailed study of the first-trimester embryo and fetus. Moving prenatal diagnosis from the mid-trimester to the end of the first trimester has obvious benefits that, along with logistic, economic, and humane advantages, open up the possibility of fetal diagnosis at a time in immunologic development when the embryo or fetus has less chance of "graft" rejection [2, 3].

Spontaneous Abortions

Despite the fact that most nonviable embryos ultimately undergo abortion, the uterus may not expel the products of conception for days and weeks [4]. Thus, it is of significant importance to correctly identify those nonviable pregnancies in order to prevent complications. Among the complications may be prolonged uterine bleeding, septic abortion, and, of equal importance, psychological disturbance to the patient and her family [4].

About 25%–30% of pregnant women experience vaginal bleeding during the first trimester. It is estimated that more than half of these patients abort spontaneously [5]. Hormonal assays alone – one of the two main modalities of evaluating the early gestation – provide us with information on the placenta, not on the embryo, and therefore lack the ability to determine accurately all the nonviable gestations. The second modality for evaluation of the early gestation is ultrasonography. Pathologic appearance of the gestational sac was considered in the past to be reliable evidence of an abnormal gestation [6]. Abdominal ultrasonography could correctly identify only 53% of the abnormal gestational sacs [6], and, by combining abdominal ultrasonography with radioimmunoassay (RIA) of human chorionic gonadotropin (hCG), the intrauterine gestation was correctly diagnosed by Nyberg et al. [7]. The same authors extended their studies to the pathologic early gestation and described the successful use of simultaneous β-hCG concentration measurements and gestational sac diameter in the identification of abnormal gestations [8]. Previous reports stressed the necessary identification of an intrauterine gestational sac in every case when the β-hCG serum concentration is 6500 mIU/ml or higher [4, 9]. This level was called the "discriminatory zone." Recently, Nyberg et al. [7] updated the discriminatory zone and demonstrated gestational sacs at a β-hCG level of 1800 mIU/ml. The comparison of the hormonal RIA results to the ultrasonographic findings revealed that only 36% of the abnormal gestations were detected by RIA by applying the discriminatory zone of 1800 mIU/ml. When the diameter of the gestational sac was used, 65% of abnormal pregnancies were identified [8]. More recently, the discriminatory level of hCG for intrauterine gestational sac detection by transvaginal scanning (TVS) was 730–1042 mIU/ml [10, 11]. By using abdominal ultrasonography, the following morphologic criteria were proposed to identify an abnormal early gestation [4]:

1. A "small-for-date" gestational sac. This characteristic requires an accurate knowledge of the ovulation date, which is usually known only in cases of ovulation induction and/or basal body temperature measurements. In most cases, the accurate ovulation date is unknown.
2. The absence of fetal heart motion by the end of the 7th week after the 1st day of the last menstrual period (LMP). This criterion also necessitates accurate pregnancy dating [4, 12].
3. An irregularly shaped, "bizarre," or unnaturally large gestational sac lacking an embryo within it, and the absence of the "double decidual contour" ("double sac" sign) [3, 4].
4. A gestational sac without an embryo within it may result from an early embryonic demise, or from a "never developed embryo" [3, 4] (the "blighted ovum").

Nyberg et al. [6] further defined the morphologic characteristics of the abnormal gestational sac by the following criteria:

1. A gestational sac of a diameter larger than 24 mm without an embryo [3, 4].
2. The gestational sac is distorted.
3. The choriodecidual "reaction" is less than 2 mm wide, and no yolk sac is present.
4. The ultrasound choriodecidual echoes are of low amplitude.
5. The contour is irregular or crenated.
6. The typical "double sac" sign is absent when the diameter of the gestational sac is ≥10 mm.
7. The sac is located in a very low position within the cavity of the lower uterine segment.

Detection of one of the first three characteristics has been associated with a 100% specificity and positive predictive value in diagnosing an abnormal pregnancy [4]. Because of the very high correlation of these ultrasound criteria with missed abortion, they are considered "major criteria" [4]. The remaining criteria are considered "minor" because their interpretation is more subjective and because they are less specific in correctly diagnosing an abnormal gestation [4].

The ultrasound detection and description of the yolk sac is also an important criterion to differentiate normal and abnormal gestation (see Chap. 11). In cases where a yolk sac is seen without an embryonic (fetal) pole, the diagnosis of missed abortion is highly suspected [2, 3, 13]. These criteria, as well as a small-for-date and "free-floating" yolk sac were suggested to be the "yolk sac sign."

Although abnormal-appearing yolk sacs have been associated with poor pregnancy outcome, the wide scatter of normal values at each gestational age seems to preclude the usefulness of yolk sac diameter in predicting many fetal anomalies [2–4]. For example, in 22 instances in which chorionic villus sampling results proved abnormal karyotypes, the yolk sac diameters were within 95% confidence levels at the time of the ultrasound scan [2]. In our experience with TVS, the yolk sac by itself could not serve as a reliable indicator of the pregnancy outcome since there were many false-positive and false negative cases.

One should always bear in mind that an abnormal-appearing early gestation may ultimately turn out to develop normally [4]. Therefore, extreme caution must be exercised, since structurally malformed early gestations can be detected by using the transvaginal probe earlier than by the abdominal ultrasonography [1–4]. In the abnormal-appearing early gestation, the decidua around the gestational sac may show irregular echogenicity, subdecidual bleeding, and fragmentation [1, 3, 4, 14] (Figs. 1–3). The gestational sac is smaller and usually detached from the decidual wall and from the chorion or appears crumpled and flat [1, 4]. The embryonic/fetal pole also usually shows signs of abnormality. Our ultrasound criteria leading to the early and relatively simple diagnosis of abnormal gestation are presented in Table 1.

The Abnormal Gestational Sac

The pseudogestational sac that occurs in approximately 8%–10% of ectopic gestations is usually visualized more precisely by TVS than by abdominal ultrasonography [1, 4]. Such a pseudogestational sac may occasionally be detected even without a pregnancy (Fig. 4). The TVS scan may be helpful in detecting early anomalies as molar gestation (Fig. 5), cysts of umbilical cord (Fig. 6), or pregnancy in the presence of intrauterine adhesions (Fig. 7). In

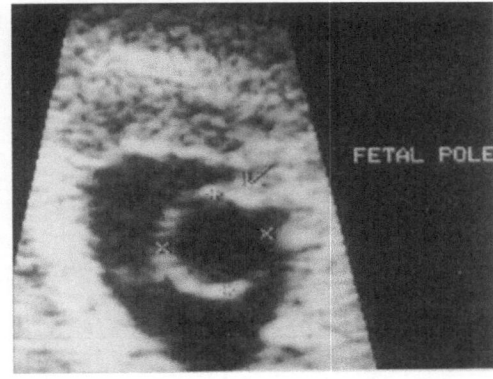

Fig. 1. A normal 6-week intrauterine pregnancy detected by TVS. Note the fetal (embryonic) pole (*arrow*)

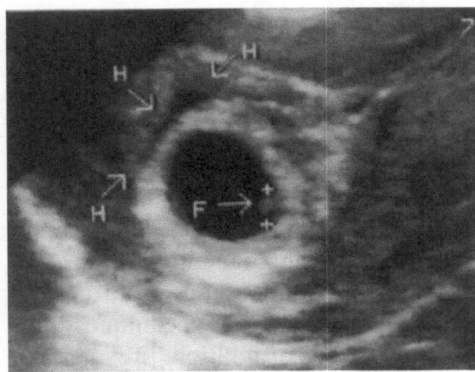

Fig. 2. A 6-week intrauterine pregnancy with fetal pole (*F, arrow*) and interdecidual hemorrhage (*H, arrows*). This pregnancy ended in missed abortion

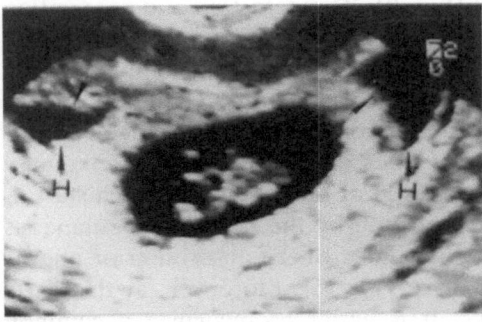

Fig. 3. A 7-week intrauterine pregnancy with decidual hemorrhage (*H, arrows*). Although the heartbeat motion is existent at this stage, the pregnancy ended in missed abortion 3 days later (disappearance of fetal heartbeats)

contrast to the snowstorm appearance of a molar pregnancy on abdominal ultrasonography, TVS detects a hydatidiform mole as multiple cysts in an amorphic material in the uterine cavity (Fig. 5) significantly earlier than by the abdominal route. The existence of an intrauterine pregnancy concomitantly with an intrauterine device (IUD) can be clearly demonstrated by TVS (Fig. 8).

Table 1. Ultrasound criteria leading to early diagnosis of abnormal gestation

Uterus and decidua	Gestational sac and membranes	Fetal pole and yolk sac
Irregular decidua	Size and date discrepance	Degenerative changes
Pathologic double ring	Empty sac 20 mm or more in diameter	No fetal heartbeat after 6½ weeks (LMP)
Subchorionic bleeding (Figs. 3, 4)	Detached membranes	Severe repeated bradycardia
Disappearance of the products of conception	Irregular or shrunken amniotic membranes	Irregular, small, or nonexistent yolk sac
	Severe oligohydramnios	

Multiple Gestational Sacs

By virtue of its ability to visualize a gestational sac as early as 2–5 days after the missed period, TVS is very suitable for early diagnosis of multiple pregnancies (twins, triplets, quadruplets, etc.). Recently, the incidence of multiple pregnancies has increased tremendously mainly due to the increased usage of in vitro fertilization (IVF) and the embryo transfer technique, but also due to the increasing popularity of ovulation induction for in vivo fertilization (by clomiphene citrate and human menopausal gonadotropin in conjunction with hCG). All these "assisted reproduction" techniques have brought about a significant increase in the number of multiple pregnancies. About 30%–40% of the early multiple sacs spontaneously "reduce" themselves to singleton pregnancies [15] (Fig. 9). The detection and recognition of this possibility would facilitate research and enable the study of the actual behavior of multiple pregnancies, which seem to be more common at the beginning of gestation than used to be appreciated from the cited incidence of multiple pregnancies. Previous studies were based on the detection of multiple sacs at 8 weeks' gestation or later. Our and others' preliminary findings suggest that a significant proportion of the gestations which start as multiple pregnancies turn out to be singleton by the so-called spontaneous fetal reduction or vanished twin phenomenon where absorbtion of one or more of the gestational sacs takes place [15, 16]. Moreover, at this early stage of gestation, i.e., 5–6 weeks, it is easier to recognize a common chorion and thus to correctly diagnose and differentiate the monozygotic twins (which bear a worse prognosis) from the dizygotic twins which have a better outcome. These minute but clinically important data are detectable by TVS examination in the first trimester and are usually not detected so early by the traditional abdominal ultrasonography. Serial TVS of the multiple gestation may often reveal the spontaneous absorption of one or more fetuses, so that by 12 weeks only one or two viable fetuses are detected.

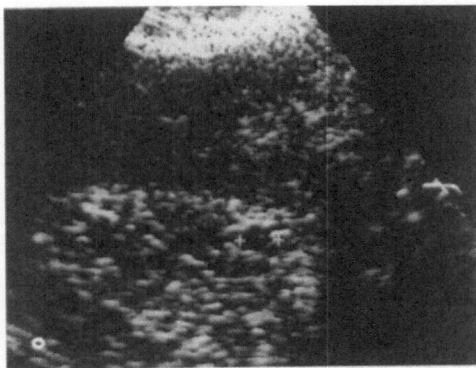

Fig. 4. A 2.5-mm intrauterine pseudo-gestational sac in a patient with polycystic ovaries on day 16 of the menstrual cycle. The serum β-hCG was zero

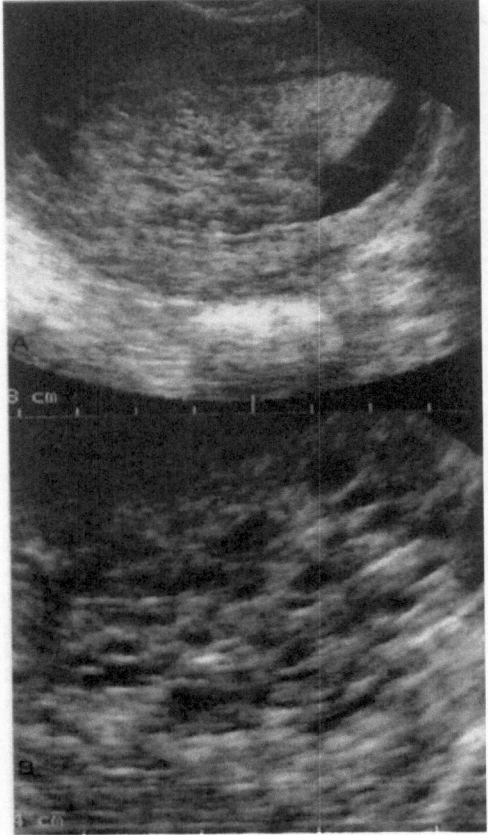

Fig. 5 A,B. An intrauterine molar gestation (8 weeks). Note multiple small vesicles at low magnification (**A**) and at high magnification (**B**)

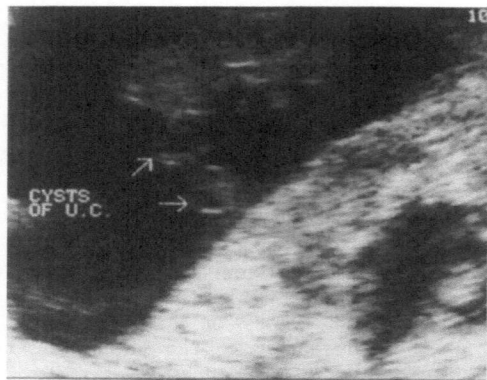

Fig. 6. A missed 12-week intrauterine pregnancy. Note cysts of umbilical cord (*U.C.*, *arrows*)

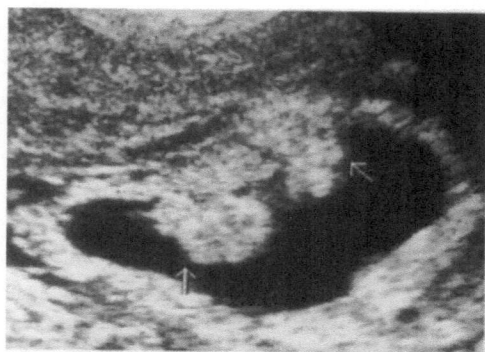

Fig. 7. A 6-week intrauterine pregnancy in the presence of intrauterine adhesions (*arrows*) proven by hysterosalpingography and by hysteroscopic vision. The pregnancy ended in an early abortion

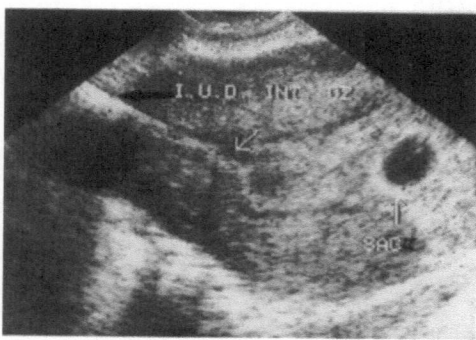

Fig. 8. A 6-week intrauterine gestation in the presence of a partially expulsed IUD. Note the intrauterine location of the gestational sac (*white arrow*) and the IUD (*black arrow*) in the cervical canal. *Left white arrow* points to the internal os. After taking out the IUD from the cervical canal the pregnancy continued normally and ended in delivery of a normal term baby

This occurrence may save iatrogenic fetal reduction only for those multiple gestations of more than three viable fetuses at 12 weeks of gestation, thus preventing the risk of losing all the fetuses as a consequence of iatrogenic reduction. This complication may occur in 10%–15% of these interventions. TVS follow-up during the 1st trimester may thus be helpful in keeping with the traditional rule of *primum non nocere*.

Ectopic Pregnancy

Ectopic pregnancy (EP) accounts for a significant rate of maternal death (26% of all maternal deaths), fetal wastage of almost 100%, and a high incidence of maternal morbidity [17, 18]. A marked increase in the recorded incidence of EP has been reported in different countries and continents [17]. In the United States the rate of EP doubled between the years 1965 and 1976 [17]. Rubin et al. [19] reported an increase in the rate of EP from 4.5 to 9.4 per 1000 pregnancies. In Sweden [20] the rate increased from 5.8 to 11.1 per 1000 gestations in the 15 years of study, and in Great Britain the rate of EP increased from 3.2 to 4.3 per 1000 live births and terminations of pregnancy [17, 18].

The average reported delay in making the correct diagnosis of EP is 3 weeks in about 14% of the cases [17, 18]. Therefore, TVS which may "make the diagnosis" of EP is a significant clinical aid to every gynecologic practitioner, 24 h a day. The main reason for the increase in the epidemiology of ectopic gestation is the earlier age of first intercourse, bringing about a higher rate of sexually transmitted diseases and unplanned conceptions at earlier ages than 50 years ago [17, 18]. Although it has been frequently quoted that one in 300 pregnancies is ectopic, Fontanilla and Anderson [21] indicated that the actual occurrence is considerably more frequent. In Baltimore, ectopic gestation occurred once in 200 gestations among white women and once in 120 gestations among black patients, nearly an 80% difference [18]. This strongly suggests that pelvic inflammatory disease is the responsible factor, and obviously consideration of any reported incidence should include the racial differences. Breen's [22] New Jersey study recorded one ectopic gestation in every 87 deliveries in an 85% black group of patients.

The study of Bobrow and Bell [23] of the Harlem Hospital in New York noted one ectopic gestation to every 64 live births, which would indicate the highest ratio ever recorded for any large American series. This approached the rate of one in 28 gestations reported from Jamaica [18]. That repeat ectopic pregnancy may recur in the remaining tube in approximately 10% of cases has been accepted by most gynecologists, although normal intrauterine gestation is much more common (approximately 25%) [18].

According to Curran's projection [24], by the beginning of the twenty-first century, 10% or more of the women at the reproductive age will become involuntarily infertile as a result of the sequelae of pelvic inflammatory disease. More than 3% of them will experience an ectopic gestation [17, 18, 24]. It is evident that it is almost impossible to prevent ectopic gestations. However, early detection and accurate diagnosis of most ectopic pregnancies will significantly reduce maternal mortality and morbidity, and contribute to conservation of future fertility. TVS is a major breakthrough in the achievement of this goal.

The literature dealing with various aspects of the ultrasound diagnosis of ectopic gestation is quite abundant [1, 17]. The diagnosis, by abdominal ultrasonography, relies primarily on the exclusion of intrauterine pregnancy in the presence of an adnexal mass and sometimes fluid in the cul-de-sac [1, 17]. More recently, it became obvious that the "classic" ultrasound findings of ectopic gestation were not always present [1, 17]. Unfortunately, abdominal ultrasonography reaches diagnostic accuracy only by the 7th postmenstrual week. The false-positive rate ranges between 3% and 30% (mean 16%) and the false-negative 2%–35% (mean 5%) [17, 25]. Since more than half of the cases of ectopic gestation are diagnosed before the 7th week, only a minority of ectopic gestations will benefit from abdominal ultrasonography [1, 17]. In contrast, diagnosis of tubal gestation by TVS is based upon the direct demonstration and imaging of a tubal ring (Fig. 10) [26, 27].

The TVS scan also rules out intrauterine pregnancy, as well as blood collection in the cul-de-sac [1, 26, 27]. Anecdotal terms describing the telltale clue include "bagel sign" and "doughnut ring" [1] (Fig. 10). In a series of 658 patients scanned by TVS for suspected ectopic pregnancy, 176 were surgically proven. There were two false-positive and seven false-negative results. Thus the sensitivity of the transvaginal approach is 96% and the specificity almost 99%. Both the positive and the negative predictive values are significantly higher than the respective values generated by abdominal ultrasonography [1, 17].

Intrauterine pregnancy can be documented with absolute reliability as early as 5 weeks after LMP, regardless of patient obesity or uterine position [17]. Although the intrauterine gestational sac is seen by TVS when the peripheral β-hCG concentration reaches the level of 500 mIU/ml [1, 17, 28], one may occasionally detect an intrauterine gestational sac at lower levels of β-hCG, as shown in Fig. 11 (one day after a β-hCG level of 166 mIU/ml). The very early visualization of embryonic and extraembryonic structures by TVS enhances the reliability of diagnosing a normal or abnormal intrauterine pregnancy. Once intrauterine pregnancy has been diagnosed, one may, with high confidence, discard the suspicion of ectopic pregnancy, since the possibility of combined intrauterine and ectopic gestation is quite remote, except for pregnancies generated by IVF or gamete intrafallopian transfer (GIFT), where such an occurrence should be borne in mind.

The endometrium, responding to the hormonal stimulation induced by the developing ectopic gestation, may be demonstrated by TVS as a highly echogenic, thickened endometrium (decidua) [17]. Eventually the ectopic and immature trophoblast ceases to secrete, leading to decidual degeneration and bleeding. This may have been what various authors described as the pseudogestational sac of ectopic pregnancy [17]. This pseudogestational sac is primarily located symmetrically in the center of the uterus, outlining its cavity [17]. In cases where an ectopic gestation is detected early enough by TVS (at a time when no decidual degeneration has occurred), no intracavital

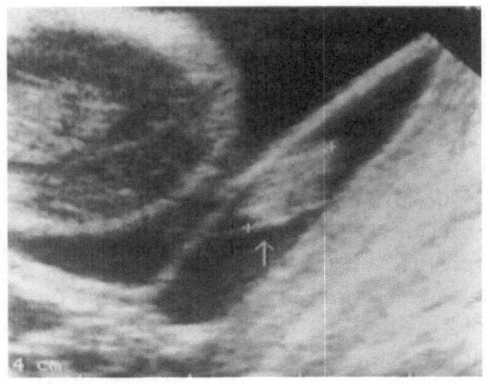

Fig. 9. Spontaneous fetal reduction. A 12-week twin pregnancy with spontaneous demise of the twin in the right sac (*arrow*). The twin in the left sac developed normally

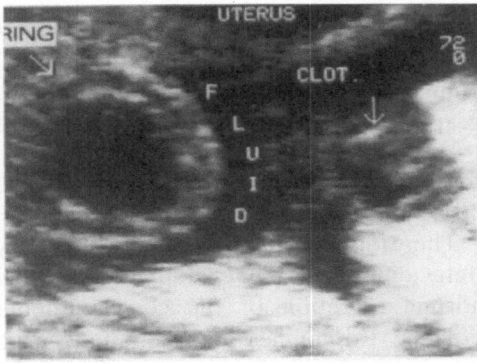

Fig. 10. An 8-week tubal pregnancy. Note tubal ring (bagle sign), surrounded by fluid and blood clot in the peritoneal cavity

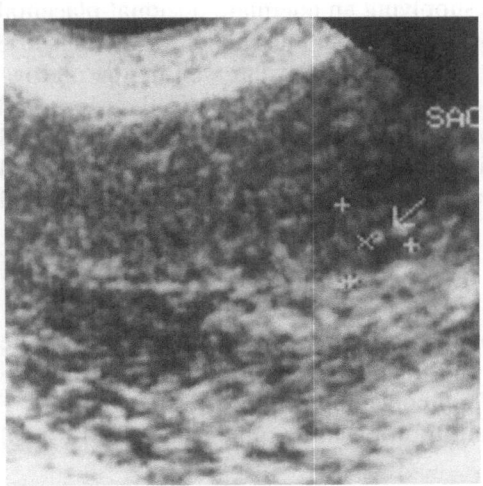

Fig. 11. A very early intrauterine gestational sac at 4 weeks and 2 days after LMP. The β-HCG concentration on the previous day was 166 mIU/ml

bleeding occurs, and therefore no pseudogestational sac is detected [17]. The appearance of the pseudogestational sac, as visualized by TVS, may precede overt vaginal bleeding or spotting by several hours or days.

In the case of a live and developing tubal gestation, the embryo will continue to grow as long as the tubal epithelium (endosalpinx) is able to sustain the implanted sac [17]. At this early stage, development of the embryo and of the extraembryonic structures is similar to that of an intrauterine pregnancy. The ectopic gestational sac with a visible heartbeat adjacent to the fetal pole is visualized by TVS (Fig. 12). The advantage of the high-frequency 6.5 MHz probe lies in its ability to visualize a viable tubal pregnancy before its rupture or before tubal abortion occurs. Embryonic structures, such as the fetal pole with heartbeats, or extraembryonic structures, such as the yolk sac and the early placenta, may be recognized (Fig. 12).

It has been noted that in cases of unruptured tubal pregnancy, whether there is fetal heart motion or not, the tubal wall containing the gestational sac was thickened, measuring 4–6 mm [17]. Rather than the term *adnexal ring* used in abdominal ultrasonography, we suggest the more descriptive term *tubal ring* [17] (Fig. 10).

The ectopic and immature trophoblast in the tube may not secrete adequate quantities of hCG, which is luteotrophic and believed to sustain a normal corpus luteum [17, 29]. Therefore, corpus luteum insufficiency occurs bringing about embryo or fetal demise [17]. Development of tubal gestation is dependent on the normal secretion of the corpus luteum and also on the invasion ability of the trophoblast into the underlying endosalpinx to ensure blood support and placental formation. Since the endosalpinx is usually incapable of supplying an adequate decidual-placental bed, the ectopic trophoblast will invade the tubal muscularis for this purpose [17]. This penetration and erosion of the tubal muscular wall by the ingrowing trophoblast in search of supporting blood supply weakens the tubal wall and may result in bleeding into the gestational sac, between the tubal wall and the sac, and eventually outside the serous lining into the peritoneal cavity. Once the gestational sac has separated from the tubal wall, several events may take place [17]:

1. Rupture of the gestational sac into the tubal lumen.
2. Absorption of the gestation (Fig. 13).
3. Abortion of the ectopic pregnancy into the peritoneal cavity through the fimbrial opening. In this case, the thickened tubal wall may be demonstrated by TVS in the bloody fluid collecting in the pelvis (Fig. 13).
4. A slow blood leak or overt tubal rupture leading to pelvic intraperitoneal fluid collection (Fig. 10).

Careful examination of the cul-de-sac may detect the presence or absence of free fluid with or without blood clots (Fig. 10). The blood clots are depicted as a bizarre-shaped mass of irregularly echogenic substance [1,

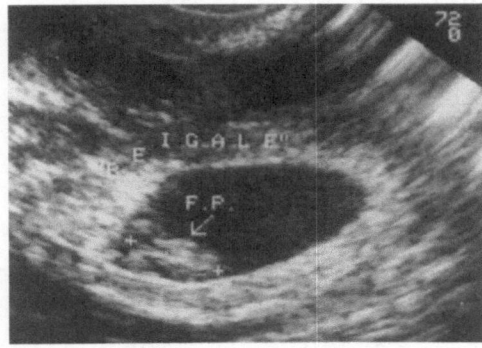

Fig. 12. A 71/2-week developing tubal pregnancy. Heartbeats were clearly observed on real-time TVS. *F.P.*, fetal pole

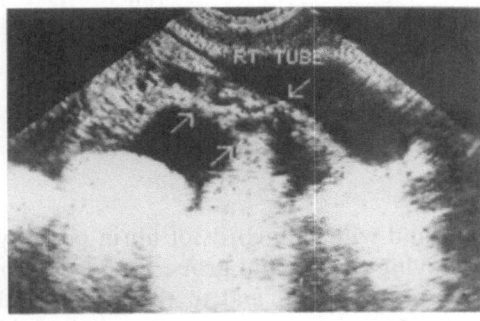

Fig. 13. A 7-week ectopic pregnancy. Note thickened tubal wall (*arrows*) surrounded by bloody fluid in the pelvic peritoneal cavity

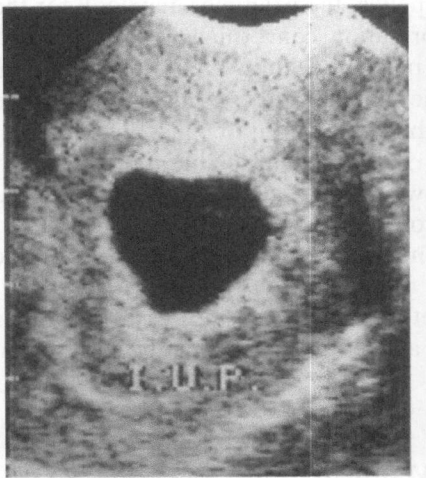
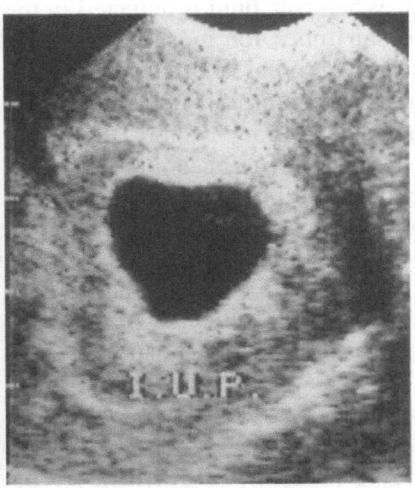

A   B

Fig. 14. An intrauterine pregnancy (*I.U.P.*) detected as a gestational sac (**A**), and a false-positive diagnosis of an ovarian ectopic pregnancy detected by a "sac" in the ovary (**B**, *arrows*). The cystic structure erroneously suspected to be a gestational sac within the ovary turned out to be a normal cystic corpus luteum of pregnancy (seen by laparoscopy)

Table 2. Fetal malformations detected in 11 fetuses by 1st-trimester TVS

System	Malformation	Cases (n)	Earliest week of detection	Outcome
Nuchal-lymphatic	Septated cystic hygroma	5	9⁵⁷	2 TOP, 2 missed, 1 normal neonate
CNS	Acrania, anencephalus, spina bifida	3	11⁶⁷	TOP
Gastrointestinal	Intraabdominal cysts	3	10	1 TOP, 2 missed
Umbilicus	Cysts of umbilical cord	3	12	1 missed, 2 normal neonates
Skeletal	Polydactyly, distorted fingers	2	11	TOP
Cardiovascular	Hypoplastic lt. heart, Fallot's tetralogy	2	12	TOP
Urinary	Horseshoe kidney, hydronephrosis	2	12	TOP
Mole	Partial mole	1	12	TOP
	Eleven fetuses	21		

17]. The clots may float freely in the fluid with thin cords of fibrin possibly being attached to their surface. By moving the vaginal probe in and out or by asking the patient to tilt her pelvis to the right and to the left several times, the blood clots may be visualized as moving within the pelvic fluid [17]. When free fluid is detected in the pelvis, the examiner may look for the fallopian tubes for more detailed information concerning the presence or absence of tubal contents [17]. The peritoneal fluid creates an optional contrast medium for visualization of the fallopian tubes, thus improving the diagnostic accuracy. It may, therefore, be very helpful to place the patient in the reversed Trendelenburg's position. If the patient is able to stand up, it may be helpful to ask her to do so and wait for 1–2 min before assuming the proper reversed Trendelenburg's position for TVS [17].

The examiner should bear in mind that a small fraction of ectopic pregnancies is located in the ovaries, uterine cervix, pelvis, or abdomen. Even with high-frequency TVS, the detection of these rare cases continues to present serious diagnostic problems and pitfalls, as shown by the case in Fig. 14.

Early Detection of Human Fetal Malformations

Ultrasonography plays a central role in modern obstetric practice. However, whether every pregnant woman should undergo an ultrasound examination may well be the most controversial question in obstetric ultrasonography at present [30]. That ultrasound examinations should be recommended when

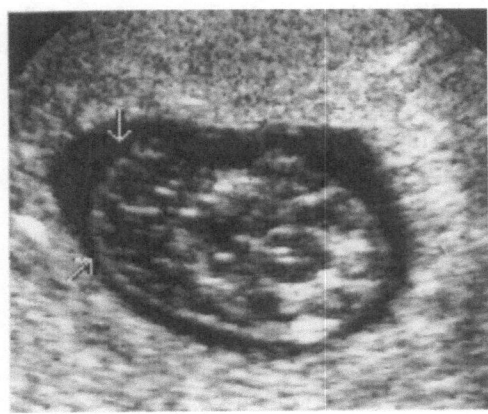

Fig. 15. A septated cystic hygroma detected by TVS in a 10-week fetus. The *arrows* point to the septated nuchal cystic lesion (transverse section)

indicated and performed with the patient's consent is not open to question [30]. The present debate concerning the use of obstetric ultrasonography has been shaped by the dichotomy: "examine all pregnancies [30, 31] or examine only when indicated" [30, 32].

Many severe and potentially lethal fetal abnormalities can be detected by early examination, as our [33] and others' [34] findings clearly demonstrate (Table 2). Out of 600 fetal screenings performed by TVS before the 13th week of gestation, we have identified 21 malformations in 11 fetuses (1.66%, Table 2). The most common fetal malformation identified during the 1st trimester was nuchal cystic hygroma (Fig. 15) [35] and, less frequently, fetal intraabdominal cysts and fetal skeletal malformations. The various fetal malformations identified by TVS during the 1st trimester and the earliest week of detection are summarized in Table 2. Patients, in our study as well as in the study of Hegge et al. [36], were much more likely to choose the option of termination when fetal abnormalities were found early. TVS screening in the 1st and early 2nd trimester offers several important advantages, as detailed below.

Early Detection

1. Early detection preserves the entire spectrum of therapeutic options including the possibility of therapeutic abortion before a tight psychologic bonding between the patient and her fetus is formed [36]; midtrimester (22–24 weeks) is often too late [36]!
2. Termination of pregnancy (TOP) is permitted, using dilatation and curettage or dilatation and evacuation (D&E), instead of the more dangerous Buero procedure practiced later in pregnancy [37].
3. Malformations, such as septated nuchal cystic hygromas or hydronephrosis, indicate a need for karyotyping. Ultrasound diagnosis at

22–24 weeks may raise technical difficulties in culturing amniotic cells, or may indicate the more dangerous chordocentesis, and may also have serious medicolegal implications if the karyotype turns out to be abnormal and TOP is considered after 24 weeks' gestation.

4. In cases of severe malformations, the early detection at 10–14 weeks may save possible invasive procedures such as mid-trimester cervical suture, where indicated, or amniocentesis.

5. Fetal sex identification may be essential in cases of multiple pregnancy to rule out homozygosity or in cases of sex-linked diseases [38].

Higher Detection Rate

Our "corrected" detection rate in the low-risk population at 9–16 weeks was 1.96% by TVS, versus 0.9% by abdominal ultrasonography at later gestational ages (19–28 weeks).

Most Fetal Malformations Are Detectable in the 1st and Early 2nd Trimester

We disagree with the accepted assumption that most fetal anomalies are detected only in the late mid-trimester. Only about 7% of the fetal anomalies appear late in pregnancy [39, 40], i.e., after 26 weeks of gestation. The overall cited incidence of these malformations is about 15:10 000 (0.15%) [41]. The vast majority of fetal anomalies (>90%) are detectable early in pregnancy, therefore postponing the ultrasound screening to 18–22 weeks will gain little in terms of detection of more malformations and will not detect those anomalies which appear after 27 weeks of gestation.

Transient Malformations

Most, if not all, malformations which can be detected at 18–22 weeks can also be detected earlier, at 9–16 weeks; however, several types of anomalies may disappear at later gestational ages. Of such "transient" anomalies, between 60%–100% disappeared at later stages of gestation. Chromosomal anomalies were detected in more than 6% (5/81) of the fetuses with transient anomalies. By postponing the first ultrasound systematic scan to 20 weeks, most of these transient malformations would not have been detected, including five of the ten cases with chromosomal anomalies. Not only cystic hygromas and hydronephrosis, but also the transient cysts of choroid plexus may be associated with dyskaryosis [42].

Higher Detection Rate of Dyskaryosis

By means of early TVS at 9–16 weeks we were able to locate six cases of trisomy 21, two cases of trisomy 18, one case of triploidy, and one case of 45

XO. The calculated incidences of these three chromosomal anomalies is comparable to the reciprocal incidences cited for each abberation [38, 41], suggesting that the vast majority of existing chromosomal abnormalities in the screened population were detected by early TVS in our series. This presents a serious argument for performing an early scan in search of those fetal anomalies, possibly associated with chromosomal abberations. Failure to do so may leave those cases of transient anomalies, undetected by later ultrasound scans, at 20 weeks of gestation. Whereas conventional indications for prenatal diagnosis (maternal age, previous child with chromosomal abnormality, parental structural anomaly, and parental balanced translocation) would, at best, allow the antenatal detection of 30% of the existing chromosomal malformations, the ultrasound diagnosis of fetal anomalies would detect four times as many chromosomal abberations [42]. In our series 6.8% (10/147) of the fetuses with anomalies detected by TVS had chromosomal abnormalities – one in 340 fetuses (10/3402). This detection rate is comparable to previous series [42].

Although our recommendation for low-risk pregnancies is to perform systematic TVS screening for fetal malformations at 13–14 weeks of gestation, one may even consider earlier ultrasound screening (at 12 weeks) for those high-risk cases such as previous fetal malformations, parental consanguinity, parental balanced translocation, etc. Detection of fetal malformations before the 13th week will enable the termination of pregnancy, when indicated, by dilatation and evacuation. Termination of pregnancy by this method will prevent, in most cases, the postabortal confirmation of the diagnosed malformation, but in those cases where TOP was postponed to the second trimester, and therefore performed by the Buero procedure, the postabortal examination did confirm the detected anomalies.

References

1. Timor-Tritsch IE, Rottem S (1988) High-frequency transvaginal sonography: new diagnostic boon. Contemp Obstet Gynecol 31:111–133
2. Green JJ, Hobbins JC (1988) Abdominal examination of the first-trimester fetus. Am J Obstet Gynecol 159:165–175
3. Mendelson EB, Bohm-Velez M, Saker M (1990) Transvaginal sonography in the abnormal first trimester. Semin Ultrasound Comput Tomogr Magu Reson 11:34–43
4. Timor-Tritsch IE, Rottem S, Blumenfeld Z (1987) Pathology of the early intrauterine pregnancy. In: Timor-Tritsch IE, Rottem S (eds) Transvaginal sonography. Elsevier, New-York, pp 109–123
5. Fantel AG, Shepard TH (1981) Basic aspects of early (first trimester) abortion. In: Iffy L, Kaminetzky HA (eds) Principles and practice of obstetrics and perinatology, vol 1. Wiley, New York, pp 553–563
6. Nyberg DA, Laing FC, Filly RA (1986) Threatened abortion: sonographic distinction of normal and abnormal gestational sacs. Radiology 158:397–400
7. Nyberg DA, Filly RA, Mahoney BS, et al. (1985) Early gestation: correlation of hCG levels and sonographic identification. AJR 144:951–954

8. Nyberg DA, Fily RA, Duarte-Filho DL, et al. (1986) Abnormal pregnancy: early diagnosis by VS and serum chorionic gonadotropin levels. Radiology 158:393–396
9. Kadar N, DeVore G, Romero R (1981) Discriminatory HCG zone: its use in sonographic evaluation for ectopic pregnancy. Obstet Gynecol 58:156–161
10. Cacciatore B, Tiitinen A, Stenman UH, Ylöstalo P (1990) Normal early pregnancy: serum hCG levels and vaginal ultrasonography findings. Br J Obstet Gynaecol 97: 899–903
11. Aleem FA, DeFazio M, Gintautas J (1990) Endovaginal sonography for the early diagnosis of intrauterine and ectopic pregnancies. Hum Reprod 5:755–758
12. Hertz JB (1984) Diagnostic procedures in threatened abortion. Obstet Gynecol 64:223–229
13. Bertrand KG, Cooperberg PL (1985) Sonographic differentiation between blighted ovum and early viable pregnancy. AJR 144:597–601
14. Stabile I, Campbell S, Grudzinskas JG (1989) Threatened miscarriage and intrauterine hematomas – sonographic and biochemical studies. J Ultrasound Med 8:289
15. Blumenfeld Z, Drugan A, Nahhas F, et al. (1987) Identical twins, spontaneous absorption of multiple fetuses and IVF. 5th International Congress of In-Vitro Fertilization and Embryo Transfer, Norfolk
16. Sulak LE, Dodson MG (1986) The vanishing twin: pathologic confirmation of an ultrasonographic phenomenon. Obstet Gynecol 68:811–815
17. Rottem S, Timor-Tritsch IE (1987) Think ectopic. In: Timor-Tritsch IE, Rottem S (eds) Transvaginal sonography. Elsevier, New-York, pp 125–141
18. Stabile I, Grudzinskas JG (1990) Ectopic pregnancy: a review of incidence, etiology and diagnostic aspects. Obstet Gynecol Surv 45:335–347
19. Rubin GL, Peterson HB, Dorfman SF, et al. (1983) Ectopic pregnancy in the United States, 1970 through 1978. JAMA 249:1725–1729
20. Westrom L, Bengtsson L, Mardh PA (1981) Incidence, trends and risks of ectopic pregnancy in a population of women. Br Med J 282:15–18
21. Fontanilla J, Anderson GW (1968) Further studies on racial incidence and mortality of ectopic pregnancy. Am J Obstet Gynecol 70:312–319
22. Breen JL (1970) A 21-year survey of 654 ectopic pregnancies. Am J Obstet Gynecol 106:1004–1010
23. Bobrow ML, Bell HG (1962) Ectopic pregnancy: a 16-year survey of 905 cases. Obstet Gynecol 20:500–508
24. Curran JW (1980) Economic consequences of pelvic inflammatory disease in the United States. Am J Obstet Gynecol 138:848–851
25. Levi S, Leblicq P (1980) The diagnostic value of ultrasonography in 342 suspected cases of ectopic pregnancy. Acta Obstet Gynecol Scand 59:29–36
26. Kivikoski AI, Martin CM, Smeltzer JS (1990) Transabdominal and transvaginal ultrasonography in the diagnosis of ectopic pregnancy: a comparative study. Am J Obstet Gynecol 163:123–128
27. Laing FC (1990) Sonographic determination of tubal rupture in patients with ectopic pregnancy: is it feasible? Radiology 177:330–331
28. Blumenfeld Z, Rottem S, Elgali S, Timor-Tritsch IE (1987) Transvaginal sonographic assessment of early embryological development. In: Timor-Tritsch IE, Rottem S (eds) Transvaginal sonography. Elsevier, New-York, pp 87–108
29. Blumenfeld Z, Nahhas F (1988) Luteal dysfunction in ovulation induction: the role of repetitive human chorionic gonadotropin supplementation during the luteal phase. Fertil Steril 50:403–407
30. Chervenak FA, McCullough LB, Chervenak JL (1989) Prenatal informed consent for sonogram: an indication for obstetric ultrasonography. Am J Obstet Gynecol 161:857–860
31. Royal College of Obstetricians and Gynaecologists (1984) Report of RCOG Working Party Routine ultrasound examination in pregnancy. Royal College of Obstetricians and Gynaecologists, London

32. Consensus Development Conference (1984) Diagnostic ultrasound imaging in pregnancy. Report of a consensus development conference sponsored by the National Institute of Child Health and Human Development, the Office of Medical Applications of Research, the Division of Research Resources and the Food and Drug Administration, Feb 6–8. National Institutes of Health, Bethesda
33. Rottem S, Bronshtein M, Thaler I, Brandes JM (1989) First trimester transvaginal sonographic diagnosis of fetal anomalies. Lancet 1:444–445
34. Cullen MT, Green J, Whetham J, Salafia C, Gabrielli S, Hobbins JC (1990) Transvaginal ultrasonographic detection of congenital anomalies in the first trimester. Am J Obstet Gynecol 163:466–476
35. Bronshtein M, Rottem S, Yoffe N, Blumenfeld Z (1989) First-trimester and early second-trimester diagnosis of nuchal cystic hygroma by transvaginal sonography: diverse prognosis of the septated from the nonseptated lesion. Am J Obstet Gynecol 161:78–82
36. Hegge FN, Franklin RW, Watson PT, Calhoun BC (1989) An evaluation of the time of discovery of fetal malformations by an indication-based system for ordering obstetric ultrasound. Obstet Gynecol 74:21–24
37. Atrash HK, Lawson HW, Smith JC (1990) Legal abortion in the US: trends and mortality. Contemp Obstet Gynecol 35::58–69
38. Bronshtein M, Rottem S, Yoffe N, Blumenfeld Z, Brandes JM (1990) Early determination of fetal sex using transvaginal sonography: technique and pitfalls. JCU 18:302–306
39. Nelson LH, Clark CE, Fishburne JI, Urban RB, Penry MF (1982) Value of serial sonography in the in utero detection of duodenal atresia. Obstet Gynecol 59:657–660
40. Kurtz AB, Filly RA, Wagner RJ, et al. (1986) In utero analysis of heterozygous achondroplasia – variable time of onset as detected by femur length measurements. J Ultrasound Med 5:137–139
41. Romero R, Pilu G, Jeanty P, Ghidini A, Hobbins JC (1988) Prenatal diagnosis of congenital anomalies. Appleton and Lange, Norwalk, pp 21, 54, 77, 236, 239, 313
42. Eydoux P, Choiset A, Le Porrier N, et al. (1989) Chromosomal prenatal diagnosis: study of 936 cases of intrauterine abnormalities after ultrasound assessment. Prenat Diagn 9:255–268

22 Embryoscopy

E.A. Reece

Introduction

Endoscopic visualization of the embryo (embryoscopy) is a new and evolving technology. The concept of direct visualization of the embryo reflects not only our ongoing quest for improved diagnostic techniques, but also the potential for direct, targeted embryonic therapy. Advancing technology has rendered the fetus and its environment accessible to the "outside" world and, therefore, to diagnosis and therapy. Among the latest of such technologies is embryoscopy, which utilizes high-resolution fiberoptic equipment for direct visualization of the embryo/fetus [1–6].

The first direct transuterine visualization of the fetus was successfully performed by Westin [5] in 1954 using a 10-mm hysteroscope. In spite of the image quality being rather limited by the insufficient optical system, fetal parts and movements could be identified. Nevertheless, even without optimal delineation of the fetal structures, the invasion of the fetal compartment had begun, and the stage was set for targeted fetal diagnosis and direct in utero therapy [5, 6].

The advent of amniocentesis for Rh isoimmunization in the 1950s followed by genetic amniocentesis in the 1960s were major milestones in the development of fetal intervention. Ian Donald introduced ultrasonography into obstetrics and gynecology in 1958 [7] and had thus paved the way for the widespread use of an extremely powerful modality of fetal visualization. The static-image ultrasound scan enabled us to learn about fetal and uterine morphology and resulted in a blossoming of descriptive literature on obstetric ultrasonography. Before long, advancing technology yielded the gray-scale dynamic ultrasound image, which added yet another dimension to our understanding. Not only did the "real-time" scanner improve visualization, but for the first time some aspects of human fetal physiology were observed in vivo. Investigators were fascinated by the "live" observation of fetal body, limbs, and breathing movements, heart motion, bladder filling and emptying, and the recording of fetal heart rate patterns. Circadian rhythms and periodicity of fetal physiologic functions were discovered and correlated with maternal physiologic status [7–14].

Prenatal diagnosis of congenital anomalies improved exponentially with the advanced resolution of modern ultrasound equipment [15]. In conjunction with electronic fetal heart rate monitoring, a rapidly evolving concept of "fetal well-being" emerged to impact significantly on the practice of obstetrics. The interest of our specialty shifted its focus from the delivery process and associated surgical skills to the quest for understanding fetal biology [1, 4].

Most physiologic and biochemical processes in the fetus were expected to result in alterations of metabolic products in the fetal blood circulation, similar to those of an adult. This concept was translated into the technique of intrapartum determination of fetal capillary scalp blood pH for detection of fetal distress. This technique became an important step towards an understanding of fetal acid-base status; however, it remained limited to parturition [4].

Initial attempts at gaining access into the fetal circulation during fetal development focused on ultrasound-guided aspiration of fetal blood from the placenta. This approach was hampered by significant difficulties in obtaining pure fetal blood specimens and by relatively low success rates. In the early 1970s, Valenti [17] and Scrimgeour [18] reported the use of 6-mm and 2.2-mm endoscopes, respectively, for examination of the human fetus. Their technique involved exposing the uterus by laparotomy and inserting the instruments through a myometrial incision. The widespread diagnostic and therapeutic use of fetoscopy in ongoing pregnancies became feasible when Hobbins and Mahoney [19] developed the Dyonics "needlescope" which employed an ultrasound-guided percutaneous transabdominal technique under local anesthesia. Direct fetal visualization allowed for the diagnosis of external structural anomalies, aspiration of fetal blood, skin, liver, and fetal tumor biopsies, as well as intravascular infusion of blood and blood products [4, 20, 22]. The foundation for fetal medicine as a subspecialty in its own right had been laid.

Major drawbacks of fetoscopy, however, were related to a fetal loss rate of 4%–8% as well as inconsistent success in obtaining fetal blood samples devoid of amniotic fluid contamination; risk-versus-benefit considerations limited the use of fetoscopy somewhat. Additionally, fetoscopy could not be used when the amniotic fluid was discolored and was less efficacious in advanced pregnancies.

With the dramatic improvement in ultrasound technology, Daffos et al. [24] were the first to substitute direct vision for ultrasound guidance of a spinal needle for percutaneous umbilical blood sampling (PUBS). This technique quickly replaced fetoscopy for many diagnostic and therapeutic indications due to its safety, simplicity, and ability to obtain 100% pure fetal blood [25]. Consequently, our understanding of fetal infectious disease, metabolism, and transplacental pharmacology was greatly enhanced, and important work on the pathophysiology of intrauterine growth retardation and the fetuses of diabetic mothers is continuing [4].

All of the above-described diagnostic and therapeutic techniques are limited to use during the second and third trimesters; the first-trimester embryo and fetus have so far received only some diagnostic and relatively little therapeutic attention. Of course, chorionic villus sampling (CVS) can be performed as early as 6 weeks' gestation and provides extremely valuable genetic information. Furthermore, the utility of CVS has recently been enhanced by techniques of DNA analysis and our growing knowledge of DNA sequence [27, 28]. However, the diagnostic capability of CVS is limited to the degree by which trophoblastic and embryonic tissues are genetically congruent. Additionally, nongenetically determined structural anomalies are beyond the diagnostic scope of CVS. First-trimester ultrasonography, particularly transvaginal scanning, can diagnose major anomalies relatively early [29–34]. It appears, however, that the limits of sound wave resolution, with currently used frequencies, have been rapidly approached, particularly when it comes to minor malformations. Furthermore, none of the above techniques can be used for treatment of diseases affecting the embryo [4].

Embryoscopy is a recently introduced technique which consists of transcervical fiberoptic endoscopy of the first-trimester embryo/fetus. It can be performed as early as 5 weeks' menstrual age. A side channel, customized for our instrument, allows for the passage of an aspiration needle for access into the embryonic blood circulation. Currently, this is the only instrument which allows direct visualization of the human embryo in vivo and access to its circulation. Multiple diagnoses of embryonic malformations have been made by embryoscopy; however, the diagnostic and therapeutic potential is just beginning to unfold [1–4].

Technique

The technical aspects of the procedure have been described elsewhere [1–4]. In brief, embryoscopy utilizes a rigid fiberoptic endoscope, 30 cm in length with diameters from 2.0 to 3.5 mm and a 0° or 30° angle lens. Under ultrasound guidance, the endoscope is passed transcervically into the extracoelomic cavity (chorionic cavity) without disturbing the amnion (Fig. 1). A complete examination of the embryo or fetus includes visualization of the head, face, dorsal and ventral walls, limbs, umbilical cord, and yolk sac. The average length of time for this procedure is 5 min, and the success rate for visualization of the fetus is above 95%. Dumez has reported a similar success rate [35–37]. He has also performed embryoscopy, undertaken to rule out head and limb anomalies, in over 50 continuing pregnancies. In his early attempts, two pregnancies were lost, but, in subsequent work, no adverse outcome was noted. It should be emphasized that overall relatively few ongoing pregnancies have so far been investigated, but this is expected to change with further improvement in the technique and its applications to a variety of conditions and circumstances (Table 1).

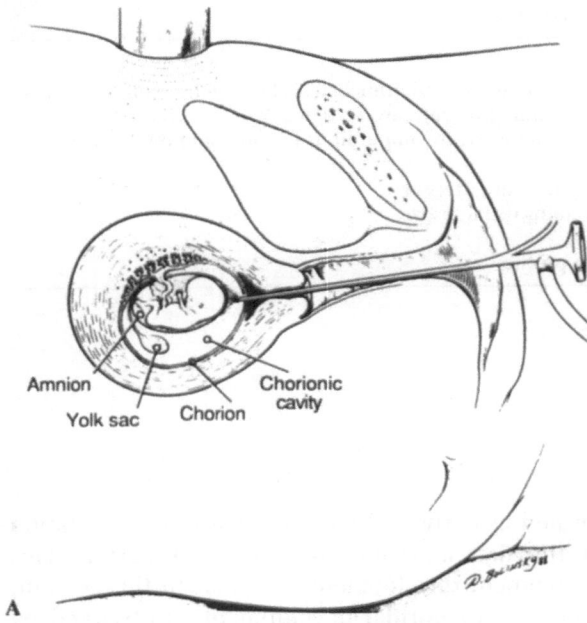

Amnion
Yolk sac
Chorion
Chorionic cavity

Fig. 1. A Embryoscope inserted transcervically and passed through the chorion into the extracoelomic space. Here the entire conceptus can be easily visualized through the amnion. Note the anatomic landmarks as indicated. **B** Ultrasound demonstration of the embryoscope (*arrow*) and the embryo (*E*) to which the endoscope is directed through the chorion (*C*) and into the chorionic space

A

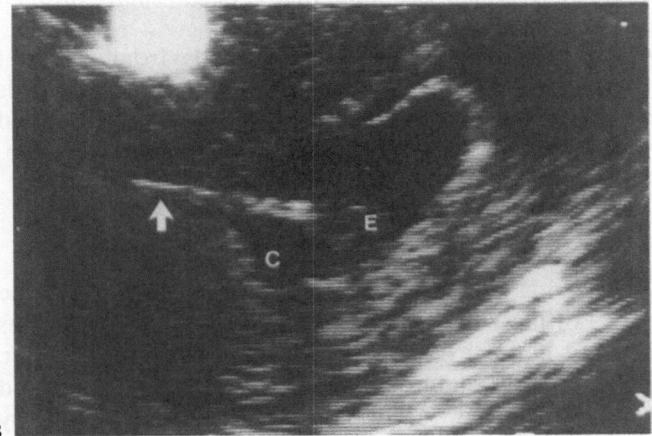

B

Table 1. Potential applications of embryoscopy. (From [1])

1. Prenatal diagnosis of structural anomalies
2. Confirmation of early ultrasound diagnosis prior to pregnancy termination
3. Early blood and/or tissue sampling
4. Embryo access for cell and/or gene therapy
5. The biology of embryonic development

Table 2. Development of head and neck

Conceptional week
  4 Otic depression, optic evagination, and flexed branchial arches can be seen
  5 The nasal pit, primitive mouth, and optic cups are recognizable
  6 The oral and nasal cavities are confluent, the upper lip is formed, and the head
    remains larger and flexed
  7 All external and internal structures are present
  8 The head is less flexed, but retains its large size
  9 The membranes fuse
 10 The face has a normal human profile

Current Applications

Morphologic Evaluation

Embryoscopy can be performed as early as 3 conceptional weeks' gestation and continued throughout the first and into the second trimester. This procedure can be used to document developmental events. In this section, we will discuss the main features of normal development as observed by embryoscopy, and anomalies which are potentially diagnosable through embryoscopy.

Following conception, the embryonic period extends from weeks 4 through 8, when all major external and internal structures develop. This is the period of greatest vulnerability to the effects of teratogens. By the end of this embryonic period, there are recognizable features of human development [1, 38–40].

The Head and Neck

Growth of the fetal body tends to occur at a disproportionate rate, with the cephalic end being ahead of the rest of the body. This disproportion begins very early, with the highest number of somites clustering in the future base of the head nurtured by a well-established blood supply. The brain and associated sense organs also develop early and remain disproportionately larger than other extracranial regions throughout embryonic life [38]. Table 2 outlines the prominent normal developmental milestones of the head and neck.

The endoscopic view of the fetal face at 6 conceptional weeks reveals a prominent forehead, widely spaced eyes, and confluent oral and nasal cavities (Fig. 2). At 8 conceptional weeks, greater facial details are seen; and by 10 conceptional weeks, normal human facial features are recognized (Fig. 3).

Malformations of the head which will have external manifestations include anencephaly, acrania, hydrocephaly, microcephaly. However, those

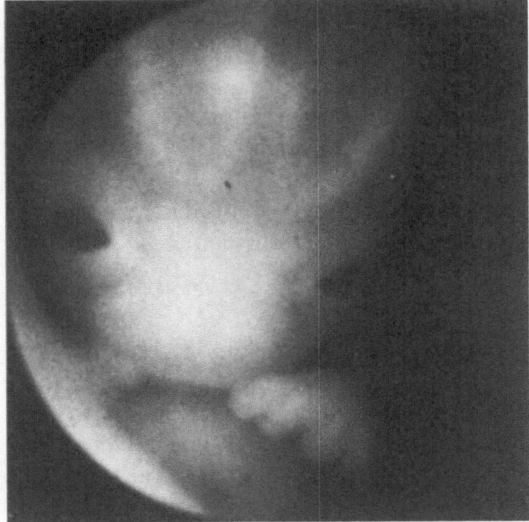

Fig. 2. Fetal face at 6 conceptual weeks

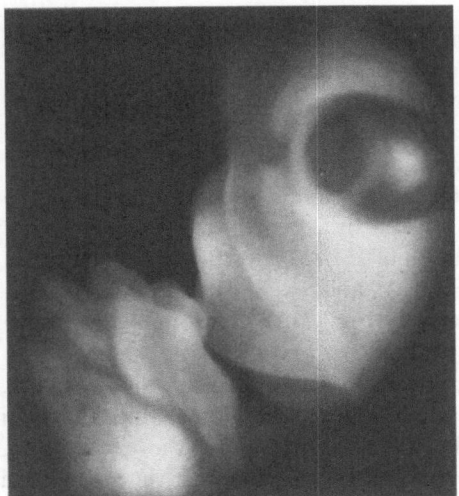

Fig. 3. Profile of fetal face at 10 conceptual weeks. The normal facial features can be recognized

considered to be most likely diagnosable in the 1st trimester by embryoscopy are anencephaly and acrania. Potentially diagnosable anomalies of the face include micrognathia and cleft lip. The latter is usually unilateral (left more often than right), but can be bilateral. The lip is often the only area involved, but the bony upper jaw may also be involved [1, 39].

The eyes are initially located laterally and far from each other, but gradually this disproportion decreases as broadening of the head occurs.

Potentially diagnosable anomalies via embryoscopy include hypotelorism or hypertolorism and the absence of one or both eyes.

The ears develop around the first branchial groove and deepen to form the external meatus by 5 conceptional weeks, then shift to a higher and more lateral position by 10 conceptional weeks. Anomalies are often related to failure of upward shifting and also seem potentially diagnosable using embryoscopy during early gestation [1, 38].

The nose develops from a complex merging of frontonasal bones with maxillary processes; however, anomalies are not often seen.

The Trunk

The early embryo begins as a layered plate that undergoes infolding and results in a hollow cylinder, the trunk. In the dorsal region, the neural tube also undergoes infolding and forms a neural tube. The cephalic part of the neural tube rapidly grows to form the forebrain, midbrain, and hindbrain vesicles. The ventricles are formed by the lumina of the neural tube. The future axial skeleton and muscles will be represented by somites located in this dorsal region. Development of the internal organs adds to the form and shape of this cylinder [39].

Development of the gut occurs at a time when the abdominal cavity is still small; hence, herniation occurs in the body stalk at about 5 weeks. The gut remains extruded until about 10 weeks when reinsertion occurs followed by complete closure of the ventral wall.

These normal developmental events can be documented by embryoscopy. The neural tube is seen with the cephalic end open at about 5 conceptional weeks of gestation (Fig. 4); but by 7 conceptional weeks, there is complete closure of the neural tube (Fig. 5). The ventral hernia is seen as early as 4 conceptional weeks (Fig. 6); and by 8 conceptional weeks, the hernia is almost completely resolved (Fig. 7). Occasionally, we have observed the gut to remain extruded after the 8th conceptional week [1].

Dorsal and ventral wall defects can also be diagnosed by embryoscopy. Omphalocele results when the abdominal wall fails to close after the 10th conceptional week, and bowels and even liver persist in the body stalk. Gastroschisis occurs when the bowel protrudes from a defect that is located below and generally to the right of the umbilicus. The liver is rarely involved. Faulty dorsal wall closure is associated with conditions such as spina bifida or rachischisis with or without saccular protrusions and herniation of neural elements [1, 38].

The Limbs

The normal development of limb buds is manifested first as lateral swellings or paddle-shaped structures in the late 4th conceptional week. Initially, arm buds are located far down on the body in the region of the lumbosacral somites. Flattening constrictions will demarcate foot and hand, leg and

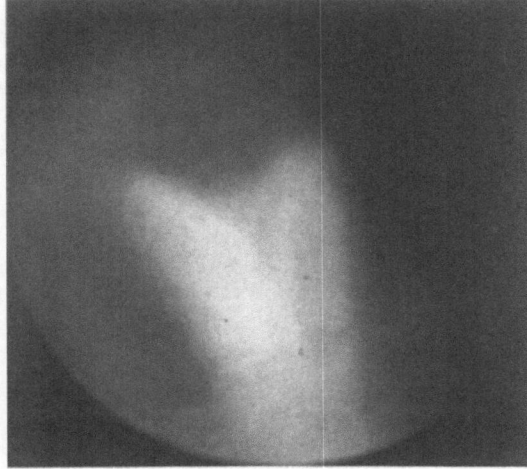

Fig. 4. Embryo at 5 conceptual weeks. The neural tube is seen with the cephalic end open

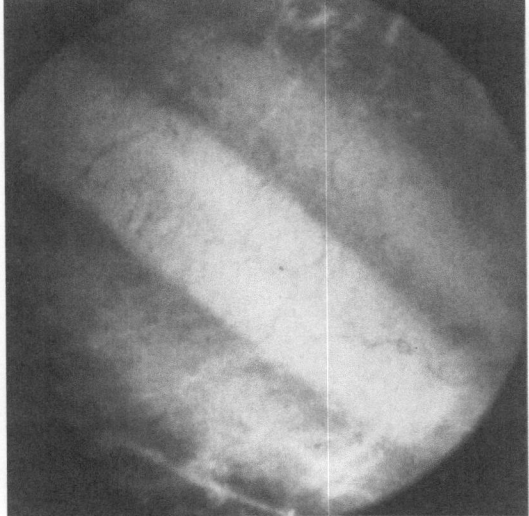

Fig. 5. The completely closed neural tube is seen at 7 conceptual weeks

thigh, and forearm and arm regions. Hand-radial ridges, suggested by grooves, predict the location of digits. The lower limbs develop at a slower rate. However, the thumbs and great toes separate early from the other digits in both upper and lower extremities [38–40]. The upper limb buds appear first, develop first, and attain normal appearance first. A disproportionate growth is maintained until the 2nd year of postnatal life when legs and arms become of equal length.

In our investigation in first-trimester pregnancies, we have been able to document these normal developmental events with embryoscopy. The hand

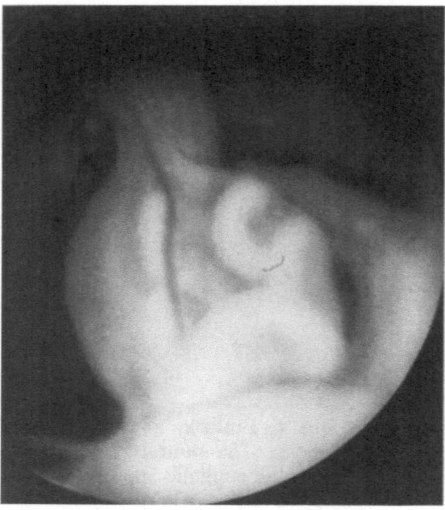

Fig. 6. The physiologic ventral hernia is seen at 4 conceptual weeks with loops of bowel contained within the umbilical stalk

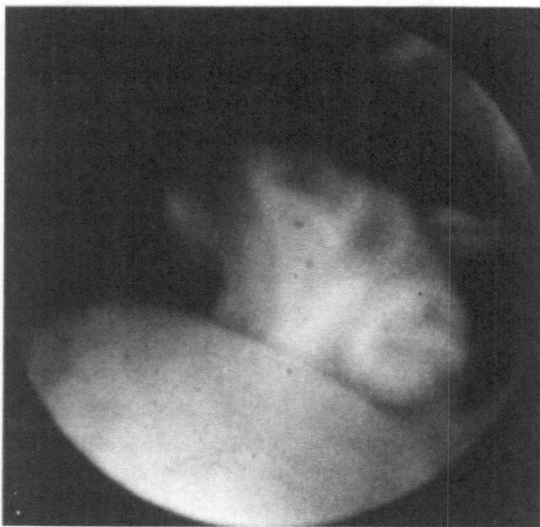

Fig. 7. By 8 conceptual weeks, the physiologic hernia is almost completely resolved with the abdominal contents re-entering into the expanded abdominal cavity

paddles are seen at 4 conceptional weeks with subtle demarcation of finger rays (Fig. 8); but by 6 conceptional weeks webbed fingers are observed (Fig. 9). At 7 conceptional weeks a fully developed hand is seen (Fig. 10). The developmental progression of the upper limbs can be as much as 2 weeks ahead of the lower extremities. Figures 11 and 12 show the fetal foot at 6 and 9 conceptional weeks with toe rays and fully developed foot, respectively.

Since limb development can be so beautifully documented with embryoscopy, it is very likely that abnormalities of hands and feet could be easily

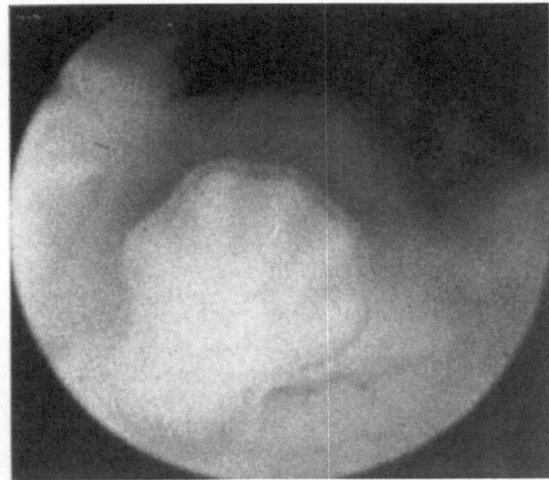

Fig. 8. An example of a hand paddle can be seen at 4 conceptual weeks with subtle demarcations of finger rays

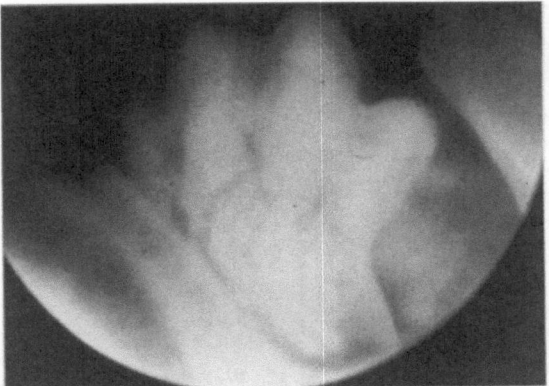

Fig. 9. At 6 conceptual weeks webbed fingers can be observed

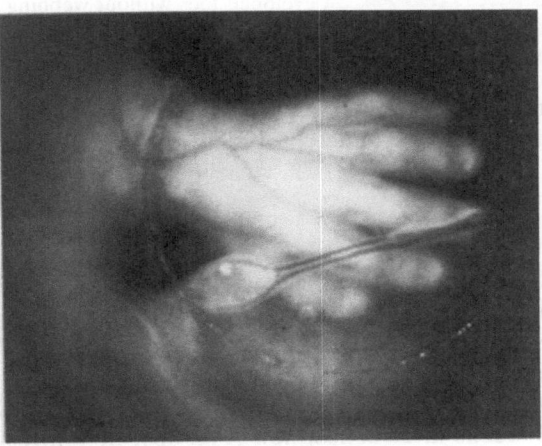

Fig. 10. At 7 conceptual weeks a fully developed hand is observed without webbing

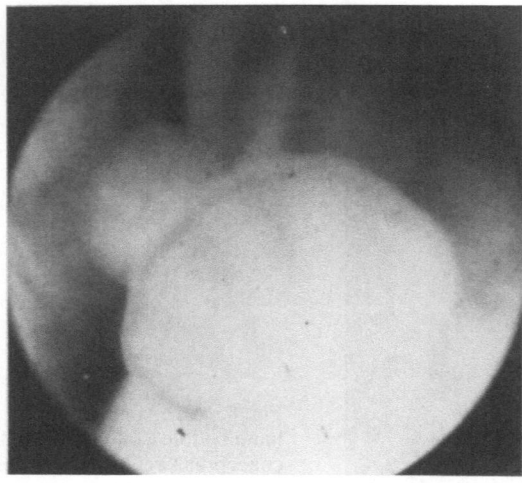

Fig. 11. Here the foot paddle at 6 conceptual weeks is demonstrated. The developmental process of the lower extremities usually lags behind the upper extremities by about 2 weeks

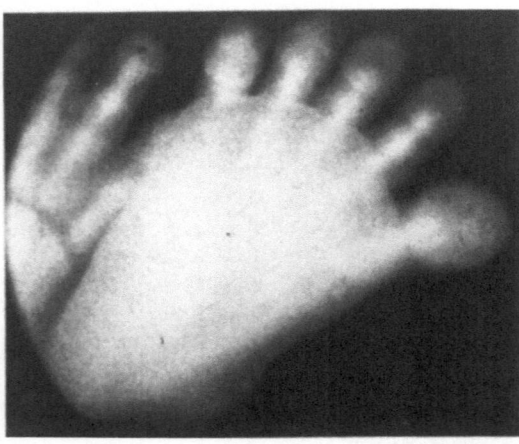

Fig. 12. Well-developed fetal foot at 9 conceptual weeks with well-developed toes without webbing

identified. For example, we have diagnosed polydactyly at 9 conceptional weeks of gestation. The potential exists to diagnose hemimelia, phocomelia, sirenomeli, missing digits, lobster claw, syndactyly, brachydactyly, or even club hand or foot [20].

Other Structures

The yolk sac has an essential role during organogenesis of providing blood cells, gonadocytes, and epithelia for the digestive and respiratory tracts. After the yolk sac invaginates and contributes to the midgut, the extruded remnant of the yolk sac becomes nonfunctional.

Embryoscopically, the yolk sac is observed to have a confluent and prominent vasculature at 3–8 weeks after conception (Fig. 13), in contrast

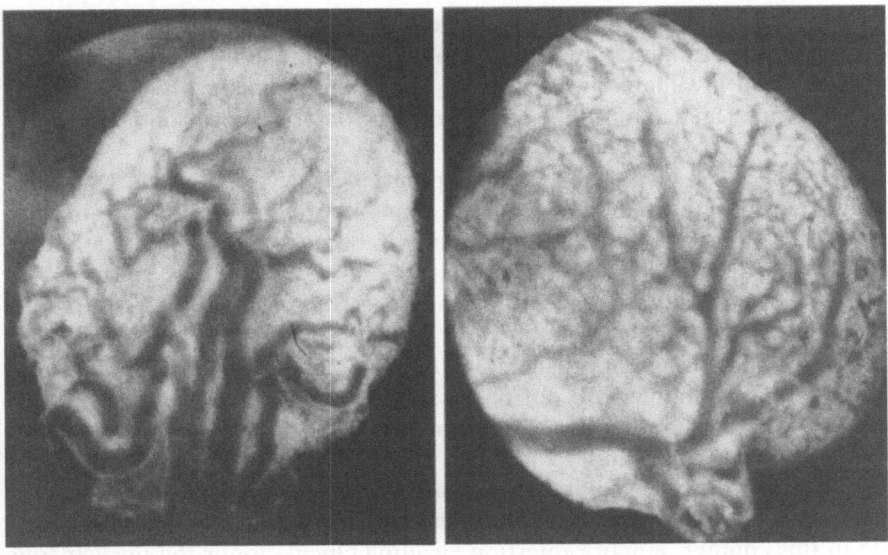

13 **14**

Fig. 13. The yolk sac at 3 conceptual weeks with confirmed vasculature and main vitelline vessels traversing the yolk sac

Fig. 14. The yolk sac at 10 conceptual weeks with blood vessels less prominent, smaller, and more numerous

to the yolk sac observed at 10 conceptional weeks where the blood vessels are smaller, more numerous and appear to be less prominent (Fig. 14). In an in vitro rat embryo model in which fetal anomalies were induced by suffusing the embryo preparation in high concentrations of glucose, the yolk sac structure and appearance were markedly altered [39, 41–44]. Some ultrasound studies have shown that yolk sac size and shape may be predictive of abnormal fetal outcomes. Others have found this relationship [45]. However, with new technologies such as embryoscopy, greater attention to the early conceptus is expected. Therefore, the relationship between yolk sac and embryo development will be explored in greater detail.

Confirmation of Early Ultrasound Prenatal Diagnosis

With the continued improvement in high-resolution ultrasound equipment, we have been able to expand our prenatal diagnostic program to examine fetuses even in the early second trimester. When congenital anomalies are diagnosed and patients choose to undergo pregnancy termination, pathologic confirmation and careful anatomic studies are performed following a dilation and extraction procedure. Verification of prenatally diagnosed anomalies is vital to accurate patient counseling. Using embryoscopy, we are now able to obtain a direct view of the embryo prior to evacuation and to document abnormalities by videotaping the examination. This technology has also been invaluable in identifying additional anomalies not recognized by ultrasound and confirming others already detected ultrasonographically [1].

Furthermore, from a conceptual point of view, embryoscopy will be of increasing importance to general human embryology. Much of our knowledge of early human development is based on the investigation of human aborti or analogies drawn from animal research. It is conceivable, however, that the very same pathology that causes a miscarriage could affect fetal growth and development as well, thereby somewhat limiting the validity of the former approach. On the other hand, stages of embryonic development in animals are not necessarily representative of human development, thereby limiting the validity of the latter approach. Embryoscopy visualizes the human embryo in vivo, unaffected by any pathology of the uterine or hormonal environment, and representative of the intrinsic capability of a given fetus for early development.

In this context, a major question awaiting elucidation is the unclear role of the yolk sac in early human development. The yolk sac provides blood cell precursors, gonadocytes, and epithelia of the digestive and respiratory tracts and has been demonstrated to be structurally altered when exposed to high glucose concentrations in embryo culture experiments. It has been suggested, therefore, that the yolk sac could play an essential role in the pathogenesis of congenital anomalies among fetuses of diabetic mothers [42–45]. Direct observation of the yolk sac by embryoscopy as well as aspiration of its contents for laboratory analysis could enhance our understanding of human malformations considerably.

The Future

Embryoscopic examination with the potential for embryonic therapy is an extension of the existing diverse prenatal diagnostic techniques as well as the beginning of a new era of first-trimester intervention. We can expect fetal tissue sampling and gene therapy to be goals of the future to which embryoscopy can be applied.

Tissue Sampling

First-trimester diagnoses of genetic diseases by organ tissue retrieval and/or embryonic blood sampling also appear within reach. The evolution of sophisticated molecular techniques using gene probes for DNA analyses makes embryoscopy seem even more attractive for the early prenatal diagnosis of many diseases.

Diagnosis can be done on the trophoblast via CVS at 9–12 menstrual weeks of gestation; and in almost 1% of cases, an additional fetal source, such as the fetal blood, is necessary to clarify initial aberrant chromosomal findings. Unfortunately, in many cases such evaluation needs to wait until the second trimester. Embryoscopic blood sampling in the first trimester may soon be feasible and would obviate this anxiety-laden waiting patients presenting after the first trimester and desiring rapid prenatal diagnosis or early fetal blood typing in a mother with isoimmunization [1].

Gene and Cell Therapy

Rapid progress has also been made in recent years towards the concept of human gene and cell therapy [46, 48]. If this concept becomes a reality, embryoscopy will permit accessibility to the human embryo at a time when embryos are immunologically "naive" and may therefore be receptive to these grafts.

Embryoscopy now opens a new era in fetal medicine. Prenatal diagnosis in the future will change considerably with more emphasis placed on the first trimester and performed directly on the embryo or on the placenta, as in CVS, rather than conducted in the second and third trimesters on the amniotic fluid and fetus. It is expected that the perfection of embryoscopy will provide access to the entire conceptus, permitting both early diagnosis as well as the potential for early treatment.

Gene or cell therapy has not yet been approved for clinical trials; however, we are at the threshold of an era in which this new therapeutic approach will probably be used in the treatment of currently incurable diseases. Obviously, the in utero application of this technology would be limited to serious genetic diseases which produce irreversible damage by the time of birth. At present, the only human tissue that can be used effectively for gene or cell transfer is bone marrow. It is expected, however, that more will be learned in the future about how to package the DNA, to purify stem cells, and to make it tissue-specific. It would then be feasible to use the intravenous route for injection of genetic material and access to the embryo and its circulatory system.

The potential feasibility of the above approach has been tested by Kantoff et al. [49] They demonstrated expression of the human adenosine deaminase (ADA) gene in the blood cells of irradiated monkeys that were reinfused with their own bone marrow cells after the cells had been treated in vitro with an ADA retroviral vector. The first attempts at postnatal human therapy of ADA deficiency are at advanced stages. Other genetic diseases, particularly those which affect the brain (e.g., Lesch-Nyhan syndrome), will require in utero intervention in order to be effective. Since in utero transplantation of donor tissue might become necessary to achieve this goal, embryoscopy could be an invaluable access tool to the early embryo/fetus, which would allow for graft acceptance by a naive immune system.

Embryoscopy, therefore, is expected to have a major impact on fetal medicine in the years to come, and prenatal diagnosis and therapy will undoubtedly change with more emphasis placed on intervention in the first trimester.

References

1. Reece EA, Hobbins JC (1991) Embryoscopy: an evolving technology for early prenatal diagnosis. In: Chapman M, Grudzinskas G, Chard T (eds) The embryo normal and abnormal development and growth. Springer, Berlin Heidelberg New York, pp 123–140
2. Cullen MT, Reece EA, Whetham J, Hobbins JC (1989) Embryoscopy: description and utility of a new technique. Am J Obstet Gynecol 162:82–86
3. Cullen MT, Reece EA, Viscarello RR, Hobbins JC (1989) Transcervical endoscopic visualization of prenatally diagnosed anomalies prior to second trimester termination. Proceeding of the Annual Meeting of the Society for Gynecology Investigation, March 1989
4. Reece EA, Rothmensch S, Whethan J, Cullen M, Hobbins JC (1991) Embryoscopy: a closer look at first-trimester diagnosis and treatment. Am J Obstet Gynecol (submitted)
5. Westin B (1954) Hysteroscopy in early pregnancy. Lancet ii (267):872
6. Galliant A, Lueken RP, Lindermann HJ (1978) New instruments and new methods. A preliminary report about transcervical embryoscopy. Endoscopy 10:47
7. Donald I, MacVicar J, Brown TG (1958) Investigation of abdominal masses by pulsed ultrasound. Lancet ii (274):1188–1194
8. Hon EH, Quilligan EJ (1967) The classification of fetal heart rate. Conn Med 31:779
9. Green JJ, Hobbins JC (1988) Abdominal ultrasound of the first trimester fetus. Am J Obstet Gynecol 159:165
10. Hon EH (1968) An atlas of fetal heart rate patterns. Harty, New Haven
11. Manning FA, Hill LM, Platt LD (1981) Qualitative amniotic fluid volume determination by ultrasound: antepartum detection of intrauterine growth retardation. Am J Obstet Gynecol 139:254
12. Patrick J, Campbell K, Carmichael L, Natale R, Richardson B (1982) Patterns of gross fetal body movements over 24 hour observation intervals during the last 10 weeks of pregnancy. Am J Obstet Gynecol 142:363–371
13. Manning FA, Lange FR, Morrison J, Harman CR (1983) Determination of fetal health: methods for antepartum and intrapartum fetal assessment. Curr Probl Obstet Gynecol 7:1
14. Beard RW, Filshie GM, Knight CA, Roberts GM (1979) The significance of the changes in the continuous fetal heart rates in the first stage of labor. J Obstet Gynecol Br Commonw 78:865
15. Romero R, Pilu G, Jeanty P, Ghidini A, Hobbins JC (1988) Prenatal diagnosis of congenital anomalies. Appleton and Lange Norwalk
16. Golbus MS, McGonigle KF, Goldberg JD, Filly RA, Callen PW, Anderson RL (1989) Fatal tissue sampling. The San Francisco experience with 190 pregnancies. West J Med 150:423–430
17. Valenti C (1972) Endoamnioscopy and fetal biopsy. A new technique. Am J Obstet Gynecol 141:561–564
18. Scrimgeour JB (1973) Other techniques for antenatal diagnosis. In: Emery AEH (ed) Antenatal diagnosis of genetic disease. New York: Churchill-Livingstone, New York, pp 40–57
19. Hobbins JC, Mahoney MJ (1974) In utero diagnosis of hemoglobinopathies. Technique for obtaining fetal blood. N Engl J Med 290:1065–1067

20. Rodeck CH, Nicolaides KH (1986) Fetoscopy. Br Med Bull 42:296–300
21. Mahoney MJ, Hobbins JC (1977) Prenatal diagnosis of chondroectoderman dysplasia (Ellis-Van Creveld syndrome) with fetoscopy and ultrasound. N Engl J Med 297:258
22. Elias S (1983) Fetoscopy in prenatal diagnosis. Clin Perinatol 10:357–367
23. Hobbins JC, Mahoney MJ (1974) Progress toward in utero diagnosis of hemoglobinopathies: technique for obtaining fetal blood. N Engl J Med 290:1065–1067
24. Daffos F, Capella-Pavlovsky M, Forestier F (1985) Fetal blood sampling during pregnancy with use of a needle guided by ultrasound. A study of 606 consecutive cases. Am J Obstet Gynecol 153:655–660
25. Hobbins JC, Grannum P, Romero R, Reece EA (1985) Percutaneous umbilical blood sampling. Am J Obstet Gynecol 152:1–6
26. Sievers S, Hagenbusch M, Eckert M (1983) Management of pregnant women with thalassemia minor. Fortschr Med 101:687–688
27. Hogge WA, Schomberg SA, Golbus MS (1986) Chorionic villus sampling: experience of the first 1000 cases. Am J Obstet Gynecol 154:2349
28. Martin AO, Simpson JL, Rosinsky B et al. (1986) Chorionic Villus sampling in continuing pregnancies. II. Cytogenetic Reliability. Am J Obstet Gynecol 154:1353
29. Hobbins JC, Grannum PAT, Berkowitz RL et al. (1979) Ultrasound in the diagnosis of congenital anomalies. Am J Obstet Gynecol 134:331
30. Bulic M, Podobnik M et al. (1987) First trimester diagnosis of low obstructive uropathy. An indication of initial renal function of the fetus. J Clin Ultrasound 6:715
31. Vergani P, Ghidini A et al. (1987) Antenatal diagnosis of fetal acrania. J Ultrasound Med 6:715
32. Hill LM, Thomas ML et al. (1988) Sonographic assessment of the first trimester fetus. A cautionary note. Am J Perinatol 7:97
33. Cutis JA, Watson L (1988) Sonographic diagnosis of omphalocele in the first trimester of fetal gestation. J Ultrasound Med 7:97
34. Timor-Tritsch IE, Warner WB, Peisner DB, Pirsone E (1989) First trimester midgut herniation. A high frequency transvaginal sonographic study. Am J Obstet Gynecol 161:831–833
35. Dumez Y (1990) Embryoscopy and congenital malformations. Proceedings of International Conference on Chorionic Villus Sampling and Early Prenatal Diagnosis, 28–29 May 1990, Athens
36. Roume J, Aubry MC, Labbe F, Dumez Y, Aubry JP, Henrion R (1985) Prenatal diagnosis of limb and digital abnormalities. Evaluation of the activity of the Port Royal University Clinic from 1979 to 1983. J Genet Hum 33:457–461
37. Dumez Y (1988) Embryoscopy and congenital malformations. Proceedings of the International Conference on Chorionic Villous Sampling and Early Prenatal Diagnosis, 28–29 May 1988, Athens
38. Moore KL (1988) The developing human. Clinically oriented embryology, 4th edn. Saunders, Philadelphia
39. Arey LB (1974) Development anatomy, 7th rev edn Saunders, Philadelphia
40. Hamilton WJ, Boyd JD, Mossman HW (1964) Human embryology, 3rd edn. Heffner, Cambridge
41. Pinter E, Reece EA, Leranth CZ, Sanyal MK, Hobbins JC, Mahoney MJ, Naftolin F (1986) Yolk sac failure in embryopathy due to hyperglycemia. An ultrastructural analysis of the yolk sac differentiation associated with embryopathy in rat conceptuses under hyperglycemic conditions. Teratology 33:363–374
42. Reece EA, Pinter E, Leranth CZ, Sanyal MK, Hobbins JC, Mahoney MJ, Naftolin F (1985) Ultrastructural analysis of malformations of the embryonic neural axis induced by hyperglycemic conceptus culture. Teratology 32:363–374
43. Reece EA, Pinter E, Leranth CZ, Garcia-Segura LM, Sanyal MK, Hobbins JC, Mahoney MJ, Naftolin F (1989) Yolk sac failure in embryopathy due to hyperglycemia: horseradish peroxidase uptake in the assessment of yolk sac dysfunction. Obstet Gynecol 74:755–762

44. Reece EA, Hobbins JC (1986) Diabetic embryopathy: pathogenesis, prenatal diagnosis and prevention. Obstet Gynecol Surv 41:325–335
45. Reece EA, Scioscia AL, Pinter E, Hobbins JC, Green JJ, Leranth CZ, Mahoney MJ, Naftolin F (1988) Prognostic significance of the human yolk sac assessed by ultrasonography. Am J Obstet Gynecol 159:1191–1194
46. Anderson WF (1984) Prospects for human gene therapy. Science 226:401
47. Reece EA, Pinter E, Leranth CZ, Sanyal MK, Hobbins JC, Mahoney MJ, Naftolin F (1985) Ultrastructural analysis of malformations of the embryonic neural axis induced by hyperglycemic conceptus culture. Teratology 32:363–374
48. Anderson WF (1985) Human gene therapy: scientific and ethical considerations. J Med Philos 10:275
49. Kantoff PW, Gillis AP, McLachlin OR et al. (1987) Expression of human adenosine deaminase in nonhuman primates after retrovirus-mediated gene transfer. J Exp Med 166:219–234

23 Prenatal Diagnosis and Invasive Techniques in the First Trimester of Pregnancy

B. Brambati

Introduction

The last 2 decades have witnessed the transfer of a number of prenatal diagnosis methods from applied research to being a routine component of genetic counseling and obstetric management. More than 100 different genetic conditions can now be diagnosed in utero by amniocentesis; chorionic villus sampling (CVS); cordocentesis; and cytogenetic, enzymatic, and DNA analysis. In addition, more than 200 fetal malformations can be detected by ultrasonography, and neural tube defects (NTD) and Down's syndrome can be screened in maternal blood in the first half of the second trimester [1]. In the meantime, the increasing demand to make reproductive decisions as early as possible and the rapid evolution of the techniques involved in genetic diagnosis have led to new approaches moving from the mid-second trimester to the preembryo stage.

Chorionic Villus Sampling

There is no doubt about the psychologic advantages of first-trimester diagnosis by CVS when compared to second-trimester procedures (amniocentesis, cordocentesis). Moreover, the complications of inducing abortion in the case of an affected fetus are significantly reduced in relation to the age of pregnancy. Though the public generally appears to have a favorable opinion regarding the option of prenatal diagnosis to prevent genetic diseases, the studies on the acceptance of a prenatal test indicate wide variation, the decision pro and contra the test depending on many individual and social factors. Several of these obstacles have been overcome by the possibility of using CVS early in the pregnancy, and, thanks to this new method, genetic diagnosis is now being performed for the first time all over in the world. Reliable information on the worldwide application of CVS cannot be gathered for a number of reasons, but mainly the lack of national programs and steering committees, and the number of uncontrolled tests being performed in private centers. Nevertheless, a rough evaluation of the tendency and total number of cases per year might be obtained from the WHO CVS

Table 1. Timetable of major CVS research steps and technical specifications

Reference	Sampling route	Sampling method	Remarks
Hahnemann [5]	TC	Endoscope and biopsy forceps	First experimental study
Anshan [7]	TC	Metallic cannual, blind insertion	First clinical study
Kazy et al. [9]	TC	Biopsy forceps, US guidance	First US use
Old et al. [10]	TC	Plastic cannula, US guidance	First DNA diagnosis
Brambati and Simoni [11]	TC	Plastic cannula, US guidance	First cytogenetic diagnosis
Smidt-Jensen et al. [12]	TA	US-fixed needle guide, double coaxial needle	First TA-CVS
Nicolaides et al. [20]	TA	Freehand US-guided needle insertion	First TA-CVS in the 2nd–3rd trimesters
Brambati et al. [16]		Freehand US-guided needle insertion	First TA-CVS in the 1st trimester by 20-gauge single needle
Brambati et al. [21]	TA	Freehand US-guided needle insertion	First TA-CVS at 6 weeks of gestation

US, ultrasound.

registry [2]: a steady yearly increase by a factor of 1.5 has been evident since 1985, and in 1991 more than 70 000 and 40 000 CVS tests are expected in the world and in Europe, respectively.

History of CVS

Historically, techniques for placental biopsy were developed for the rapid diagnosis of hydatidiform mole and as a diagnostic aid for placental pathology in 1958 [3] and 1966 [4]. However, its genetic value started to be evaluated in the late 1960s in Denmark in parallel with the early development of mid-trimester amniocentesis (Table 1). The experiments were done only in patients undergoing voluntary abortion [5]. In the absence of real-time ultrasound, the procedure had to be guided by direct vision, using a straight 5.0-mm diameter endoscope: the instrument was inserted through the cervix and passed through the decidua covering the ovum, the villi were identified by direct vision and sampled by a biopsy forceps. Another Scandinavian group used a similar method in preabortion patients [6]. Both these studies reported a high number of complications which would have led to the loss of pregnancy, and a low success rate of karyotyping was obtained. In 1975, a group of Chinese investigators reported an aspiration technique to obtain placental tissue for sexing by sex chromatin [7]. They used a 3-mm metal cannula inserted transcervically without optical guidance and stopped after any soft resistance was noted: chorionic tissue was aspirated by a syringe attached to a smaller coaxial tube which was advanced an additional 0.5–1.0 mm. The procedure was demonstrated to be highly successful and relatively safe, but no investigator in the Western countries was able to repeat a similar performance [8]. The key to success in sampling came with the application of ultrasound to guide the sampling device: Kazy et al. [9] in the Soviet Union sampled chorionic tissue through the cervical canal using ultrasound to guide the biopsy forceps. In the last 26 cases sampling was done for sexing by both X and Y chromatin stains. However, CVS could be introduced into clinical activity only after the works of Old et al. [10] in London and of Brambati and Simoni [11] in Milan. The high efficacy and the low risk of transcervical (TC) chorionic tissue aspiration by a thin plastic catheter under ultrasound guidance and the reliability of direct karyotyping and molecular analysis methods were extensively evaluated. Since then, several variations of the original devices (biopsy forceps and plastic catheter) have been used. A substantial methodological contribution was further given by Smidt-Jensen et al. [12] who suggested transabdominal (TA) sampling as an alternative to the TC method. Using this method, the potential risk of ascending intrauterine infection could be avoided.

General Methodological Considerations and Prerequisites for a CVS Program

CVS is an outpatient procedure, requiring neither preoperative preparation nor postoperative care. In the case of rhesus incompatibility, Coombs' test should be performed before CVS, and, if no antibodies are present, a dose of at least 50 µg of anti-D immunoglobulin should be administered within 72 h of sampling. A full bladder is not required, but may be advantageous either to enlarge the angle of a pronounced antiflexion of the uterus or to push intestinal loops away from the anterior uterine wall. The operating room should contain an examination table, facilities for scrubbing, and an inverted microscope for the immediate identification and weight evaluation of the chorionic tissue sampled. An ultrasound machine with a 5-MHz scanner rather than a conventional 3.5-MHz one is recommended to guarantee an accurate study of the uterus and its contents, embryo-fetal biometry and cardiac activity, and a reliable guidance of the sampling device; the high-resolution ultrasound equipment becomes mandatory when sampling earlier than 9 weeks of gestation. Although the placental site is generally easily identifiable by a characteristic strong and thick reflecting zone surrounding the gestational sac, the identification of the umbilical cord insertion on the placenta is sometimes very useful or even essential to localize a suitable area for CVS.

The minimal amount of chorionic tissue which is reliable for the different diagnostic processes (direct karyotyping, culture, DNA analysis)

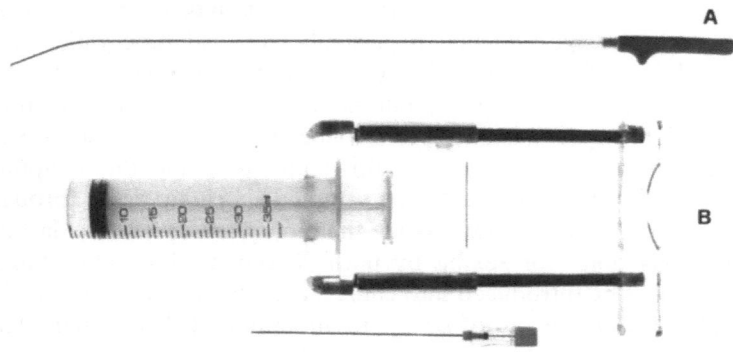

Fig. 1. A TC aspiration system: 26-cm plastic cannula housing a bendable metallic obturator with blunt distal end. B TA aspiration system: a 20-gauge spinal needle and a 30-ml syringe mounted on a syringe holder

should be established in advance with the laboratory, and the quality and quantity of each specimen needs an immediate evaluation under the microscope to decide whether to stop sampling or not. The sampling team (sonographer and/or operator) should be trained to do the first evaluation, in this way saving time for the laboratory staff. No more than two sampling attempts should be permitted in the same session [13], and, in the case of failure, the reasons should be analyzed and any further decision discussed with the patient.

It is generally advisable to aspirate the tissue sample in a nutrient medium supplemented with heparin to avoid the inclusion of villi in maternal blood clots. However, if DNA analysis is to be performed, a saline solution is recommended to avoid any interference of heparin with the DNA extraction process.

CVS Between 8 and 12 Weeks of Gestation

Although several sampling methods were considered in the early phase of clinical experience, only two of them are currently used: TC aspiration using a thin plastic catheter [14, 15] and TA needle biopsy [12, 16, 17].

TC Aspiration

TC aspiration is carried out by a freehand ultrasound-guided catheter via the cervical canal. The catheter most frequently used is Trophocan (Portex Ltd., England), a polypropylene cannula, 1.5 mm in outer diameter, and 21 cm or 26 cm long, provided with a shaped metallic obturator with a blunt distal end (Fig. 1). The patient is placed in the gynecologic position and after external genital toilet, the cervix and vagina are visualized with a speculum and cleaned with a broad-spectrum antiseptic solution (e.g., povidone-iodine). It has been suggested that the anterior lip of the cervix should be grasped with a tenaculum to be able to modify the angle between the cervical canal and a too anteflexed uterus by a traction applied to the cervix. However, in our experience the use of the tenaculum is not necessary because the same favorable effect on uterine anteflexion can generally be obtained by filling the bladder; moreover, tenaculum insertion is frequently painful, and in rare cases cardiovascular collapse has also been reported as an effect of vasovagal reaction. The catheter is slightly curved by the operator to conform to the angle needed to approach the selected sampling site. The use of a metallic probe with a rounded tip of about 3 mm in diameter (hysterometer) has proven to be very useful to provide information on the cervical canal and its angle with the uterus painlessly and safely, and clearly to fix the internal cervical os on the ultrasound display. The catheter is then slowly inserted through the cervical canal and stopped at the internal cervical os: from this point further movement should only be undertaken under ultrasound guidance to avoid any damage to the ovular sac and maternal decidua. The end of the catheter should enter deep into the placenta and should be positioned next to the chorionic plate. The obturator is removed without disturbing the position of the catheter, and a 20-ml syringe containing 3–4 ml Hank's medium or saline solution is attached. About 10 ml negative pressure is applied, and the catheter is slowly removed from the placenta under negative pressure.

Transabdominal Chorionic Villus Sampling

TA needle biopsy is carried out by a freehand ultrasound-guided technique using a disposable 20-gauge spinal needle, 90 mm long (Fig. 1). The skin is sterilized with a topical application of povidone-iodine solution and the area draped in a sterile fashion. Because the transducer is placed 3 cm or more away from the insertion site, it does not need to be protected by a sterile glove nor have sterile contact gel to avoid contamination (Fig. 2). The needle is introduced without local administration of anesthetics and should be oriented so that the longest placental axis can, as far as possible, be brought into the needle pathway. The needle orientation and the insertion point are directly related to both the distance of the placenta from the maternal surface and the limited length of the needle available for sampling. If a 20-gauge spinal needle longer than 9 cm is used, its higher elasticity should be held in due consideration because of a less efficient control of its direction. When the needle tip is clearly visible within the placenta (Fig. 3) the stylet is removed, and a 30-ml syring containing 3–4 ml Hank's solution or saline solution is attached. The syringe is set in a holder to make the aspiration easier, and the chorionic tissue is aspirated by combining repeated

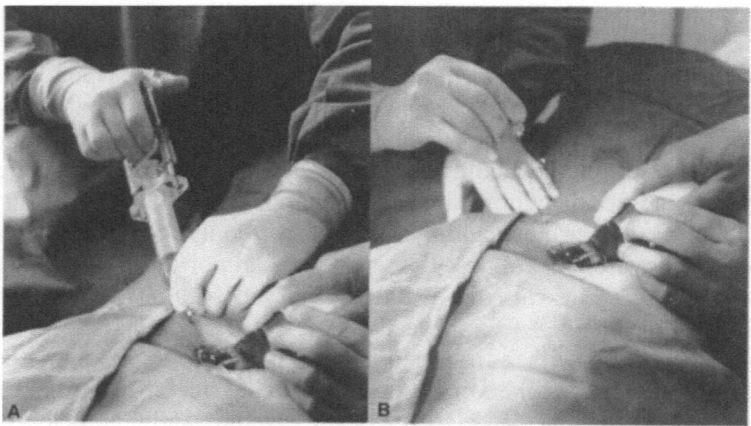

Fig. 2A,B. Main steps of TA-CVS. **A** The most favorable scanning plane for needling is selected; **B** aspiration of villi is performed by means of a syringe holder

suctions of about 20-ml depression and slow to-and-fro movement of the needle tip along the axis of the placenta. The sampling system is withdrawn by keeping a fully depressed syringe.

CVS in Twin Pregnancy

Documentation of two plancetal sites in a twin pregnancy does not substantially change methodological problems arising out of singleton pregnancy. When placentae are fused, it is easy in the early first trimester to distinguish between dichorionic and monochorionic pregnancy because chorion laeve is still substantially present all around gestational sac and makes it easier to recognize the thick septum between the sacs at ultrasound investigation. However, visualization of only one placental site and dichorionic sacs does not permit any distinction between monozygotic and dizygotic twins. This uncertainty requires specific technical criteria to sample chorionic tissue from each of the two gestational sacs reliably [18]. The most valuable criteria are to limit aspiration to the placental area next to the umbilical cord insertion and to avoid passing through the placenta of the second twin. In the case of a fused placenta, TA sampling has proved to be the most successful approach, although in some instances the combined use of both TA and TC sampling has been most profitable. Moreover, when two identical sexes are provided by cytogenetic analysis and no differences have been found in relation to the specific risk indication, the reliability of sampling may be proven only by chromosome polymorphism studies and/or use of minisatellite DNA probes that detect highly polymorphic regions of the human genome [18, 19]. Furthermore, if identical results follow the

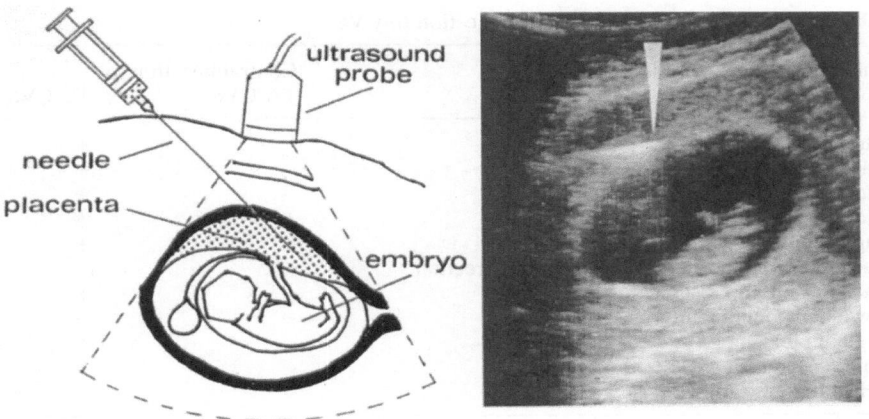

Fig. 3. Diagram and ultrasound description of TA-CVS

latter genetic studies, the patient should be counseled to wait for confirmation by amniocentesis or cordocentesis.

Early (Before 8 Weeks) and Late (After 12 Weeks) CVS

The TA method has also proven to be a valuable approach for sampling chorionic tissue before and after the traditional 8–12 weeks' gestation period [20–24]. The viability of the pregnancy can easily be demonstrated by ultrasound detection of embryo heart activity from the end of the 5th week onward, and in more than 90% of cases the chorion frondosum site can be well defined by a 5-MHz abdominal probe at 6–7 weeks. In addition, before sampling, a more detailed investigation of the uterine content can be obtained by a 6.5-MHz vaginal probe, and this is certainly helpful to establish clearly the relationship between placenta and embryo. The embryo pole is a useful indicator of the placental site because at this time the very short umbilical cord keeps it next to the chorionic surface of the placenta [25]. Early sampling can be peformed by a 22-gauge spinal needle thanks to the smaller size of chorionic villi, whereas in the second and third trimesters an 18-gauge spinal needle is recommended because of the larger size of the villi.

The feasibility of CVS before 8 weeks and the use of rapid diagnostic tests (direct karyotyping, enzymatic analysis on fresh tissue, DNA analysis by polymerase chain reaction) offer the opportunity, if selective abortion is required, to interrupt pregnancy by medical methods (oral antiprogestins together with vaginal prostaglandins) with obvious advantages for the patient [26].

Table 2. Clinical and anatomic contraindication to CVS

Anatomic/clinical conditions	Contraindication to TA-CVS	TC-CVS
Vaginal infection	−	+
Inaccessible cervical canal	−	+
Excessive uterine anteflexion	−	+
Miomas	+	+
Vaginismus	−	+
Placental site opposite to cervical canal direction	−	+
Placental site too distant	+	+
Bowels surrounding uterus	+	−
Vaginal bleeding	−	+
Pending cerclage	−	+
RH isoimmunization	+	+

+, Contraindication.

The extension of CVS beyond the first trimester has proven very convenient as an alternative method to amniocentesis or cordocentesis in three specific situations: (a) to avoid any delay in fetal diagnosis by waiting for amniocentesis at 15–16 weeks; (b) to obtain fetal diagnosis in time by rapid karyotyping for medicolegal reasons; (c) to acquire genetic information in patients at risk and marked by a technically unfavorable condition for amniotic fluid or fetal blood sampling (e.g., severe oligohydramnios).

Contraindications to CVS

A number of clinical or anatomic conditions may be identified as contraindications to one of the two sampling approaches or to both of them (Table 2). A recent history or evidence of vaginal infection is unanimously accepted as an unfavorable condition for TC sampling and, together with vaginismus and an inaccessible cervical canal, is considered a clear indication for an alternative approach, i.e., TA CVS or amniocentesis. In cases of an anteflexed uterus, a too pronounced angle with the cervix could make it too risky to attempt to reach the anterior or posterior placenta via the cervical canal, while, if the uterus is retroflexed, intestine loops could make it too hazardous for TA sampling. A gestational age of more than 12 weeks is in principle considered to be an unfavorable condition for TC aspiration because thereafter the adhesion of the gestational sac membranes to the decidua parietalis becomes firmer, and the placenta is frequently far from the internal cervical os. Myomas are potential obstacles to sampling procedures by deviating the sampling device from the correct path. A major concern is caused by rhesus-isoimmunized patients: microtransfusion from the fetus to the mother has been demonstrated in the majority of CVS cases by an increased maternal alpha-fetoprotein (AFP) level after sampling

Table 3. Diagnostic indications to CVS

Chromosome anomalies
Inborn errors of metabolism
DNA analysis (some of the diseases diagnosable by DNA analysis)

Autosomal dominant	Polycystic kidney disease
	Retinoblastoma
	Huntington's chorea
	Myotonic dystrophy
	Neurofibromatosis
Autosomal recessive	α-Thalassemia
	β-Thalassemia
	Sickle cell anemia
	Cystic fibrosis
	Adrenogenital syndrome
	Phenylketonuria
X-linked diseases	Duchenne's muscular dystrophy
	Becker's muscular dystrophy
	Hemophilia A
	Hemophilia B
	Adrenoleukodystrophy
	Retinitis pigmentosa
Other indications	Rubella virus
	Cytomegalovirus
	Toxoplasma
	Blood grouping
	Paternity

[27, 28]; therefore the potential risk of CVS should be carefully evaluated and periodical determinations of the antibody titer should be done after sampling. When the distance of the placenta from the insertion point clearly exceeds the length of the sampling device, contraindication to both sampling methods is obvious. History or evidence of vaginal bleeding is a very puzzling condition: the effect of CVS on pregnancy outcome is still unknown, but in these cases TA sampling is in principle preferable to avoid any potential risk of iatrogenic ascending infection. In our experience, in carrying out TA CVS in patients with a history of bleeding, but a normal pregnancy, no relationship has been found with fetal loss rate [14].

Indications for CVS

Direct karyotyping and DNA analysis by restriction fragment length polymorphisms made chromosomopathies and hemoglobinopathies the first specific diagnostic indications of chorion biopsy [10, 11]. The determination of normal enzymatic levels in villi [29, 30] also made possible first-trimester diagnosis of metabolic diseases in pregnancies at risk (Table 3). More recently information about blood grouping and some infectious agents (i.e., rubella virus, toxoplasmosis, cytomegalovirus) have also become available

Table 4. Technical and clinical aspects of TA and TC methods

	CVS methods	
	TA	TC
Simplicity (methodology, technology, perioperative cares)	High	High
Efficacy >10 mg villi by single attempt (%)	96.9	96.4
Contraindications (%)	3.1	15.8
Success at first attempt (%)	97.5	92.5
Success at second attempt (%)	99.6	99.2
Costs Spinal needle ($)	2	
TROPHOCAN catheter ($)		15
Total time of the procedure (min)	10–15	15–25
Training time	Short	Long
Risks Maternal	Very rare	Very rare
Early complications (%)	3.8	6.4
Total abortion rate (<28 weeks) (%)	4.0	4.8
Acceptability	High	High

[31–34]. A major advantage of using chorionic tissue for prenatal diagnosis is the high yield of substrate obtained for analysis. On the other hand, the inconvenience of maternal decidual contamination may be easily removed by accurate selection of the material in experienced hands. Moreover, inconclusive diagnosis, due to inadequate enzyme or DNA source, possible maternal contamination, laboratory difficulties, chromosomal mosaicism, gives the opportunity for a later confirmation by the analysis of fetal cells obtained by amniocentesis or fetal blood sampling. All chromosomal anomalies and inborn errors of metabolism, previously diagnosed on cultured amniocytes, are now diagnosable in fresh or cultured chorionic tissue (Table 3). In addition, with the rapidly progressing human genome mapping it is foreseeable that the diagnosis of an unlimited number of hereditary diseases will be available in the near future.

Efficacy of CVS

As has been demonstrated by randomized trials, both TA and TC approaches should be considered highly efficient sampling methods [35]: success was obtained in more than 90% of cases at the first insertion and in more than 99% of cases at the second insertion; no more than two insertions have been attempted. Although chorionic tissue weight distribution of TA needling appears significantly shifted towards lighter values when compared with TC aspiration, specimens of at least 10 mg or more have been obtained in about 95% of cases by a single insertion in both series (Table 4). TA-CVS proved an easier method to learn and the operator attained the highest skill by the first 100 cases, instead of 300 cases when the TA method was used [36]. In case of failure at the second insertion, it is advisable to stop any

Table 5. Early and late complications after CVS

	TA-CVS (%)	TC-CVS (%)
Early complications		
Bleeding	2.3	5.6
Peritoneal reaction	0.3	–
Uterine cramping	1.6	0.3
Uterine infection[a]	–	0.2
Late complications		
Vaginal bleeding (abruptio placentae/placenta previa)	0.3	0.9
Preeclampsia/hypertension	1.2	2.3
Premature delivery	8.1	7.1
Birthweight <2500	5.2	4.8
Malformations[b]	2.4	1.6

[a] Uterine infection cases were reported only before randomized study.
[b] Including cases diagnosed in utero.

further attempt and evaluate the opportunity to wait for amniocentesis or to repeat CVS some days later. In the latter case it is preferable to use only the TA route with the aim of reducing as much as possible ascending bacterial infection of the uterine cavity by further catheter passage through the vagina and cervical canal.

Early and Late Complications

Early complications occur more often after TC sampling (Table 5), and bleeding is the most frequent clinical feature [37]. In TC-CVS bleeding is usually observed immediately after sampling and for less than 1 day, whereas it frequently appears later after TA-CVS and lasts 2 or more days.

Mild localized peritonitis has been very occasionally observed after TA-CVS, and the etiopathogenetic causes have been ascribed either to the transfixion of bowels with shedding of intestinal bacteria in the peritoneal cavity or to the inflammatory effect of bleeding from the uterine wall. Ascending infection of the uterine cavity following the introduction of the catheter was suggested as potential danger by culturing bacteria from the tip of the catheter [38, 39]. Septic abortion has been documented in a number of published cases, and in two of them maternal shock complicated intrauterine infection [14, 39–41]. Bleeding, fever, and uterine cramps were reported 1–2 weeks after sampling.

With the exception of uterine infection, all of the other early complications cleared up without any consequence for the mother and the fetus, and no relationship was demonstrated with fetal loss rate.

Clinical and obsteric complications of the late second and third trimesters did not show any increase [37]: vaginal bleeding (abruptio placentae/placenta previa), pre-eclampsia, hypertension, premature delivery, and birthweight have been reported in the range of the general population.

Fetal Loss After CVS

In recent randomized trials no statistically significant difference in fetal loss rate was found between amniocentesis and TC-CVS [42], or between TC-CVS and TA-CVS [37]. Therefore, one can infer that for both TA-CVS and TC-CVS methods the expected fetal loss rate following the sampling procedure is not unlike that for amniocentesis, i.e., about 1%.

Laboratory Aspects of CVS

It appears that all the enzymatic analyses of inborn metabolic defects previously diagnosed on cultured amniocytes are also diagnosable on villi. Frequently, by using fresh chorionic tissue, the result can be obtained much more rapidly and efficiently than with amniotic fluid because of the large amount of substrate available. Similarly, a great advantage of using chorionic tissue for molecular diagnosis is the high yield of DNA obtained for analysis. Although maternal contamination has been claimed to be a major potential pitfall, appropriate sampling and accurate tissue selection obviate this inconvenience. Moreover, in a recent study of prenatal diagnosis of β-thalassemia, the reliability of dot blot analysis with DNA amplification was evaluated against the traditional oligonucleatide analysis, and in all cases the diagnoses were concordant [43].

Villi are usually processed for cytogenetic analysis in two separate ways: (a) by the direct method; and (b) by the long-term tissue culture method [44]. The direct method provides rapid results and minimizes the effect of maternal contamination; tissue culture is better able to offer an optimal morphologic analysis of chromosomes by high-resolution banding. The direct method is, in principle, more suitable to extend fetal diagnosis there where a sterile condition cannot be assured and poor technical resources are available. Both methods are highly successful and accurate (Table 6); however, the percentage of chromosomal anomalies confined to the placenta ("false positive") is slightly higher in the direct method series and the very rare false-negative results in cytogenetic diagnosis have been reported when only the direct method has been used [45–48]. A workshop on cytogenetic problems of first-trimester diagnosis [49] analyzed and compared CVS series performed by only direct or long-term culture methods or by both, and concluded that each of these approaches is efficient and accurate and per se advisable for clinical use. More recently, in the United States collaborative study [50], a very high cytogenetic laboratory success rate and diagnostic accuracy was found with both the direct and culture methods of analysis.

Table 6. Success rate and diagnostic accuracy in the first-trimester cytogenetic analysis of chorionic tissue

Reference	Cases (n)	Success rate (%)	False positive (%)	False negative (%)	Method of analysis
Hogge et al. [48]	1000	>99.0	1.2	None	Long-term culture
Green et al. [45]	940	99.8	1.3	None	Direct
Sachs et al. [46]	1034	99.7	0.2	None	Direct
Simoni and Brambati [44]	2413	99.3	1.03	0.04	Direct
Ledbetter et al. [50]	3884	99.6	0.9	0.12	Direct
	3777		0.4	None	Long-term culture

Maternal cell contamination was confirmed to be rare in direct preparations but occurred in 1.9% cases of long-term culture. However, a higher frequency of incorrect predictions of fetal cytogenetic status was observed with the direct method than with the culture method. When both methods were used, no incorrect predictions were observed.

Chromosomal anomalies confined to the placenta have been related to intrauterine growth retardation [51]. The biologic effect of the above chromosomal condition on the fetal development is still under investigation, and hypotheses have been formulated on the basis of the clinical observations ranging from normal intrauterine development to multiple malformations, mild or severe growth retardation, intrauterine death [52–55]. Then, if a false-positive cytogenetic result is suspected, it is advisable: (a) to suggest diagnostic control by amniocentesis or cordocentesis; and (b) thereafter, if confinement of chromosomal anomaly to the placenta is demonstrated, to start intensive surveillance of fetal development.

Conclusions

Between 8 and 12 weeks CVS can be equally successfully performed either by the TA or TC route. However, better efficiency of the diagnostic service is assured only when both methods are available.

TA sampling offers the possibility to anticipate genetic diagnosis before 8 weeks or to extend it to the 2nd and 3rd trimesters of pregnancy.

CVS has proved to be a very efficient procedure, and its fetal loss risk has been demonstrated to be as low as in amniocentesis. No adverse effects were found on fetal growth during intrauterine development and at birth. Laboratory failure and diagnostic accuracy rates are comparable to the values obtained by amniotic fluid processing. The recent progress on DNA analysis make genetic diagnosis possible in a very short time and offers the opportunity to investigate an unlimited number of hereditary diseases.

Amniocentesis in the 1st Trimester

The usual time for amniocentesis and chromosome or biochemical analysis is 15–17 weeks of gestation. At that time the procedure is technically simple, relatively safe, and provides a sufficient number of viable cells to allow successful culture. The recent Danish randomized trial [56] reported an average fetal loss rate of 1% (95% confidence limits 0.3% – 1.5%), a statistically significant difference between amniocentesis patients and controls in the rate of respiratory distress syndrome (1.1% versus 0.5%) and infant pneumonia (0.7% versus 0.3%), and a culture failure rate of 0.56%.

However, the later in pregnancy genetic diagnosis is performed, the more difficult the situation becomes both psychologically and surgically. Therefore, attempts have been made over the past few years to move amniocentesis closer to the first trimester to make the procedure a more acceptable competitor for the increased privacy afforded by CVS [57–66]. Recent advances in real-time ultrasound and tissue culture techniques have made possible the development of earlier amniotic fluid sampling.

There are wide variations in the gestational age of study populations, and in some cases amniocentesis has been done as early as 8 weeks. Although fetal loss rate has been frequently reported close to the midtrimester values, the results are difficult to evaluate because the studies are usually small and grossly unbalanced by an excess number of 13- and 14-week-old pregnancies, real first-trimester procedures being present in very limited numbers. As with later amniocentesis, the procedure is performed by inserting a 22-gauge needle into the amniotic sac under continuous ultrasound surveillance. However, a number of anatomic and physiologic peculiarities of the late first trimester should be considered in order to anticipate potential technical difficulties and fetal risks.

The chorion forms a sac, the chorionic sac, within which the embryo and its amnion and the yolk sac are suspended by the connecting stalk. The extraembryonic coelom, which derives from the fusion of the coelomic spaces in the extraembryonic mesoderm, separates chorionic and amniotic membranes. The extraembryonic coelom is filled with a clear mucinous and highly viscous material and is far greater than the amniotic cavity until the end of the 8th week. However, because of the steady expansion of the amniotic sac, the extraembryonic coelom is gradually reduced to a thin layer by the 12th week and disappears by the 13th–14th weeks, leaving the amniotic membrane attached to the chorion.

In the first trimester, the amniotic sac might be compared to a balloon inflated at low pressure, and this light pressure gradient between amniotic and extra-amniotic cavity explains the flag-waving of the amniotic membrane caused by fetal movements and observed during ultrasound examination (Fig. 4). Moreover, before the 10th week, the fetal crown-rump length is equivalent to the amniotic sac diameter, and the amniotic fluid pocket

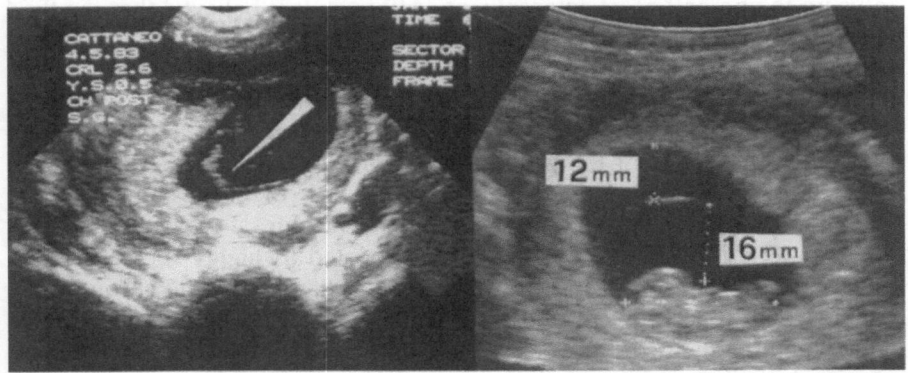

A   B

Fig. 4. A Floating amniotic membrane (*arrow*) at 9.3 weeks of gestation. **B** Extraembryonic coelom and amniotic cavity measurements in relationship with the embryo at 9.3 weeks of gestation

Table 7. Risks and technical problems of amniocentesis before 12 weeks' gestational age

Sampling pitfalls
  Total amount of amniotic fluid might be critical
  "Tenting" of the amniotic membrane might result because of the low pressure in the
    amniotic sac
  Amniotic fluid pockets free of fetal parts are rather small
Potential adverse biologic effects
  Developmental abnormalities
  Direct fetal damage
  Fetal loss
  Chronic amniotic fluid leakage
  Lung hypoplasia and respiratory distress syndrome
Potential drawbacks in cytogenetic analysis
  Too small a number of viable cells
  Higher rate of pseudomosaicisms

surrounding the fetus does not exceed 1–1.5 cm (Fig. 4). These anatomic characteristics clearly indicate the difficulties expected when entering the amniotic sac and ensuring fetal safety (Table 7).

Before the 10th week the amniotic fluid resembles a transmembranous dialysate of the maternal serum and at that stage the role of the embryo is negligible [67, 68]. By the 10th week a new pathway gains importance: the kidney starts to function and significant fetal urinary excretion is detectable after the 13th week [69]. Moreover, the fetal skin may be a pathway for the exchange of water and basic biologic substances up to the 20th week. The volume of amniotic fluid rises from a mean value of 29.7 ml at 10 weeks to 58 ml at 12 weeks with a range of 18–33 ml and 35–86 ml, respectively [70–76]. Therefore, the removal of 1 ml of amniotic fluid per week of gestation to provide a sufficient amount of cells for culture could deprive the fetus up to one-half of its environment. Although it is believed that the fluid aspirated can be replaced in few hours, the significant increase of respiratory complications reported in midtrimester amniocentesis in human pregnancy [56, 77] and the lung hypoplasia observed after early amniocentesis in monkeys [78] suggests the need to accumulate more information to assess the long-term effects of the procedure. Until then, no reliable data can be provided to the patient about the potential risk of early amniocentesis, which must be considered an investigational procedure.

Amniotic fluid cell culture has been reported as highly successful only after the 12th week [64], and the mean time to have complete chromosome analysis and report results did not significantly differ from midtrimester amniocentesis; before 12 weeks a high number of culture failures was observed, while much more time was required to provide results [60, 64]. Acetylcholinesstherase levels cannot be used before the 12th week of gestation to detect open NTD because of the high number of false-positive results [79], whereas from the 10th week 17-HO-progesterone level determination has proved to be as reliable as in midtrimester amniocentesis to diagnose 21-hydroxylase deficiency-type congenital adrenal hyperplasia [80].

Preimplantation Genetic Diagnosis

Currently, genetic diagnoses are limited to the postimplantation period and are performed by CVS and amniocentesis. However, there are couples at risk of hereditary disease who cannot accept pregnancy termination for emotional, moral, or religious reasons, and therefore show great interest in preimplantation diagnosis to select for implantation only those embryos that are not at risk of carrying genetic disease.

Successful preimplantation diagnosis requires not only an easy access to gametes or preembryo, but also reliable and safe microbiopsy procedures and analytic techniques sufficiently sensitive and specific to permit the analysis of the genetic material of a single cell.

Two main technical approaches have been suggested for recovery and manipulation of oocytes and/or preembryos (Fig. 5): preovulatory oocyte pick-up by needle aspiration followed by in vitro fertilization, and nonsurgical lavage of the uterine cavity after natural conception [81, 82]. Material for genetic diagnosis can be provided either by first polar body biopsy of the mature oocyte [83, 84], or single cell aspiration of the six- to eight-cell stage zygote [85], or trophectoderm herniation and biopsy of expanded blastocyst [82] (Fig. 6).

Two diagnostic techniques are currently under investigation to be used for molecular analysis in the human preembryo: polymerase chain reaction (PCR) [85] and in situ hybridization [86–88]. PCR allows amplification of specific DNA sequences by a factor of a millionfold or more and therefore

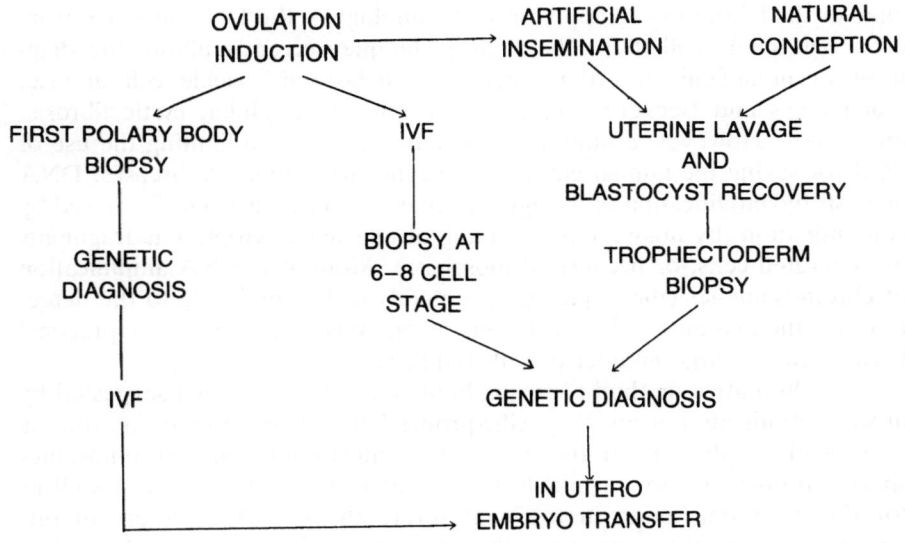

Fig. 5. Main approaches to preimplantation genetic diagnosis

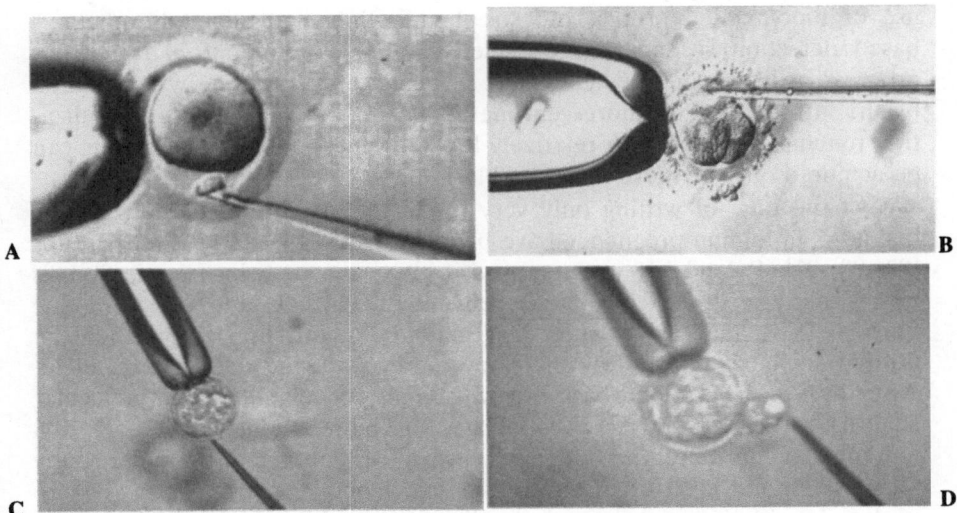

Fig. 6A–D. First polar body removal (**A**), blastomere aspiration at four-cell stage (**B**), and trophectoderm biopsy at the expanded blastocyst stage (**C**, **D**). (Courtesy of Y. Verlinsky, Chicago, USA)

enables the laboratory to apply probe technology to the DNA message from even one single cell [89]. The PCR technique currently allows the diagnosis of monogenic disorders such as β-thalassemia, sickle cell disease, Duchenne's and Becker's muscular distrophy, hemophilia, cystic fibrosis, and others. However, a number of potential pitfalls are limiting the use of PCR for sexing the human embryo or diagnosing hereditary diseases. DNA analysis by amplification of a single cell may result in misdiagnosis caused by contamination, by inadvertent sampling of anucleate cytoplasmic fragments or generated cells, or because of mosaicism. Moreover, DNA amplification of chromosome-specific sequences is unlikely to be applicable to the detection of aneuploidies, while cytogenetic analysis does not seem practical because of the large number of cells required.

An alternative method of sexing human embryos has been suggested by in situ hybridization using Y-specific probes [86–88]. In situ hybridization is also widely applicable for the detection of aneuploidy since chromosome-specific probes are now available for one-half of the autosomes as well as for the X chromosome; in the near future the use of a battery of different probes would make it possible to evaluate the presence of an extra-chromosome among the most frequent aneuploidies. However, the use of radiolabeled probes requires several days to obtain diagnosis and would then necessitate cryopreservation of biopsied embryos. More recently the use of biotynylated probes and streptavidin-linked alkaline phosphatase-based detection systems has provided results in less than 24 h. Further progress in increasing efficiency and reducing detection time was obtained by the introduction of fluorescent methods. Nevertheless, in situ hybridization remains a system with relatively low efficiency, and reliable results can be acquired only by more than one blastomere.

At the time of writing only very few cases of preimplantation genetic diagnosis in human pregnancy have been published by two research groups [83, 90, 91]. Polar body was removed in two cases at risk of cystic fibrosis and α_1-antitrypsin deficiency, respectively [83, 90]. In the case of cystic fibrosis, PCR analysis revealed that polar body was homozygous for the normal allele, demonstrating that the oocyte contained the affected allele. After IVF one blastomere was removed at the six-cell stage, and molecular analysis revealed that the pre-embryo was homozygous for delta-F 508 mutation, confirming the diagnosis previously obtained by polar body analysis. In the case of α_1-antitrypsin deficiency, five polar bodies were analyzed and two oocytes containing the unaffected gene were fertilized and the embryos transferred to the uterus, but no pregnancy was established.

In couples at risk of X-linked disease and undergoing IVF one blastomere was aspirated from an eight-cell stage zygote [90]. The sex was determined by DNA amplification of a Y-specific sequence. Female embryos only were transferred in five women and resulted in two twin and one singleton pregnancy. The diagnostic accuracy was confirmed by CVS. A further case of preimplantation diagnosis for gender determination has been

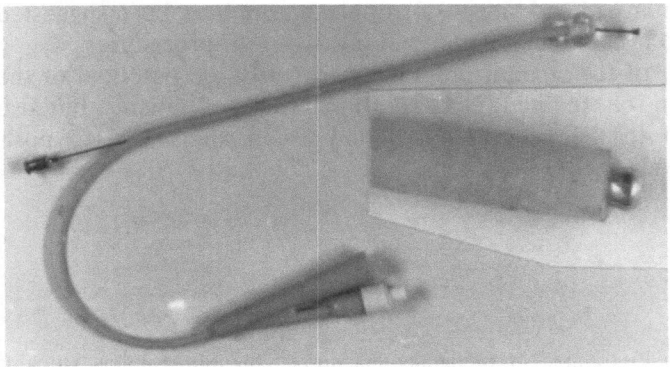

Fig. 7. Uterine lavage system made by a collecting balloon, silicon cannula, and coaxial flushing metallic catheter with a blunted extremity; the balloon is inflated when just passing the internal cervical os

successfully reported in a woman with a previous child with fragile X syndrome: two diploid female embryos were transferred, but pregnancy ended in spontaneous miscarriage at 7 weeks' gestational age [92].

Until now no cases of genetic diagnosis by blastocyst-stage biopsy after uterine lavage have been reported. The inconvenience associated with DNA amplification is expected to be less critical than for the previously described diagnostic procedures because of the larger amount of material available for processing. On the basis of the previous experience reported for transferring embryo from fertile donors to infertile recipients, one can expect to successfully obtain blastocyst by uterine lavage at the 4th–6th days after LH peak in about 40% of cases [93, 94]. The percentage could be further increased to more or less 50% by optimizing the efficiency of the washing system.

The removal of some of the trophectoderm cells from the expanded blastocyst has been successfully obtained in primates, rabbits, mice, and cattle, and no detrimental effects have been described in the offspring [95–99]. More recently, trophectoderm biopsy has also been carried out in human blastocysts [100]. A slit was made in the zona pellucida and herniation of 10–30 cells was obtained after 18–24 h in about 66% of cases, while hatching occurred in 38% of cases, comparing very favorably with the 18% observed in the control series of nonmanipulated embryos.

For a preimplantation genetic program flushing of the uterine cavity (Fig. 7) after natural conception should be considered as the most suitable way for a number of reasons [101]. Couples at high genetic risk are generally young and fertile, and it seems obvious to offer them genetic diagnosis without significantly interfering with the quality of their reproductive and sexual life. Therefore, spontaneous oocyte maturation, normal and free intercourse and natural conception, together with blastocyst recovery by nonsurgical uterine lavage and biopsy of trophectoderm have been suggested as an alternative to cycle manipulation and invasive IVF procedures.

The final point of the current program is to verify the potential of the method, not only in selected and relatively small groups of patients, but also in being extended to the general population as an alternative to post-implantation methods.

References

1. Weaver DD (1989) Catalog of prenatally diagnosed conditions. Johns Hopkins University Press, Baltimore
2. Jackson LG (1989) CVS Newsletter. 31 January 1989, Philadelphia
3. Acosta-Sison H (1958) Diagnosis of hydatiform mole. Obstet Gynecol 12:205–208
4. Alvarez H (1966) Diagnosis of hydatiform mole by transabdominal placental biopsy. Am J Obstet Gynecol 95:538
5. Hahnemann N (1974) Early prenatal diagnosis: a study of biopsy techniques and cell culturing from extra-embryonic membranes. Clin Genet 6:294–306
6. Kullander S, Sandhal B (1973) Fetal chromosome analysis after transcervical placental biopsies in early pregnancy. Acta Obstet Gynercol Scand 52:355–359
7. Anshan (1975) Fetal sex prediction by sex chromatin of chorionic villi cells during early pregnancy. Chin Med J [Engl] 1:117–126
8. Hoewell DH, Loeffler FE, Coleman DV (1983) Assessment of a transcervical aspiration technique for chorionic villus biopsy in the first trimester of pregnancy. Br J Obstet Gynaecol 90:196–198
9. Kazy Z, Rozovsky IS, Bakharev VA (1982) Chorion biopsy in early pregnancy: a method of early prenatal diagnosis for inherited disorders. Prenat Diagn 2:39–45
10. Old JM, Ward RHT, Karagozlu F, Petrou M, Modell B, Weatherall DJ (1982) First trimester fetal diagnosis for haemoglobinopathies: three cases. Lancet 2:1414–1416
11. Brambati B, Simoni G (1983) Fetal diagnosis of trisomy 21 in the first trimester of pregnancy. Lancet 1:586
12. Smidt-Jensen S, Hahnemann N, Jeansen PKA, Therkelsen AJ (1984) Experience with transbdominal fine needle biopsy from chorionic villi in the first trimester-An alternative to amniocentesis. Nordic Meeting on Medical Genetics, Febr 17–19, Helsinki
13. WHO Consultation on First Trimester Fetal Diagnosis (1986) Risk evaluation in chorionic villus sampling. Prenat Diagn 6:451–456
14. Brambati B, Oldrini A, Ferrazzi E, Lanzani A (1987) Chorionic villus sampling: an analysis of the obstetric experience of 1000 cases. Prenat Diagn 7:157–169
15. Wapner RJ, Jackson L (1988) Chorionic villus sampling. Clin Obstet Gynecol 31(2):328–344
16. Brambati B, Oldrini A, Lanzani A (1987) Transabdominal chorionic villus sampling: a freehand ultrasound-guided technique. Am J Obstet Gynecol 157:134–137
17. Brambati B, Lanzani A, Oldrini A (1988) Transabdominal chorionic villus sampling. Clinical experience of 1159 cases. Prenat Diagn 8:609–617
18. Brambati B, Oldrini A, Simoni G, Terzoli GL, Romitti L, Rossella F, Ferrari M (1984) First trimester fetal karyotyping in twin pregnancy. J Med Genet 21:58–60
19. Hill AVS, Jeffreys AJ (1985) Use of minisatellite DNA probes for determination of twin zygosity at birth. Lancet 2:1394–1395
20. Nicolaides KH, Southill PW, Rodeck CH, Warren RC (1986) Why confine chorionic villus sampling to the first trimester? Lancet 1:543
21. Brambati B, Tului L, Simoni G, Travi M (1988) Prenatal diagnosis at 6 weeks. Lancet 2:397

22. Holzgreve W, Miny P, Basaran S, Fuhrmann W, Belle FK (1987) Safety of placental biopsy in the second and third trimester. N Engl J Med 317:1159
23. Pijpers L, Jahoda MGJ, Reuss A, Wladimiroff JW, Sachs ES (1988) Transabdominal chorionic villus biopsy in second and third trimesters of pregnancy to determine fetal karyotype. Br Med J 297:822–823
24. Monni G, Olla G, Rosatelli C, Cao A (1990) Second trimester placental biopsy versus amniocentesis for prenatal diagnosis of beta-thalassemia. N Engl J Med 322:60–61
25. Brambati B, Tului L, Simoni G, Travi M (1991) Genetic diagnosis before the eighth gestation week. Obstet Gynecol 77:1–4
26. UK Multicentre Trial (1990) The efficacy and tolerance of mifepristone and prostaglandin in first trimester termination of pregnancy. Br J Obstet Gynaecol 97:480–486
27. Brambati B, Guercilena S, Bonacchi I, Oldrini A, Lanzani A, Piceni L (1986) Feto-maternal transfusion after chorionic villus sampling: clinical implications. Hum Reprod 1:37–40
28. Fuhrmann W, Altland K, Kohler A, Holzgreve W, Jovanovic V, Rauskolb R, Pawlowitzki IH, Miny P (1988) Feto-maternal transfusion after chorionic villus sampling. Hum Genet 78:83–85
29. Simoni G, Brambati B, Danesino C, Rossella F, Terzoli GL, Ferrari M, Fraccaro M (1983) Efficient direct chromosome analysis and enzyme determinations from chorionic villi samples in the first trimester of pregnancy. Hum Genet 63:349–357
30. Grebner EE, Wapner RJ, Barr BA, Jackson L (1983) Prenatal Tay-Sachs diagnosis by chorionic villi sampling. Lancet 2:286–287
31. Kanhai HH, Gravenhorst JB, Veer MBR, Mass CJ, Bevetstock GC, Bernini LF (1984) Chorionic biopsy in management of severe rhesus isoimmunization. Lancet 2:157–158
32. Terry GM, Ho Terry L, Warren RC, Rodeck CH, Cohen A, Rees KR (1986) First trimester prenatal diagnosis of congenital rubella: a laboratory investigation. Br Med J 292:930–933
33. Amirhessami-Aghili N, Lahijani R, Manalo P, St Jeor S, Tibbitts FD, Hall MR (1989) Persistence of human cytomegalovirus DNA sequences without cell-virus homology in human placental explants in culture. Int J Fertil 34:411–419
34. Verhofstede C, van Renterghem L, Plum J, Vanderschueren S, Vanhaeserbrouck P (1990) Congenital toxoplasmosis and TORCH. Lancet 2:622–623
35. Brambati B, Oldrini A, Lanzani A, Terzian E, Tognoni G (1988) Transabdominal versus transcervical chorionic villus sampling: a randomised trial. Hum Reprod 3:811–813
36. Brambati B, Lanzani A, Tului L (1990) Transabdominal and transcervical chorionic villus sampling: efficiency and risks evaluation of 2411 cases. Am J Med Genet 35:160–164
37. Brambati B, Terzian E, Tognoni G (1991) Randomized clinical trial of transabdominal versus transcervical chorionic villus sampling methods. Prenat Diagn 11:285–293
38. Scialli AR, Neugebauer DL, Fabro SE (1985) Microbiology of the endocervix in patients undergoing chorionic villi sampling. In: Fraccaro M, Simoni G, Brambati B (eds) First trimester fetal diagnosis. Springer, Berlin Heidelberg New York, pp 69–73
39. Brambati B, Matarrelli M, Varotto F (1987) Septic complication after chorionic villus sampling. Lancet 1:1212–1213
40. Blackmore KJ, Mahoney MJ, Hobbins JC (1985) Infection and chorionic villus sampling. Lancet 2:339
41. Barela A, Kleinman GE, Goldtich IM, et al. (1986) Septic shock with renal failure after chorionic villus sampling. Am J Obstet Gynecol 154:1100–1102
42. Canadian Collaborative CVS-Amniocentesis Clinical Trial Group (1989) Multicentre randomised clinical trial of chorionic villus sampling and amniocentesis. Lancet 1:1–6

43. Ristaldi MS, Pirastu M, Rosatelli C, et al. (1989) Prenatal diagnosis of beta-thalassemia in Mediterranean populations by dot blot analysis with DNA amplification and allele specific oligonucleotide probes. Prenat Diagn 9:629–638
44. Simoni G, Brambati B (1989) Fetal karyotype diagnosis by first trimester chorionic villus sampling. In: Reed GB, Claireaux AE, Bain AD (eds) Diseases of the fetus and newborn. Chapman and Hall, London, pp 627–639
45. Green JE, Dorfmann A, Jones SL, et al. (1988) Chorionic villus sampling: experience with an initial 940 cases. Obstet Gynecol 71:208–212
46. Sachs ES, Jahoda MGJ, Kleijer WJ, Pijpers L, Galjaard H (1988) Impact of first trimester chromosome, DNA and metabolic studies on pregnancies at high genetic risk: experience with 1000 cases. Am J Med Genet 29:293–303
47. Simoni G, Terzoli G, Rossella F (1990) Direct chromosome preparation and culture using chorionic villi: an evaluation of the two techniques. Am J Med Genet 35:181–183
48. Hogge WA, Schonberg SA, Golbus MS (1986) Chorionic villus sampling: experience of the first 1000 cases. Am J Obstet Gynecol 154:1249–1252
49. Fourth International Conference on Chorionic Villi Sampling and Early Prenatal Diagnosis, Athens, May 1988
50. Ledbetter DH, Martin AD, Verlinsky Y, et al. (1990) Cytogenetic results of chorionic villus sampling: high success rate and diagnostic accuracy in the United States Collaborative Study. Am J Obstet Gynecol 162:495–501
51. Kalousek DK, Dill FJ (1983) Chromosomal mosaicism confined to the placenta in human conceptions. Science 221:665–667
52. Verp MS, Rosinsky B, Sheikh Z, Amarose AP (1989) Non-mosaic trisomy 16 confined to villi. Lancet 2:915–916
53. Schwinger E, Seidl E, Klink F, Rehder H (1989) Chromosome mosaicism of the placenta-a cause of developmental failure of the fetus? Prenat Diagn 9:639–647
54. Hashish AF, Monk NA, Lovell-Smith MPF, et al. (1989) Trisomy 16 detected at chorion villus sampling. Prenat Diagn 9:427–432
55. Johnson A, Wapner RJ, Davis G, Jackson L (1990) Mosaicism in CVS: an association with poor perinatal outcome. Obstet Gynecol 75:573–577
56. Tabor A, Madsen M, Obel EB, Philip J, Bang J, Norgaard-Pedersen B (1986) Randomized controlled trial of genetic amniocentesis in 4606 low-risk women. Lancet 1:1287–1293
57. Stripparo L, Buscaglia M, Longatti L, et al. (1990) Genetic amniocentesis: 505 cases performed before the sixteenth week of gestation. Prenat Diagn 10:359–364
58. Johnson A, Godmilow L (1988) Genetic amniocentesis at 14 weeks or less. Clin Obstet Gynecol 31:345–352
59. Hanson FW, Zorn EM, Tennat FR, Marionos S, Samuels S (1987) Amniocentesis before 15 weeks' gestation: outcome, risks and technical problems. Am J Obstet Gynecol 156:1524–1531
60. Elejalde BR, de Elejalde MM, Acuna JM, Thelen D, Trujillo C, Karrman M (1990) Prospective study of amniocentesis performed between weeks 9 and 16 of gestation: its feasibility, risks, complications and use in early genetic prenatal diagnosis. Am J Med Genet 35:188–196
61. Henry G, Peakman D, Winkler W, O'Connor K (1985) Amniocentesis before 15 weeks instead of CVS for earlier prenatal cytogenetic diagnosis. Am J Hum Genet [Suppl] 37:A219
62. Luthardt FW, Luthy DA, Karp LE, Hickok DE, Resta RG (1985) Prospective evaluation of early amniocentesis for prenatal diagnosis. Am J Hum Genet [Suppl] 37:A222
63. Godmilow L, Weiner S, Dunn LK (1987) Genetic amniocentesis between 12 and 14 weeks gestation. Am J Hum Genet [Suppl] 41:A275
64. Rooney DE, MacLachlan N, Smith J, Rebello MT, Loeffler FE, Beard RW, Rodeck C, Coleman DV (1989) Early amniocentesis: a cytogenetic evaluation. Br Med J 299:25

65. Arnovitz KS, Priest JH, Elsas LJ, Strumlauf E (1988) Amniocentesis prior to 14.5 weeks gestation: experience in 142 cases. Am J Hum Genet [Suppl] 43:A232
66. Nevin J, Nevin NC, Dornan JC, Sim D, Armstrong MJ (1990) Early amniocentesis: experience of 222 consecutive patients, 1987–1988. Prenat Diagn 10:79–83
67. Ostergard DR (1970) The physiology and clinical importance of amniotic fluid: a review. Obstet Gynecol Surv 25:297–319
68. Finegan J-AK (1984) Amniotic fluid and midtrimester amniocentesis: a review. Br J Obstet Gynaecol 91:745–750
69. Abramovich DR (1968) The volume of amniotic fluid in early pregnancy. J Obstet Gynaecol Br Commonw 75:728–731
70. Fuchs F (1966) Volume of amniotic fluid at various stages of pregnancy. Clin Obstet Gynecol 9:449–460
71. Lind T, Hytten FE (1970) Relation of amniotic fluid volume to fetal weight in the first half of pregnancy. Lancet 1:1147–1149
72. Harrison RG, Malpas P (1953) The volume of human amniotic fluid. J Obstet Gynaecol br Emp 60:632–639
73. Haswell GL, Morris JA (1973) Amniotic fluid volume studies. Obstet Gynecol 42:725–732
74. Monie IW (1953) The volume of the amniotic fluid in the early months of pregnancy. Am J Obstet Gynecol 66:616–625
75. Queenan JT, Thompson W, Whitefield CR, Shah SI (1972) Amniotic fluid volume in normal pregnancies. Am J Obstet Gynecol 114:34–38
76. Wagner G, Fuchs F (1962) The volume of amniotic fluid in the first half of human pregnancy. J Obstet Gynaecol Br Commonw 69:131–136
77. Report to the Medical Research Council by their Working Party on Amniocentesis (1978) An assessment of the hazard of amniocentesis. Br J Obstet Gynaecol 85, Suppl 2
78. Hislop A, Fairweather DV (1982) Amniocentesis and lung growth: an animal experiment which clinical implications. Lancet 2:1271–1272
79. Muller F, Oury JF, Boué A (1989) First-trimester amniotic fluid acetylcholinesterase electrophoresis. Prenat Diagn 9:173–175
80. Raux-Demay M, Mornet E, Boué J, Couillin P, Oury JF, Ravise N, Deluchat C, Boué A (1989) Early prenatal diagnosis of 21-hydroxylase deficiency using amniotic fluid 17-hydroxyprogesterone determination and DNA probes. Prenat Diagn 9:457–466
81. Penketh RJA, McLaren A (1987) Prospects for prenatal diagnosis during preimplantation human development. Baillierès Clin Obstet Gynaecol 1:747–764
82. Edwards RG, Hollands P (1988) New advances in human embryology: implications of the preimplantation diagnosis of genetic disease. Hum Reprod 3:549–556
83. Verlinsky Y, Ginsberg N, Lifchez A, Valle J, Moise J, Strom CM (1990) Analysis of the first polar body: preconception genetic diagnosis. Hum Reprod 5:826–829
84. Monk M, Holding C (1990) Amplification of a beta-haemoglobin sequence in individual human oocytes and polar bodies. Lancet 1:985–988
85. Handyside AH, Pattinson JK, Penketh RJA, Delhanty JDA, Winston RML, Tuddenham EGD (1989) Biopsy of human preimplantation embryos and sexing by DNA amplification. Lancet 1:347–349
86. Penketh RJA, Delhanty JDA, van de Berghe JA, Finkestone EM, Handyside EH, Malcolm S, Winston RML (1989) Rapid sexing of human embryos by non-radioactive in situ hybridization: potential for preimplantation diagnosis of X-linked disorders. Prenat Diagn 9:489–500
87. Jones KW, Singh L, Edwards RG (1987) The use of probes for the Y chromosome in preimplantation embryo cells. Hum Reprod 2:439–445
88. West JD, Gosden JR, Angell RR, Hastie ND, Glasier AF, Thatcher SS, Baird DT (1987) Sexing human pre-embryo by DNA-DNA in situ hybridisation. Lancet 1:1345–1347

89. Li A, Gyllensten UB, Cui X, Saiki RK, Erlich HA, Arnheim N (1988) Amplification and analysis of DNA sequences in single human sperm and diploid cells. Nature 335:414–419
90. Handyside AH, Kontogianni EH, Hardy K, Winston RML (1990) 344:768–770
91. Strom CM, Verlinsky Y, Milayeva S, et al. (1990) Preconception genetic diagnosis of cystic fibrosis. Lancet 2:306–307
92. Milayeva S, Verlinksy Y, Enriquez G, et al. (1990) Successful preimplantation diagnosis for gender determination. J In Vitro Fert Embryo Transfer 7:197
93. Buster JE, Bustillo M, Modi IA, et al. (1985) Biologic and morphologic development of donated human ova recovered by nonsurgical uterine lavage. Am J Obstet Gynecol 153:211–217
94. Formigli L, Formigli G, Roccio C (1987) Donation of fertilized uterine ova to infertile women. Fertil Steril 47:162–165
95. Betteridge KJ, Hare WCD, Singh EL (1981) Approaches to sex selection in farm animals. In: Brackett BG, Seidel GE, Seidel SM (eds) New techologies in animal breeding. Academic, New York, pp 109–125
96. Gardner RL (1971) Manipulations on the blastocyst. Adv Biosci 6:279–296
97. Gardner RL, Edwards RG (1968) Control of the sex ratio at full term in the rabbit by transferring sexed blastocysts. Nature 218:346–349
98. Rowson LEA, Moore RM (1966) Development of sheep conceptus during the first fourteen days. J Anat 100:777–785
99. Summers PM, Campbell JM, Miller MW (1988) Normal in-vivo development of marmoset monkey embryos after trophectoderm biopsy. Hum Reprod 3:389–393
100. Dokras A, Sargent IL, Ross C, Gardner RL, Bar low DH (1990) Trophectoderm biopsy in human blastocysts. Hum Reprod 5:821–825
101. Brambati B, Formigli L (1991) Uterine lavage for preimplantation genetic diagnosis. In: Verlinsky Y, Kuliev A (eds) Preimplantation diagnosis. Plenum Press, New York, pp 165–174

Treatments

24 Medical Therapy

E.R. Barnea

Introduction

The epidemiology and etiology of early pregnancy dysfunction (EPD) has been discussed in Chap. 15. From that discussion it becomes clear that early pregnancy dysfunction is not a homogenous group, but it encompasses a wide range of known and unknown etiologies. During the last decade, important investigations have been carried out in order to understand various aspects of human reproduction. New and effective regimens for inducing ovulation have been introduced, and pregnancies can now be achieved through assisted reproductive technologies. Recent data have shown that the "bottleneck" of these new technologies is the establishment of a successful pregnancy. Ovulation, fertilization, and embryo formation are achieved in most cases; however, implantation and pregnancy development are less frequently obtained. In spite of major advances in reproduction, prevention of early pregnancy complication and treatment in cases of early pregnancy dysfunction remain almost unexplored. Further-more, the subjects are actually considered as highly controversial.

The period of time between implantation and the 9th week of gestation is the most critical period of gestation. During this time most early pregnancy dysfunction, congenital malformations, or failures will take place. Thereafter, the rate of spontaneous abortion decreases to less than 5% (see Chap. 15). Based on the concept that a pregnancy that tends to fail should be allowed to do so, in many medical centers there is a policy not to treat threatened abortion, except for bed rest. This no-touch approach assumes that the embryo must be defective and therefore it should be left to its destiny. Now, if the pregnancy continues untreated, can we assume that the conceptus is normal?

It is obvious that as more early pregnancy losses have been diagnosed in recent years (see Chap. 15), new and effective measures have to be put forward in order to reduce the rate of early pregnancy complications. Therefore, a new unbiased look is needed on this critical aspect of human reproduction. The purpose of this chapter is to review the latest information in this field and to introduce some new thoughts for managing early pregnancy dysfunction.

Habitual Abortions

It is of note that in spite of its rarity (<1% of patients) the subject of habitual abortion (HA) has received major attention. A large number of investigations have been carried out which have stimulated a more scientific approach to the problem, leading to the development of numerous treatment modalities. However, as a public health concern, the rate of early pregnancy dysfunction in the general population is much more important than that of HA, and it can be estimated that 15%–20% of early pregnancies will fail. This distorted view is illustrated by the following example: if a woman has had two miscarriages in a row in the 1st trimester, she is not considered to be at risk for a subsequent EPD and is usually encouraged to conceive again naturally. If three consecutive pregnancy failures occur, then the problem is investigated more seriously and the woman is considered as a habitual aborter. This approach should be modified: for a mother-to-be any loss is important and should not be minimized. Epidemiologic studies have also shown that the likelihood of early pregnancy dysfunction and loss increases with each subsequent pregnancy loss.

This strict view is based on large cytogenetic studies performed on early abortuses where a high rate of chromosomal malformations (up until 60%) was found (see Chaps. 16, 18, 23). However, these data are questionable since habitual aborters who are successfully treated have a low malformation rate. This strongly suggests that the figures quoted previously were much too high and did not reflect reality. In addition, new diagnostic techniques (see Chaps. 20, 23) may help to answer whether the embryo is normal before deciding whether or not a pregnancy should be treated. Among them, transvaginal ultrasonography and some biochemical markers are very useful in the early diagnosis of a viable pregnancy. Obviously, future studies will be needed to make this important differential diagnosis. Embryos who are destined to fail because of anomalies should not respond to any therapy.

General Management of Early Pregnancy Dysfunction

Figure 1 illustrates a suggested flowsheet for managing early pregnancy dysfunction. It can be observed that emphasis is placed on identification of the site of the disorder. The diagnostic methods presently available encompass history, physical examination, biochemical measures, ultrasonography, and, more recently, Doppler techniques.

The safety of drugs in early trimester therapy is of legitimate concern since this is the time period when most malformations can happen. This was emphasized in the late 1970s when progestin use was suggested to be associated with the VACTREL syndrome and virilization of the female fetus [1]. Based on these suggestions the Food and Drug Administration (FDA)

- IF PREGNANCY SUSPECTED : measure hCG
- if positive : perform vaginal ultrasound (TVS)

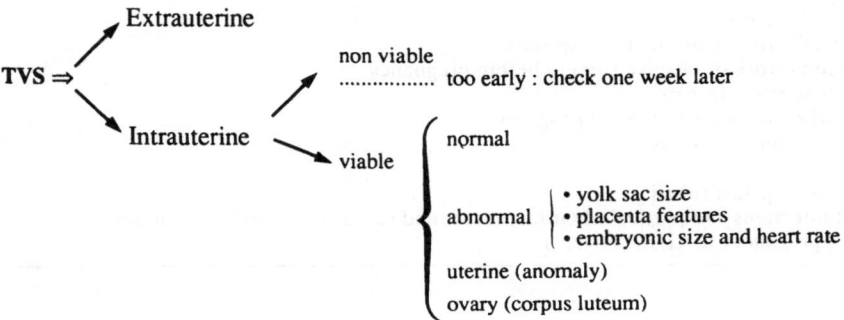

Fig. 1. Flowchart for diagnosing early pregnancy dysfunction

in the United States has prohibited the use of these drugs in early pregnancy. However, more recent studies have indicated that natural progesterone and its close derivatives are not teratogenic. Therefore these drugs are still currently used for treating corpus luteum dysfunction through early pregnancy.

Preconceptional Prevention of EPD

The least frustrating and best approach is preconceptual prevention. By correcting a specific condition the patient apporaches pregnancy with a sense of cautious optimism which may also help for the success of the pregnancy. The patient must be aware of any problem which has to be identified and corrected (Table 1). Thus, as all that can be done is done, there must be acceptance of whatever the outcome may be. Moreover, there is also a certainty that by this approach there is no risk for the developing embryo of being exposed to medications during early pregnancy. Though it is prominent, this approach is unfortunately also the least used. Management should be the same for patients who are attempting to get pregnant for the first time. Obviously, the knowledge of any endocrinologic reproductive problems is paramount (see Chap. 15). Table 2 illustrates the various measures that may be proposed to correct these disorders.

Existing Systemic Diseases and General Medications

A woman should be able to become pregnant if she is healthy and really wishes to do so. Any systemic disease should be diagnosed and treated in

Table 1. Preconceptional prevention: general aspects

Correct and treat systemic illnesses, treat collagen disorders and autoimmune disease
Control diabetes mellitus
Stop medications if possible
Change and adjust medications if needed
Avoid irradiation
Avoid adverse environmental exposure
Do dental work or elective surgery before pregnancy
Eliminate stress factors
Help with an adequate dietary program
Avoid strenuous exercise
Stop smoking
Immunize against rubella
Treat infections and particularly urinary tract and sexually transmitted diseases
Avoid pregnancy if AIDS carrier

Table 2. Preconceptional prevention: reproductive aspects

Favor regular cycles: improve ovulation, diagnose any dysfunction
Treat hyperandrogenism and polycystic ovaries
Correct thyroid disorders
Correct adrenal disorders
Lower hyperprolactinemia
Treat endometriosis
Correct uterine anomalies
Wait 3 months after stopping oral contraception
Wait 2 months after removing IUD
Wait 3 months after a previous pregnancy loss
Improve sperm quality
Avoid male toxin exposure

order to lower the risk for early pregnancy dysfunction. Patients who suffer from systemic disorders should have their medications evaluated and the least harmful for pregnancy selected. If this is done properly, then patients should be given time to adjust to them in order to avoid side effects. In particular, psychiatric, antiepileptic, or antihypertensive drugs should receive special attention. The intake of tranquilizers and any other similar medications should be minimized. However, only a few drugs are clearly teratogenic, and most are agents which have no real indication.

It has been shown that in cases of diabetes mellitus preconceptional control with insulin leading to strict normoglycemia reduces the rate of congenital malformations. Excessive dietary measures and strenuous exercise must also be avoided. Recent studies have shown that an adverse environment, including cigarette smoking, is associated with a high rate of early second trimester pregnancy wastage. It has been demonstrated that the early placental function, as expressed by xenobiotic metabolic enzyme activity and hormone secretion, is affected by exposure to tobacco, fumes, neuroleptics, tranquilizers, and opiates (see Chaps. 7, 15). It is evident that exposure of the patient to irradiation (hysterosalpingogram or any X-ray during the second half of the cycle) must be avoided.

A sizeable proportion of patients take medications inadvertently before the first delay of menses; others consume them regularly (e.g., tranquilizers, opiates, or alcohol). Preconceptional rehabilitation should be applied with proper help and guidance. The enormous responsibility of becoming pregnant while using drugs should be emphasized, and proper contraception should be advocated until the problem is solved. If, however, the patient becomes pregnant while still taking drugs, this influence is clearly a high risk, and the pregnancy should be followed very carefully from the beginning.

Reproductive System Abnormalities

Regular cycles with ovulation at midcycle are optimal for conception and are likely to be associated with low early pregnancy dysfunction rates. In contrast, patients with irregular, too long or too short cycles have a lower rate of conception and are likely to have a higher rate of early pregnancy dysfunction.

Disorders related to thyroid and adrenal conditions should be corrected promptly, and only after that should pregnancy be allowed. It has been shown that certain fertility drugs not only correct infertility disorders, but may also lower the rate of early pregnancy dysfunction. This is the case in the treatment of polycystic ovaries. Furthermore, patients with a short luteal phase are at risk for subsequent pregnancy loss. Ovulation induction with gonadotropins and in vitro fertilization/embryo transfer (IVF/ET) procedures also have an increased abortion rate. Therefore, in these cases, luteal supplementation with either human chorionic gonadotropin (hCG) or progesterone should be carried out. Patients with known uterine synechia or large septum should be operated on before attempting pregnancy.

The need for a delay after oral contraception or the removing of an intrauterine device (IUD) is not emphasized enough. In addition, it is mandatory to wait a couple of months before contemplating a new conception after a previous pregnancy failure.

Male exposure to some toxins was recently suggested as a cause of increased rates of congenital malformations. The involvement of the male partner obviously needs further investigation. Whether the patient is immunized against rubella should be checked and, if necessary, she should be given immunization, a simple but very important preventive measure. Nowadays, with the widespread of AIDS, preventive testing of the couple to avoid vertical transmission should also be contemplated in high-risk groups.

We suggest that a preconceptual check list should be given to the patient by the gynecologist; this could be of great help in identifying the problems, and preventive measures could be subsequently applied if needed.

Table 3. Therapy for preventing and treating early pregnancy dysfunction

1. Local	Bed rest
	Avoidance of intercourse
	Cerclage
2. General	Gonadotropins (+hCG) and closely related molecules
	Progesterone
	Corticosteroids
	Dopamine agonists
	Aspirin + dipyridamole
	Heparin
	Reduction of stress factors
3. Immune therapy	

Postconceptional Prevention of EPD

Preconceptual prevention of EPD involves identification of patients who are at risk of pregnancy loss in the very early stages of pregnancy. The management includes either pre- and post-, or only postconceptional measures. Treatment will be needed from the earliest stages of pregnancy on, and such a pregnancy is thus considered as being at high risk. This condition is obviously less favorable than preconceptional prevention.

Existing Systemic Disease and General Medications

Adequate control of systemic disease can lower the risk of abortion. Tight control of diabetes mellitus, through diet and insulin therapy before and throughout pregnancy, and avoidance of an adverse environment are other illustrative examples. Avoidance of smoking, alcohol, and drugs is always more difficult to achieve at once, while it is feasible in the prepregnancy period.

Reproductive System Abnormalities

The decision whether to treat or not will often be made after conception has occurred. In documented short luteal phase, progesterone replacement therapy can be initiated early. The prophylactic value of progesterone for patients at risk of early pregnancy loss was recently examined in a prospective randomized study [2]. A clear benefit of progesterone was demonstrated: the abortion rate was only 3/50 in the treated group as compared to 14/50 when no supplementation was given. Corticosteroid treatment of female embryos with 21 hydroxylase deficiency can prevent the formation of anatomic defects of the external genitalia [3]. Cerclage in the late 1st trimester is also most useful in reducing the rate of pregnancy loss associated with incompetent cervix; it is also valuable in cases of HA [4].

Ad Hoc Treatment of Early Pregnancy Dysfunction

The third possibility is ad hoc intervention when the problem is identified and already ongoing. This is the most difficult and obviously least desirable but sometimes a necessary approach. Treatment is often not based on identifying a specific etiology as is the case in pre- or sometimes in post-conceptual prevention (Table 3). Therefore, the approach is essentially empiric and based on clinical symptoms, ultrasound, and assay of biologic fluids. Treatment is always an immediate necessity because the patient is anxious. In these cases the outcome remains uncertain since the approach is not disorder specific. This makes the patient even more apprehensive.

Diagnostic Clues for Early Pregnancy Dysfunction

Vaginal bleeding is the most important clinical symptom of a threatening abortion. The frequency of first-trimester vaginal bleeding among 670 women aborting with chromosomally normal conceptuses and 219 women with chromosomally abnormal losses was compared with that of 3089 women delivered at term [5]. Vaginal bleeding early in gestation was predictive of pregnancy outcome in that moderate or heavy bleeding was associated with a fourfold risk of loss of either chromosomally normal or abnormal conceptions. Spotting or slight bleeding was associated with a 2.7-fold risk of loss of a chromosomally normal conception, but not with the loss of chromosomally abnormal conceptions. The magnitude of the problem is great since 15% of women have vaginal bleeding in the first trimester, and 30%–50% of these will subsequently abort.

Since the causes of early pregnancy dysfunction are numerous, it seems likely that therapy could be more successful if treatment could be targeted rather than using empiric methods as is presently done. The use of ultrasound techniques could be very useful. Development of embryo- specific markers of well-being are probably the step in the right direction (see Chap. 20).

The current forms of therapy for early pregnancy dysfunction are rather limited. Table 3 lists the treatment modalities normally employed. Once early pregnancy dysfunction is diagnosed, the cornerstone of management has, for many years, been immediate bed rest for a few days until vaginal bleeding and pain subside, with continued rest for 3 or more days after the symptoms have stopped. There is a standard recommendation to avoid intercourse during that time. It is assumed that uterine contractions are produced by seminal prostaglandins and during orgasm which may lead to placental separation. This is effective in cases of premature labor and is widely utilized for both the prevention and treatment of this condition (together with tocolytics). In some early cases, these two measures are enough to stop bleeding and to help pregnancy progress with increasing circulating progesterone levels.

Medical Treatment

Progesterone

Older studies demonstrated the indispensability of corpus luteum integrity for the maintenance of early pregnancy before 8 weeks. Studies performed on patients after ovum donation have also demonstrated that progesterone together with estradiol are the only hormones required to support pregnancy. Isolated estradiol deficiency has not been considered as an etiologic factor in pregnancy loss.

Progesterone can be administered by various routes: i.m., orally, or through vaginal suppositories. It has not been established which is better, natural progesterone or its close derivatives such as dihydrogesterone. Some authors advocate the use of the vaginal route because of reliable absorption and consistent plasma levels. The optimal plasma levels are probably those of the late luteal phase or early normal pregnancy, i.e., 20–25 ng/ml. Actual tissue levels are probably most important, and, in order to achieve elevated local levels, higher peripheral levels might be required as recently suggested [6].

Certain progestins such as medroxyprogesterone acetate have a negative feedback effect on the corpus luteum function and thus lower progesterone output. This, together with their potential teratogenic effect, precludes their use. Recently the concept that progesterone values <15 ng/ml are associated with inevitable abortion was challenged [7]: in one study, progesterone replacement prevented abortion in 70% of treated cases (19 out of 27). However, the issue is not settled, and there is no firm belief as regards progesterone action. However, progesterone can effectively reduce the HA rate when combined with bed rest.

Indications for Therapy

Absolute. The main indication is clear-cut for corpus luteum deficiency. In these cases treatment is continued until 8–9 weeks, when the luteoplacental shift takes place. On the same basis, replacement is needed following corpus luteum cyst rupture or removal in early pregnancy. Low progesterone levels, lower than 15 ng/ml in early pregnancy, are an indication for treatment [8]. Finally, progesterone replacement is also indicated following embryo replacement in patients who have had ovum donation.

Relative. Therapy is also indicated in cases of early pregnancy bleeding or abdominal pain related to uterine contractions associated with a viable embryo. One cannot predict when bleeding will be of sufficient magnitude to induce abortion. There are also patients who are pregnant with an IUD in place and will continue the pregnancy after removal of the device or even if the device cannot be removed. Progesterone may also be included as a preventive measure in the following cases: after IVF/ET, HA, multiple gestations, death of one twin followed by bleeding, and in cases of pregnancy following inadvertent exposure to gonadotropin-releasing hormone (GnRH) analogues. Lastly, patients with a fibroid or septate uterus may also benefit from such a therapy because of possible incorrect implantation.

Mechanism of Progesterone Action

The mechanisms by which progesterone exerts its beneficial effects are probably multiple and not well known. They could include the following:

1. Progesterone is responsible for the formation of the secretory endometrium with associated decidual reaction and secretion of PEG, pregnancy-associated plasma protein A (PAPP-A), and other proteins (Chap. 2).
2. It helps to stabilize the lysosomes present in endometrial cells, thereby preventing menstruation [9].
3. It prevents phospholipase C activation, thus blocking prostaglandin release.
4. Progesterone may help in the process of tubal transport.
5. It may also act as an immunosuppressor: it lowers the levels of interleukin 2, a monokine and a mediator of transplantion rejection, and increases those of PAPP-A, an immunomodulator.
6. It diminishes uterine irritability and blocks uterine contractions perhaps by lowering the density of oxytocin reception [10]. These contractions are especially prone to develop in cases of bleeding due to the separation of decidua or placenta from the uterus. Recent work [11] has demonstrated that the abortificient RU-486 leads to abortion by forming a retrodecidual and placental hematoma which is the first sign of pregnancy demise.
7. Progesterone acts as an antiestrogen: it blocks estrogen action, lowers endometrial estradiol density, stimulates the conversion of estradiol to estrone by 17β-OH-dehydrogenase, and favors estrogen sulfation [12].
8. A direct effect of progesterone on placental function is also possible. In physiologic concentrations progesterone reduces hCG production and secretion, perhaps leading to the physiologic plateau in hCG secretion which is noted at 8–9 weeks of gestation.
9. Progesterone promotes prostaglandin E_2 (PGE_2) production by placental macrophages [13].

In summary, the local and general effects of progesterone are very complex. New modalities of administration should be studied, in particular, pulses mimicking the in vivo conditions would be of interest: progesterone is secreted by the placenta and the corpus luteum in such a mode (see Chap. 8).

Human Chorionic Gonadotropin

The efficacy of hCG in managing threatening abortion, either isolated or habitual, was reported earlier in noncontrolled studies. Recently, a placebo-controlled study [14] showed encouraging results in a group of HA patients in whom no specific reason for the problem was found or who had low progesterone levels. According to this study, the results of a large multicenter study with a conclusive answer should be forthcoming shortly.

Indications for Therapy

Absolute. hCG is a widely used agent for promoting fertility. Because of its resemblance in form and function to luteinizing hormone (LH), hCG is used to induce ovulation at midcycle. It is effective in supporting corpus luteum function and correcting corpus luteum deficiency. A number of controlled studies [15] have shown that hCG administration during the luteal phase increases pregnancy rates, especially following gonadotropin-induced ovulation. It may also reduce the rate of very early pregnancy loss.

Relative. There is a great similarity between the suggested role of progesterone and that of hCG. The latter probably has a role in the treatment of HA. Its effectivity in increasing pregnancy rate following assisted reproductive technologies is not settled, though it is used for such a purpose in many centers. This mode of therapy could be useful following the death of one twin. Patients with threatening abortion might benefit from hCG treatment. However, it prevents the use of serum hCG assay for monitoring the pregnancy status. Measurement of free β-hCG can overcome this problem. Lastly, there could be cases where the placenta apparently secretes low amounts of hCG: such pregnancies could be rescued by hCG administration. It must be remembered in this respect that abortion may be related to poor implantation or a defective placenta, but not necessarily to a defective embryo (see Chap. 16).

Administration

The usual form of administration of hCG is the i.m. route and the scheme of administration is highly variable. Injection of 2500 units every 3 days, repeated three times is the preferred modality during the luteal phase, after a 5000- or 10000-units shot given at midcycle [16]. Several other regimens have also been advocated, such as 1000 units or 1500 units every few days. In early pregnancy the use of 5000 units hCG twice a week is advocated. In addition, newer forms of administration should also be contemplated. We have demonstrated that the pattern of hCG secretion during early pregnancy is pulsatile and we have reported that the origin of this pulsatile mode is placental (see Chap. 8). Therefore, a pulsatile administration of the hormone could be evaluated for the developing pregnancy at risk. Such a mode of administration is currently utilized for ovulation induction and should be given in the luteal phase and in early pregnancy.

Mechanism of Action

The mechanisms by which hCG supports early pregnancy are not clear. Some hypotheses can, however, be put forward:

1. It supports the corpus luteum and can prolong its lifespan for 1 week by binding to LH receptors. Consequently, progesterone and estradiol levels increase.
2. Recently specific binding sites for hCG were identified in the endometrium [17]: in vitro, epithelial cells primed by medroxyprogesterone acetate increased their prolactin production.
3. Prior to implantation the embryo can produce hCG which could serve as a signal for maternal recognition [18].
4. Its immunosuppressive effect is not clearly demonstrated since this is produced only at very high concentrations.
5. Recently specific binding sites for hCG were found at the trophoblast cell membrane, confirming previous studies demonstrating that hCG stimulates adenylate cyclase locally, a strong indication for a paracrine role of hCG [19].

Dopamine Agonists

The value of dopamine agonists in correcting hyperprolactinemia is well established. This condition leads to menstrual abnormalities including anovulation and luteal phase defect. Patients with corpus luteum defect caused by high prolactin levels benefit from dopamine agonists which promptly correct the condition [20]. During pregnancy, the decidua is an important source of prolactin. However, this is not affected by exposure to dopamine agnoists. We have recently found that prolactin added to dynamic placental cultures reduced hCG secretion by explants. Through a paracrine influence (increase in local progesterone secretion?) (see Chap. 8) [21], prolactin was also shown to increase myometrial contractility. Such an activity may cause bleeding and placental separation.

Treatment by dopamine agonists in early pregnancy is not settled. Pregnancies associated with hyperprolactinemia have a substantial rate of spontaneous abortion. Some advocate the use of dopamine agonists for habitual aborters, others use it only to support early pregnancy in hyperprolactinemic patients. The earlier concern raised regarding the teratogenicity of such medications is now settled as no adverse effect was detected in a large population.

Anticoagulants and Immune Therapy

Anticoagulants are used in cases of HA (primary or secondary) and are effective in preventing severe growth retardation and fetal death later in gestation (see Chap. 26) [22]. These medications are started towards the end of the first trimester and are continued until the third trimester. Among them, low-dose aspirin (100 mg/day), which favors prostacycline over thromboxane production, is useful, together with dypiridamole 225 mg/day [23]. Others [24] have advocated the use of heparin 5000 units daily for treating secondary HA. The drug may help prevent the formation of placental vascular thrombosis.

Corticosteroids were suggested to be useful in the treatment of habitual aborters. They may be effective either alone or in combination with anti-coagulants. The immunosuppressive effect of corticoids may aid in pregnancy maintenance. The use of these drugs is not settled, and studies are currently being conducted to evaluated their effectiveness. The doses could be 10–30 mg prednisone a day. In certain cases treatment with very high i.v. doses of the steroid has been shown to prevent pregnancy loss. Morevoer, for women with antisperm antibodies short high doses, i.e., 96 mg methyl prednisolone could prevent the recurrence of abortion [25]. Very recently, early corticoid administration proved effective in preventing the virilization of the female fetus in cases of 21 hydroxylase deficiency [3]. This is an example of a well-directed specific therapy for the first-trimester embryo.

Gonadotropin

Patients with polycystic ovaries have increased early pregnancy losses due to fertilization of aged gametes [26]. This is effectively reduced by the use of ovulation induction with Metrodin, a gonadotropin containing only follicle-stimulating hormone (FSH) [27, 28]. We also have some personal experience in treating HA patients with gonadotropins. We have a case report of a patient previously treated with every available therapy (including husband and pooled leucocytes) for nine consecutive losses and who was given gonadotropins for ovulation induction. Conception followed and the patient delivered healthy twins at term. Could this be last-chance treatment? The rationale behind this approach could be that adequate ovulation and luteal phase lead to good embryonal development. Multiple pregnancies are sometimes the sole chance for accurate gamete selection. This is also supported by the finding that in IVF the transfer of more than one embryo increases the chances of success for pregnancy.

Other Supportive Therapies

A whole variety of other drugs were utilized to support early pregnancy at risk. Most of them were reported only sporadically and no controlled studies were carried out. Among them are antibiotics such as tetracycline which are used to treat suspected chlamydial infection [29]. Other compounds are serotonin blockers, multiple vitamins, trace elements, and homeopathic drugs. There is no scientific reason for their use.

Surgical Means to Prevent and Treat Early Pregnancy Dysfunction

Surgical means are used in the presence of intrauterine abnormalities such as synechia, polyps, submucous myomas, and large septum which may lead to EPD or loss. The diagnosis should be made by hysterosalpingogram followed by hysteroscopy between pregnancies. Treatment can be planned according to the severity of the disorder, i.e., hysterocopical removal, dilation and curettage, or surgery. In cases of Asherman's syndrome (severe intrauterine adhesions), balloon or IUD insertion followed by administration of large doses of estrogens is to be preferred. Cervical lesions (tears, conizations, etc.) induce a risk of abortion. Cerclage may be proposed once the viability of the embryo is established. In addition, it has been suggested that habitual aborters may benefit from such a procedure as a preventive measure. Finally, the decision to carry out such a cerclage procedure may be taken if cervical shortening or opening is observed. The best way to document such possibilities is by the use of transvaginal ultrasound when the length of the cervix and the degree of internal os opening can be accurately determined.

Unavoidable Early Pregnancy Dysfunction

The last category is the inactive approach where the disorder is not treatable and therefore it is left to its destiny to abort alone or to be evacuated by surgical means. This is the least favorable category, and it should be explained to the patient that the situation is unavoidable, but that the perspective for a future pregnancy remains fair.

All the available information on the aborting pregnancy should be recorded. This is especially important because there is a strong tendency to forget the circumstances surrounding the event. The information regarding this pregnancy should be available in order to be able to perhaps manage the next pregnancy better, or even to apply preconceptional measures if it is at all possible.

Until recently it was customary to carry out dilation and curettage in all patients with incomplete or missed abortion. However, this practice has changed, and nowadays early pregnancies with declining hCG levels are left to abort spontaneously. Obviously, monitoring of the serum hCG levels and ultrasound examination are required to confirm that the uterine cavity has been completely emptied. In cases of missed abortion lasting more than 4 weeks, a coagulation profile is required.

The decision whether to terminate a high-risk pregnancy has to be evaluated on an individual basis. In some cases the decision is clear-cut because of the proven teratogenicity of ingested drugs. We must not forget that the retinoic acid derivatives are teratogenic for a long time after they are taken.

References

1. Herbst AL (1981) Diethyl stilbesterol and other sex hormones during pregnancy. Obstet Gynecol 58 Suppl 35
2. Check JW, Wu C, Adelson H (1985) Decreased abortions in HMG-induced pregnancies with prophylactic progesterone therapy. Int J Fertil 30(3):45–47
3. Speiser PW, Laforgia N, Kato K, Pareira J, Khan R et al. (1990) First trimester prenatal treatmet and molecular genetic diagnosis of congenital adrenal hyperplasia (21-hydroxylase deficiency). J Clin Endocrinol Metab 70:838–848
4. Edmonds DK (1988) Use of cervical cerclage in patients with recurrent first trimester abortion. In: Beard RW, Sharp F (eds) Early pregnancy loss mechanisms and treatment. Springer, Berlin Heidelberg New York, pp 411–416
5. Strobino BA, Pantel-Silverman J (1987) First-trimester vaginal bleeding and the loss of chromosomally normal and abnormal conceptions. Am J Obstet Gynecol 157:1150–1154
6. Dickey RP (1984) Evaluation and management of threatened habitual first trimester abortion. Adv Clin Obstet Gynecol 2:7
7. Check JW, Winkel CA, Check ML (1990) Abortion rate in progesterone treated women presenting initially with low trimester serum progesterone levels. Am J Gyn Health 4:33–34
8. Huang K (1986) The primary treatment of luteal phase inadequacy: progesterone versus clomiphene citrate. Am J Obstet Gynecol 155:824–830
9. Henzl MR, Smith RE, Voost G, Tyler ET (1972) Lysosomal comcept of menstrual bleeding in humans. J Clin Endocrinol Metab 34:860–864
10. Nisselson R, Flouret G, Hechter O (1978) Opposing effects of estradiol and progesterone on oxytocin receptors in rabbit uterus. Proc Natl Acad Sci USA 75: 204
11. Shalev J, Menashe J, Blankstein J, Pariente G, Nebel L, Serr DM (1989) Mechanism of abortion by RU 486. International meeting on pregnancy loss, Tel-Aviv, 1989
12. Tseng L, Gurpide E (1975) Effect of progestins on estradiol receptor levels in human endometrium. J Clin Endocrinol Metab 41:402–406
13. Yagel S, Hurwitz A, Rosen B, Keizer (1987) Progesterone enhancement of prostaglandin E_2 production by fetal placental macrophages. Am J Reprod Immunol Microbiol 14:45–48
14. Harrison RF (1988) Early recurrent pregnancy failure: treatment with human chorionic gonadotropin. In: Beard RW, Sharp F (eds) Early pregnancy loss mechanisms and treatment. Springer, Berlin Heidelberg New York, pp 421–428
15. Jovich JL, Edirisinghe WR, Cummins JM (1991) Evaluation of luteal support therapy in a randomized controlled study within gamete intrafallopian transfer program. Fertil Steril 55:131–139
16. Buvat J, Marcolin J, Juetterd C, Herbault JC, Louvet AL, Dehaene JL (1990) Luteal support after luteinizing hormone-releasing hormone agonists for in vitro fertilization: superiority of human chorionic gonadotropin over oral progesterone. Fertil Steril 53:490–495
17. Levin JH, Tonetta RA, Lobo RA (1990) HCG enhances progestin stimulation of prolactin production by human endometrial stromal cells in culture: evidence for trophoblast-endometrial paracrine interaction. 38th Annual Meeting of the Pacific Coast Fertility Society, 1990
18. Hay DL, Lopata A (1989) Chorionic gonadotropin secretion by human embryos in vitro. J Clin Endocrinol Metab 67:1322–1324
19. Menon KMJ, Jaffe RB (1973) Chorionic gonadotropin-sensitive adenyl cyclase in human term placenta. J Clin Endocrinol Metab 36:1104–1107
20. Crosingani PG, Ferrari C, Scarduelli C, Picciotti MC, Caldara R, Malinverni A (1981) Spontaneous and induced pregnancies in hyperprolactinemic women. Obstet Gynecol 58:708–713

21. Barnea ER, Fares, F, Kol S (1989) Stimulatory effect of prolactin on human placental progesterone secretion at term in vitro: possible inhibitory effect on oestradiol secretion. Placenta 10: 37–43
22. McIntyre JA, Coulam CB, Faulk WP (1989) Recurrent spontaneous abortion. Am J Reprod Immunol 21: 100–104
23. Walsh SW (1989) Low-dose aspirin: treatment for the imbalance of increased thromboxane and decreased prostacyclin in preeclampsia. Am J Perinatol 6:124–132
24. Gleicher N (1986) Pregnancy and autoimmunity. Acta Haematol 76:68–77
25. Bronson RA (1990) Sperm antibodies. Immunol Allergy Clin North Am 10:165–184
26. Homburg R, Armar NA, Eshel A, Adams J, Jacobs HS (1988) The influence of serum luteinzing hormone concentrations on ovulation, conception, and early pregnancy loss in patients with polycystic ovary syndrome. Br Med J 297:1024–1026
27. Remorgida V, Venturini PL, Anserini P, Salerno E, de Ceccol (1991) Use of combined exogenous gonadotropins and pulsatile gonadotropin-releasing hormone in patients with polycystic ovarian disease. Fertil Steril 55:61–65
28. Homburg R, Eshel A, Armar NA, Tucker M, Mason PW, Adams J, Kilborn J, Sutherland JA, Jacobs HS (1989) One hundred pregnancies after treatment with pulsatile gonadotropin hormone releasing hormone to induce ovulation. Br Med J 298:809
29. Center for Disease Control (1985) Chlamydia trachomatis. MMWR 34 (Suppl 3)

25 Surgical Therapy

R.L. Fischer and R.J. Wapner

Introduction

Surgical procedures on the fetus are currently available in the second and third trimesters. Intrauterine fetal blood transfusions, as well as percutaneous drainage of effusions and decompression of cysts, are performed at most perinatal centers. Other procedures, such as correction of diaphragmatic hernias and placement of diversion catheters in hydrocephalus and obstructed urinary tracts are more experimental and are performed in only a few select centers. Although performance of such procedures during the first trimester has theoretical advantages, experience is nonexistent. The advent of embryoscopy, as reviewed in Chap. 22, may offer surgical access in the first trimester (e.g., umbilical blood sampling and gene therapy); however, embryoscopy is still in its infancy. Substantially improved ultrasonography may also spark attempts at first-trimester invasive fetal treatments, but at present this feasibility remains unknown.

Presently, first trimester surgery is limited to a relatively few ultrasound-guided procedures. Two of these, multifetal pregnancy reduction and removal of a retained intrauterine device, seek to improve perinatal outcome by removing potential sources of preterm labor and delivery. A third technique, selective termination of a genetically or anatomically abnormal fetus within a multiple gestation, offers patients a management alternative to total pregnancy termination or birth of a handicapped child. This chapter will explore the history, technique, and perinatal risks of these procedures.

Multifetal Pregnancy Reduction

The spontaneous occurrence of multiple gestations is relatively uncommon. In the United States, the incidence of naturally occurring twins is 1.05% – 1.35%, while spontaneous triplets are encountered every 6000–10000 pregnancies [1]. With the advent of increasingly sophisticated ovulation induction techniques, rates of multiple gestations have risen markedly. It is estimated that the rate of a twin gestation is 5%–10% with clomiphene citrate, and as high as 10%–40% with human menopausal gonadotropin (hMG) [2, 3]. Three or even more fetuses may occur in 2%–3% of hMG inductions, resulting in a condition some have referred to as high-order or grand multiple gestations. In vitro fertilization (IVF) and gamete intra-fallopian transfer (GIFT), with the transfer of numerous preembryos or gametes, also contribute to an increased occurrence of high-order multiple gestations.

Multifetal pregnancies present an increased risk of both maternal and perinatal complications, including preterm labor and delivery, pre-eclampsia, prolonged bedrest and hospitalization, deep venous thromboses, post-partum hemorrage, and the need for Cesarean section. Additionally, there are social and financial considerations for parents of multiple gestations which are further compounded when one or more of their children are handicapped as a result of extreme prematurity.

The perinatal morbidity and mortality of three or more fetuses is difficult to determine accurately as the numbers in reported series are relatively small and are subject to reporting bias (Table 1). While large series of grand multiple gestations are lacking, it appears that gestational age at delivery is inversely proportional to the number of fetuses [4, 5]. The potential long-term morbidity of severe prematurity is well known and includes bronchopulmonary dysplasia, intraventricular hemorrhage, hydro-cephalus, retrolental fibroplasia, and developmental delay.

Until recently, the couple with a grand multiple gestation faced a serious dilemma: continue a pregnancy with almost certain adverse outcome, or terminate the entire gestation. For couples with long-standing infertility, this decision was especially painful.

As reproductive technology became more frequently utilized and such high-order multiple gestations became more frequent, efforts were made to improve outcome in these grand multiple gestations by reducing the number of viable fetuses. Early work was dedicated to developing the safest approach for such procedures. Initially, Dumez and Oury [6] reported 15 cases of pregnancy reduction using transcervical gestational sac aspiration at 8–11 weeks. A 50-ml syringe was linked to a suction curette which was introduced through the cervix. Under continuous ultrasound guidance, the catheter was placed just under the most accessible embryo. Sudden suction was applied, and the gestational sac with the contained embryo was aspirated. The trophoblast was left in situ. The initial report [6] was encouraging, with a miscarriage rate of only 13.3%.

Concerned about the hypothetical risks of hemorrhage and infection, other investigators began utilizing transabdominal techniques previously applied to 2nd-trimester twins containing a genetically abnormal or anomolous fetus. Known by various terms, such as selective reduction, selective termination, selective survival, and selective birth, these procedures entailed termination of the abnormal twin while allowing the normal one to remain in utero. Earlier techniques included intracardiac needle

Table 1. Pregnancy outcome of published series of multiple gestations

Reference	Fetuses	Patients (n)	Gestational age at delivery								Uncorrected PNM (per 1000)	Dates
			<28 Weeks (n)	(%)	28–31 Weeks (n)	(%)	32–36 Weeks (n)	(%)	≥37 Weeks (n)	(%)		
Caspi et al. [4]	Twins	21	3	14	2	10	7	33	9	43	167	1968–1975
Keith et al. [63]a,c	Twins	588	24	4	31	5	140	24	372	63	72	1971–1975
Medearis et al. [64]b	Twins	2831	105	4	145	5	800	28	1781	63	116	1972–1976
Persson and Grennert [65]a	Twins	136	0	0	2	2	32	24	102	75	44	1973–1978
Australia IVF Collaborative Group [66]	Twins	169	8	5	10	6	63	38	86	52	N/A	1979–1988
Savona-Ventura and Grech [67]a	Twins	190	7	4	11	6	44	23	128	67	98	1963–1985
Itzkowic [68]	Triplets	59	6	10	8	15	29	49	15	25	232	1946–1976
Syrop and Varner [69]	Triplets	20	3	15	[12			60]	5	25	216	1946–1983
Michlewitz et al. [70]	Triplets	15	1	7	0	0	12	80	2	13	133	1954–1976
Daw [71]	Triplets	14	1	7	0	0	8	57	5	36	310	1958–1977
Holcberg et al. [72]	Triplets	31	5	16	6	19	16	52	4	13	312	1960–1979
Caspi et al. [4]d	Triplets	5	0	0	2	40	3	60	0	0	200	1968–1975
Ron-El et al. [73]d	Triplets	19	[7			37]	[12			63]	N/A	1970–1978
Australia IVF Collaborative Group [66]	Triplets	32	1	3	11	36	18	58	1	3	N/A	1979–1985
Newman et al. [74]	Triplets	198	10	5	36	18	127	64	24	12	66	1985–1986
Vervliet et al. [75]	Triplets	15	3	17	3	20	10	67	1	7	N/A	1985–1988
Caspi et al. [4]d	Quadruplets	3	1	33	1	33	0	0	1	33	583	1968–1975
Ron-El et al. [73]d	Quadruplets	6	0	0	2	33	1	17	3	50	N/A	1970–1978
Lipitz et al. [76]	Quadruplets	8	2	25	0	0	6	75	0	0	313	1975–1989
Gonen et al. [77]	Quadruplets	5	0	0	5	100	0	0	0	0	100	1978–1988
Vervliet et al. [75]	Quadruplets	6	2	33	1	17	3	50	0	0	N/A	1985–1988
Collins and Bleyl [78]	Quadruplets	71	9	13	25	35	34	48	3	4	147	1980–1989
Lipitz et al. [76]	Quintuplets	2	0	0	2	100	0	0	0	0	0	1975–1989
Gonen et al. [77]	Quintuplets	1	0	0	1	100	0	0	0	0	0	1978–1988
Lipitz et al. [76]	Sextuplets	1	0	0	1	100	0	0	0	0	167	1975–1989

Note: includes series providing gestational age at delivery.

a Data modified to fit gestational age grouping.

b Gestational age at delivery for "uncomplicated" twin pregnancies only. PNM refers to all 3594 twin pregnancies from 1972 to 1976.

c Gestational age at delivery not stated in 21 twin pregnancies.

d Studies from same hospital with overlapping periods (1970–1975). Studies may be overlapping.

puncture with either traumatic disruption [7], fetal exsanguination [8], air embolization [9], or calcium gluconate injection [10], as well as hysterotomy [11] and fetoscopically guided umbilical vein puncture with air embolization [12, 13].

Since the above procedures sought specifically to target a genetically or anatomically abnormal fetus for termination, the terms "selective reduction/termination" were considered appropriate. However, when utilized in a nonselective manner in an attempt to improve overall perinatal outcome, these designations were felt to be inaccurate and misleading [14]. Rather, the term "multifetal pregnancy reduction" (MFPR) was recommended as a more accurate description of these procedures.

The initial use of the transabdominal approach for MFPR occurred in 1986, when Kanhai et al. [15] reported a quintuplet to twin reduction at 10 weeks of gestation by needling the cardiac region of each embryo with a 20-gauge spinal needle. Although the procedure had to be repeated on one embryo 1 week later due to continued cardiac activity, the patient carried uneventfully to term, delivering two healthy infants.

These initial reports were followed by numerous other series of pregnancy reductions performed by various techniques. Table 2 includes all series of first-trimester MFPR involving five or more patients. With the exception of the earlier reports of Dumez and Oury [6] and a few transcervical attempts by Berkowitz and Lynch [14], most investigators employed the technique of fetal pericardiac potassium chloride injection. The quantity of potassium chloride required ranged from 0.2–4.5 ml of a 2 mEq/ml potassium chloride solution. No cases of maternal toxicity from the potassium chloride were observed, and there were no reports of maternal disseminated intravascular coagualtion, despite the prolonged retention of the demised fetus(es). Antibiotics were occasionally employed prophylactically [16, 17], while the lack of use was not associated with an increased infection rate [18].

The loss rate associated with MFPR ranged from 0% to 40%. The largest series reported was the multicenter registry of Dumez et al. [19] which contained 399 cases. The early loss rate occurring within 4 weeks of the procedure was 6%, while losses after 4 weeks (late losses) were 12%. The total loss rate was 17%, with a higher loss rate (both early and late) noted in pregnancies starting with five or more fetuses. The gestational age at delivery of continuing pregnancies was inversely related to both the starting and final number of fetuses. When categorized by the latter, the average gestational age at delivery of triplets was 33 weeks, for twins 35.7 weeks, and for singletons 38 weeks, which appears to be consistent with previously published norms [4].

In most series, the route for MFPR by intrathoracic potassium chloride injection was transabdominal, usually with one puncture per fetus terminated. This technique employed a 20- or 22-gauge needle, which was directed into the fetal thorax. Successful fetal entry, often facilitated by a quick and sudden motion, could be confirmed by observing "passive fetal

Table 2. Pregnancy outcome of published series of multifetal pregnancy reductions

Reference	Year	Patients (n)	Route	Technique	Range (weeks)	Fetuses post reduction	Pregnancy loss <24 weeks (n) (% of total)		On-going preg-nas	Gestational age at delivery (weeks)						Comments
							(n)	(% of total)		28–31 (n) (%)		32–36 (n) (%)		≥37 (n) (%)		
Dumez and Oury [6]	1986	15	Transcervical	Gestational sac aspiration	8–11	Triplets 1 / Twins 10 / Single 4	2	13.3	4	1	11.1	4	44.4	4	44.4	1 vanished twin
Berkowitz et al. [24]	1988	3	Transcervical	Gestational sac aspiration	9–10	Twins 2	1	33.3[a]	0	0	0	0	0	2	100	[a] One pregnancy loss due to hemorrage during reduction
		9	Transabdominal	Intrathoracic KCl	9–13	Triplets 1 / Twins 8	3	33.3	0	1	16.7	5	83.3	0	0	One fetus lost after PUBS at 22 weeks
Shalev et al. [21]	1989	10	Transabdominal	Intrathoracic KCl	10–13	Quads 1 / Triplets 3 / Twins 6	4	40	0	1	16.7	2	33.3[b]	3	50	[b] Spontaneous abortion of one embryo, delivery of remaining fetus at 35 weeks
		10	Transvaginal	Intrathoracic KCl	8–11	Triplets 2 / Twins 8	1	10	1	0	0	2	25.0	6	75.0	
Evans et al. [79]	1990	2	Transabdominal	Traumatic cardiac disruption	9–12	Triplets 1 / Twins 21	5	22.7	0	2	11.8	15	88.2 >35 weeks		>35 weeks	Spontaneous abortion of one embryo
		20	Transabdominal	Intrathoracic KCl												
Gonen et al. [77]	1990	6	Transvaginal	Intrathoracic KCl	8–13	Triplets 2 / Twins 2 / Single 2	1	16.7	2	0	0	2	66.7	1	33.3	One patient with both transabdominal and transvaginal reduction

Study	Year	No.	Approach	Technique	Weeks	Plurality									Comments
Lynch et al. [16]	1990	85	Transabdominal	Intrathoracic KCl	10.5–13[d]	Triplets 4 / Twins 80 / Single 1	8	9.4	32	4	8.8	25	55.6	16	35.6	[d] One reduction at 9.5 weeks
Tabsh [17]	1990	40	Transabdominal	Intrathoracic KCl	11–13[e]	Triplets 1 / Twins 38 / Single 1	0	0	12	2 7 <33 weeks / 16 57 33–36 weeks				10	36	[e] One reduction at 15 weeks
Wapner et al. [18]	1990	46	Transabdominal	Intrathoracic KCl	9–13	Triplets 1	0	0	0	1[f]	100	0	0	0	0	[f] Two infants died from prematurity, one infant with IVH
						Twins 32	1	3.1	0	1	3.2	13	41.9	17	54.8	
						Single 13	1	7.7	0	0	0	3	25.0	9	75.0	
Dumez et al. [19]	1991	399	Transabdominal	Intrathoracic KCl	9–14	Quads 1 / Triplets 9 / Twins 262 / Twins 38	17 (6% < 4 weeks, 12% ≥ 4 weeks)		89	Triplets – average 33 weeks / Twins – average 35.7 weeks / Single – average 38.0 weeks						Multicenter registry (may include [16, 18, 24, 79])

PUBS = Percutaneous umbilical blood sampling; IVH = intraventricular hemorrhage.

movement" in response to movement of the needle hub [20]. The fetus(es) chosen for termination was based on technical accessibility by the spinal needle. In most cases, the fetus overlying the cervical os was avoided to prevent potential rupture of a necrotic sac and subsequent ascending infection. Initially, the needle was left in place long enough to demonstrate cardiac asystole and then removed. However, owing to numerous reports of immediate resumption of cardiac activity, most authors elected to leave the needle in place for at least 2–3 min after asystole was observed in the event that additional potassium chloride might be required [16, 18]. Additionally, after death was confirmed, some investigators withdrew the needle slightly and removed amniotic fluid, which reduced the amount of tissue to be resorbed [17, 18].

An alternate technique occasionally employed for MFPR utilized the transvaginal approach. Initially, a 16- or 18-gauge needle was directed under transvaginal ultrasound guidance into the gestational sac. A smaller-gauge needle was then inserted through the larger one (double-needle technique) and then directed into the fetal thorax. In one small series of 20 patients, the loss rate was 10% with the transvaginal approach, compared to 40% with the transabdominal route [21]. The primary advantage of the transvaginal approach was the improved ultrasound visualization, leading proponents of this technique to suggest that termination could be accomplished as early as 7–8 weeks [20–22]. However, others recommended delaying MFPR until 12–13 weeks to allow for potential spontaneous reduction of some of the fetuses [23, 24]. As the phenomenon of "vanishing twins" may occur as frequently as 21.2% after fetal heart activity has been documented [25], delaying the technique might avoid unnecessary procedures. Although few performed MFPR after 13 weeks out of concern for the potential increased risk of disseminated intravascular coagulation with greater tissue loads, this has not been observed.

Owing to sporadic reports of delayed resumption of fetal cardiac activity, most authors repeated the ultrasound examination, ranging from 30 min to 1 week following MFPR. Repeating the scan on the day of the procedure would allow for re-injection at a time when identification of the initially injected fetus would be certain.

A widely debated issue has involved the number of fetuses that should be left following MFPR, as well as the number of fetuses that would justify MFPR. Most series of MFPR used twins as their endpoint for reduction, unless otherwise requested by the patient. While morbidity and mortality associated with singletons is lower than for twin gestations, a reduction to a singleton would leave no allowance for spontaneous fetal loss [6, 21, 23]. Additionally, the perinatal risks involved in twin gestations are felt by many to be acceptable (see Table 1), and hence reduction to a singleton may be seen as more of a social or financial decision. A similar rationale was used by those who would not offer MFPR for a twin gestation. Others have honored such requests, albeit relunctantly, on the grounds of the legal right of a woman's choice to abortion [18]. Recently, some have even questioned the need for MFPR on triplets [26], in light of advances in neonatal care and improved perinatal outcome. Although perinatal survival with triplets has dramatically improved, the incidence of perinatal and neonatal complications remains higher than with singleon gestations, despite newer tocolytic therapy and home uterine contraction monitoring. At present, however, where abortion is legal, this issue ultimately remains a personal decision between the patient and her physician.

Neural tube defect (NTD) screening in the remaining fetuses is likely to be affected by MFPR, due to the release of alpha-fetoprotein (aFP) from the deceased fetuses. Grau et al. [27] evaluated 32 women who had undergone first-trimester MFPR to a twin gestation. Twenty-two had maternal serum (MS) aFP screening, and all but one had elevated levels, defined as greater than 4.5 multiples of the median (MOM). The average elevated MS-aFP level was 9.4 MOM, with a range of 5.0–19.0 MOM. No correlation was observed between either the number of fetuses terminated (range, one to three) or the termination-to-testing interval (range, 1–7 weeks), and the the MS-aFP level. Amniocentesis was performed on 27 women, 16 for advanced maternal age and 11 for elevated MS-aFP. Of the 53 amniotic fluid samples (one fetus had an encephalocele and was selectively terminated without amniocentesis), 13 (24.5%) had amniotic fluid (AF) aFP levels greater than 2.0 MOM, and one (2%) was positive for acetylcholinesterase. No fetuses were found to have an NTD at birth. It was concluded that MS-aFP screening following first-trimester MFPR is not useful in the detection of NTD, and that AF-aFP levels must be interpreted with caution. High-resolution ultrasonography in the second trimester would hence appear to be the most accurate means of NTD detection in women undergoing MFPR.

There are a number of actual and theoretical risks associated with MFPR, in addition to entire pregnancy loss. Disseminated intravascular coagulation (DIC) in the mother has been reported in cases of prolonged retention of a spontaneously occurring intrauterine fetal demise both in singleton [28] and multiple gestations [29]. The etiology for the DIC was felt to be secondary to thromboplastin, which was released from the degenerating conception and entered the maternal circulation. In all series of MFPR reported in Table 2, no cases of DIC were observed. This may be due to the relatively small necrotic tissue load following 1st-trimester reductions.

Similarly, thrombotic complications, such as multicystic encephalomalacia, hydranencephaly, renal cortical necrosis, and small bowel atresia, have been reported in fetuses whose co-twins have died [30]. In the majority of cases, these complications occurred in monozygotic and, presumably, monochorionic gestations. Since vascular anastamoses exist in over 85% of monochorionic placentas, thromboplastin from the necrotic conception could cross to the living fetus, leading to DIC and vascular disruption of various organs. In most series of MFPR, when a monochorionic gestation was observed in a grand multiple gestation, either both fetuses were terminated, or the twin gestation was left undisturbed. Since monochorionic gestations have significant inherent perinatal risks, intentionally leaving this situation would appear unwise if avoidable. Golbus et al. [31] reported, among their experience with MFPR, five cases of monochorionic twins in which one fetus was terminated. Four of the co-twins died within 4–12 h. While acute release to tissue thromboplastin may be etiologic, another likely possibility is a sudden hemodynamic shift across anastomotic placental vessels due to the asystolic twin, resulting in hypovolemia, hypotension, and sudden death of the co-twin. The only successful case in the series of Golbus et al. [31] was performed by hysterotomy, which led them to suggest that if one fetus of a monochorionic twin pair had to be terminated, it should be performed either by hysterotomy or umbilical cord ligation.

Boulot et al. [32] reported two separate cases of "anencephaly-like malformations" in remaining fetuses after 1st-trimester MFPR by intra-thoracic hypertonic saline injections. In both cases, "normal" embryos had been observed by transvaginal ultrasound prior to MFPR, although chorionicity was not mentioned. The anomalous fetuses were later identified at the time of repeat ultrasound evaluation at 17–18 weeks. Selective termination was performed on both these fetuses and normal karyotypes identified. The two singleton pregnancies were subsequently delivered at 39 and 32 weeks. While the cause of these anomalous fetuses might be attribut-able to the hypertonic saline or prolonged retention of a dead uterine co-inhabitant, this might also represent a previously unrecognized abnor-mality, as multiple gestations have a higher incidence of encephalocele and anencephaly [33].

Another theoretical concern with MFPR is an alteration of hormone levels as a result of embryo destruction. O'Keane et al. [34] serially followed human chorionic gonadotropin (hCG), estradiol (E-2), and progesterone levels in a pregnancy reduced in stages from quintuplets to twins by intra-embryonic sodium chloride injection at 8.5 and 9.5 weeks. A fourfold decrease in hCG was observed by 12 weeks, to a level below that expected for twins. In contrast, estradiol levels increased, despite the reduction of fetuses that are thought to be responsible for formation of estrogen precursors. Progesterone levels fell slightly by 12 weeks, but were still elevated over normal twin levels. The pregnancy continued uneventfully to 37 weeks. Although this represents only one case, it would appear that by 9.5 weeks the remaining placental/fetal units are able to maintain adequate hormonal levels to sustain the pregnancy.

Intense debate has arisen over the medical/ethical significance of MFPR [35–37]. To some, it represents a significant scientific breakthrough that allows patients with no other hope for normal pregnancy outcome to deliver healthy infants. To others, especially those who oppose elective abortion, it represents an abuse of medical technology with unwarranted destruction of healthy fetuses. Blaskiewicz [38] commented that the perinatal loss rate was far greater with MFPR if one included the fetuses that had been intentionally terminated. Advocates of MFPR countered with the obvious reduction of neonatal morbidity as a result of prolongation of pregnancy, as well as reduced maternal morbidity from grand multiple gestations.

From a legal perspective, if abortion is legal, MFPR may simply represent a "variation of first-trimester termination of pregnancy" [39]. Exceptions may exist. In England, the Voluntary Abortion Act of 1967 defines lawful conditions for terminating a pregnancy. Strictly speaking, MFPR does not terminate the entire pregnancy; hence, the Abortion Act may not apply [40].

All sides appear to agree that the ideal solution to this medical/ethical/ legal dilemma would be to limit the number and extent of grand multiple gestations. For women undergoing ovulation induction, this could be accomplished by close surveillance of estradiol levels, and follicle number and growth. Additionally, limitation of the number of preembryos and oocytes transferred in IVF and GIFT, respectively, would reduce the incidence of such high-order multiple gestations. Toward that end, the Voluntary Licensing Authority in England has placed a limit on embryos and oocytes transferred to three, with four allowable under exceptional clinical conditions [41].

Owing to the relative infancy of MFPR, there are no studies to evaluate its long-term effects on both the surviving fetuses and the mother. Developmental studies on the infants, as well as psychological profiles on women who have terminated wanted fetuses, are necessary to determine the appropriateness of continuing MFPR. The registry established by Dumez et al. [19] will also aid in determining overall loss and prematurity rates. For the present time, it would seem appropriate for individuals experienced in ultrasound-guided procedures to continue to offer MFPR to women with grand multiple gestations in an attempt to reduce ultimate perinatal morbidity and mortality. As with any surgical procedure, the risks and benefits of MFPR, as well as alternative therapies, should be thoroughly discussed with the patients. Ethical and emotional issues should be explored and openly discussed before, during, and even after MFPR has been performed [42].

Selective Termination

With the increased frequency of first-trimester prenatal diagnosis, an increasing number of women are discovering genetic abnormalities by 13 weeks of gestation. In multiple gestations, the identification of one abnormal fetus would allow for "selective" termination of the affected fetus, while leaving the unaffected sibling(s) undisturbed. Mulcahy et al. [43] were among the first to perform first-trimester selective termination in a hemophilia carrier with a twin gestation. Chorionic villus sampling (CVS) at 8.5 weeks revealed discordant sexes. Two weeks later, amniocentesis was performed on the male fetus, and additional aspiration removed 20 ml of bloody fluid containing fetal tissue. Brambati et al. [44] reported a quadruplet-to-twin reduction in a couple who were both carriers for beta-thalassemia. CVS was initially performed on two fetuses at 8 weeks, resulting in the discovery of one doubly heterozygous fetus and one single heterozygote. The first was selectively terminated by intrathoracic potassium chloride, while CVS was performed on the remaining two fetuses. Both were single heterozygotes. A second fetus was then terminated due to favorable location. The remaining two fetuses were delivered at term, and their heterozygote status was confirmed postnatally.

Wapner [18] reported four cases of first-trimester selective termination for genetic abnormalities. Two of the four remaining singleton gestations were delivered at term, one was a small-for-gestational-age infant at 33 weeks, and one pregnancy was lost at 14 weeks. This loss occurred in a monochorionic twin gestation where the affected fetus had a cystic hygroma and a mosaic karyotype of 45 X/46 XY. Karyotype of the other twin following its demise also revealed a similar mosaic pattern that had not been initially identified. Since these four selective terminations were performed up to 3 weeks after CVS, and since reversal of fetal positions has been reported [45], the importance of detailed mapping of both fetal and placental positions at the time of initial genetic testing was stressed to assure that the correct fetus was being subsequently terminated. Alternatively, a repeat CVS utilizing rapid karyotyping of the cytotrophoblast would allow for confirmation of a genetic abnormality within 24 h prior to the scheduled termination.

Technically, first-trimester selective termination is not different from MFPR, except for the necessity for accurate identification of the abnormal fetus. In some cases, the affected fetus may overly the cervical os or be less accessible, possibly increasing the potential risk of the procedure. However, first-trimester selective termination allows the patient a viable alternative to entire pregnancy loss or delivery of an infant with a known genetic or anatomic abnormality.

Retrieval of Retained Intrauterine Device

Although use of the intrauterine device (IUD) has declined in the United States, it is still a commonly used form of birth control in other countries, especially in the Third World. It is estimated that there are 2 million IUD users in the United States, with approximately 50 million users worldwide [46]. With a failure rate of two of three per 100 users per year, a significant number of women will conceive with the IUD in situ.

Both maternal and fetal complications have been reported in pregnancies with a retained IUD. At least 15 cases of maternal death and 219 cases of septic abortion have been reported with the Dalkon Shield, which has since been voluntarily withdrawn [47, 48]. The incidence of spontaneous abortion with a retained IUD in five independent series encompassing 1672 pregnant IUD users remained remarkably constant, ranging from 42.6% to 56.8% [49–53]. The rate of preterm labor ranged from 5% to 17.4%. Additional complications, such as abruptio placentae, septicemia, and fetal demise, have also been observed [54]. It was postulated that higher tissue levels of prostaglandin could predispose to preterm labor, while placental implantation over the imbedded IUD could lead to a higher incidence of abruptio placentae.

The practice of IUD removal at the time of pregnancy recognition has reduced, although not eliminated, the risk of adverse perinatal outcome. The rate of spontaneous abortion following IUD removal fell to 20.3% – 25.7%, while the incidence of preterm labor was also reduced to 2.5% – 4.3% [49, 52, 53]. Foreman et al. [55] found that removal of the IUD in the 1st trimester of pregnancy was not associated with an increase in 2nd trimester fetal loss, whereas retention of the IUD resulted in a tenfold higher risk of spontaneous abortion, intrauterine fetal death, or delivery of an infant weighing less than 750 g.

These and other series have led to the recommendation that the IUD should be removed whenever encountered during pregnancy [46]. In most cases, endometrial blastocyst implantation will cause the nonembedded IUD to be extra-amniotic in location. When the IUD string is visualized, simple traction will usually be successful. However, Shalev et al. [56] reported 13 IUD extractions, six of which were observed sonographically. In three cases in which the IUD was implanted lateral or cephalad to the gestional sac, traction on the string resulted in distortion of the sac. These IUDs were ultimately extracted by "gentle, intermittent traction up to two minutes." In contrast, when two lateral/cephalad IUDs were extracted without ultrasound visualization, both pregnancies aborted within 7 days of the procedure. Eight cases of caudally implanted IUDs were removed without complications, whether or not it was accompanied by ultrasonography.

When the IUD tail cannot be identified at the external os during pregnancy, three possibilities exist: unrecognized expulsion, uterine perforation, or string retraction into the uterine cavity due to growth of the gestation. Koetsawang et al. [53] observed that the incidence of string retraction was related to gestational age. At 8 weeks, 98.8% had a visible tail. This rate fell to 19.1% and 2.1% at 9–12 and >12 weeks, respectively.

A variety of methods for IUD localization in the nongravid state are available to the clinician, including blind probing with a uterine sound or hook, hysteroscopy, anteroposterior and lateral radiographs, or ultrasonography. In pregnancy, however, the latter would appear to be the least invasive and hence most advisable [57].

In the past, when the IUD was identified in the uterine cavity and no tail was visualized, a dilemma existed. The patient could continue the pregnancy, risking serious sequelae to her baby and even herself, or she could terminate the pregnancy. It was out of this dilemma that alternatives were sought.

Stubblefield et al. [58] removed a Copper 7 IUD from a 7-week gestation under ultrasound guidance. After cleansing the vagina and cervix, a tenaculum was placed on the cervix. An alligator forceps with a diameter of 3 mm was then introduced through the cervical canal and was guided under ultrasound visualization to the IUD in the lower uterine segment. The IUD was grasped with the forceps and gently withdrawn. Antibiotics were given prophylactically for 5 days, and the patient ultimately delivered at term without complications. Three additional patients were similarly managed at 7–8 weeks of gestation. All had lower uterine segment IUDs (one Copper 7, one Lippes Loop, and one Progestasert), all were removed uneventfully, and all three delivered at term.

Another series of ultrasound-directed IUD retrievals was reported by Shalev et al. [59] who used a technique similar to the one described above. IUD extraction was successful in all 16 attempts, which occurred at 7–10 weeks of gestation. The IUD position was noted to be "high" in two cases. Of 13 completed pregnancies, four electively terminated, one aborted (anterior IUD), and eight "healthy" infants were delivered at undisclosed gestations.

Wagner et al. [60] compared four IUD removal modalities in 54 patients: blind probing with small forceps, ultrasound-guided extraction, fluoroscopic-guided retrieval, and carbon dioxide hysteroscopic-guided removal. Success rates were 56.3%, 66.7%, 87.5%, and 100%, respectively. Rates of pregnancy "injury" (described as rupture of membranes or hemorrhage resulting in pregnancy loss) occurred in 50.0%, 58.3%, 12.5%, and 22.2% of cases, respectively. Four injuries occurred with fundally implanted IUDs.

Assaf et al. [61] also reported favorable outcomes following removal of retained copper IUDs by carbon dioxide hysteroscopy prior to 12 weeks of gestation. Successful removal occurred in 17 of 21 women. Two patients miscarried within 2 days; the remainder either delivered at term or were ongoing pregnancies.

Owing to the unknown effects of carbon dioxide on the developing embryo, an alternative hysteroscopic technique was proposed by Gustavii et al. [62] who employed saline to distend the uterine cavity. Successful extraction occurred in all seven women who were scheduled for first-trimester termination. "Copious" bleeding was encountered in one patient, but due to the rinsing action of the saline instillation, the bleeding subsided. At the present time, it is unknown whether the light associated with hysteroscopy could have any adverse effects on fetal visual development.

If IUD removal is unsuccessful, the patient should be counselled regarding possible adverse pregnancy outcomes, and termination of pregnancy offered. Although no fetal malformations have been reported in association with a retained IUD [57], deformations can occur. Continuing pregnancies should be followed closely for signs of preterm labor, bleeding, and ascending infection. The latter must be aggressively treated in order to prevent such serious sequelae as septicemia, shock, and maternal death.

Conclusion

The first-trimester invasive procedures described demonstrate that significant manipulation of the early gestation is feasible without significant untoward effects on the pregnancy. While utilization of the presently available procedures is limited, the knowledge gained will be helpful in the future. Although it is unlikely that first-trimester correction of fetal structural malformations will be possible in the forseeable future, other fetal therapy, such as fetal blood sampling and gene insertion, may benefit from these early ultrasound-guided ventures into the first-trimester uterus. Although a great deal of further investigation remains in order to develop more sophisticated techniques of fetal therapy, a small step has been taken in this direction.

References

1. Guttmacher A (1953) The incidence of multiple births in man and some other uniparae. Obstet Gynecol 2:22
2. Schenker JG, Yarkoni S, Granat M (1981) Multiple pregnancies following induction of ovulation. Fertil Steril 35:105–123
3. American College of Obstetricians and Gynecologists (1988) Medical induction of ovulation. ACOG, Washington (Technical bulletin, no 120)
4. Caspi E, Ronen J, Schreyer P, Goldberg MD (1976) The outcome of pregnancy after gonadotrophin therapy. Br J Obstet Gynaecol 83:967–973
5. Gonen R, Heyman E, Asztalos EV, Ohlsson A, Pitson L, Shennan A, Milligan J (1990) The outcome of triplet, quadruplet, and quintuplet pregnancies managed in a perinatal unit: obstetric, neonatal, and follow-up data. Am J Obstet Gynecol 162:454–459
6. Dumez Y, Oury JF (1986) Method for first trimester selective abortion in multiple pregnancy. Contrib Gynecol Obstet 15:50–53
7. Aberg A (1978) Cardiac puncture of fetus with Hurler's disease avoiding abortion of unaffected co-twin. Lancet 2:990–991
8. Kerenyi TD, Chitkara U (1981) Selective birth in twin pregnancy with discordancy for Down's syndrome. N Engl J Med 304:1525–1527
9. Petres R, Redwine F (1981) Selective birth in twin pregnancy (Letter). N Engl J Med 305:1218–1219

10. Antsaklis A, Politis J, Karagiannopoulos C, Kaskarelis D (1984) Selective survival of only the healthy fetus following prenatal diagnosis of thalassaemia major in binovular twin gestation. Prenat Diagn 4:289–296
11. Beck L, Terinde R, Rohrborn G, Claussen U, Gebauer H, Rehder H (1981) Twin pregnancy, abortion of one fetus with Down's syndrome by sectio parva, the other delivered mature and healthy. Eur J Obstet Gynecol Reprod Biol 12:267–269
12. Rodeck CH, Mibashan RS, Abramowicz J, Campbell S (1982) Selective feticide of the affected twin by fetoscopic air embolism. Prenat Diagn 2:189–194
13. Redwine FO, Hays PM (1986) Selective birth. Semin Perinatol 10:73–81
14. Berkowitz RL, Lynch L (1990) Selective reduction: an unfortunate misnomer. Obstet Gynecol 75:873–874
15. Kanhai HHH, van Rijssel EJC, Meerman RJ, Bennebroek Gravenhorst J (1986) Selective termination in quintuplet pregnancy during first trimester (Letter). Lancet 1:1447
16. Lynch L, Berkowitz RL, Chitkara U, Alvarez M (1990) First-trimester transabdominal multifetal pregnancy reduction: a report of 85 cases. Obstet Gynecol 75:735–738
17. Tabsh K (1990) Transabdominal multifetal pregnancy reduction: report of 40 cases. Obstet Gynecol 75:739–741
18. Wapner RJ, Davis GH, Johnson A, Weinblatt VJ, Fischer RL, Jackson LG, Chervenak FA, McCullough LB (1990) Selective reduction of multifetal pregnancies. Lancet 335:90–93
19. Dumez Y, Evans M, Wapner R, Lynch L, Dommergues M, Goldberg J, Johnson M, Golbus M, Berkowitz R (1991) Efficacy of multifetal pregnancy reduction (MFPR): collaborative experience of the world's largest centers. 11th Annual Meeting of the Society of Perinatal Obstetricians, San Francisco
20. Birnholz JC, Dmowski WP, Binor Z, Radwanska E (1987) Selective continuation in gonadotropin-induced multiple pregnancy. Fertil Steril 48:873–876
21. Shalev J, Frenkel Y, Goldenberg M, Shalev E, Lipitz S, Barkai G, Nebel L, Mashiach S (1989) Selective reduction in multiple gestations: pregnancy outcome after transvaginal and transabdominal needle-guided procedures. Fertil Steril 52:416–420
22. Itskovitz J, Boldes R, Thaler I, Bronstein M, Erlik Y, Brandes JM (1989) Transvaginal ultrasonography-guided aspiration of gestational sacs for selective abortion in multiple pregnancy. Am J Obstet Gynecol 160:215–217
23. Evans MI, May M, Drugan A, Fletcher JC, Johnson MP, Sokol RJ (1990) Selective termination: clinical experience and residual risks. Am J Obstet Gynecol 162:1568–1575
24. Berkowitz RL, Lynch L, Chitkara U, Wilkins IA, Mehalek KE, Alvarez E (1988) Selective reduction of multifetal pregnancies in the first trimester. N Engl J Med 318:1043–1047
25. Landy HJ, Weiner S, Corson SL, Batzer FR, Bolognese RJ (1986) The "vanishing twin": ultrasonographic assessment of fetal disappearance in the first trimester. Am J Obstet Gynecol 155:14–19
26. Lorenz J, Terry J (1988) Selective reduction of multifetal pregnancies (Letter). N Engl J Med 319:949–950
27. Grau P, Robinson L, Tabsh K, Crandall BF (1990) Elevated maternal serum alpha-fetoprotein and amniotic fluid alpha-fetoprotein after multifetal pregnancy reduction. Obstet Gynecol 76:1042–1045
28. Pritchard J, Ratnoff O (1955) Studies of fibrinogen and other hemostatic factors in women with intrauterine death and delayed delivery. Surg Gynecol Obstet 101:467–477
29. Skelly H, Marivate M, Norman R, Kenoyer G, Martin R (1982) Consumptive coagulopathy following fetal death in a triplet pregnancy. Am J Obstet Gynecol 142:595–596
30. Hoyme HE, Higginbottom MC, Jones KL (1981) Vascular etiology of disruptive structural defects in monozygotic twins. Pediatrics 67:288–291

31. Golbus MS, Cunningham N, Goldberg JD, Anderson R, Filly R, Callen P (1988) Selective termination of multiple gestations. Am J Med Genet 31:339–348
32. Boulot P, Hedon B, Deschamps F, Laffargue F, Viala JL, Humeau C, Arnal F (1990) Anencephaly-like malformation in surviving twin after embryonic reduction (Letter). Lancet 335:1155–1156
33. Windham G, Sever L (1982) Neural tube defects among twin births. Am J Hum Genet 34:988–998
34. O'Keane JA, Yuen BH, Farquharson DF, Wittmann BK (1988) Endocrine response to selective embryocide in a gonadotropin-induced quintuplet pregnancy. Am J Obstet Gynecol 155:364–367
35. Weiner J (1988) Selective first-trimester termination in octuplet and quadruplet pregnancies: clinical and ethical issues (Letter). Obstet Gynecol 72:821–822
36. Diamond E (1988) Selective reduction of multifetal pregnancies (Letter). N Engl J Med 319:950
37. Zaner R, Boehm F, Hill G (1990) Selective termination in multiple pregnancies: ethical considerations. Fertil Steril 54:203–205
38. Blaskiewicz R (1990) Transabdominal multifetal pregnancy reduction: report of 40 cases (Letter). Obstet Gynecol 76:735–736
39. Hobbins JC (1988) Selective reduction – a perinatal necessity? N Engl J Med 318: 1062–1063
40. Howie PW (1988) Selective reduction in multiple pregnancy. Br Med J 297:433–434
41. Brahams D (1987) Assisted reproduction and selective reduction of pregnancy. Lancet 2:1409–1410
42. American College of Obstetricians and Gynecologists (1991) Multifetal pregnancy reduction and selective fetal termination. ACOG, Washington (Committee opinion, no 94)
43. Mulcahy M, Roberman B, Reid S (1984) Chorion biopsy, cytogenetic diagnosis, and selective termination in a twin pregnancy at risk of haemophilia (Letter). Lancet 2:866–867
44. Brambati B, Formigli L, Tului L, Sindai G (1990) Selective reduction of quadruplet pregnancy at risk of B-thalassaemia. Lancet 336:1325–1326
45. Benke PJ (1982) Twins discordant for Down's syndrome. Clin Genet 21:104–106
46. American College of Obstetricians and Gynecologists (1987) The intrauterine device. ACOG, Washington (Technical bulletin, no 104)
47. Monif GRG, d'Allessandri RM, Khakoo RA, Kluge RM (1977) Fatal sepsis associated with an intrauterine device and pregnancy. South Med J 70:249–250
48. Christian CD (1974) Maternal deaths associated with an intrauterine device. Am J Obstet Gynecol 119:441–444
49. Tatum HJ, Schmidt FH, Jain AK (1976) Management and outcome of pregnancies associated with the Copper T intrauterine contraceptive device. Am J Obstet Gynecol 126:869–879
50. Vessey MP, Doll R, Johnson B, Peto R (1974) Outcome of pregnancy in women using an intrauterine device. Lancet 1:495–498
51. Lewit S (1970) Outcome of pregnancy with intrauterine devices. Contraception 2:47–57
52. Alvior GT Jr (1973) Pregnancy outcome with removal of intrauterine device. Obstet Gynecol 41:894–896
53. Koetsawang S, Rachawat D, Piya-Anant M (1977) Outcome of pregnancy in the presence of intrauterine device. Acta Obstet Gynecol Scand 56:479–482
54. Wiles PJ, Zeiderman AM (1974) Pregnancy complicated by intrauterine contraceptive devices. Obstet Gynecol 44:484–490
55. Foreman H, Stadel BV, Schlesselman S (1981) Intrauterine device usage and fetal loss. Obstet Gynecol 58:669–677
56. Shalev J, Greif M, Ben-Rafael Z, Itzchak Y, Serr DM (1982) Continuous sonographic monitoring of IUD extraction during pregnancy: preliminary report. AJR 139:521–523

57. Fallon JH (1985) Pregnancy with IUD in situ. Kans Med 86:322–324
58. Stubblefield PG, Fuller AF Jr, Foster SC (1988) Ultrasound-guided intrauterine removal of intrauterine contraceptive devices in pregnancy. Obstet Gynecol 72:961–964
59. Shalev E, Edelstein S, Engelhard J, Weiner E, Zuckerman H (1987) Ultrasonically controlled retrieval of an intrauterine contraceptive device (IUCD) in early pregnancy. JCU 15:525–529
60. Wagner H, Schweppe K, Kronholz H, Beller F (1980) Möglichkeiten der Extraktion von Intrauterinpessaren bei eingetretener Schwangerschaft. Med Welt 31:1317–1320
61. Assaf A, El Tagy A, El Kady A, El Agezy H (1988) Hysteroscopic removal of copper-containing intrauterine devices with missing tails during pregnancy. Adv Contracept 4:131–135
62. Gustavii B, Edvall H, Nordenskjöld F (1984) Removal of IUDs in early pregnancy. Acta Obstet Gynecol Scand 63:571–572
63. Keith L, Ellis R, Berger GS, Depp R (1980) The Northwestern University multihospital twin study. I. A description of 588 twin pregnancies and associated pregnancy loss, 1971 to 1975. Am J Obstet Gynecol 138:781–787
64. Medearis AL, Jonas HS, Stockbauer JW, Domke HR (1979) Perinatal deaths in twin pregnancy. Am J Obstet Gynecol 134:413–421
65. Persson PH, Grennert L (1979) Towards a normalization of the outcome of twin pregnancy. Acta Genet Med Gemellol (Roma) 28:341–346
66. Australia IVF Collaborative Group (1988). In-vitro fertilization pregnancies in Australia and New Zealand, 1979–1985. Med J Aust 148:429
67. Savona-Ventura C, Grech ES (1988) Multiple pregnancy in the Maltese population. Int J Gynecol Obstet 26:41–50
68. Itzkowic D (1979) A survey of 59 triplet pregnancies, Br J Obstet Gynaecol 86:23–28
69. Syrop CH, Varner MW (1985) Triplet gestation: maternal and neonatal implications. Acta Genet Med Gemellol (Roma) 34:81–88
70. Michlewitz H, Kennedy J, Kawada C, Kennison R (1981) Triplet pregnancies. J Reprod Med 26:243–246
71. Daw E (1978) Triplet pregnancy. Br J Obstet Gynaecol 85:505–509
72. Holcberg G, Biale Y, Lewenthal H, Insler V (1982) Outcome of pregnancy in 31 triplet gestations. Obstet Gynecol 59:472–476
73. Ron-El R, Caspi E, Schreyer P, Weinraub Z, Arieli S, Goldberg MD (1981) Triplet and quadruplet pregnancies and management. Obstet Gynecol 57:458–463
74. Newman R, Hamer C, Miller M (1989) Outpatient triplet management: a contemporary review. Am J Obstet Gynecol 161: 547–553
75. Vervliet J, de Cleyn K, Renier M, Janssens P, Buytaert P, Gerris J, Delbeke L (1989) Management and outcome of 21 triplet and quadruplet pregnancies. Eur J Obstet Gynecol Reprod Biol 33:61–69
76. Lipitz S, Frenkel Y, Watts C, Ben-Rafael Z, Barkai G, Reichman B (1990) High-order multifetal gestation – management and outcome. Obstet Gynecol 76:215–218
77. Gonen Y, Blankier J, Casper R (1990) Transvaginal ultrasound in selective embryo reduction for multiple pregnancy. Obstet Gynecol 75:720–722
78. Collins M, Bleyl J (1990) Seventy-one quadruplet pregnancies: management and outcome. Am J Obstet Gynecol 162:1384–1392
79. Evans MI, Fletcher JC, Zador IE, Newton BW, Quigg MH, Struyk CD (1988) Selective first-trimester termination in octuplet and quadruplet pregnancies: clinical and ethical issues. Obstet Gynecol 71:289–296

26 Immunologic Therapy
for Recurrent Spontaneous Abortion

C.B. Coulam

The mammalian conceptus represents both an autograft and an allograft to its host mother. To allow an antigenically unique individual to survive as a graft, the mammal mother must adapt her immunologic response during pregnancy. When this adaptive mechanism is disturbed, reproductive problems ensue. The role of immunity in abnormal human pregnancies came under intense investigation with studies of recurrent spontaneous abortion [1]. An immunologic etiology for recurrent spontaneous abortion has been proposed for many couples with otherwise unexplained reproductive wastage [1, 2]. Since the first reports of successful pregnancy following leukocyte transfusions in 1981 [3, 4], interest in the possibility that immunotherapy might prevent recurrent spontaneous abortions has been generated. Therapy is based on the concept that successful pregnancy requires or is markedly enhanced by maternal allogeneic recognition [5]. However, skepticism toward immunotherapy has been growing because evidence supporting the rationale for treatment has been lacking and results of clinical trials using immunotherapy have been conflicting.

Data providing support for the rationale for treatment are rapidly accumulating. However, recent results of clinical trials describing the efficacy of various forms of immunotherapy do not support the notion that immunotherapy with leukocyte immunization is a useful treatment for couples with recurrent spontaneous abortion. It is the purpose of this chapter to present the evidence supporting the rationale for immunotherapy in the treatment of recurrent spontaneous abortion and to summarize the literature describing results of immunotherapy for treatment of recurrent spontaneous abortion.

Rationale for Immunotherapy

Research in reproductive biology has led to the concept that successful pregnancy requires maternal immunologic recognition of the trophoblast [5]. In the absence of such recognition or in the absence of an appropriate response to such recognition, recurrent pregnancy loss will occur. Previously, doubt had been cast on an immunologic cause of recurrent pregnancy loss because of lack of evidence of an immunologic response to the pregnancy by the mother. More recently evidence for such a response is rapidly accumulating in several areas. That maternal humoral immunologic response to pregnancy occurs has been demonstrated by experiments documenting the presence of idiotypic (Ab1) and antiidiotypic (Ab2) antibodies directed toward trophoblastic antigens in normally pregnant women [6]. Immunohistologic studies have shown that placentae from normal pregnancies manifest immunopathologic responses known as villitis which are similar to immunopathologic responses existing in transplanted human organs [7]. Data demonstrating that human pregnancy initiates coagulation and fibrinolysis, just as in other allogeneic reactions such as renal allografts, have been reported [8]. That these immunologic responses are involved in the mechanisms of recurrent spontaneous abortion are supported by the following observations: (a) some individuals suffering from recurrent spontaneous abortion contain Ab1 but not Ab2 directed toward trophoblastic antigens [9]; (b) placentae from recurrent spontaneous abortions show an increased among of villitis compared with normal 1st-trimester placentae in some instances and a decrease in others [10]; (c) and some abnormal pregnancies are associated with increase in coagulation compared to normal pregnancies [6].

Idiotype-Antiidiotype Network

The role of the idiotype-antiidiotype network in immunoregulation and its presence during pregnancy have been described [6]. The results of investigations describing Ab1 and Ab2 in normal and abnormal pregnancies has been used to build a pathophysiological model for recurrent spontaneous abortions.

The Model

In normal pregnancy the mother's immune system recognizes a trophoblastic antigen and produces an Ab1 antibody to the antigen as well as an Ab2 antibody [6]. Ab2 is required for successful pregnancy because it downregulates maternal allogeneic rejection reactions and it forms complexes with Ab1. In the absence of Ab2, the pregnancy is lost. Ab2 can be lacking either because a women is a poor auto-antiidiotype producer or she fails to produce Ab1 and thus is not stimulated to make Ab2. If a woman is a poor auto-antiidiotype producer and produces Ab1 but not Ab2, her Ab1 could bind to the placenta directly or be transported across the placenta via immunoglobin G (IgG) crystallizable fragment (Fc) receptors and enter the fetus. If a woman does not immunologically recognize her pregnancies (as a result of genetic compatibility with trophoblastic antigens), she fails to produce an Ab1 antibody to these trophoblast antigens and thus fails to produce Ab2. In the absence of Ab2, maternal allogeneic reactions are not down-regulated, and innate killer cell responses can attack the blastocyst.

A concept is emerging that the production of protective antibodies occurs in response to an antigen present not only on the trophoblast, but also in seminal plasma [11]. In this way immunization could occur at the time of mating, making the environment at the time of implantation of the blastocyst favorable for successful pregnancy. These data provide a rationale for immunization trials using trophoblast vesicles and seminal plasma in the treatment of recurrent spontaneous abortion [12]. The antigen(s) responsible for stimulating maternal immunologic recognition and production of a protective response is not known. Thorough testing on panels of normal adult and fetal tissues has shown that specificity solely for trophoblast remains elusive. In a WHO-sponsored workshop, only five of 45 submitted monoclonal antibodies (mAbs) showed minimal cross-reactivity with non-trophoblastic cells and were reactive with trophoblast [13]. These five mAbs identified low molecular weight cytokeratins, CD46, placentae alkaline phosphatase (PLAP), an epithelial membrane antigen (HMFG1) and truncated class I major histocompatibility complex (MHC) antigens (W6/32). To support the hypothesis that maternal immunization begins at insemination, the antigen responsible for maternal recognition should be present in semen and should be polymorphic. Trophoblast-lymphocyte cross-reactive (TLX) antigens whose mAbs cross-react with CD46 antigens and PLAP are present in seminal plasma and they are polymorphic [14–16].

Data Supporting the Model

Results from several experiments using sera from normal pregnant women and women who are poor auto-antiidiotype producers support the model of idiotype-antiidiotype responses during pregnancy being important in the outcome or success of that pregnancy [6]. By using the Ab1 isolated from a recurrent aborter who was a poor auto-antiidiotype producer it was possible to demonstrate the existence of an Ab1-Ab2 complex in the blood of normal primigravid women. To do this, the lymphocytotoxic Ab1 from the recurrent aborter was attached to a solid phase matrix and used to absorb the primigravida's serum. The selection bias here is that the aborter's Ab1 must be cytotoxic to the primigravida's husbands lymphocytes. Thus, the primigravida's Ab1-Ab2 complex equilibrium would be dissociated by the solid phase Ab1 binding to the primigravida's Ab2 leaving an excess of the primigravida's Ab1 in the supernate. To obtain the primigravida's Ab2, the solid phase Ab1 column with its bound Ab2 was resuspended and incubated with the primigravida's husband lymphocytes. The Ab2 is competitively removed and recovered in the supernate. It was also demonstrated that the Ab2 isolated from the normal primigravida could neutralize or block the Ab1 cytotoxicity of the poor auto-antiidiotype producer [6]. These observations support the model of unchecked immune responses generated during the pregnancies of a subset of recurrently aborting women.

Another approach using the poor auto-antiidiotype producer involved the production of heterologous Ab2. In this experiment a highly specific rabbit Ab2 was generated to Ab1 from a poor auto-antiidiotype producer and used to study Ab1 cross-reactive idiotypic (CRI) antibodies in the blood of multiparous, normal pregnant, nonpregnant, and poor auto-antiidiotype producers [6]. Specificity of the heterologous Ab2 antibody was confirmed by showing reactivity with the Fab and $F(ab^1)_2$ IgG fragments from the immunizing Ab1. Moreover, the data showed that the Ab2 reagent was negative with the IgG isolated from nontransfused, nontransplanted males, but positive with some normal pregnant and multiparous women as well as with a proportion of the secondary aborting women (recurrent abortions after a live birth or stillbirth). The importance of these observations is that the rabbit Ab2 could detect CRI-Ab1 antitrophoblast antibodies in both gravid and multiparous women, despite the failure to demonstrate the presence of these Ab1 antibodies by other traditional laboratory tests. These findings strengthen the hypothesis that women respond immunologically to allotypic trophoblast antigens and that the idiotype-antiidiotype regulation of these responses is characteristic of normal pregnancies.

Immunopathology of Placentae

For at least 20 years it has been known that placentae from normal pregnancies manifest evidence of immunopathology similar to that shown in transplanted human organs. Normal placentae have villi with histologic evidence of chronic inflammation, known as areas of chronic villitis of unestablished etiology [7]. These areas are composed predominantly of helper (i.e., $CD4^+$) T lymphocytes and activated (i.e., HLA-DR positive) macrophages, suggesting that they are areas of immunopathology. The number of areas of chronic villitis is increased in placentae from pregnancies complicated by infection, intrauterine growth retardation, maternal immunologic disease, and some cases of recurrent spontaneous abortion. Immunochemical studies show that some recurrent spontaneous aborters fail to mount an immunologic response to their pregnancies. These individuals typically present clinically as primary aborters (no pregnancies progressing beyond 20 weeks of gestation). The results of quantitative immunohistologic studies indicate that significantly more villitis areas are found in placentae from secondary aborters compared with normal placentae. Fewer areas of villitis are found in placentae from primary aborters compared with normal placentae. A large difference between the percentage of affected villi in placentae from primary and secondary aborters has been reported [10]. These findings offer an immunopathologic correlate for the clinical diagnosis of recurrent spontaneous abortion, and they support the concept of immunologic hyporesponsiveness in primary and a hyperresponsiveness in secondary aborters.

Hemostasis

Coagulation and fibrinolysis are initiated as a result of allogeneic reactions such as renal allografts [8]. The fetus and its extraembryonic membranes represent an allograft to its host mother. Fibrinogen and many other clotting factors are present in human placentae, but very little fibrin is found in normal chorionic villi. This is because the endothelial surface of fetal stem vessels contains the thrombomodulin anticoagulant pathway. However, this pathway is sensitive to cytokines released during immunologic reactions. For example, interleukin 1 released from activated macrophages causes a down-regulation of thrombomodulin from the endothelial plasma membrane, and this results in the presence of tissue factor on the endothelial surface which activates factor VII and sets in motion the coagulation cascade. Thus, fibrin deposition is found in those villi that make allogeneic contact with maternal tissues in the basal plate, and in areas of villitis of unestablished etiology where tissue factor has been identified on fetal stem vessel endothelial cells [8].

It has been shown that areas of villitis are foci of immunopathology, and that these areas are more common in placentae from secondary recurrent spontaneous aborters [7, 10]. Since there is evidence that cytokines are released in areas of immunopathology, one would predict that placentae from secondary recurrent spontaneous aborters should contain more fetal stem vessels with evidence of coagulation, and that is what has been reported [7, 8, 10]. Thus, it appears that both the immunopathology and coagulation data in placentae from mothers with recurrent spontaneous abortions, particularly secondary recurrent spontaneous abortions, manifest multiple immunologic reactions. It is not clear what caused these reactions, but the most likely candidate is allogeneic responses analogous to that observed in renal transplants.

Diagnosis of Recurrent Pregnancy Loss

To be effective, treatment should be directed toward the cause of recurrent pregnancy loss. Therapy is based on the concept that successful pregnancy requires an appropriate response to allogeneic recognition [5]. Appropriate responses result in the production of auto-antiidiotypic antibodies and the release of interleukin-1. Interleukin-1 down-regulates endothelial thrombomdulin and converts endothelium from a thromboresistant to a thrombogenic state, thus promoting the deposition of fibrin which secures the blastocyst to the implantation site [8].

Since effective treatment of recurrent pregnancy loss related to an immunologic event is dependent upon whether the fetal wastage is the result of lack of maternal immunologic recognition or lack of appropriate immunologic response to the recognition of the conceptus, it is important to distinguish the two entities diagnostically. Tools available for diagnosis are limited and have included HLA typing of mating couples, assaying for maternal antipaternal lymphocytotoxic antibodies, mixed lymphocyte cultures, and tests for maternal autoantibodies.

HLA Typing

Diagnostic tools to identify those individuals who recurrently abort because of lack of maternal immunologic recognition are lacking. In the absence of specific reagents, HLA antigen sharing has been used as a nonspecific predictor of maternal immunologic recognition. Studies describing increased sharing of MHC antigens among couples with a history of recurrent spontaneous abortion compared with normally reproducing couples have been both confirmed and refuted [17]. In support of an association between MHC and reproduction success, reproduction has been found to be better in couples who share fewer HLA antigens compared with those who share two or more, and indexes such as number of children and interchild intervals seem to vary as a function of the degree of HLA sharing among Hutterite couples who do not practice contraception [31].

The significance of shared MHC antigens between partners experiencing recurrent spontaneous abortion is not clear. One interpretation of the role of MHC antigens in human reproduction is that it is not the HLA markers that are important, but it is their association with the trophoblast antigens that generates maternal recognition and protection that is important, and that compatibility between mother and blastocyst results in failure of such recognition. Thus, shared HLA may be an index of shared trophoblast antigens. This interpretation has yet to be proven. Alternatively, reproductive wastage in MHC-sharing couples could be occurring via different (nonimmunologic, i.e., genetic) mechanisms.

Lymphocytotoxic Antibodies

Maternal antipaternal lymphocytotoxic antibodies are rarely detected in women experiencing recurrent spontaneous abortions who have no pregnancies progress beyond 20 weeks of gestation, yet such antibodies are commonly identified in women who recurrently abort after having a livebirth or a stillbirth. These antibodies are broadly reactive with lymphocytes from many different donors, and they are removed by absorption with trophoblast membranes or with tissues that express TLX antigens, such as platelets.

Different tests for lymphocytotoxic antibodies can produce different results on the same sample. For example, complement fixation assays only detect complement-fixing antibodies. Complement-independent assays, such as antibody-dependent cellular cytotoxicity, are sometimes positive when complement fixation assays are negative, and vice versa. In addition, lymphocytotoxic antibodies in recurrent aborters are sometimes present with inhibitors that are anticomplementary. This problem can be overcome if the lymphocytes are washed free of the patient's serum after incubation. This step removes the inhibitors and allows complement fixation to proceed. A similar inhibitor has been identified in a complement-independent assay.

There is much speculation about the nature of the immunogen responsible for the antibody production as well as controversy regarding the significance of the presence of the antibodies. Since maternal antipaternal antibodies are associated with a history of successful pregnancies in 20% of primiparas and 35%–64% of multiparas, the presence of antipaternal cytotoxic antibodies in women experiencing recurrent spontaneous abortion is difficult to interpret. Several studies have reported that preexisting antipaternal cytotoxic antibodies predict good outcome in subsequent pregnancies [19].

Despite the controversy regarding the significance of the presence of maternal antipaternal lymphocytotoxic antibodies in predicting pregnancy outcome, the presence of such antibodies is usually considered to be a contraindication to the use of leukocyte immunotherapy. There are practical reasons, such as anaphylaxis, to expect that patients with preformed lymphocytotoxic antibodies could respond adversely to immunotherapy with leukocytes. More information is needed about the character of these antibodies.

Mixed Lymphocyte Culture Reactions

One of the best examples of cell-mediated immunity in pregnancy is the mixed lymphocyte culture (MLC) reaction. In this reaction, lymphocytes from two different individuals are mixed together and cultured under conditions that permit measurement of their DNA metabolism, which is an index of the intensity with which one cell reacts to the other. As a model of pregnancy, the father's cells are sufficiently irradiated to disallow their ability to respond immunologically, but their capacity to stimulate the mother's lymphocytes is retained. This is called a one-way MLC reaction.

One-way MLC reactions between pregnant wife and husband proceed normally in the mother's nonpregnant or third-party serum or plasma, but are blocked by plasma from the pregnant mother. The most widely accepted explanation for this blockage of allogeneic recognition is that it is caused by blocking antibodies in the pregnant mother's blood. These antibodies have specificities for allotypic trophoblast antigens on the father's lymphocytes. They can be eluted from placentae and blocked by allogeneic and xenogeneic antibodies to human trophoblast membrane antigens.

Convincing support of a role for blocking factors in pregnancy comes from clinical investigations done in unexplained spontaneous abortions. Many women with primary recurrent spontaneous abortions do not produce a factor in their blood which blocks in vitro models of cell-mediated immunity between lymphocytes from the mating partners. In some cases, this deficiency can be overcome by immunizing the wife with lymphocytes [4].

The production of blocking factors subsequent to immunotherapy has been reported as a good prognostic indicator for pregnancy success. The meaningfulness of the finding of a blocking factor to MLC is not clear. Furthermore, the absence of this blocking factor occurs in normally responding women when they are not pregnant. Thus, the utility of assaying for an MLC-blocking factor in the evaluation of women with recurrent spontaneous abortion is questionable.

Autoantibodies

The study of autoantibodies is an emerging area in pregnancy research. An increase in antiphospholipid antibodies has been reported in women experiencing reproductive wastage. Antiphospholipid antibodies are autoantibodies detected by standard tests for syphilis, lupus anticoagulant test (activated partial thromboplastin time, kaolin clotting time, Russell's viper venom), or by enzyme-linked immunoabsorbent assay for anticardiolipin. Retrospective studies associate the lupus anticoagulant and anticardiolipin antibodies with recurrent fetal loss [20–21]. Affected women usually have high titers of antibodies over an extended period of time. The mechanism of fetal loss in women with these autoantibodies is unknown, although placental vessel thrombosis resulting in infarction and placental insufficiency has been proposed.

While many unanswered questions remain regarding the nature of autoantibodies directed toward phospholipid, the presence of maternal antibodies to phospholipids is of obvious clinical importance in the management of high-risk patients. Individuals with high antibody titers to phospholipids would not be expected to respond to immunotherapy using leukocytes or other immunogens to stimulate an immune response.

Review of Immunotherapy

Recurrent spontaneous abortion can result from lack of allogeneic recognition or inappropriate response to such recognition. The therapeutic approach to each of these is different.

Lack of Allogeneic Recognition

Diagnostic tools to identify those individuals who recurrently abort because of lack of maternal allogeneic recognition are deficient. In the absence of specific reagents, HLA antigen sharing between mating pairs has been used as a nonspecific predictor of maternal allogeneic recognition [17]. While sharing two or more HLA antigens between mates has occurred more frequently among populations of couples experiencing recurrent spontaneous abortion compared with normally reproducing couples, HLA sharing between individual couples with a history of recurrent spontaneous abortion has not been predictive of pregnancy outcome with or without immunotherapy [17]. Recently, progesterone receptor-positive CD8$^+$ lymphocytes have been shown to be a marker for allogeneic recognition [18]. In the presence of progesterone, these cells secrete a 34-kDa suppressor substance the concentration of which has been associated with pregnancy outcome. If this observation can be repeated, progesterone receptor-positive CD8$^+$ lymphocytes induced by allogeneic stimulation will be a welcome diagnostic tool to identify those individuals who recurrently abort because of a failure in allogeneic recognition.

Since the first reports in 1981 [3, 4], indicating successful pregnancies could occur after immunotherapy, several immunization regimens have been introduced [19, 23]. In addition to using leukocytes from either partners or third-party donors for immunization as originally reported, immunization with trophoblast vesicles and donor seminal plasma have also been reported [12]. The intravenous route has been used for both third-party and partner immunization with leukocytes as well as trophoblast vesicles; the intradermal route for inoculation with a partner's leukocytes; and the vaginal route for immunization with seminal plasma. Table 1 outlines and Table 2 summarizes the results of these studies which have appeared in the literature [21]. All of the different models of immunization are associated with a viable birth rate of 70%–80% in the uncontrolled observations. Three placebo-controlled randomized trials using paternal leukocytes have been published and the results of these trials are presented in Table 3 [24]. Only one of the three controlled studies published has indicated a beneficial effect of paternal leukocytes compared with control groups [23]. This study reported 78% successful pregnancies in recurrently aborting women immunized with paternal leukocytes, compared with 37% in recurrently

Table 1. Immunotherapy review

Selection criteria	RSA type	Immunogen	Route	Outcome		
				Total (n)	Delivered (n)	(%)
3+ SAB Ab neg	1	Spouse MNC	i.v., s.c., i.d.	16	12	75
		Autologous MNC	×1	23	8	35
	2	Spouse MNC		6	5	82
		Autologous MNC		4	2	50
Pre-CON Post-CON	1 + 2	Spouse MNC	i.v., s.c., i.d.	77	59	77
		Autologous MNC	×1	75	53	71
3+ SAB	1 + 2	Spouse MNC	i.v., s.c., i.d.	74	43	58
		None	×1	25	8	32
3+ SAB	1 + 2	Spouse MNC	i.d. ×2 pre-pregnancy	37	26	70
3+ SAB	1 + 2	Spouse MNC	i.d. ×2	53	42	79
		None	pre-pregnancy	49	28	57
3+ SAB 2 or > SH HLA	1	3–5 Buffy coats	i.v. ×2 s.c. ×9	26	23	88
3+ SAB	1	Third Party MNC Buffy coats	Transfusion ×3	26	23	88
	2	Suffy coats		21	11	52
3+ SAB Ab Neg	1	Trophoblast membranes	100 mg i.v. ×1	11	6	55
	2			10	9	90
3+ SAB	1	Spouse MNC	40 ml ID. ×2	20	12	60
3+ SAB	1 + 2	Spouse MNC	i.v., s.c., i.d.	39	21	54
		Autologous MNC	×1	49	23	47
		Third party MNC		11	5	45
3+ SAB	1 + 2	Spouse MNC	i.v., s.c., i.d.	21	13	62
		Saline	×1	25	19	76

RSA, recurrent spontaneous abortion; SAB, spontaneous abortion; MNC, mononuclear cells; Ab, Antibody; CON, conception; SH HLA, shared HLA antigens.

aborting women receiving their own leukocytes. Results from the other placebo-controlled trials revealed viable pregnancy rates in the control groups of 58%–67% compared with 60%–75% in the treated groups [24]. Differences between control and treated groups were not significant. The significance of a placebo effect is important since supporting psychotherapy alone for recurrent abortion has claimed similar success rates. Epidemiologic studies have also indicated that the chance for a successful pregnancy after three previous miscarriages without treatment is 60%.

The controversy that now exists regarding interpretation of the efficacy of immunizations with paternal leukocytes revolves around the results in the nontreated groups [24]. The results in the nontreated groups will depend upon the number of confounders for risk of recurrence of pregnancy loss in

Table 2. Summary of the results of studies in the literature describing the efficacy (live birth rate) of various types of immunotherapy for women with recurrent spontaneous abortion

Immunotherapy	Pregnancy outcome		
		Live births	
	(n)	(n)	(%)
Mononuclear cells			
Spouse	549	369	67
Third party	84	62	74
Autologous	76	33	45
Trophoblast membranes	21	15	71
Seminal plasma or placebo	44	22	50
Immunglobulin i.v.	25	20	80
Saline	25	19	76
None	99	44	44

Table 3. Summary of results of pregnancy outcome from three placebo-controlled clinical trials using paternal mononuclear cells to immunize women with a history of recurrent spontaneous abortion

Reference	Spouse MNC		Control		
		Delivered		Delivered	Type
	(n)	(%)	(n)	(%)	
Mowbray et al. [23]	22	77	27	37	Autologous MNC
Cauchi (in [24])	21	62	25	76	Saline
Ho (in [24])	39	62	49	47	Autologous MNC
	82	62	101	51	

the women studied. A number of potential factors that predict an increased risk of another pregnancy loss have been reported. These have included maternal age, number of previous abortions or other adverse pregnancy outcome, history of infertility or delay in conception, previous live-born child, sharing of HLA antigens among mating couples, and the presence of lymphocytotoxic antibodies, antinuclear antibodies, and mixed lymphocyte reaction-blocking factor in maternal serum. Of these potential factors, only number of previous miscarriages, history of adverse pregnancy outcomes, and a history of subfertility have been confirmed. Maternal age as a risk factor for recurrence of pregnancy wastage has been both confirmed and refuted. Since the populations in the three controlled studies using paternal leukocytes for treatment of recurrent pregnancy loss were not homogeneous for risk factors for recurrence, pregnancy outcomes reported could represent population differences. To answer the question of who would benefit from paternal leukocyte immunization in the treatment of recurrent spontaneous abortion, a study in which the treated and control populations are homogeneous with respect to known risk factors for recurrence of pregnancy loss should be performed.

Not only has the efficacy of leukocyte immunization treatment of recurrent pregnancy loss been questioned, but also the safety of leukocyte transfusions remains controversial. The association of acquired immunodeficiency syndrome with transfusion of blood products has complicated the use of this therapy. Other considerations include risks of maternal sensitization to leukocyte, erythrocyte, or platelet antigens; perinatal host-versus-graft disease; and transfusion-related risks including infections. Intrauterine growth retardation has been reported after intradermal immunization with paternal leukocytes.

Alternative approaches to inducing the necessary maternal immunologic responses are being evaluated. One study uses isolated trophoblast microvillous membranes [12] and another employs seminal plasma [25]. Both of these materials contain TLX and CD46 antigens, and mAbs to both of the antigens have been shown to be cross-reactive. Another approach is to provide passive immunity by administering gamma-globulin [26, 27]. These treatments have the advantage of lacking intact cells and nuclear material. If they are effective, they have the potential of offering couples suffering from lack of allogenic recognition a safer option for treatment.

Inappropriate Response to Immunologic Recognition

In addition to maternal allogeneic recognition of the blastocyst, successful pregnancy requires an appropriate immunologic response [1]. An inappropriate response to maternal immunologic recognition is associated with pregnancy failure. Since the blastocyst is made up of gene products from both the mother and the father, it can act as an autograft as well as an allograft. If the maternal immunologic response to the blastocyst is not appropriate and the problem involves a maternally derived antigen, an autoimmune cause of recurrent spontaneous abortion is diagnosed. Alternatively, if failed or inappropriate response to a paternally derived antigen is found, an alloimmune cause of recurrent spontaneous abortion is diagnosed.

Treatment for Inappropriate Autoimmune Recognition. Antiphospholipid antibodies are associated with autoimmune conditions characterized by vascular abnormalities [20–22, 28, 29]. Individuals in whom these antibodies are observed have adverse pregnancy outcomes, including maternal thrombosis, severe preeclampsia, and early and late fetal death. Because the vascular lesions and the range of pregnancy outcomes are similar to those observed in abortions associated with poor auto-antiidiotype production, a role for hemostasis and fibrinolysis in the pathophysiology of recurrent pregnancy loss has been postulated. There are very few published cases of viable infants delivered of untreated women who were clearly documented to have lupus anticoagulant activity prior to or early in pregnancy. Preterm delivery is one approach to managing pregnancies complicated by circulatory anticoagulants, but this is usually preceded by attempts to decrease antibody titers to phospholipids by immunosuppression. Experience has accumulated with the use of prednisone, and more refractory patients have been given azathioprine or cyclophosphamide. Aspirin is given concomitantly with immunosuppression. This is done to inhibit thromboxane synthetase with the expectation of reducing the threat of small-vessel thrombosis. Heparin has also been used to treat women with antiphospholipid antibody syndrome to combat the thrombotic events which have been reported in these women. No studies with aspirin or immunosuppressive agents have been controlled, and the results are descriptive. Definition of the efficacy of the various treatments awaits results of randomized clinical trials.

Treatment for Inappropriate Alloimmune Recognition. Women who are diagnosed as poor auto-antiidiotype producers are not treated with immunization, but no agreement exists about how these patients should be treated. Twice-daily subcutaneous injections of 5000 units of heparin have been used successfully to prevent secondary abortions in women displaying the presence of antipaternal lymphocytotoxic antibodies [25]. The rationale for this therapy is drawn from laboratory studies of sera from poor auto-antiidiotype producers before and after solid-phase heparin absorption. This procedure revealed significant decreases in lymphocytotoxic antibody activity after heparin absorption. Clinical trials are in progress where the poor auto-antiidiotype producers are treated with intravenous immunoglobulin (IVIG) which is thought to contain both Ab1 and Ab2 antibodies to trophoblast antigens [26].

Risks of Immunotherapy

Several immunotherapeutic regimens have been used, but, as with all new forms of treatment, the benefits must be weighed against the risks. Untreated pregnancies that progress beyond the 28th week in women with a history of recurrent miscarriage have more complications than pregnancies in women without such a history. In a study of 97 mothers who gave birth to 118 babies, 72 had a viable first pregnancy before experiencing three or more miscarriages. This group of secondary aborters had an increase in small-for-gestational-age (SGA) infants, preterm births, and perinatal mortality rates. Others have also reported an increase in preterm delivery,

SGA infants, and congenital anomalies among offspring of secondary aborters [30].

The risks to the fetus in primary aborters is, by definition, 100% loss. No published data have described pregnancy outcome among women who experienced three or more spontaneous abortions and then produced a viable infant without treatment. Information describing perinatal and pediatric follow-up of children of treated primary aborters shows that complications are fewer than those expected from untreated secondary aborters, but more than those expected from normally reproducing couples. Clearly, controlled investigation with better experimentally and clinically defined objectives is required before objective judgments can be drawn as to optimal therapeutic procedures.

Conclusion and Future Directions

Recurrent spontaneous abortion occurs in 2%–5% of reproducing couples, making this complication of pregnancy as frequent or more frequent than pregnancy complicated by diabetes. Efficacious treatment for recurrent spontaneous abortion is needed. However, the consensus of the Workshop on Unification of Immunotherapy Protocols held during the Tenth Annual Meeting of the American Society for Immunology of Reproduction in Chicago, Illinois, June 20–23, 1990, was: (a) proof of efficacious treatment for recurrent spontaneous abortion does not exist; and (b) controversial data describing the efficacy of immunotherapy in the treatment of recurrent spontaneous abortion have been reported. Until these important questions can be answered, it was recommended that immunotherapy should be performed only at centers that have studies designed to help elucidate what is efficacious treatment for recurrent spontaneous abortion.

To answer the question of safe and efficacious treatment for recurrent spontaneous abortion future studies must focus on identifying markers that will predict those patients that will be expected to respond to immunotherapy. as well as identifying the specific immunomodulator.

References

1. Coulam CB (1986) Unexplained recurrent pregnancy loss (Epilogue). Clin Obstet Gynecol 29:999–1004
2. Hill JA (1990) Immunological mechanisms of pregnancy maintenance and failure: a critique of theories and therapy. Am J Reprod Immunol 22:33–41
3. Taylor C, Faulk WP (1981) Prevention of recurrent abortion with leukocyte transfusions. Lancet 2:68–70

4. Beer AE, Quebbeman JF, Ayers JW, Haines RF (1981) Major histocompatibility complex antigens, maternal and paternal immune responses and chronic babitual abortions in humans. Am J Obstet Gynecol 141:987–999
5. Faulk WP, McIntyre JA (1981) Trophoblast survival. Transplantation 31:1–5
6. Torry DS, Faulk WP, McIntyre JA (1989) Regulation of immunity to extraembryonic antigens in human pregnancy. Am J Reprod Immunol 21:76–81
7. Labarrere CA, McIntyre JA, Faulk WP (1990) Immunohistologic evidence that villitis in human normal term placentas is an immunologic lesion. Am J Obstet Gynecol 162:515–522
8. Faulk WP (1989) Placental fibrin. Am J Reprod Immunol 19:132–135
9. Chaouat G, Lanker D (1988) Vaccination against spontaneous abortion in mice: protection against spontaneous abortion by preimmunization with an antiidiotypic antibody. Am J Reprod Immunol Microbiol 16:146–150
10. Labarrere CA (1989) Allogeneic recognition and rejection reactions in the placenta. Am J Reprod Immunol 21:94–99
11. Thaler CJ (1989) Immunologic role for seminal plasma in insemination and pregnancy. Am J Reprod Immunol 21:147–150
12. Johnson PM, Chia KV, Hart CA, Griffith HB, Francis WJA (1988) Trophoblast membrane infusion for unexplained recurrent miscarriage. Br J Obstet Gynecol 95:342–347
13. Anderson DJ, Johnson PM, Alexander NJ, Jones WR, Griffin PD (1987) Monoclonal antibodies to human trophoblast and sperm antigens: report of two WHO-sponsored workshops. J Reprod Immunol 10:231–257
14. Ballard LL, Bora NS, Yu GH, Atkinson JP (1988) Biochemical characterization of membrane cofactor protein of the complement system. J Immunol 141:3923–3929
15. McLaughlin PJ, Lewis-Jones DI, Hutchinson GE, Johnson PM (1986) Placental-type alkaline phosphatase in human seminal plasma from fertile and infertile males. Fertil Steril 46:934–937
16. Kajino T, Torry DS, McIntyre JA, Faulk WP (1988) Trophoblast antigens in human seminal plasma. Am J Reprod Immunol Microbiol 17:91–95
17. Coulam CB, Moore SB, O'Fallon WM (1987) Association between major histocompatibility antigen and reproductive performance. Am J Reprod Immunol Microbiol 14:54–58
18. Szekeres-Bartho J, Kinsky R, Varga P, Chaouat G (1990) Immunomodulatory effects of progesterone and the outcome of pregnancy. Am J Reprod Immunol 22:95–96
19. Cowchock SF, Smith JB, David S, Sher J, Batzer F, Corson S (1990) Paternal mononuclear cell immunization therapy for repeated miscarriage: predictive variables for pregnancy success. Am J Reprod Immunol 22:12–17
20. De Wolf F, Carreras LO, Moermann P, et al. (1982) Decided vasculopathy and extensive placental infarction in a patient with repeated thromboembolic accidents, recurrent fetal loss, and a lupus anticoagulant. Am J Obstet Gynecol 142:829–834
21. Lockshin MD, Druzin MC, Goei S, Quamar T, Magid MS, Jovanovic L, Ferenc M (1985) Antibody to cardiolipin as a predictor of fetal distress or death in pregnant patients with systemic lupus erythematosus. N Engl J Med 313:1322–1326
22. Lubbe WF, Butler WS, Palmer SJ, Liggins GC (1983) Fetal survival after prednisolone suppression of natal lupus anticoagulant. Lancet 1:1361–1363
23. Mowbray JF, Gibbings C, Liddell H, Reginald PW, Underwood JL, Beard RW (1985) Controlled trial of treatment of recurrent spontaneous abortion by immunization with paternal cells. Lancet 1:941–943
24. Coulam CB, Smith JB, Branch DW (1991) Report on the Workshop for the Unification of Immunotherapy Protocols held at the 10th Annual Meeting of the Society for Immunology of Reproduction, June 20–23, 1990, Chicago, Illinois. Am J Reprod Immunol (in press)
25. McIntyre JA, Coulam CB, Faulk WP (1989) Recurrent spontaneous abortion. Am J Reprod Immunol 21:100–104

26. Mueller-Eckhardt G, Heine O, Neppert J, Kunzel W, Mueller-Eckhardt C (1989) Prevention of recurrent spontaneous abortion by intravenous immunoglobulin. Vox Sang 56:151–154
27. Coulam CB, Peters AJ, McIntyre JA, Faulk WP (1990) The use of intravenous immunoglobulin for the treatment of recurrent spontaneous abortion. Am J Reprod Immunol 22:78–85
28. Scott JR, Rote NS, Branch D (1987) Immunologic aspects of recurrent abortion and fetal death. Obstet Gynecol 70:645–656
29. Triplett DA, Harris EM (1989) Antiphospholipid antibodies and reproduction. Am J Reprod Immunol 21:123–131
30. Reginald PW, Beard RW, Chapple J, Forbes PB, Liddell HS, Mowbray JF, Underwood JL (1987) Outcome of pregnancies progressing beyond 28 weeks gestation in women with a history of recurrent miscarriage. Br J Obstet Gynaecol 94:643
31. Ober CL, Hauck WW, Kostyu DD, O'Brien E, Elias S, Simpson JL, Martin AO (1985) Adverse effects of HLA-DR antigen sharing on fertility: a cohort study in a human isolate. Fertil Steril 44:227

Future Guidelines

27 Human and Animal Models: Limitations and Comparisons

G.J. Burton

Introduction

Ethical considerations, coupled with the relative inaccessibility of the uterine lumen, severely limit the opportunities to examine closely or to manipulate experimentally the crucial events taking place during the first 12 weeks of gestation in humans. Researchers have therefore turned to animal models, or recently to in vitro culture systems in order to investigate many of the processes involved. However, the extreme diversity displayed amongst mammals in terms of the reproductive strategies employed, and in particular in the type of placentation that has evolved, means that questions must be asked concerning the validity of animal models. Equally, as evidence emerges of the complexity of the dialogue that takes place between the blastocyst and endometrium at the time of implantation, the limitations of in vitro systems are being exposed.

This chapter therefore attempts to highlight some of the major species differences and other factors that must be taken into account when considering data obtained from model systems. By necessity of space, it will be restricted to an appraisal of the common laboratory mammals and domestic species, but within this group a wide range of gestation lengths obviously exists, and so the "first 12 weeks of gestation" will therefore be interpreted as denoting the stages of preimplantation development, implantation and embryogenesis.

Underlying many of the differences observed is the fundamental relationship between the trophoblast and the uterine tissues. In the majority of species the chorion, defined as the trophoblast and its supporting layer of extra-embryonic mesoderm, is simply apposed to the uterine epithelium. There is no destruction or invasion of the maternal tissues, and so the conceptus remains within the uterine lumen throughout gestation. Such a relationship is usually described as central implantation, although the use of the term "implantation" in these species is to an extent a misnomer [1, 2]. Nevertheless, this form of interaction is seen, for example, in the domestic pig, the horse and all ruminants.

A limited degree of invasion is demonstrated by the carnivores, but the process is most extensive in representatives of four apparently unrelated orders of mammals, namely the insectivores, rodents, bats and primates. In these, the blastocyst attaches to the endometrium at an early stage and rapidly becomes completely embedded in the uterine wall during a process known as interstitial implantation.

Clearly the mode of implantation, whether it be non-invasive or invasive, will have a profound influence on the development of both the blastocyst and the endometrium prior to implantation, and on the subsequent development of the placenta. These aspects of the feto-maternal relationship will be dealt with in turn in these two broad categories of mammals.

Preimplantation Development of the Blastocyst

Following fertilisation in the ampullary portion of the oviduct, the embryos of most species enter the uterine lumen at a roughly equivalent stage of development, namely the four to eight cell stage some 2–4 days post-conception. In humans and the laboratory rodents this usually coincides with formation of the blastocyst, whereas in the non-invasive species the development of the blastocyst is a later event, occurring between days 5 and 7 post-conception.

Ultrastructurally the early blastocysts of most mammals are remarkably similar, but later dramatic changes take place in the blastocysts of those species displaying non-invasive implantation. First, at about day 10 the trophoblast cells overlying the inner cell mass, the polar trophoblast or cells of Rauber degenerate and thus expose the epiblast layer of the developing germ disc to the contents of the uterine lumen. In anticipation of this event, the epiblast cells develop microvilli on their free surfaces, whilst the intercellular spaces become sealed by junctional complexes. Amnion formation occurs by folding of the somatopleure at the margins of the germ disc over the dorsal surface of the disc and is a comparatively late event, final fusion occurring on day 17 post-conception. The term "pleuramnion" has been used to describe an amnion formed in this way to distinguish it from a cavitation amnion which develops by delamination from the inner cell mass. The latter process is almost exclusively restricted to those species in which the blastocyst attaches to the endometrium by the embryonic pole, and in which the polar trophectoderm remains intact throughout. These blastocysts go on to develop some form of invasive implantation.

Secondly, commencing around day 11 there is a phase of rapid elongation. This phenomenon is seen to its maximal extent in the domestic pig, where the blastocyst may achieve lengths of 1 m or more, and to a lesser extent in the sheep and cow. Curiously, it is absent from the horse where the blastocyst remains almost perfectly spherical. The reasons for these differences are unclear. Comparative studies amongst the primates suggest that interstitial implantation may restrict the possibilities for blastocyst expansion for, once embedded in the uterine wall, it is completely surrounded by maternal tissues [3]. On the other hand, it could be argued that blastocyst expansion is advantageous in centrally implanting species in that it will effect a more extensive and immediate fetomaternal contact. The process of elongation may be very rapid in the pig with rates of 3–5 cm/h, and, whilst initially due to cellular remodelling, it is dependent in its later stages on hyperplasia [4].

Throughout this period the blastocyst is reliant upon the uterine secretions for its metabolic requirements. Prior to hatching, the zona pellucida may play a regulatory role in the transport of substances for, whilst in some species, such as the laboratory rodents, it appears freely permeable to macromolecules such as ferritin and peroxidase, in others, including the pig, it excludes them from the trophoblast. After hatching these macromolecules are readily endocytosed by the trophoblast of all species and may pass either into the lysosomal system or via a vesicle-mediated pathway into the blastocoele. It should be noted that the mechanism of hatching also varies between species, the zona pellucida being digested away from the inside by the trophectodermal cells or from the outside by proteolytic enzymes secreted by uterine glands, or by a combination of both processes.

Cues for Implantation

Despite these striking differences in morphology, there seem to be general similarities in the dialogues that take place between the blastocysts and their respective endometria immediately prior to attachment. Investigations of the phenomenon of delayed implantation have clarified many of the endocrinological events taking place at this crucial time. Delayed implantation may be either facultative, where it is associated with maternal lactation and helps to synchronise reproduction with care of the young, or obligatory, where it is part of the overall reproductive strategy of seasonal breeders. Facultative delay is seen in the mouse and rat, when high levels of prolactin inhibit the release of luteinising hormone. When lactation ceases, this negative feedback is removed, oestrogen levels rise and implantation ensues. This may be mimicked experimentally by ovariectomising rats in early pregnancy and rendering maintainance doses of progesterone. The blastocyst remains in the uterine lumen until a dose of exogenous oestrogen is administered.

In obligatory delay an oestrogen surge again seems to provide the cue to implantation. For example, mating in the roe deer takes place during late July or early August, but the blastocyst remains free in the uterine lumen for the next 5 months. Just after the winter solstice, plasma oestrogen levels rise, there is an abrupt discharge of secretions from the uterine glands, and elongation of the blastocyst takes place [5]. Central implantation soon follows, and the young are born in May.

Although such studies reveal the essential role of a 'nidatory' surge of oestrogen in many species, it is apparently not required in the rabbit, guinea pig, hamster or sheep, even though oestrogen is secreted during the luteal phase in several of these. The hormone appears to act by stimulating the release of uterine secretions which, besides maintaining the blastocyst and serving to lyse the zona pellucida in some species, may have other diverse and important roles.

Denker [6] has emphasised the essential paradox of implantation, namely that two epithelia come into intimate contact and often form attachments to each other. This transgresses the normal rules of cell biology, and so some modification of each epithelial surface must take place in order to render it "receptive" to the other. This could involve either the expression of cell adhesion molecules de novo or the modification of existing surface molecules. Glycosidases contained in the uterine secretions may, for example, remove terminal sugars, whilst the action of other enzymes, such as sialidase, might result in increased cell permeability. Several growth factors, notably insulin-like growth factors I and II and epidermal growth factor, have also been identified in these secretions and could exert a profound influence on the development of both the embryo and the endometrium [7].

Although the trophoblast is known to bind and endocytose numerous proteins, there is also traffic in the feto-maternal direction. Across a number of diverse species, such as the sheep, rabbit, pig, roe deer and human, thin-walled bulbous protrusions have been observed arising from the apical surface of the endometrial cells. Whilst it has been suggested these have a secretory role, there is also considerable evidence that they are engaged in the uptake of macromolecules from the uterine lumen. The structures have therefore been termed "pinopodes", and their function may be to transport signals released by a closely juxtaposed trophoblast to the uterine epithelium. Pinopodes are seen during the luteal phase of the non-pregnant cycle in the human, and so the presence of a blastocyst is not essential for their formation. Indeed, it is clear that they may be a second result of the nidatory oestrogen surge, for they attain maximal development and endocytic activity following the administration of oestrogens in ovariectomised pregnant rats [8]. It is interesting to note that in those species where placental attachment is restricted to localised sites, for example, the sheep, pinopodes are only formed at the areas of future close apposition [9].

Luteal or nidatory oestrogens clearly have important roles to play in the regulation of implantation, and it is intriguing that in those species in which the corpus luteum is unable to synthesise oestrogens, for example, the pig, the trophoblast may take over this role.

Spacing of Embryos and the Siting of Attachment

Trophoblastic oestrogens may also play other roles than those outlined above. Thus, in both the horse [10] and the pig [11] they are thought to be the signal for the maternal recognition of pregnancy, whilst in the latter species they may also be responsible for spacing of embryos within the uterine horns. Prior to attachment in this species, the embryos become evenly spaced throughout both uterine horns, and this takes place before either blastocyst elongation or uterine horn growth has occurred. The process is not simply due to myometrial activity since silastic beads only become spaced if they are coated with oestrogens. It is thought that the hormone may influence the release of either histamines or prostaglandins in the vicinity of the blastocyst, resulting in local alterations in smooth muscle activity.

Migration throughout the uterine horns is also seen in the rat and mouse, but in these species there is a strict orientation of the blastocyst around the circumference of the uterine horn so that the initial attachment takes place on the antimesometrial side. Species in which the blastocyst attaches to the antimesometrial rather than to the mesometrial or central regions of the horn tend to possess a trophoblast that is highly invasive during the early stages of development. How this initial recognition is achieved is not known at present, but it seems to have little bearing on the final location of the definitive placenta which is mesometrial in all rodents [1].

In other species, particularly the ruminants, the sites of attachment and the location of the placenta are dictated by the presence of uterine specialisations known as caruncles. These are non-glandular areas of the endometrium, visible in the non-pregnant uterus as roundish elevations, the size, number and exact shape of which is species specific [2]. In the sheep they are made particularly conspicuous by the presence of melanocytes and, numbering between 80 and 100 per uterus, they are arranged in four rows along each uterine horn. By comparison, in the cow there are 70–140 caruncles, but in some species of deer there may be as few as four per uterine horn. Placental villi only develop opposite these caruncles, forming a composite unit known as a "placentome" or "cotyledon" (Fig. 1). These are located over all parts of the chorionic sac and are 1–2 cm in diameter in a small ruminant such as the sheep, and 5–8 cm in diameter in a large ruminant such as the cow.

The presence of such discrete and restricted areas of placentation in the sheep has provided fetal physiologists with several important benefits. It allows for easy access to the fetus via the intervening areas of unspecialised chorion, so that catheters may be implanted for chronic experiments without disrupting the placenta. In addition, by removing caruncles from the non-pregnant uterus prior to conception, it is possible to restrict the extent of the

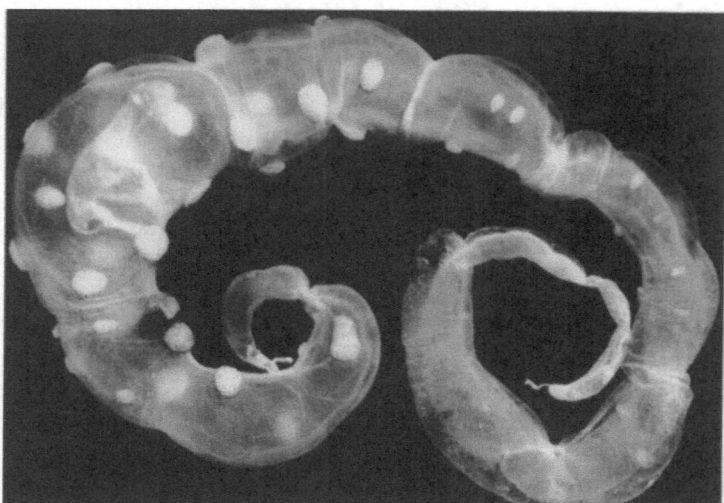

Fig. 1. The fetal membranes of a calf during the first trimester of pregnancy (crown-rump length of fetus = 85 mm). The development of villi is restricted to discrete areas, referred to as "cotyledons", scattered over the surface of the elongated chorionic sac. These arise opposite the non-glandular caruncles of the uterus and represent the principal sites of placental exchange. The presence of the fetus, contained within the amnion, separates the chorionic sac into two limbs. The fetus lies at the base of the pregnant horn of the bicornuate uterus, whilst the longer limb of the sac extends into the non-pregnant horn. The allantois can be clearly seen extending to the tip of each limb

mature placenta to a precise predetermined degree. This enables the correlation between placental and fetal growth to be accurately assessed, and the reserve capacity of the placenta to be determined experimentally.

Interaction with the Uterine Epithelium

The most simple arrangement is found in those species displaying central or non-invasive implantation, and so this category will be described first.

Non-invasive Implantation

The classical example of this type of implantation is provided by the domestic pig, and the early stages were recently reviewed by Stroband and van der Lende [12]. Attachment begins around days 13–14, in the region of the developing germ disc, and then spreads gradually to the tips of the elongating blastocyst over the next few days. This coincides with the outgrowth of the allantoic sac which, as it fills with fluid, may distend the

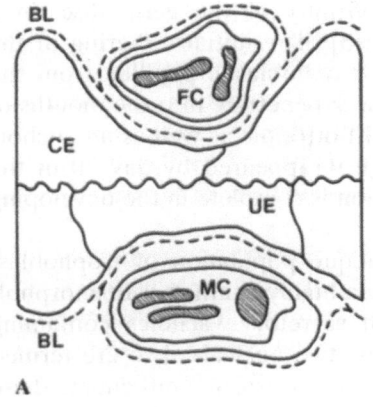

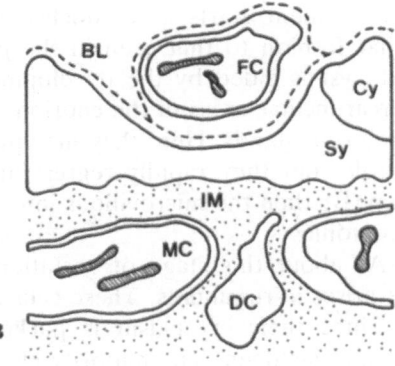

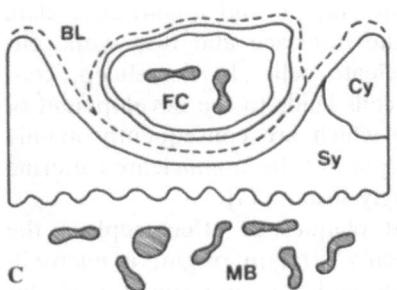

Fig. 2A–C. Different histological types of chorio-allantoic placentae recognised. **A** Epitheliochorial; **B** endotheliochorial; **C** haemochorial. *BL*, basal lamina; *CE*, chorionic epithelium; *Cy*, cytotrophoblast; *DC*, decidual cell; *FC*, fetal capillary; *IM*, interstitial membrane; *MB*, maternal blood space; *MC*, maternal capillary; *Sy*, syncytiotrophoblast; *UE*, uterine epithelium. (Based on [2])

blastocyst and so bring the two epithelial surfaces into contact. Initially there is a loss of microvilli from the apical surface of the trophoblast cells at the sites of attachment, a process which may allow for either a more intimate contact between the membranes or for an increase in membrane surface area through remodelling. Microvilli then reform on the trophoblastic surface, and by day 18 there is extensive interdigitation with those of the uterine epithelium. Indeed, cytoplasmic processes from the trophoblast may penetrate between adjacent uterine cells, and there is some debate as to whether these processes may reach the basal lamina [13].

Opposite the mouths of the uterine glands specialisations of the chorionic sac known as "areolae" develop. In these regions the trophoblast cells are transformed from cuboidal to columnar and phagocytose the glandular secretions. Otherwise the materno-fetal interface is represented throughout gestation by the simple microvillous interdigitation, strengthened by the presence of junctional complexes linking the two epithelia. There is thus no destruction of the maternal tissue, and so this relationship is classified histologically as epitheliochorial (Fig. 2).

The sheep, cow and horse also fall into this epitheliochorial category, but each shows unique species variations. In the ruminants the process of attachment starts at caruncles in the vicinity of the germ disc, in a similar fashion to that seen in the pig [9, 13]. The initial tethering of the blastocyst is aided by the development of trophoblastic papillae from the intercaruncular areas of the chorion, and these penetrate into the mouths of the uterine glands. Here they may play a histiotrophic as well as an anchoring role, but they rapidly regress and have disappeared by day 20 in the sheep. By then the microvillous interdigitation is complete in the developing placentomes.

At about this stage of gestation, a unique population of trophoblast cells arises in ruminants. These cells display a highly characteristic morphology since their cytoplasm is packed with secretory vacuoles containing placental lactogen. The majority also possess two nuclei and so are termed "binucleate cells". Although formed in the trophoblastic epithelium, these cells do not take part initially in the materno-fetal interface. However, there is now convincing morphological, autoradiographic and quantitative data demonstrating that they migrate across the interface and fuse with cells of the uterine epithelium to form trinucleate cells. In the sheep, continued migration and fusion of binucleate cells leads to the development of localised multinucleated syncytial plaques which are consequently mainly fetal in origin. Initially the plaques are separated by uninucleated uterine epithelial cells, but these become increasingly scarce [14].

With increasing time, these syncytial plaques in effect replace the uterine epithelium. Despite being essentially fetal in origin, a microvillous interdigitation persists between them and the apposing unicellular trophoblast. It has been suggested that one of the functions of the migration is to transport secretory granules of ovine placental lactogen across the materno-fetal interface. The immunological consequences of the process must not be forgotten, however, for it is likely that the basal surface of the syncytial plaques will be of mixed maternal and fetal origin. Despite this, there does not appear to be any significant or consistent maternal lymphocytic response to the tissue.

By contrast, in the cow and deer binucleate cell migration and fusion take place, but, once their granules are released, the resultant trinucleate cells die and are phagocytosed by the trophoblast. The uterine epithelium thus remains essentially cellular in these species [15].

In the small ruminants, therefore, there is an intriguing migration of fetal cells into the maternal tissues at the sites of the uterine caruncles. The intercaruncular areas of the uterine epithelium remain unicellular and are not involved in this process. A similarly localised migration of fetal cells into the endometrium takes place in the horse, but in this case it does not involve the region participating in placental exchange.

As mentioned previously, the horse blastocyst has a characteristic spheroidal shape and remains free in the uterine lumen for a considerable period. It is held within the lumen, at the base of one of the uterine horns, by a marked increase in myometrial tone that develops during the 3rd week of gestation. Attachment to the uterine epithelium is not achieved until day 42 post-conception.

During this period the allantoic sac grows out and expands towards the abembryonic pole of the blastocyst. Opposite its leading edge a restricted band of trophoblast extending around the entire circumference of the blastocyst becomes thrown into a series of folds, first visible at about day 25. With subsequent development, the trophoblast cells at the apices of the folds proliferate, and the whole band, now known as the "chorionic girdle", moves progressively towards the abembryonic pole of the blastocyst as the allantois enlarges. Then, quite suddenly, between days 36 and 38 of pregnancy, the girdle cells leave the blastocyst and individually migrate through the uterine epithelial cells, breach the basal lamina and come to lie as aggregations within the endometrial stroma. These aggregations form a discontinuous ring around the base of the pregnant horn and are known as the "endometrial cups". They have a limited life span, becoming necrotic and finally being sloughed from the endometrium between days 120 and 150 of gestation.

Within the cups the fetal cells are generally binucleate, and, since they secrete high levels of pregnant mare's serum gonadotrophin, it has again been suggested that this localised invasion is a means of transporting a hormone across the materno-fetal interface. The immunological aspects of this migration have been extensively researched [16] since a pronounced lymphocytic and plasma cell accumulation takes place around the periphery of the cups and plays an active role in their final destruction. Interestingly, the cells of the chorionic girdle express a high density of conventional class 1 major histocompatability complex antigens just prior to their invasion of the endometrium. Subsequently, the cells of the endometrial cups continue to express these antigens, but at a much lower level, whereas expression generally cannot be detected over the remaining non-invasive areas of the chorion [17].

Invasive Implantation

In the second broad category of mammals, i.e. those displaying invasive implantation, there is a more widespread penetration of the maternal epithelium by the trophoblast. This may vary in its extent and depth so that, whilst in some species such as the dog or rhesus monkey most of the blastocyst remains exposed to the uterine lumen, in others such as the mouse and man it becomes completely embedded within the uterine wall. On the basis of electron-microscopic observations, three types of epithelial invasion were indentified in the classic study of Schlafke and Enders [18]. They are as follows:

1. The "displacement type" (for example, rat, mouse). In this, the initial attachment of the unicellular trophoblast to the uterine epithelium

stimulates degeneration and sloughing of the latter. The basal lamina is thus exposed, but penetration and erosion of this layer appears to be the result of activity in the underlying endometrial cells rather than in the advancing trophoblast.

2. The "fusion type" (for example, rabbit). Here, as the name implies, there is a fusion between the trophoblast cells and those of the uterine epithelium to form a multinucleated syncytium of mixed maternal and fetal origin. Unlike the situation in the sheep, this syncytium is highly invasive and rapidly penetrates both the basal lamina and the endometrial stromal tissue.

3. The "intrusion type" (for example, ferret, dog and cat). In this process there is an initial attachment and adhesion between the two unicellular epithelia achieved by microvillous interdigitation and the formation of junctional complexes. This is followed by the development of a syncytial trophoblast which appears to be entirely of fetal origin and which penetrates between the uterine epithelial cells to invade the endometrial stroma.

Clearly, these very early stages of implantation are of vital importance to the establishment of a pregnancy and so they are subject of much research effort. In order to observe them in closer detail, attempts have been made to reproduce them in vitro using a variety of culture systems [19]. The earliest of these involved laying blastocysts onto strips of rabbit endometrium maintained on agar in an organ culture system. Keeping such a complex tissue as endometrium fully viable in culture is a difficult task, however, and the non-specific way in which the blastocysts attached, coupled with the unresponsiveness of the system to hormone stimuli raised doubts over the validity of the results obtained.

Subsequent studies therefore attempted to simplify the system by using monolayers of either uterine epithelial or stromal cells. These have the advantage that the viability of the preparations is more certain, but their very simplicity presents a number of important drawbacks. The lack of an underlying endometrial stroma means that, once the trophoblast has penetrated between the cultured cells, its behaviour may become very atypical. Clearly, it cannot be influenced by any factors produced in vivo by the endometrium, but in addition it may be abnormally affected by the competing adhesive properties of the culture dish. In order to remove the latter complication, mouse blastocysts have been cultured with fragments of uterine epithelium in hanging drops of culture medium [19].

Here the epithelial fragments form small hollow spheres with the apical surfaces of the cells facing outwards. These spheres are brought into contact with the blastocysts by the gentle curvature of the hanging drop, and there is no competing substrate for adhesion. Adhesive interactions between the two epithelia have been observed within 1–2 days of culture, with an initial interdigitation of microvilli progressing to close membrane-to-membrane apposition in an identical fashion to that seen in vivo.

One problem common to both monolayer and hanging drop culture systems is that contact between trophoblast and uterine tissue is only established at isolated points on the surface of the blastocyst. This is unlike the situation in vivo, where the endometrium generally closes around the blastocyst and covers it on all aspects, even in those species displaying non-invasive central implantation. In a more three-dimensional approach, Hohn and Denker [20] constrained rabbit blastocysts and endometrial fragments within dialysis tubes. The endometrial fragments were precultured under progestagenic stimulation in order to ensure regeneration of a complete epithelial lining all around the fragments, so preventing atypical direct contact between the trophoblast and endometrial stroma. Although this system does provide for greater physical contact between the blastocyst and the uterine fragments, the additional presence of the dialysis membrane may hinder the diffusion of metabolites and nutrients, for increased rates of cell degeneration were observed within the uterine tissues. Nonetheless, in cases where both sets of tissue remained apparently healthy, the attachment of the trophoblast to and subsequent invasion of the endometrium followed a very similar course to that seen in vivo.

Despite these shortcomings, in vitro systems offer many possibilities for future research into implantation. Whilst the monolayer cell culture and hanging drop systems are best suited to investigations of the earliest stages of attachment and adhesion, by incorporating the presence of endometrial stroma the three-dimensional models allow for studies of trophoblast invasion. Early attempts to manipulate these events by the introduction of steroid hormones, serum factors and enzymes into the culture media have so far failed to yield consistent results. This is most likely due to inconsistencies in the methodologies employed, but, as experience increases, these models should prove valuable research tools.

Most studies have been based on laboratory rodents, but the results of Lindenberg et al. [21] from monolayer cultures suggest that human blastocysts implant by the intrusion method.

Species Differences in Trophoblastic Invasiveness

Marked differences in the interaction between trophoblast and endometrium thus exist both between and within the various orders of mammals. This raises the intriguing question of whether this reflects differences inherent in the trophoblast, or whether the activities of the tissue are modified in different ways by the various uterine epithelia. Several pieces of evidence suggest that it is almost certainly the latter case.

It has been known for many years that trophoblast behaves atypically when placed in abnormal locations. For example, the usually non-invasive trophoblast of the domestic pig becomes highly invasive when transplanted to ectopic sites, even if these are lined by epithelia [22]. This is accompanied by a conversion of the normally unicellular trophoblast into multinucleated syncytial masses.

Equally, it is well established that mouse blastocysts, which will readily invade non-uterine tissues such as muscle and kidney, are only able to implant into the endometrium during its "receptive phase". The early experiments of Cowell [23] demonstrated that this restriction can be overcome by removing the uterine epithelium, allowing the blastocyst to implant at any phase of the oestrous cycle. More recent evidence from the in vitro model systems has demonstrated that blastocysts preferentially attach to damaged or degenerating epithelial cells, suggesting that under normal conditions the uterine epithelium actively prevents blastocyst attachment.

As mentioned previously, changes in cell receptivity must take place in order to allow attachment between the two epithelia [6]. Alterations in the lectin affinity or membrane expression of glycoproteins have been put forward as a basis for these, special attention focusing upon heparan sulphate proteoglycans and lactosaminoglycans [24]. Whether there are species variations in the expression of such molecules or their receptors, and whether this could be the basis of the differences in trophoblastic behaviour observed remains to be determined, however.

Development of the Yolk Sac and Allantois

Before considering subsequent stages of implantation, mention must briefly be made of the ways in which the yolk sac and allantois develop, for these have profound influences not only on placental transfer, but also on the further differentiation of the trophoblast, particularly in the laboratory rodents.

The chorion is intrinsically avascular, and so for effective placental exchange it must be vascularized by vessels developing in the extra-embryonic mesoderm associated with either the yolk sac or the allantois. Yolk sac or chorio-vitelline placentation is widely seen amongst marsupials and is generally considered to be the more primitive type of placentation in mammals. In all eutherian mammals the definitive placenta is vascularised from the allantois, even though development of the sac may be rudimentary in some species, for example, in humans. However, a transient chorio-vitelline phase is often seen during early gestation in many eutherian species such as the pig, dog and horse. In later development the outgrowth of the fluid-filled allantois strips the yolk sac away from the chorion to a greater or lesser degree, depending on the species concerned [1, 2].

Oddly, development of a highly elaborate chorio-vitelline placenta which persists until term is seen amongst the laboratory rodents. In, for example, the rat, rabbit and mouse, the yolk sac is far larger than the allantois and fills most of the extra-embryonic coelom. It develops to such a degree that the developing embryo sinks into the yolk sac and so becomes surrounded by two layers of yolk sac endoderm. The outer or parietal layer of endoderm is discontinuous, non-vascularised and lies directly on the inner surface of the trophoblast. Along with the trophoblast it secretes an acellular extra-embryonic membrane unique to rodents and insectivores, i.e. Reichert's membrane, which is initially interposed between the two layers of the yolk sac. However, soon both the outer wall of the yolk sac and the overlying trophoblast disintegrate, leaving the inner vascularised wall of the yolk sac exposed directly to the uterine lumen, except for its covering of Reichert's membrane [1, 2, 25]. This configuration is termed an "inverted yolk sac placenta" and represents an important pathway for antibody transfer later in gestation. A similar situation occurs in the guinea pig, except that it develops so early that the intermediate stages described above are generally not observed. In humans the yolk sac is very rudimentary and never establishes contact with the chorion.

Further Trophoblastic Proliferation and Invasion

During invasive implantation, once the trophoblast has penetrated the uterine epithelium, its behaviour will be heavily modified by factors released from the cells of the endometrial storma and by molecules present in the extracellular matrix. In carnivores there is restricted erosion of the maternal tissues, with the endothelium of the endometrial capillary network remaining intact. Sheets of syncytiotrophoblast come to lie adjacent to these vessels, separated from them by only a layer of basal lamina-type material, and so this relationship is classified histologically as endotheliochorial (Fig. 2).

More extensive destruction of maternal tissue is seen in the laboratory rodents and higher primates. In these species, the trophoblast is bathed directly by maternal blood, and so the placentae are classified as haemochorial (Fig. 2). There are, however, major differences in the development and proliferation of the trophoblast within this group [26].

The laboratory rodents do not exhibit the widespread infiltration of the endometrium by cytotrophoblast cells that occurs in humans, but instead proliferation of the trophoblast is restricted to the polar region of the blastocyst. In part, this reflects differences in the way the blastocysts implant, but is also due to the development of the inverted yolk sac placenta outlined above.

The blastocysts of these species come to lie in a crypt on the anti-mesometrial wall of the uterine horn, with the inner cell mass orientated towards the uterine lumen. The polar trophoblast cells proliferate to form a large cone of cells termed the "ectoplacenta" or "Trager" [25]. This is directed towards and subsequently invades the mesometrial wall of the uterus. In the rat and mouse, the ectoplacenta is itself invaded by the allantoic mesoderm in such a way that it is split into a deep and superficial layer. The superficial layer, known as the "trophospongium", is essentially a remnant of the ectoplacenta, but does undergo some differentiation and may retain a proliferative activity. On a comparative basis, it is equivalent to the trophoblastic shell or basal plate of the human placenta [1]. The deeper layer develops into the chorio-allantoic placenta, with an effective maternal and fetal circulation, and is referred to as the "zona intima".

In the other main laboratory rodent, the guinea pig, the situation is made more complex due to a different pattern of allantoic growth. A small central area of the trophospongium in contact with the endometrium undergoes separate development to form a specialisation known as the sub-placenta. From this region streamers of trophoblast cells, many forming syncytial giant cells, penetrate into the endometrium in a similar fashion to the interstitial invasion seen in humans.

Major differences therefore exist in the way that trophoblast invasion occurs, even in those species displaying haemochorial placentation. One important feature common to humans and the laboratory rodents is, however, that retrograde migration of trophoblast takes place into the maternal arteries supplying the developing placenta [26]. This migration is vital for the normal functioning of the vessels later in gestation.

The details of how this trophoblastic invasion is controlled and limited are only beginning to be clarified, but it is clear that the endometrial stroma plays a very important role.

Endometrial Response to Trophoblastic Invasion

In response to the breaching of the uterine epithelium by trophoblast, the endometrial stroma undergoes a series of profound changes which are collectively termed the "decidual response". Such changes are not observed in species possessing epitheliochorial placentae, but are seen to a limited degree in association with endotheliochorial placentae. Maximal development occurs in conjunction with haemochorial placentation, although even in this category major species differences are seen. Thus, there is a profound response in the rat and mouse, but it is virtually absent in the rabbit. Amongst the higher primates, a strong correlation exists between the depth of invasion and the extent of the decidual response, for in the rhesus monkey, where implantation is much more superficial than in humans, the response is localised to the area immediately beneath the blastocyst [27].

Decidualisation involves considerable morphological and biochemical changes. Most research has been based on the laboratory rodents, but again there are differences between these species and humans. Most importantly, the decidual reaction occurs in the rat and mouse in direct response to blastocyst implantation, whereas in humans it appears to anticipate this event, for changes are seen during the luteal phase of even non-pregnant cycles. An intermediate situation is found in the guinea pig where predecidual changes take place prior to implantation.

One important component of the human decidua is the granulated lymphocyte, and details are beginning to emerge of the complex dialogue that may take place between these cells, the decidual cells and the invading trophoblast.

Granulocytes are also a conspicuous component of the decidua of both the rat and mouse, and interestingly have also recently been described within the endometrial stroma of the sheep. The equivalence and inter-relationships between these cell populations in difference species are still very uncertain, however [28].

The Form of the Mature Placenta

From this brief account it can be seen that a tremendous diversity of placental types exists amongst even eutherian mammals. Not only do they vary in overall shape, for example, discoid, zonary or cotyledonary, but they also vary histologically in the number of tissue layers separating the maternal and fetal circulations.

Laboratory rodents are often used as models for research since they display haemochorial placentation and are readily available. In view of this, two other notes of caution must be mentioned and borne in mind when extrapolating data from one species to another.

First, the nature of the trophoblast varies between even haemochorial placentae [29]. In humans and the guinea pig, there is an outer layer of syncytiotrophoblast derived from the discontinuous cytotrophoblast layer, and this arrangement has been termed "haemomonochorial". By contrast, in the rabbit the trophoblast is two layered, the outer layer being syncytial and the inner being a mixture of syncytial and cellular elements. This species has thus been classified as "haemodichorial". In the rat, mouse and hamster, a third layer of trophoblast exists, and so these species are considered "haemotrichorial". Here the outer layer remains cellular, whilst the middle and probably the inner layers are syncytial in nature.

Lastly, placentae are classified as villous or labyrinthine depending on their architecture [1, 2]. The villous situation occurs when the chorion develops into a number of finger-like projections displaying varying degrees of lateral branching. These may be bathed by maternal blood in the haemochorial placentae of humans and the rhesus monkey, or invested by the walls of maternal crypts in the epitheliochorial placentae of the sheep or horse. The majority of placentae fall into the labyrinthine category, in which the pattern of trophoblast invasion is such that it completely surrounds the maternal blood channels. The trophoblast may be likened to the substance of a sponge with the maternal blood channels being represented by the interconnecting holes or spaces. These blood channels may still be lined with endothelium in, for example, the cat or dog, or simply contain free maternal blood, as in the rat, mouse and guinea pig. A wide variety of intermediate forms exists between these two extremes, and more comprehensive comparative texts should be consulted for further information.

Conclusion

By now it should be apparent that no one species of laboratory or domestic mammal displays an identical pattern of implantation and placental development to that seen in humans. Researchers interested in different aspects of this intriguing process will have differing requirements and must select a species accordingly. The possible limitations of that model must be borne in mind at all times, however, and extreme caution must be exercised when extrapolating data from one species to another, even though they may appear to be closely related.

References

1. Mossman HW (1987) Vertebrate fetal membranes. Macmillian, London
2. Steven DH (ed) (1975) Comparative placentation. Essays in structure and function. Academic, London
3. Luckett WP (1974) Comparative development and evolution of the placenta in primates. Contrib Primatol 3:142–234
4. Geisert RD, Brookbank JW, Roberts RM, Brazer FW (1982) Establishment of pregnancy in the pig. II. Cellular remodelling of the porcine blastocyst during elongation on day 12 of pregnancy. Biol Reprod 27:941–955
5. Aitken RJ, Burton J, Hawkins J, Kerr-Wilson R, Short RV, Steven DH (1973) Histological and ultrastructural changes in the blastocyst and reproductive tract of the roe deer, *Capreolus capreolus*, during delayed implantation. J Reprod Fertil 34:481–493
6. Denker H-W (1990) Trophoblast-endometrial interactions at embryo implantation: a cell biological paradox. In: Denker H-W, Aplin JD (eds) Trophoblast invasion and endometrial receptivity. Plenum, New York, pp 3–29 (Trophoblast research, vol 4)
7. Simmen RCM, Simmen FA (1990) Regulation of uterine and conceptus secretory activity in the pig. J Reprod Fertil [Suppl] 40:279–292
8. Leroy F, van Heock J, Bogaert C (1976) Hormonal control of pinocytosis in the uterine epithelium of the rat. J Reprod Fertil 47:59–62
9. Guillomot M, Flechon J-E, Wintenberger-Torres S (1981) Conceptus attachment in the ewe: an ultrastructural study. Placenta 2:169–182
10. Heap RB, Hamen M, Allen WR (1982) Studies on oestrogen synthesis by the preimplanation equine conceptus. J Reprod Fertil [Suppl] 32:343–352

11. Geisert RD, Zavy MT, Moffatt RJ, Blair RM, Yellin T (1990) Embryonic steroids and the establishment of pregnancy in pigs. J Reprod Fertil [Suppl] 40:293–305
12. Stroband HWJ, van der Lende T (1990) Embryonic and uterine development during early pregnancy in pigs. J Reprod Fertil [Suppl] 40:261–277
13. King GJ, Atkinson BA, Robertson HA (1982) Implantation and early placentation in domestic ungulates. J Reprod Fertil [Suppl] 31:17–30
14. Wango EO, Wooding FBP, Heap RB (1990) The role of trophoblast binucleate cells in implantation in the goat: a quantitative study. Placenta 11:381–394
15. Wooding FBP (1982) The role of the binucleate cell in ruminant placental structure. J Reprod Fertil [Suppl] 31:31–39
16. Allen WR (1982) Immunological aspects of the endometrial cup reaction and the effect of xenogeneic pregnancy in horses and donkeys. J Reprod Fertil [Suppl] 31:57–94
17. Donaldson WL, Zhang CH, Oriol JG, Antczak DF (1990) Invasive equine trophoblast expresses conventional class 1 major histocompatibility complex antigens. Development 110:63–71
18. Schlafke S, Enders AC (1975) Cellular basis of interaction between trophoblast and uterus at implantation. Biol Reprod 12:41–65
19. Morris JE, Potter SW (1990) An in vitro model for studying interactions between mouse trophoblast and uterine epithelial cells. In: Denker H-W, Aplin JD (eds) Trophoblast invasion and endometrial receptivity. Plenum, New York, pp 51–69 (Trophoblast research, vol 4)
20. Hohn H-P, Denker H-W (1990) A three-dimensional organ culture model for the study of implantation of rabbit blastocyst in vitro. In: Denker H-W, Aplin JD (eds) Trophoblast invasion and endometrial receptivity. Plenum, New York, pp 71–95 (Trophoblast research, vol 4)
21. Lindenberg S, Hyttel P, Sjogren A, Greve T (1989) A comparative study of attachment of human, bovine and mouse blastocysts to uterine epithelial monolayer. Hum Reprod 4:446–456
22. Samuel CA, Perry JS (1972) The ultrastructure of pig trophoblast transplanted to an ectopic site in the uterine wall. J Anat 113:139–149
23. Cowell TP (1969) Implantation and development of mouse eggs transferred to the uteri of non-progestational mice. J Reprod Fertil 19:239–245
24. Carson DD, Wilson OF, Dutt A (1990) Glycoconjugate expression and interactions at the cell surface of mouse uterine epithelial cells and preimplantation – stage embryos. In: Denker H-W, Aplin JD (eds) Trophoblast invasion and endometrial receptivity. Plenum, New York, pp 211–241 (Trophoblast research, vol 4)
25. Kaufman MH (1983) The origin, properties and fate of trophoblast in the mouse. In: Loke YW, Whyte A (eds) Biology of trophoblast. Elsevier, Amsterdam, pp 23–68
26. Pijnenborg R, Robertson WB, Brosens I, Dixon G (1981) Trophoblast invasion and the establishment of haemochorial placentation in man and laboratory animals. Placenta 2:71–92
27. Enders AC, Welsh AO, Schlafke S (1985) Implantation in the rhesus monkey: endometrial responses. Am J Anat 173:147–169
28. Bulmer JN, Ritson A, Pace D (1990) Endometrial leukocytes in human pregnancy. In: Denker H-W, Aplin JD (eds) Trophoblast invasion and endometrial receptivity. Plenum, New York, pp 431–451 (Trophoblast research, vol 4)
29. Enders AC (1965) A comparative study of the fine structure of the trophoblast in several haemochorial placentas. Am J Anat 116:29–68

28 Classification of Abortion Material; Myth or Reality?

H. Fox

Pathologists always prefer to examine material within the framework of a classification, and those studying tissues from spontaneous abortions are no exception. Most pathologists usually receive, however, only placental and decidual tissue from cases of abortion, and therefore any classification suitable for widespread application has to be based largely on placental morphology. It is, perhaps, fortuitous that this pragmatic necessity is bolstered by a widespread view that placental changes offer the best basis for classification of abortion material. Any useful classification of abortion material in terms of placental morphology should fulfill two criteria:

1. The classification should be based on simple histological examination of haematoxylin and eosin-stained sections and should not necessitate the use of techniques such as morphometry or electron microscopy.
2. There should be a correlation between the types of placental abnormality recognised in the classification and the presumed cause of the abortion.

It is with this second criterion that most difficulties are encountered. It is, of course, well recognised that karotypic abnormalities of the fetus contribute substantially to early fetal loss, and there have been many claims of an abilty to correlate placental morphology with specific chromosomal abnormalities or, less ambitiously but more realistically, with an abnormal, as opposed to a normal, karotype. If these claims are true, then they could serve as a valid, though not necessarily clinically useful, basis for a classification of abortion material.

Is it, in fact, true that karotypic abnormalities are associated with specific histological changes in the placental villi? There is general agreement that a partial hydatidiform mole is usually, though by no means invariably, associated with a fetal triploidy, but it has also been claimed that there are specific placental changes associated with trisomies 13, 16, 18, 20 and 21 and with XO monosomy [1–7]. Rehder et al. [8] have examined these claims and correlated cytogenetic data with placental morphology in 200 cases of abortion. They divided the placentas into five groups, one of which was equivelant to a partial mole and one to a hydropic abortion; the other three groups comprised placentas showing varying degrees of hydropic and post mortem change. There was an 80% incidence of chromosomal abnormalities amongst the partial hydatidiform moles, but an approximately 50%–65% incidence of such anomalies in the other four groups. The authors concluded that, with the probable exception of a fetal triploidy, it was not possible to identify any morphological criteria to allow for either the diagnosis or the exclusion of a chromosomal abnormality of any type.

Minguillon et al. [9], as a result of a similar study, also concluded that the predictive value, in terms of recognising a chromosomal abnormality, of placental villous histology was inadequate and indeed no higher than would be anticipated by chance. Novak et al. [10] had previously come to the same comclusion and were of the opinion that, with the exception of triploidy, villous morphology was an inaccurate and insensitive indicator of chromosomal anomalies.

It is only fair to point out that Rockelein et al. [11, 12] have disagreed with these negative findings on the basis of their light and scanning electron-microscopic findings: their conclusions currently represent, however, a minority view, and it is highly probable that a histological diagnosis of a chromosomal abnormality, based on placental villous morphology, is extremely inaccurate.

Rushton [13–15] has proposed a less ambitious but more realistic classification of abortion material, again based largely upon placental morphology. His classification may be paraphrased along the following lines:

Group 1. A considerable proportion of the placental villi show hydropic change.

Group 2. Most villi are non-hydropic and show post mortem changes such as stromal fibrosis and sclerosis of the villous fetal vessels.

Group 3. The villi show no evidence of hydropic or post mortem change and are of normal appearance for the length of the gestational period.

Rushton [13–15] has emphasised that each of these groups is, in terms of the aetiology of the abortion, heterogenous, and that the appearances reflect more the stage of pregnancy at which fetal death occurred. Nevertheless, it would be reasonable to assume that the placentas in group 1 are from cases of early fetal death, that in such cases the fault is more likely to be intrinsic to the fetus than to lie in the maternal environment and that a karotypic abnormality is statistically the most likely cause. By contrast, placentas in group 3 are from relatively late abortions in which it is probable that the fetus died because the mother aborted, the fault leading to abortion being more likely to be in the maternal environment than in the fetus itself. Placentas in group 2 occupy an intermediate position in which it is less easy to draw conclusions.

How useful is this classification in practice? Houwert-de Jong et al. [16, 17] used this classification to examine material from cases of recurrent and sporadic abortion. They found that the pattern of abnormalities was identical in these two groups and that, furthermore, they could not correlate the histopathological findings with the outcome of subsequent pregnancies. These studies probably require further amplification, but nevertheless their clear implication is that a classification of the type proposed by Rushton is unlikely to be of any practical value in the study and management of cases of recurrent abortion.

If classification of abortion material fails to identify the recurrent aborter and has no predictive value for subsequent pregnacies, is there any point in continuing to use any form of morphological classification? The answer, regrettably, is probably no. Obstetricians request pathological examination of material from abortions for two thoroughly practical reasons:

1. To confirm that an intrauterine pregnancy has actually been present
2. To exclude the possibility of a hydatidiform mole

Satisfactory answers to both these questions can usually be given, though it must be pointed out that the diagnosis of an intrauterine pregnancy rests solely upon the finding of a placental site reaction, and that there is considerable interobserver variation in the diagnosis of a partial hydatidiform mole. There appears to be no obvious benefit to be gained, in routine practice, to go beyond this by attempting to classify the abortion material in morphological terms, this applying with equal force to both sporadic and recurrent cases of abortion.

This conclusion may appear to be a nihilistic one, but there is little merit in perpetuating a pointless activity which appears to provide no information of diagnostic or prognostic value and which does not influence the clinical management of the patient.

References

1. Philippe E, Boué J (1969) Le placenta dans les aberrations chromosomiques léthales. Ann Anat Pathol (Paris) 14:249–266
2. Philippe E (1973) Morphologie et morphométrie des placentas d'aberration chromosomique létale. Rev Fr Gynecol Obstet 68:645–649
3. Honore LH, Dill FJ, Poland BJ (1976) Placental morphology in spontaneous human abortions with normal and abnormal karotypes. Teratology 14:151–166
4. Geisler M, Kleinebrecht J (1978) Cytogenetic and histologic analysis of spontaneous abortion. Hum Genet 45:239–251
5. Göcke H, Muradow I, Cremer H (1982) Morphologische und zytogenetische Befunde bei Frühaborten. Verh Dtsch Ges Pathol 66:141–146
6. Gocke H, Schwanitz G, Muradow I, Zerres K (1985) Pathomorphologie und Genetik in der Frühschwangerschaft. Pathologe 6:249–259
7. Muntefering H, Dallenbach-Hellweg G, Ratschek M (1988) Pathologische-anatomische Befunde bei der gestörten Frühschwangerschaft. Gynakologe 21:262–272
8. Rehder H, Coerdt W, Eggers R, Klink F, Schwinger E (1989) Is there a correlation between morphological and cytogenetic findings in placental tissue from early missed abortions? Hum Genet 82:377–385
9. Minguillon C, Eiben B, Bahr-Porsch S, Vogel M, Hansmann H (1989) The predictive value of chorionic villus histology for identifying chromosomally normal and abnormal spontaneous abortions. Hum Genet 82:373–376

10. Novak R, Agamonalis D, Dasu S, Igel H, Platt M, Robinson H, Shehata B (1988) Histological analysis of placental tissue in first trimester abortions. Pediatr Pathol 8:477–482
11. Rockelein G, Schroder J, Ulmer R (1989) Korrelation von Karotyp und Plazentamorphologie beim Frühabort. Pathologe 10:306–314
12. Rockelein G, Ulmer R, Schwille R (1990) Surface and branching of placental villi in early abortion: relationship to karotype. Virchows Arch [A] 417:151–158
13. Rushton DI (1981) Examination of products of conception from previable human pregnancies. J Clin Pathol 34:819–835
14. Rushton DI (1987) Pathology of abortion. In: Fox H (ed) Haines and Taylor: obstetrical and gynaecological pathology, 3rd edn. Livingstone, Edinburgh, pp 1117–1148
15. Rushton DI (1988) Placental pathology in spontaneous miscarriage. In: Beard RW, Sharp F (eds) Early pregnancy loss: mechanisms and treatment. Springer, Berlin Heidelberg New York, pp 149–157
16. Houwert-de Jong MH (1989) Habitual abortion: views and fact-finding. Thesis, University of Utrecht
17. Houwert-de Jong MH, Bruinse HW, Eskes TKAB, Mantingh A, Termijtelen A, Kooyman CD (1990) Early recurrent miscarriage: histology of conception products. Br J Obstet Gynaecol 97:533–535

29 Should Early Pregnancy Dysfunction Be Treated or Adequately Diagnosed, Identified, and Prevented?

T.K.A.B. Eskes, R.P.M. Steegers-Theunissen, and M. Wouters

Introduction

Nowadays spontaneous abortion and recurrent abortion constitute the most common problem in obstetric and gynecologic practice. Spontaneous abortion draws the attention of the clinician because of the fact that reproductive loss is increasingly important in women's expectations of pregnancy, and also because there is a significant shift towards pregnancy at a later age. When failure of pregnancy occurs, a couple will raise questions about pathogenesis and future outlook, often leaving the clinician behind with the unsatisfactory feeling of lack of knowledge. This chapter will describe recent developments in the clinical approach towards early pregnancy, and its early failure, using recent scientifically based knowledge, i.e., spontaneous and recurrent abortion.

Definition and Prevalence of Spontaneous Abortion

Definition of Abortion

The World Health Organization (WHO) stated that the concept of abortion should be clear: spontaneous pregnancy loss before 22 weeks of gestation and/or fetal weight of less than 500 g. However, this definition hampers both reproductive science and clinics for the simple reason that causes (and possible treatment) of reproductive loss are quite different in the various phases of pregnancy – the fertilized ovum, tubal transport and nidation, the period of organogenesis in the first 8 weeks, and the fetal period until 16–20 weeks, leading to late "abortion." When something goes wrong just a few days after the expected menstrual period, problems can be expected in the fertilization process. In the organogenetic period all sorts of influences on organ anlage and on further development of organs are at stake and depend on the specific period of pregnancy.

Prevalence of Early Abortion

The given prevalence, incidence, or frequency of early pregnancy loss depends on the methods of detection of pregnancy and the periods under study. In a prospective study [44] using human chorionic gonadotropin (hCG) levels of more than 0.025 ng per milliliter of blood in a fertile population, a 22% pregnancy loss was found before clinical detection of pregnancy. The total rate of pregnancy loss after implantation was 31%. There was no evidence that pregnancies immediately after an early loss were at higher risk than other pregnancies. It is therefore not remarkable to observe that also zygote loss is a time-related process, and that over 60% of conceptuses will be lost before 12 weeks' gestation, the majority occurring before the missed menstrual period [8].

Threatening Abortion

Symptoms

Threatening abortion is usually diagnosed by vaginal blood loss during early pregnancy. Nowadays the first clinical step is, after a detailed history, the performance of an ultrasound examination to detect a viable or nonviable pregnancy, i.e., the presence or absence of an embryo. Clinical signs such as loss of pregnancy symptoms are predictive of decreased chorionic hormonal activity.

Hormone determinations can quantitate the viability of the trophoblast. Reduced β-hCG concentrations in serum are the most common of those hormone determinations. Also high levels of CA-125 can predict spontaneous abortion [28]. It goes without saying that a good clinician always has to "think ectopic" because this type of pregnancy is increasing in frequency and can mimic intrauterine pregnancy at an early stage.

Suggested Treatment

Progesterone

Much has been said on the lack of progesterone, either in the luteal phase of the cycle or in early pregnancy, as a cause of spontaneous abortion. It is more likely, however, that the low secretion rate of progesterone is a symptom of a defect in the maturation, fertilization, and implantation of a nonoptimal oocyte. Therefore it is not surprising that progestagen therapy is of no proven value [38]. Furthermore, 19-norsteroid progestagens can cause virilization of the female fetus and are hazardous to use.

Reynders et al. [34] investigated the clinical and endocrine effects of progestagen therapy in early pregnancy in a double-blind randomized trial in

64 patients who had a viable fetus at 6 weeks' gestation and an increased risk of miscarriage. Using 17-α-hydroxyprogesterone caproate (17-OHP^{-c}) or placebo between 7 and 12 weeks' gestation 17-OHP, prolactin, thyroxin, and thyroxin-binding globulin rose significantly. The relationship between these endocrine parameters and fetal outcome was not clear, and so the authors concluded that 17-OHP^{-c} could not be recommended to prevent miscarriage.

Human Chorionic Gonadotropin

The vital role of hCG is widely accepted in the early stages of pregnancy as it stimulates the corpus luteum. The exogenous administration of hCG therefore seemed a logical approach to therapy. Harrison [15] treated 32 patients with a history of habitual abortion with hCG intramuscularly up to 16 weeks of gestation. No hCG-treated patient aborted, versus seven out of 10 patients in the placebo group. These results, in which 70% of patients in the control group aborted with viable fetus, called for another randomized trial which is currently under way.

All pharmacotherapy for threatening abortion is loaded with the serious history of medication in obstetrics of thalidomide (Softenon) and diethylstilbestrol (DES); the latter then even needed 20 years after exposure to demonstrate the adverse effects of vaginal carcinoma and infertility. This does not hold for the treatment of medical diseases like diabetes, thyroid disease, and lupus where some medication which is potentially hazardous for the offspring has to be taken. In rhesus-negative women anti-D prophylaxis is necessary during early pregnancy blood loss to prevent sensibilization and hydrops fetalis.

When a nonviable pregnancy is diagnosed, curettage is necessary to restore uterine and ovarian function as well as to provide the possibility of studying the products of conception.

Outcome of Pregnancy after Threatened Abortion and Viable Pregnancy

The effect of vaginal bleeding in the first half of pregnancy on fetal outcome has been studied retrospectively [3] and prospectively [12, 17, 20]. All these studies reported a suboptimal outcome, such as preterm birth and its neonatal sequellae such as respiratory difficulties and hyperbilirubinemia. Placenta previa and manual removal of the placenta also occurred more frequently.

Post-abortion (Ectopic) and Future Outlook

The postabortion period is important to support women following their loss, to accumulate data for the pathogenesis of this loss, and to plan a design for a future pregnancy. The risk of subsequent pregnancy losses is estimated to vary between 23.7% and 32.2% [11] with no definite correlation with the number of abortions.

The Products of Conception

When spontaneous abortion occurs, examination of expelled tissues by a gynecologic pathologist is necessary to document that a miscarriage has occurred. Because in most cases of spontaneous abortion there is a lack of data, the chorionic villi (and occasionally the embryo) have to tell us "the story." Rushton [35] described three types of histopathology of chorionic tissue:

Type I. The majority of villi show microscopic hydatidiform change, the villi being large and edematous. Trophoblastic inclusions can be found.

Type II. The majority of villi show postmortem fibrotic changes.

Type III. The majority of villi show a normal aspect, containing fetal vessels in which normoblasts can often be found.

Type I villi showed the highest correlation with chromosomal abnormalities of the conceptus. This is important because chromosome cultures of the conceptus are not always available. Quantitative morphology of chorionic villi demonstrated that the vascular density of chorionic villi was three to four times lower in cases of embryonic death or blighted ova than in controls [24].

Ectopic pregnancy can also be seen as an early reproductive loss. This condition can be compared to the spontaneous intrauterine abortion in those cases in which tubal causes of ectopic pregnancy are lacking [7]. Schermers [37] made a prospective study of 38 women with an ectopic pregnancy. Salpingitis was found in only 23.5%. Low hCG values were found in cases with a low quantity of chorionic villi. Low progesterone values correlate with pathologic ova. Systemic follow-up demonstrated abnormalities of the menstrual cycle (35%), glucose metabolism (25%), congenital uterine malformations (18%), and hypothyroidism (13%). Pathologic ova were found in 78% of cases. This finding closely resembles data of Poland and Yen [31] in specimens of spontaneous abortion.

Medical Disease and Spontaneous Abortion

Epilepsy

Relatively high rates of spontaneous abortion in early pregnancy have previously been reported for women with epilepsy [1]. Lack of adjustment of the gestational age at which the spontaneous abortion occurred and

Table 1. Characteristics of epilepsy and maternal folate concentrations in the 4th postconceptional week

		Spontaneous abortion ($n = 20$)		Normal outcome ($n = 48$)	
		Median	Range	Median	Range
Age	(years)	28	23–35	28	19–36
Primiparous	(n)	11		21	
Duration of epilepsy	(years)	18.5	6–31	12[a]	0–25
Duration of anticonvulsant therapy	(years)	15.5	6–31	11[b]	0–25
Folate					
Serum	(nmol/l)	12	2–590	13[c]	4–500
Erythrocytes	(nmol/l)	530	110–1400	500[c]	57–2100

[a] $p < 0.01$, Wilcoxon two-sample test.
[b] $p < 0.05$, Wilcoxon two-sample test.
[c] $n = 46$.

absence of a control group do not permit many conclusions about the relative risk of spontaneous abortion. We therefore prospectively studied 247 women with epilepsy and 181 controls enrolled before conception [41]. The abortion rate (up to 16 weeks) did not differ between groups, and no differences in vitamin-folate levels in serum and erythrocytes were seen in the 4th postconceptional week.

The spontaneous abortion group had a significantly longer duration of the epilepsy and a longer use of anticonvulsant therapy. This suggests that women with epilepsy should not delay pregnancy (Table 1).

Women who have taken antiemetic drugs containing vitamins or minerals have fewer abortions than women who did not take these drugs [21]. Our finding that a derangement in homocysteine metabolism (see below) could be an explanation for the pathogenesis of spontaneous abortion and that prevention of this inborn error of metabolism might be possible with specific vitamin therapy [42] deserves further study.

Diabetes

Mills et al. [25] found a 16% pregnancy loss in women with insulin dependency in whom pregnancy was identified within 3 weeks of conception. This was virtually the same rate as in a control group. Pregnancy loss in the diabetic group was associated with higher glycosylated hemoglobin values, suggesting a relationship with poor diabetic control. Mulder [26] could not substantiate a statistically significant relationship between the quality of maternal periconceptional glucose control and the degree of fetal growth delay.

During early pregnancy in women with type I diabetes, the embryo and fetus are smaller than normal [29]. Embryos and fetuses of women with type I diabetes often show early growth delay, delayed emergence of movement patterns, and disturbances in the development of behavioral states [27].

The pathogenesis of this deviation of normal reproductive outcome in case of diabetes and pregnancy still exists and needs to be clarified. Attention has to be focussed on the external and internal environment of the mother, including metabolism, and genetic predisposition. The finding of Simpson et al. [39] that DR 3^+ and DR 4^+ diabetic women have an increased risk for congenital malformations (and abortion?) is of interest.

Lupus Erythematosus

In lupus erythematosus (LE) the abortion rate is higher compared with the general population [10]. Antibodies to phospholipids are associated with recurrent spontaneous abortions, and an immune complex deposition was observed on the trophoblastic membrane [13]. Efficacy of treatment with prednisone (40–60 mg daily) or low-dose aspirin (80 mg daily) is claimed in the literature to be of benefit. Multicenter randomized trials with placebo are lacking, however.

Ovulation Disorders and Ovulation Induction

Luteinizing hormone (LH), under the influence of luteinizing hormone releasing hormone (LHRH) is crucial for ovum maturation. Oocyte maturation starts during fetal life. The number of oocytes reaches a nadir in the 6th month of pregnancy [2]. The first meiotic division is arrested in utero at the diplotene (or dictyate) stage. During the menstrual cycle in extrauterine life, the entry of LH into the preovulatory follicle at midcycle completes the first meiotic division. About 36 h later the released oocyte is mature and ready for fertilization, a time process which is within rather strict limits. From studies with donor insemination [14] it is known that, when the interval between ovulation and insemination is extended, poor rates of fertilization as well as a higher abortion rate occur.

The Key Role of LH

Abnormal secretion of LH can directly affect the maturation process of oocytes. Patients with polycystic ovarian disease who require treatment with clomiphene to induce ovulation are at increased risk of early spontaneous abortion [23, 45]. Pituitary suppression with buserelin followed by pure follicle-stimulating hormone significantly reduces the risk of spontaneous abortion in women with polycystic ovarian disease and primary recurrent spontaneous abortions [19].

Regan et al. [33] demonstrated a relationship between hypersecretion of LH, infertility, and miscarriage. They studied 193 women prospectively; there were 26 nulliparous and 167 multiparous women before conception. One blood sample was obtained in the early follicular phase of the cycle. Of the 147 women with LH concentrations of less than 10 IU/l (normal LH group), 130 (88%) conceived, whereas only 31 (67%) of the 46 women with LH values of 10 IU/l or more (high LH group) did so.

Of the 193 women, 161 became pregnant and 35 of those had a miscarriage. In the normal LH group 15 pregnancies (12%) ended in miscarriage, whereas in the high LH group 20 (65%) did so ($p < 10^{-8}$).

The finding of Regan et al. [33] can identify women before conception who have an endocrine abnormality and a substantial risk of infertility and miscarriage. Treatment (or prevention) also seems possible, for instance, with analogues of LHRH and is therefore worthwhile.

The Growth Model

Exalto et al. [9] developed a growth model for early pregnancy. Prospectively, ultrasound and hormonal parameters could be followed at weekly intervals during pregnancy. In 31 women with recurrent abortion (three or more), recurrence occurred in eight (25%).

In all cases trophoblast function (hCG; human placental lactogen: hPL) decreased followed by steroid production of the corpus luteum (17-β-estradiol, 17-α-hydroxyprogesterone, and progesterone). The diameter of the "pregnancy ring" was not predictive for spontaneous abortion. In two cases there was evidence for insufficiency of the corpus luteum and hypothyroidism, respectively. In this growth model a pregnancy can be studied prospectively, enabling the clinician to detect the exact moment of disturbance.

The Role of Transvaginal Ultrasound

Using high-frequency transvaginal ultrasonography Timor-Tritsch et al. [43] found that the gestational sac could first be detected in the 5th postconceptional week. Visualization of the heart could be done as early as the 6th week. Vaginal ultrasound shows promise in the evaluation of human embryogenesis [36].

Recurrent Abortion

There is no generally agreed definition of recurrent abortion. Two, three, or more consecutive abortions are used as criteria for primary abortion, and a live-born child before, in between, or after as secondary abortion.

Identifiable Causes

It seems likely that the cause of the majority of 1st-trimester spontaneous abortions is chromosomal anomalies, mainly of trisomic or triploid composition, in the conceptus [5]. The pathogenesis of these chromosomal abnormalities occurring during conception are, however, unknown. In approximately 5% of cases, parental chromosomal rearrangements such as balanced translocation and pericentric inversion (among others) can be found [32]. Clinicians should initiate a diagnostic evaluation after two spontaneous 1st-trimester abortions. Evaluation of a couple with recurrent pregnancy loss consists of chromosome analyses, hysteroscopy, the study of the menstrual cycle, HLA typing, lymphocytotoxic antibodies, mixed lymphocyte cultures, auto-antibodies (to nuclear antigens and phospholipids), and diseases such as diabetes, thyroidism, and lupus erythematosus. Individuals with two or more consecutive pregnancy losses in the first trimester have about a 35% chance of having a spontaneous abortion in the next pregnancy. So even without "treatment," 65% of the women can have a successful pregnancy [18].

Nonidentifiable Causes

The evaluation of causes of recurrent abortion will fail to determine abnormalities in 30%−40% of such couples. This means that more studies must be initiated on early pregnancy failure.

Therapy for Recurrent Abortion

Therapies for recurrent abortion are numerous, but many are of no proven value. Also, many of the abnormalities found can be associated but not causally related to the condition. Endocrine adjustments for thyroid function or metabolic diseases such as diabetes seem obvious, however. Anatomic uterine anomalies and cervical incompetence are associated with a mid-trimester loss and late abortion. An anatomic uterine defect can also include a loss of function [16].

White cell immunization has been recommended in patients with recurrent abortion. However, at the VIIth World Congress on Human Reproduction (Helsinki 1990) disappointing results were reported, and this form of treatment is now thought to be ethically inappropriate. Houwert-de Jong [18] described 44 couples with a history of three or more consecutive spontaneous abortions of unknown etiology. She found no aberrant immunologic reactivity, and without treatment 62% of the women delivered an infant in the first subsequent pregnancy. The histopathogic status of recurrent aborters showed a 61% incidence of abnormal villi suspect for fetal chromosomal anomalies compared with 58.5% in a control group

[18]. It was therefore suggested that fetal abnormalities are the etiology of recurrent abortion in the majority of cases. It therefore seems that "tender loving care" seems a more appropriate "therapy" for recurrent abortion, especially in cases of unidentifiable causes.

Future Developments

The Estrogen Receptor Gene

Obstetric histories of breast cancer patients with a variant estrogen receptor gene were compared with those of breast cancer patients with the wild-type gene. The women with the variant gene had a significantly higher incidence of spontaneous abortion [22], leading to the suggestion that the variant estrogen gene protein may have altered bioactivity and may not be able to maintain pregnancy.

Derangement of Homocysteine Metabolism

Maternal folate deficiency has been implicated in the genesis of spontaneous abortion. Folate as well as vitamins B_6 and B_{12} are involved in the metabolism of methionine and its intermediate product homocysteine. We hypothesized that recurrent abortion could be due to an inborn error of methionine metabolism. On the 21st day of the menstrual cycle, a standardized oral methionine loading test [4] was performed in 14 women with unexplained recurrent abortion. All women had normal liver and kidney functions, and fasting glucose and total cholesterol levels. Fasting levels of whole blood vitamin B_6, serum vitamin B_{12}, and red cell folate were not significantly different in the study group and controls. Surprisingly, 6 h after methionine loading four patients had serum peak levels of total homocysteine exceeding the mean plus twice the standard deviation in the control group. This suggests heterozygosity for homocystinuria. The same mechanism was found in women with an offspring with neural tube defects [42], suggesting that an early open neural tube defect could be a first step in the mechanisms of spontaneous abortion.

Pathologic homocysteinemia can be normalized with high dose vitamin B_6 or folic acid. It will therefore be intriguing to relate the positive effect of the use of multivitamin preparations in the periconceptional period to its beneficial effect upon a pre-existing methionine tolerance.

Summary

Early pregnancy wastage is a time-related phenomenon. In the majority of cases, pathology of the fertilized ovum is present. Treatment of threatening abortion is not advocated, but diagnosis, identification and the study of the conceptus is essential for a future outlook. Morphology of the trophoblast can differentiate between normal fetal circulation or suggest chromosomal abnormalities. Spontaneous abortion may have the same underlying mechanisms as some cases of ectopic pregnancy or recurrent abortion.

In cases of epilepsy women should not delay pregnancy because there is a statistical correlation between the duration of the disease (and the use of anticonvulsants) and the occurrence of spontaneous abortion. In cases of diabetes meticulous control in the periconceptional period seems essential. Infertility in ovulation disorders with abnormal LH levels can be treated with pituitary suppression before conception. Recurrent aborters who are carriers of a derangement of methionine-homocysteine metabolism can be treated with specific vitamin supplementation as to be proven in a clinical trial, currently undertaken.

References

1. Annegers JF, Baumgartner KB, Hauser WA, Kurland JT (1989) Epilepsy, anticonvulsants and the risk of spontaneous abortion. Epilepsia 29:451–458
2. Baker TG (1963) A quantitative and cytological study of germ cells in human ovaries. Proc R Soc Biol 158:417–433
3. Batzofin JH, Fielding WL, Friedman EA (1984) Effect of vaginal bleeding in early pregnancy on outcome. Obstet Gynecol 63:515–518
4. Boers GHJ, Smals AGH, Trijbels JMF, Fowler B, Bakkeren JAJM, Schoonderwaldt HC, Kleijer WJ, Kloppenborg PWC (1985) Heterozygosity for homocystinuria in premature peripheral and cerebral occlusive arterial disease. N Engl J Med 313:709–715
5. Boué J, Boué A, Lazar P (1975) Retrospective and prospective epidemiological studies of 1500 karyotyped spontaneous human abortions. Teratology 12:11–26
6. Dansky LV, Anderman E, Rosenblatt D, Sherwin AL, Andermann F (1987) Anticonvulsants, folate levels and pregnancy outcome: a prospective study. Ann Neurol 21:176–182
7. Eskes TKAB, van Oppen ACC (1984) Ectopic pregnancy: not only a tubal disease. Eur J Obstet Gynecol 18:391–394
8. Exalto N, Rolland R (1985) The nature of pregnancy wastage. In: Rolland R, Heineman MJ, Hillier SG, Vemer H (eds) Excerpta Medica, Amsterdam, pp 303–311
9. Exalto N, Rolland R, Eskes TKAB, Vooys GB (1982) Early pregnancy. Boehringer/ Ingelheim, Alkmaar
10. Fraga A, Mintz G, Orozco JH, Orozco J (1973) Systemic lupus erythematosus: fertility, fetal wastage and survival rate with treatment. A comparative study. Arthritis Rheum 16:541
11. Fraser FC, Fainstat TD (1951) Causes of congenital defects. A review. Am J Dis Child 82:593–603

12. Funderburk SJ, Guthrie D, Meldrum D (1980) Outcome of pregnancies complicated by early vaginal bleeding. Br J Obstet Gynaecol 87:100–105
13. Grennan DM, McCormick JN, Wojtacha D, Carty M, Behan W (1978) Immunological studies of the placenta in systemic lupus erythematosus. Ann Rheum Dis 37:129
14. Guerrero R, Rojas OI (1975) Spontaneous abortion and aging of human ova and spermatozoa. N Engl J Med 293:573–575
15. Harrison RF (1985) Treatment of habitual abortion with human chorionic gonadotrophin: results of open and placebo controlled studies. Eur J Obstet Gynecol Reprod Biol 20:158–168
16. Hein PR, Stolte LAM, Janssens J, Braaksma JT, Kars-Villanueva EB, van der Harten JJ, de Jong PA (1974) The motility of the nonpregnant congenitally malformed uterus. Eur J Obstet Gynecol Reprod Biol 4(2):51–60
17. Herz JB, Heisterberg L (1985) The outcome of pregnancy after threatened abortion. Acta Obstet Gynecol Scand 64:151–156
18. Houwert-de Jong MH (1988) Habitual abortion: views and fact binding. Thesis, University of Utrecht
19. Johnson P, Malcolm Pearce J (1990) Recurrent spontaneous abortion and polycystic ovarian disease: comparison of two regimens to induce ovulation. Br Med J 300:154–156
20. Jouppila P, Koivisto M (1974) The prognosis in pregnancy after threatening abortion. Ann Chir Gynaecol Fenn 63:439–444
21. Koller S (1982) Chancen der Abortprophylaxe in der Frühschwangerschaft. Geburtshilfe Frauenheilkd 42:204–212
22. Lehrer S, Sanchez M, Song HK, Dalton J, Levine E, Savoretti P, Thung SN, Schachter B (1990) Oestrogen receptor B-region polymorphism and spontaneous abortion in women with breast cancer. Lancet 335:622–624
23. Maggregor AH, Johnson JE, Bunde CA (1968) Further clinical experience with clomiphene citrate. Fertil Steril 19:616–620
24. Meegdes BHLM, Ingenhous R, Peeters LLH, Exalto N (1988) Early pregnancy wastage: relationship between chorionic vascularization and embryonic development. Fertil Steril 49:216–220
25. Mills JL, Simpson JL, Driscoll SG, et al. (1988) Incidence of spontaneous abortion among normal women and insulin-dependent diabetic women whose pregnancies were identified within 21 days of conception. N Engl J Med 319:1617–1623
26. Mulder EJH (1990) Fetal behaviour. Studies on normal and diabetic pregnancy. Thesis, University of Groningen
27. Mulder EJH, Visser GHA, Bekedam DJ, Prechtl HFR (1987) Emergence of behavioural states in fetuses of type-1 diabetic women. Early Hum Dev 15:231–252
28. Nowroozi K, Check JH, Adelson HG (1990) Periderm abnormalities may be predicted by elevated CA-125 levels during first few weeks following conception. European Society of Human Reproduction and Endocrinology. Congress, Milan
29. Pedersen JF, Mølsted-Pedersen L (1979) Early growth retardation in diabetic pregnancy. Br Med J 1:18–19
30. Poland BJ (1984) Recurrent spontaneous abortion (review). Eur J Obstet Gynecol Reprod Biol 16:369–375
31. Poland BJ, Yen BH (1978) Embryonic development in consecutive specimens from recurrent spontaneous abortions. Am J Obstet Gynecol 130:512
32. Portnoi MF, Joye N, van den Acker J, Morlier G, Taitlemite JL (1988) Karyotypes of 1142 couples with recurrent abortion. Obstet Gynecol 72:31–34
33. Regan L, Owen EJ, Jacobs HS (1990) Hypersecretion of luteinising hormone, infertility, and miscarriage. Lancet 336:1141–1144
34. Reynders FJL, Thomas CMG, Doesburg WH, Rolland R, Eskes TKAB (1988) Endocrine effects of 17 alpha-hydroxyprogesterone caproate during early pregnancy: a double-blind clinical trial. Br J Obstet Gynaecol 95:462–468
35. Rushton D (1981) Examination of products of conception from previable human pregnancies. J Clin Pathol 34:819–835

36. Schatz R (1991) Transvaginal sonography in early human pregnancy. Thesis, Erasmus University, Rotterdam
37. Schermers JP (1984) Ectopic pregnancy. Thesis, Free University, Amsterdam
38. Shearman RP, Garrett WJ (1963) Double blind study of the effect of 17-hydroxyprogesterone caproate on abortion rate. Br Med J 1:292–295
39. Simpson JL, Mills JL, Ober C, Holmes LB, Knopp R, Jovanovic L, Metzger BE, Arond JH, Brown Z, Comley M (1990) DR 3 + and DR 4 + diabetic women have increased risk for anomalies: evidence for genetic susceptibility in diabetic embryopathy (Abstr 391). Society for Gynecologic Investigation, St Louis
40. Smith JB, Cowchock FS (1988) Immunological studies in recurrent spontaneous abortion: effects of immunization of women with paternal mono nuclear cells on lyphocytotoxic and mixed lymphocyte reaction blocking antibodies and correlation with sharing of HLA and pregnancy outcome. J Reprod Immunol 14:99–113
41. Steegers-Theunissen RPM, Renier WO, Eskes TKAB (1991) Epilepsy, spontaneous abortion and vitamins. (to be published)
42. Steegers-Theunissen RPM, Boers GHJ, Trijbels FJM, Eskes TKAB (1991) Derangement of homocysteine metabolism and neural-tube defect. N Engl J Med 324:199–200
43. Timor-Tritsch IE, Farina D, Rosen MG (1988) A close look at early embryonic development with the high-frequency transvaginal transducer. Am J Obstet Gynecol 159:676–681
44. Wilcox AJ, Weinberg CR, O'Connor JF, Baird DD, Schlatterer JP, Canfield RE, Armstrong EG, Nisula BC (1988) Incidence of early loss of pregnancy. N Engl J Med 319:189–194
45. Yen SSC (1980) The polycystic ovary syndrome. Clin Endocrinol (Oxf) 12:177–208

30 Theological Views on Assisted Reproductive Technologies

J.G. Schenker

Introduction

Religion and science have been interrelated since the beginning of human history. The most striking change in the last 2 decades has been the secularization of bioethics. The identifiable religious influence on bioethics subsequently seemed to decline. The field of bioethics has moved from one dominated by religious and medical tradition to one now increasingly shaped by philosophical, social, and legal concepts.

Nowadays, religious groups still exert influence on the civil authorities in the field of reproduction, such as prevention of procreation, and in issues such as abortion and infertility therapy. Recent development in science and technology in the field of reproduction raise new religious questions which do not always have clear answers. The role of theology in bioethics is first of all to clarify for the different religious communities what their attitude to the new developments in the new fields of reproduction should be.

Since religious groups have been active in pressing their bioethical concepts on the public arena in different parts of the world, it is of importance to those who practice assisted reproduction to learn about the religious attitudes related to the problem of infertility and its therapeutic approach.

Attitude of the Modern Religions to the Problem of Infertility and Therapy

Jewish View

The Jewish attitude towards infertility can be learned from the fact that the first commandment of God to Adam was "Be fruitful, and multiply" (Genesis, 1:28). This is expressed in the Talmudic saying from the second century: "Any man who had no children is considered as a dead man." This attitude arises from the Bible itself and refers to 1700 years before the modern era, from the words of Rachel, who was barren, to her husband Jacob: "Give me children, or else I die" (Genesis, 30:1).

The infertile couple should be diagnosed and treated as a single unit. The medical treatment is different for men and women. When evaluating an infertile couple according to the Halakha (Jewish law), one should first evaluate the female factor. If pathology is found, one may proceed to investigate the male factor: inadequate or abnormal production, ejaculation, and deposition of spermatozoa.

According to the Jewish law, the preferred method of seminal fluid analysis should be the postcoital test, which is based on examination of motile sperm in a mucus sample collected several hours after coitus. If the results are inconclusive or abnormal (repeated several times), ejaculate should be collected following coitus interruptus, using a special condom. Examination of the semen for infertility work-up is not included in the prohibition "spilling one's seed," and, if the methods mentioned above are not possible because of mechanical, technical or emotional reasons, some rabbis permit the collection of an ejaculate induced by masturbation.

Christian View

The major branches of Christianity are the Roman Catholic Church, the Eastern Orthodox Church, and the Protestant Churches. According to the Christian view, there is no absolute right to parenthood. It is very important, but marriage does not have to have children for it to be a valid marriage. A Christian infertile couple should not indulge in behavior that undermines marriage or family or that is unacceptable in their efforts to become parents. Christians view parenting not as reproduction but as procreation. The difference relates to the purpose and act of parenting. The reason for procreation is to act as an agent of God – given creativity for the sake of God's purpose as discerned in Jesus Christ. Creating new life on God's behalf (procreation) means that the parents do not own their children as objects and children to not exist for their parents' needs.

The right to live is a fundamental one. It is the root of all rights and privileges of a human being. In the Roman Catholic Church there are principles which are very important and which guide the believers. The first principle is related to the protection of human being from its very beginning – from conception. The second is that procreation is inseparable from the psychoemotional relation of parents. This means that procreation is not performed by the physician; the physician has to be in the position to help the parents achieve conception, but not the one who is a "baby maker." The third principle is related to the personal norm of human integrity and dignity and it should be taken into consideration in medical decisions, especially in the field of infertility treatment.

Islamic View

According to the Islamic view, attempts to cure infertility are not only permissible but even a duty. The Qur'an, as well as the Old Testament, tell

of Abraham's and Zakariya's view that to have progeny is a great blessing from God. The pursuit of a remedy for infertility is therefore quite
legitimate and should not be considered as rebellion against the fate decreed
by God. Prophets of God who were childless incessantly asked their Lord to
give them children. But the treatment of infertility should by no means
trespass outside the legitimacy as ordered by God. To have offspring is
considered a blessing from God and fulfills an instinctive need. The duty of
the physician is to help the barren couple to achieve successful fertilization,
conception, and delivery of a baby.

Buddhist View

According to Buddhism there are three factors necessary for the rebirth of a
human being: the female ovum, the male sperm, and the karma. This
Karma energy is set forth by the dying individual at the moment of his or
her death. Father and mother only provide the necessary physical material for the formation of the preembryo. Marriage within Buddhism is
not of high priority to that found in monotheistic religions like Judaism,
Christianity, and Islam. Sexual activity is not especially encouraged. Marriage is considered as the second-best institution after monastic life.

Traditionally, Buddhism has imposed strict ethics on priests, while it
has taken relatively lenient attitudes towards lay people. This means that
Buddhist priests allow lay people to do whatever they want to do as long as
they do not harm others in a concrete way. This leads us to an idea that we
do not have to accept infertility as it is. If medical treatment for infertility is
available, we can make use of it. Generally speaking, having no children will
be a greater threat to a marital relationship than the practice of modern
technology of infertility. Historically, we see a lot of societies where people
do not pay respect to those women who cannot bear a child. Taking this into
consideration, it is very difficult for any society to find a rational reason to
reject a woman's desire to have a child. Therefore, any technology to
achieve this goal, according to Buddhism, is morally acceptable, and treatment should be given to unmarried as well as to married couples.

It should be mentioned that Buddhism has never been organized around
a central authority, therefore Buddhists of all types in various countries have
been comparatively individualistic and even their scriptures are not rigid.

Hindu View

Hinduism exerts its influence from the power of its thought over society and
not by formal institutional authority. The important concepts of the Hindu
religion related to the problem of infertility are:

1. The soul is eternal and indestructable.

2. Regarding the concept of rebirth, the soul (atman) is viewed as not having a definite beginning and a definite end. It has a continuous process.
3. Marriage is considered sacred and permanent.
4. Male infertility is not a cause for divorce.
5. The emphasis on reproduction is not just on having children, but on having a male offspring.
6. It is a religious duty to provide a male offspring. Therefore, the wife of a sterile male could be authorized to have intercourse with a brother-in-law or another member of the husband's family for the purpose of having a male offspring.

Assisted Reproductive Technologies

Jewish View

The various aspects of the "test-tube baby" are of considerable interest in the rabbinic literature (responsa). The basic fact which allows in vitro fertilization (IVF) and embryo transfer (ET) to be considered in the rabbinic literature at all is that the oocyte and the sperm originate from the wife and husband, respectively. The attitude favored by the Jewish religious authorities with regard to IVF and ET is based on the commandment of procreation mentioned in the Bible.

What are some of the delineating factors which would nevertheless withhold Jewish law from proceeding with IVF and ET? Some of the rabbis take a strict position and suggest that the legal and biologic ties are severed with the removal of the egg. Since the environment is sustained by medical intervention, using different culture media, this might change the biologic and legal status of the child. From the Jewish religious point of view, the Chief Rabbis of Israel, one from the Ashkenazi sector (European origin) and one from the Sephardic sector (oriental origin), support IVF and ET procedures. Jews living outside Israel are generally subjected to the laws of the country in which they live, except in cases where they wish or are required to obey the Jewish traditional personal status regulations. In such cases, the rules applicable in the state of Israel will also be applied by local rabbinic authorities, when such exist and are recognized.

Christian View

Roman Catholic Church. The Vatican's statement on IVF is very clear. It does not accept IVF as a method of procreation. In 1959 Pope Pius XII declared that attempts of artificial human fecundation in vitro must be rejected as immoral and absolutely unlawful. The arguments of the Roman Catholic Church against the practice of IVF are: (a) IVF involves disregard for human life; (b) IVF separates the human procreation from human sexual intercourse. The Vatican's instructions on respect for human life made a significant contribution to the discussion of the practice of new reproductive technologies. It states that noncoital technologies are morally illicit. The instructions call for the "reform of morally unacceptable civil laws for the correction of illicit practice" regarding assisted reproductive technologies. It also advocates making it a crime.

Eastern Orthodox Church. The Eastern Orthodox Church supports medical and surgical treatments for infertility. Nevertheless, IVF and other assisted reproductive technologies are absolutely rejected.

Protestant Churches. The Protestant Churches accept traditional treatment of infertility. Assisted reproductive technologies are acceptable only if the gametes are from the married couple and the procedure avoids damage to the preembryo.

Anglican Church. The practice of IVF, gamete intrafallopian transfer (GIFT), zygote intrafallopian transfer (ZIFT), etc. using gametes from a married couple is morally accepted. There is controversy regarding the use of donor sperm and oocytes.

Islamic View

The procedure of IVF and ET is acceptable by Islam. It can be practiced only if it solely involves husband and wife and if it is performed during the span of their marriage. According to the Islamic view, the fusion of sperm and egg, a step further than the sexual act, should take place only within the legal marriage contract.

Buddhist View

The Buddhists in different countries do not have an authorized view on the new reproductive technology. IVF has been practiced in Japan since 1982 and is also practiced in other countries with a Buddhist population. According to Buddhism, there are three necessary factors for the rebirth of a human being: the ovum, the sperm, and karma energy. This karma energy is set forth by the dying individual at the moment of his or her death. The problem that arises is whether the oocyte and the sperm have to be in their natural environment or not. If the crucial factor in the formation of human being is the karma energy, it is not necessary for the sperm and the ovum to be in their natural environment and therefore IVF should be acceptable. On the other hand, if these three entities have to be in their natural environment, the new technology of reproduction becomes effectively unacceptable. The practice of IVF raises another dilemma from the Buddhist point of view since it is involved in procreation of more preembryos than the number that is implanted in the uterus.

Hindu View

According to Hinduism, it is a religious duty to provide a male offspring. There is no objection to masturbation or spilling of seed. Assisted reproductive technology is practiced in India, and, even though there is no information in the literature on the Hindu's attitude to this practice, it seems that this new technology is not forbidden.

Artificial Insemination by Donor

Artificial insemination by donor (AID) is indicated in cases of uncurable male infertility or when the husband is a carrier of a serious inherited disease or abnormality. It is extensively used throughout the world, and today thousands of infertile couples are experiencing the satisfaction of raising children who were conceived through an AID program. Fresh or frozen semen from a donor has been used in assisted reproduction in different countries. The practice of AID is not morally acceptable to all infertile couples or their physician.

Jewish View

The various attitudes of the different rabbinic scholars to the way religion should be applied in the changing world are analyzed and discussed with regard to the legal codes and responsa literature (written opinion given by qualified authorities on questions or aspects of Jewish law).

For many centuries rabbis have discussed the principles involved in artificial insemination from a donor. The discussions are based on ancient sources in the Talmud and the codes of Jewish law. From the fifth century, Talmudic passages show that procreation without intercourse was recognized as possible by the sages of old – with regard to the midrashic legend of Ben Sira's birth.

The legend is that the prophet Jermiah went to the bath house, where his semen entered the bath water. Soon after, Jermiah's daughter came and had a bath and became pregnant by her father, and this resulted in the birth of Ben Sira, who was recognized as a legitimate child. This legend has since been quoted many times in medical literature and in rabbinic responsa dealing with AID. Some rabbinic scholars do not believe the legend that Ben Sira's birth followed a conception since concubito.

Another ancient text which indicates the possibility of conception occurring without sexual intercourse is mentioned by Rabbi Elisah of Corbell in his work Haggahot Semak of the thirteenth century: "A woman may lie on her husband's sheets but should be careful not to lie on sheets on which another man has slept, lest she becomes impregnated by his semen."

In the modern era, rabbinic authorities have extensively discussed the practice of AID. The discussions have focused on two issues: (a) is it permissible, according to Jewish law, to perform AID, or is the very act a transgression to adultery; (b) what is the status of the AID offspring?

All Jewish legal experts agree that AID, using the semen of a Jewish donor, is forbidden. It is the severity of the prohibition that is debatable. The question is whether AID constitutes adultery, which is strictly forbidden by the Torah (the Pentateuch), or whether the injunction stems from other sources – mainly the legal complications of the birth of AID offspring – as most of the experts hold. Some rabbinic authorities permit AID when the donor is a non-Jew. This eliminates some of the legal complications in regard to the personal status of the offspring. If the donor of the semen is a gentile, the child is pagan (blemished); if the child is a female, she is forbidden to marry a Cohen (priest).

AID, according to Jewish law, is prohibited for a variety of reasons: incest, lack of geneology, and the problem of inheritance. Semen donors for AID, as well as the physicians who use the semen, are violating the severe prohibition against masturbation.

Some Halakhic authorities prohibited a married woman who underwent the procedure of AID from continuing to live with her husband. However, most rabbis state that, as intercourse is not involved, the woman is not guilty of adultery and is therefore not forbidden to cohabit with her husband. If the woman underwent AID without her husband's knowledge and consent, she must accept a divorce and forfeit any rights to financial support, including her Ketubah and alimony.

AID may not be used to fulfill the rrequirement of levirate (the obligation a man has to marry his deceased brother's wife). The child conceived through AID is considered by many rabbinic scholars to have the status of a "mamzer," a bastard. The mark of bastardy severely limits the offspring's prospects for marriage in keeping with Jewish law, and this implies a severe functional handicap from a social point of view. Some rabbis consider the offspring to be legitimate, as Ben Sira was considered, while others consider it as a "safek mamzer."

There are various views regarding the legal relationship that exists between the semen donor and the child born as a result of AID. Some rule that no relationship exists, others that the child should be regarded as the donor's child, with all the legal complications of incest, inheritance, levirate marriage, etc. The majority opinion is that the donor has not fulfilled the mitzvah (obligation) of procreation by fathering an AID child. The practice of AID is accepted by part of the Jewish population in Israel and, according to the regulations of the Ministry of Health, it is allowed under special circumstances.

Although it is difficult to see fertilization with a donor's sperm as an act of adultery, there may still be a legal prohibition against IVF of an oocyte with a donor's sperm. This prohibition does not affect an unmarried woman, as long as the possibility of bastardy is excluded. Those scholars who oppose the IVF procedure, claiming that there is no paternal relationship in the IVF and ET procedures, may give an advantage to the offspring resulting from the use of donor sperm in the IVF technique. If there is no kinship, the offspring cannot be regarded as the product of an unacceptable genetic union and is thus at least as good for society as the offspring of a non-Jew and can marry a Jewess. In this case, the paternity of a non-Jewish genetic father is not recognized, while the status of the offspring is that of an ordinary Jew, almost without exception.

Christian View

The Roman Catholic Church has condemned AID for married, as well as unmarried women. The Vatican's instructions demand strict connection between procreation and intercourse. Artificial insemination involves separation between "the goods and meaning of marriage" (which means separation between unitive and procreative). Concerning AID, the instructions suggested that the AID process damages personal family relations as well as the offspring. The practice of artificial insemination is also rejected on the grounds that it is based on masturbation and that AID process is an adulterous act. The instruction asserts that "What is technically possible is not, for that very reason, morally admissible."

The Greek Orthodox Church opposes the practice of AID on the basis that it is an adulterous act. All forms of artificial insemination which involve masturbation are morally dubious. The desire to procreate does not justify what is morally illicit.

For the Protestant Churches AID is mortally illicit and, at best, morally questionable.

The Anglican Church allows semen collection by masturbation for artificial insemination by husband and for IVF. However, considerable controversy surrounds the use of sperm from a donor.

Islamic View

The practice of AID is strictly condemned by Islamic law. If the husband's infertility is beyond cure, the infertility should be accepted; according to Islamic law, AID is considered to be adultery. It enhances the chances of inadvertent brother-sister marriage in a community and it violates the legal system of inheritance. The procedure also entails the lie of registering the offspring of a man who is not the real father and therefore leads to confusion regarding lines of geneology whose purity is of prime importance to Islam. Therefore AID is not practiced at all by Moslems.

Hindu View

According to the Hindu view, in the case of male infertility the wife can be authorized to have intercourse with a brother-in-law or another member of the husband's family in order to obtain conception of a male offspring (only after 8 years of infertility or 11 years when she delivers only female offspring). According to the above statement, this may lead to the conclusion that sperm donation can be practiced according to the Hindu view, with the restriction that the sperm donor has to be a close relative of the husband.

Buddhist View

According to Buddhism, donation of sperm is not prohibited, but it is suggested to refrain from this procedure as much as possible. The reasons for it are as follows: (a) the parents, in general, may feel difficulties in taking care of a child who does not have their own genes, especially when a malformed child is born by means of gamete donation; (b) there is a danger that a donation of sperm from a third party would involve commercialization and put our society into problems; (c) donation of sperm will lead to eugenics, which may be reflected by strong social discrimination.

A Buddhist sutra says that we must not seek what we do not have, meaning that we should exclude a third person's intervention in the process of reproduction, because once the donation is accepted, there will be many social problems to be settled regarding the process of reproduction. Unfortunately, we human beings do not have wisdom enough to overcome them. Therefore, donation should be limited. However, a child who was already born following AID should be accepted as the legitimate child of the social father who has given consent to this practice. However, the child also has a right to know its genetic father when it reaches maturity. Therefore, the child's right to know his or her biologic parents should be accepted.

Ovum Donation

Successful transfer of in vitro fertilized donor oocytes has extended our ability to treat female infertility due to primary or secondary ovarian failure and in cases of genetic abnormalities. There should not be any ethical

problems with oocyte donation in societies where AID is acceptable. Oocyte
donation has an advantage over AID as both parents contribute to the birth
of the child.

Jewish View

In the case of egg donation or embryo donation, the problem that arises is
who should be considered the mother, the donor of the oocyte or the one in
whose uterus the embryo develops, the one who gives birth. In the case one
of the woman is Jewish and the other is not, the problem of the status of the
child, whether he or she is a Jew or not, will arise. Since, according to
Jewish law, the religious status of the child is determined by its mother, and
the child obtains the religion of its mother. This interesting subject has an
apparent precedent in literature. According to ancient tradition found in the
Talmud and midrashim, Dinah, the daughter of Leah and Jacob, was first
conceived in Rachel's womb, and Joseph, the son of Jacob and Rachel was
first conceived in Leah's womb, but in the end they were exchanged so that
the male embryo which was in Leah's womb was removed and implanted in
that of Rachel and the female embryo was removed from Rachel's womb
and implanted in Leah's. The result was that Leah gave birth to a daughter
called Dinah and Rachel gave birth to a son called Joseph. This description
is based on the Talmud Tactate Berachot 60:A, though in the Bible Dinah is
considered the daughter of Leah and Joseph the son of Rachel.

Contrary to the situation after artificial insemination, where the father
of the child is the sperm donor, with regard to motherhood in the case of
ovum or embryo donation, there is divisible parenthood – ownership of the
egg and the environment in which the embryo is conceived.

Jewish law states that the child is related to the one who finished its
formation – the one who gave birth. A judgement is found in the Bible that
states that when a person who starts an action, but does not complete it, and
another person comes along and completes it, the one who completed the
action is considered to have done all of it (Sota 13:8).

Christian View

According to Christianity, in reality the origin of a human person is a "result
of an act of giving." The Vatican's instruction does not accept the donation
of gametes to an infertile couple as an act of generosity. It states that
conception by gamete donation (oocyte) can damage the personal family
relations, as well as the offspring and the society. Oocyte donation is
forbidden by the main branches of Christianity: Roman Catholic, Eastern
Orthodox, and Protestant.

Islamic View

Ovum donation is similar to sperm donation and involves intervention of a third party other than the husband and wife. Therefore it would not be permitted in Islam. Donation of embryos is also prohibited by Islamic law. It should be known that adoption of a child is also prohibited by Islam.

Hindu and Buddhist Views

According to Hinduism and Buddhism, it is suggested that oocyte donation can be practiced on the same grounds as sperm donation.

Surrogacy

The practice of surrogacy, partial or complete, raises several ethical, legal, and social problems. It is practiced only in a few countries.

Complete Surrogacy. In cases of infertility due to uterine factors, when the woman is unable to carry the embryo, the ovum is fertilized with the sperm of her husband, after which it is implanted in the uterus of another women who gives birth as a "surrogate mother."

Partial Surrogacy. Partial surrogacy is when the surrogate mother is inseminated with another woman's husband's sperm and therefore donates her oocyte and leases her uterus. Following birth the child is given to the infertile couple for adoption.

Jewish View

The Jewish religion does not forbid the practice of surrogate motherhood. According to Jewish law, if partial surrogacy is practiced – a strange woman is inseminated by sperm of a man and she completes the pregnancy on agreement, the child born should be handed over to the owner of the sperm. In case of full surrogacy, when the embryo is transplanted to another woman, the question is not resolved, as was discussed in the case of ovum donation. From the religious point of view, the child will belong to the father who gave the sperm and to the mother who gave birth.

Christian View

The practice of surrogate motherhood is not accepted by the Christian religion, i.e., Roman Catholic, Eastern Orthodox, Protestant, or Anglican.

The objection is on the basis that surrogate motherhood is contrary to the unity of marriage and to the dignity of the procreation of the human person.

Buddhist View

There is no prohibition in Buddhism to the practice of surrogacy, but it may raise complications regarding the family ties, and legal and moral aspects.

Hindu View

According to Hinduism, there is no prohibition to the practice of surrogacy, but, as in Buddhism, it may raise dilemmas regarding family ties, and legal and moral issues. A special problem can arise when the surrogate mothers deliver a female offspring since, according to Buddhism, there is an obligation to deliver a male offspring.

Cryopreservation of Preembryos

Cryopreservation of preembryos is at present practiced almost routinely in IVF programs. Even though pregnancy and live birth rates following this procedure are still very low, cryopreservation solves problems of spare embryos and increases the total results of IVF and ET.

Jewish View

The freezing of the preembryo raises the basic question of whether cryopreservation, which stops the development and growth of the embryo, does not cancel all the rights of the preembryo's father. With regard to the mother, the problem is simplified, since the embryo is later transferred into her uterus and will renew the mother-embryo relationship. If we declare that maternity is decided by completion of the development of the embryo into a fetus and neonate, the woman into whose uterus the embryo is transferred for implantation has the right to motherhood. With regard to the relationship to the father, whose main function is of fertilization of the oocyte in order to form the preembryo, the period of freezing may cause a severing of the relationship between the child and his father. Freezing of sperm and preembryo can be permitted only when all the measures are taken to ensure that the father's identity will not be lost.

Christian View

Roman Catholic Church. The freezing of embryos, even when carried out in order to preserve the life of an embryo constitutes an offence against the respect due to the human being, by exposing them to great risks of death or harm to their physical integrity, and depriving them, at least temporarily, of maternal shelter and gestation, thus placing them in a situation in which further offences and manipulations are possible (Congregation for the Doctrine of Faith, Vatican, 1978).

Protestant and Greek Orthodox Churches. Cryopreservation is not an acceptable procedure.

Islamic View

According to Islamic law, freezing of the preembryo is acceptable.

Buddhist View

Buddhism accepts cryopreservation of preembryos due to the following reasons:

1. It saves some of the human embryos, that would otherwise be lost, for future use.
2. The patient's burden is reduced psychologically, financially, as well as physically.
3. The freezing of preembryos should be acceptable only during the period when the ovum donor is at a reproductive age and should not be acceptable after the death of one of the parents or the loss of the reproductive ability of the oocyte donor.

Preembryo Research

Academic and medical benefits may be achieved from preembryo and embryo research. But at the same time, most people share fears of threatening social results from the free unrestricted research on potential human beings. Benefits from human preembryo and embryo research may be achieved in the following medical fields:

1. Improvement in clinical results of assisted reproductive technologies
2. Diagnosis of genetic aberrations and possible therapy
3. Contraceptive research

4. Therapeutic use of embryonal tissue for transplantation to adults in life-threatening medical conditions

Jewish View

1. It should be allowed to create and use an in vitro preimplantation embryo for fertility research if there are real chances that the sperm owner may benefit and have a child as a result of this research.
2. Jewish law forbids destruction and use of a preembryo as long as it has a potential implantation ability.
3. An in vitro blastocyst that has hatched from its zona pellucida and has lost its implantation potential man be kept for continuing research.
4. It is prohibited to use a postimplantation preembryo for research, unless the research is essential for saving the embryo's life.
5. The arbitrary limit of 14 days approved for research on preembryos by some ethical committees is not recognized by Jewish law.

Christian View

Christianity recognizes the preembryo as a human being from the stage of conception. Any research on the preembryo is forbidden.

Islamic View

Some Islamic scholars may accept performing research on excess embryos resulting from IVF in order to increase their "ILM" (knowledge). It may be permissible in cases when it is for the sake of the individual embryo.

Buddhist View

Buddhists believe that there is a continuity from life to life through many reincarnations. According to many Buddhist scholars, experimentation on preembryos is acceptable.

Hindu View

Hinduism does not view that the soul (atman) has a specific beginning or a specific end. Research on preembryos may be permissible when it is to help the infertile couple and it serves the dharma (religious and moral duty) of the physician.

Conclusion

The practice of assisted reproductive technology has been shown to be beneficial to the infertile couple. Further developments in these techniques, based on clinical experience and basic research, may broaden our knowledge on the field of reproduction. However, the practice of these new technologies raises medical, ethical, legal, and social problems. Since religion still has a great influence on the populations in the different parts of the world, those who consider applying these novel techniques have to be knowledgeable regarding the different religious attitudes to them.

31 Extracorporeal Embryonic Development: Juridical Views

F. Pierre and J.H. Soutoul

Although embryonic development covers a very short period (12 weeks), its situation at the beginning of life, between political, social, and religious influences, and the explosion of reproduction techniques in the last decade place this subject in a position which is continuously changing. Central to the current controversy about early human embryo research is the question of when the embryo acquires moral (and juridical) status. This debate was initially opened in relation to laws concerning abortion in different countries. However, compared to the embryo in utero, the extracorporeal status has a specific vulnerability that needs increased juridical protection.

Most developed countries are currently highlighting the very early stages of embryo development in reports and/or laws concerning the use of human prenates in human artificial procreation and research. In countries where these laws exist (UK, Spain), the time of development of the primitive streak is considered as the limit for research (i.e., 14 days after fertilization or the so-called preembryonic developmental period) [3]. In contrast, the only laws regarding the moral and juridical status of the embryo up to 12 weeks of gestation is concerned with abortion, and extrapolation on an extracorporeal embryo is very difficult.

Juridical Uncertainties as a Result of Difficulties in Definition

There is no legal status for an embryo in most countries, even if it is suggested that it is a person or even a potential person. A distinction should be made between the legal personhood beginning at birth and the ethical problem of the status of an embryo (based on fundamental questions which are influenced by political and philosophical considerations). The extracorporeal situation confounds the problem even more as there is a suggestion that embryo could be a "property." The biological parents have no commercial rights in most countries other than the "power of life or death" within precise limits: they have no right to sell or purchase, but they do have the right to decide whether or not to cryoconservation, donate, or transfer the embryo (Art. L.670 of the Braibant Project in France; the same also applies in Spain). The French report (Braibant Project) specifies that donation is not understood from a juridical point of view (i.e., from the patrimonial aspect), but in its usual sense so that the notion of "ownership" is eliminated.

In all existing laws or reports (in the United Kingdom, Sweden, Norway, Victoria State in Australia, Art. L.671 of the Braibant Project in France), in vitro fertilization is reserved for the treatment of human infertility. In a few countries (Spain, Southern Australia, Germany, and in France as the Braibant Project is not yet a law) it can also be used to treat or prevent genetic or hereditary disease. Paradoxically, the deliberate destroying of surplus embryos within the 14-day limit is not a criminal act of abortion in all countries which accept in vitro fertilization, provided that the other embryos were used to cure the infertility of the "owner." For example, in Southern Australia State (see Appendix), this right can be re-examined by the "owner" every year. The attitudes in different countries depend on the way the extracorporeal embryo is considered, and, except for the United Kingdom, these debates are reserved for the embryo obtained by in vitro fertilization, with no possibility to use an embryo obtained by lavage in most countries.

In all countries (except Ireland and Liechtenstein), as soon as in vitro fertilization is accepted, the rights of the embryo rely on its moral status under the influence of political and religious considerations. These considerations include: (a) the protection of the embryo as a human being from the time of fertilization; (b) the denial of any moral status to the pre-embryonic stage (i.e., 14 days after fertilization); and (c) the personification at birth when an independent life and human relationships are possible.

Does Human Personality Begin with Fertilization?

In Louisiana State (USA), "any person who intentionally causes the fertilization of a human ovum by human sperm outside the body shall, with regard to the human being thereby produced, be deemed to have the care and custody of a child . . ." (Act No. 964, July 14, 1986, State of Louisiana, Regular Session, p. 1744). However, few countries agree with this opinion, and the majority do not grant legal status to the human embryo in vitro (i.e., during the 14 days after fertilization). Thus, even in the absence of legal rights, there is no denying that the embryo constitutes the beginning of human life, "a member of the human family" [4]. Therefore, whatever the attitude, every country has to examine which practices are compatible with the respect of that dignity and the security of human genetic material.

And Just What Is a Pre-embryo?

The definition of a pre-embryo is difficult [2, 3]. For its defendants (UK, USA), it is the period at the end of which one can be sure that there will be a single biological individual as segmentation is finished. No alternatives (hydatidiform mole, twinning, etc.) other than a single primitive streak can develop after this 14-day period. During this short period, in the absence of a constitutional right to life, the use of human embryos is uncertain, and restriction by project or laws is very different in each country. More fundamental rules rather than strict control of methodology should be introduced in order not to stop medical progress.

And What about the Embryonic Period?

The embryonic period is considered to be from 2 to 12 weeks of gestation by defendants of the pre-embryo (USA, UK), and as soon as fertilization has taken place by others. Both the common law and the civil law give legal protection to the patrimonial interests of the conceptus in utero; this is sometimes extended to extrapatrimonial interests but only in the case of a live birth. This could apply to in vitro embryos if transfer is performed after the father's death, and the subsequent pregnancy ends in a live birth.

Is There an Apparent Consensus on Certain Points?

As previously mentioned, the status of the human embryo is not juridically defined and relies on the political, social, and religious influences in each country.

In many countries there have been special reports and laws on the subject of medically assisted conception and related areas over the last decade (see Appendix). Although no overall consensus has been achieved on this subject, some points regarding the use of embryos reveal the gradual emergence of certain common areas of agreement with regard to the techniques themselves, as was recently very well described by Knoppers and Le Bris [4]: (a) access to these techniques should be limited to heterosexual married couples or to those living in stable unions; (b) clinics and physicians offering these techniques should be subject to medical supervision and regulation; (c) paternity and maternity should be provided for by law and this for all birth technologies (the mother is the woman who carries the embryo until it is born, and the father the husband if married or the man living in stable union at that time, unless it is shown that he did not consent; (d) medical records should be kept and medical records concerning participants should be confidential; (e) embryonic life in vitro should be limited to 14 days; (f) storage of gametes or embryos should be subjected to time limitations; (g) post-mortem implantation should be prohibited; (h) commercial surrogacy agencies or intermediaries should be prohibited; (i) the consent of the participants should be obtained, and standard conditions of donation should be imposed; (j) reproductive technologies should be free from commercialization; (k) no sex selection of embryos (except for sex-linked diseases) and no eugenic selection should be attempted; and (l) extreme forms of genetic engineering (cloning, creation of chimeras, parthenogenesis, interspecies fertilization, etc.) should be prohibited.

However, some major differences are still evident: (a) the question of the remuneration of embryo donors which not allowed at all in some countries such as France, Spain, and Louisiana State, but seen as "a reasonable payment for the time, pain, risk and inconvenience" in others [1]; (b) limitation of the number of children by a donor; (c) the eventual access by the child to information on the donor and the kind of information (genetic or nominative); (d) the way to keep registers (what kind, how long, by whom); (e) the donation and conservation of, and experimentation with a human embryo (by whom, for what, how long, when; if at all); and (f) the genetic diagnosis on embryos.

It is reassuring to see that projects and laws in different countries are in agreement as regards family law and the institution of matrimony so that the filiation of children born after in vitro fertilization is clear (see above). In addition, a child cannot attempt to modify this filiation for any reason in relation to its origin (donation of embryo or gametes, in vitro fertilization, different genetic origin, etc.). Maternal filiation cannot be questioned; nor can paternal filiation if the father's consent to the assisted reproduction technique was acquired when the embryo was transferred (Family Law Amendment Act, 1983, which is a unique federal decision on assisted reproduction in Australia; Sweden and the UK have similar laws).

In France, even though in the last decade some jurisdictive decisions have agreed to the denial of paternity inspite of initial consent (Court of Appeal, Toulouse, September 21, 1987, and others), the Braibant Project would like to eliminate this possibility. In Germany, a federal decision on April 7, 1983 accepted the possibility of denying paternity, even after initial consent, but very few courts have accepted this principle in their final decisions. Even though it is not applied, the biological origin seems to be a stronger notion in this country.

In most countries both the common law and the civil law are actually in agreement on this fundamental point. This clear confirmation of the juridical situation of in vitro fertilized embryo was essential to confirm the filiation of child to be born . In this respect, while meeting the ethical principles, the law is working in favor of scientific progress and favoring the rights of children above the rights to have children.

Two Persistent Discrepancies

The Power of Science Over the Human Being

A clear definition is necessary to protect the fundamental right to physical integrity of any human being, from conception to death. In France, there is no law, but the National Ethical Committee looks at every project and embryo donation is rarely allowed. In the Braibant Project, embryo donation for research could be possible for 7 days post-fertilization, but only if the initial aims as regards the infertility of the parents are achieved before this time. The time could be prolonged to 14 days on the advice of the Ethical Committee (Art. L.672 of the Braibant Project). This committee should have approved the methodology and aim of the project, which cannot include genetic manipulation (Art. L.673 of the Braibant Project). However, there is currently a strong probability that the Braibant Project could be abandoned and replaced by the recent Lenoir Report, although this is still not official (April 1991).

In the United Kingdom (Human Fertilization and Embryology Act, 1990, Chapter 37, Section 3), although interspecies fertilization is not allowed, there is a wide scope for experimentation and investigation or treatment of the pre-embryo, including genetic manipulation if this used "to increase knowledge about the causes of congenital disease. . . . Any offence to these limitations is liable on conviction or indictmtent to imprisonment for a term not exceeding 10 years or a fine or both. . . ."

In Australia (Law No. 10, Part III, Section I, Art. 14, March 10, 1988, in Southern Australia) there is a similar attitude; "research on human material shall be approved by the Authority, and whatever the method . . . it shall not be prejudicial to the embryo." The only punishment could be a fine in the case of an offence.

Anonymous Donors or Not?

The assertion that egg donation should remain anonymous and that no information should be given about the donors' identification is true in France, Norway, and Southern Australia. In this last country, it is amazing to see how great the punishment could be (i.e., imprisonment) in this case compared to other offences. In other countries, the situation is different: information about the genetic identification (federal recommendation in Australia; Sweden) or nominative information can be obtained after the child has reached the age of 16 years in Germany, 18 years in United Kingdom, or under certain circumstances in the United States [1].

It is clear that if harmonization is difficult on such fundamental points, it constitutes a stumbling block in the attempt to initiate a European project (CAHBI; see Appendix), as happens in countries comprising different states having independent legal systems. It does, however, seem to be a necessity: "Easy means of transportation and communication would allow citizens to practice procreative tourism in order to exercise their personal reproductive choices in other less restrictive countries or states" [4]. It is difficult to find a balance between respecting these new technologies (research on the embryo, new procreation methods) when they are beneficial for the children to come and the temptation to protect this potential human being by very restrictive laws that could stop this research.

Conclusion

Even though many points are still unclear, the recent laws (or projects), which draw their inspiration from Roman laws, are trying to protect the embryo's dignity, even when the embryo is extracorporeal; for example, the problems of filiation have found a homogeneous solution in most developed countries. Amazingly, in France these ethical aspects of embryo protection as a person were already confirmed in the law concerning legal abortion [Law 75-17, January 17, 1975].

More than laws on what should or should not be done in embryo research or genetic diagnosis, it is important to have precise legislation on the way physicians and clinics who are working with/on/for embryos should be subject to supervision and regulation. If this is successful, it would not stop the prodigious scientific advance with respect to our duty towards the unborn. No law or jurisprudential system could cope with the very fast progress in reproduction, and this attitude could assure a great juridical stability with respect to each potential human being.

Appendix

- In the United States there is no federal consensus, but 25 states have adopted statutes that apply explicitly to fetal research [1]. The states are very different in their approach, and, apart from ten (e.g., Act No. 964, July 14, 1986, State of Louisiana), most of them are not precise.
- In Australia, no federal law exists despite the project of the *Reproductive Technology of the Ashe Committee: RTAC report* published in 1985. However, two states have very precise laws: Infertility Medical Procedures Act in Victoria State (No. 10163, November 20, 1984) and Reproduction Technology Act in Southern Australia (No. 10, March 10, 1988).
- In Europe, five countries have specific laws concerning new reproduction technologies: Law No. 35, November 22, 1988, *Sanidad – Tecnicos*

de Reproduccion Assistada in Spain [6]; Human Fertilization and Embryology Act, November 1, 1990, in the United Kingdom; Law No. 745, October 24, 1990, *Entwurf eines Gesetzes zum Schutz von Embryonen* applied since January, 1991, in Germany; Law No. 68, June 12, 1987, in Norway; and Law No. 711, June 14, 1988 in Sweden.
- Other countries, such as France (with the draft of Braibant law, March, 1989), Italy, and Denmark, have projects and reports.
- For religious reasons, Ireland and Liechtenstein have no project or will to discuss this subject.
- A special European Committee (CAHBI) has been formed (Strasbourg, 1989) to discuss the unification of projects.

References

1. American Fertility Society (1990) Ethical considerations of the new reproductive technologies. Fertil Steril [Suppl 2] 53
2. Dawson K (1988) Segmentation and moral status "in vivo" and "in vitro": a scientific perspective. Bioethics 1:1–14
3. Jones HW, Schrader C (1989) And just what is a pre-embryo? Fertil Steril 52:189–191
4. Knoppers BM Le Bris S (1990) Recent advances in medically assisted conception: legal, ethical and social issues. WHO's Scientific Group on Recent Advanced in Medically Assisted Conception, April, 2–6, Geneva
5. Lejeune J (1990) L'enceinte concentrationnaire. Fayard, Paris
6. Peinado JA, Russell SE (1990) The Spanish law governing assisted reproduction techniques: a summary. Hum Reprod 5:634–636
7. Schenker JG (1990) Research on human embryos. Eur J Obstet Gynecol 36:267–273

32 Medical Views on Assisted Procreation

M.H. Valkenburg, M. Camus, A.C. van Steirteghem, and P. Devroey

In Vitro Fertilization and Embryo Transfer Procedures

Introduction

In vitro fertilization and embryo transfer (IVF-ET) is a procedure initially developed to treat women with severe tubal disease or absent fallopian tubes following surgery. The fertilization of oocytes outside the womb in the laboratory and the replacement of in vitro cultured embryo(s) to the uterus give a previously sterile woman and possibility to conceive. The first pregnancy by IVF-ET was established in 1976, which was, however, ectopic [1]. Subsequently, the first successful pregnancies and births were described in 1978 (Louisa Brown) and 1980 [2, 3]. Later IVF-ET was also applied to treat male subfertility, unexplained infertility, and endometriosis where other traditional methods had failed [4–6]. In 1984 new applications were opened for women without ovaries when Lutjen et al. [7] published results of a successful replacement of donor embryos into the uterus of an agonadal recipient. In 1983 Trounson et al. [8] published details of a pregnancy after oocyte donation in a patient with repeated fertilization failures. Current indications for the clinical use of IVF are given in Table 1. So far, it has been estimated that between 40 000 and 50 000 children have been born around the world from assisted procreation.

Ovarian Stimulation

Steptoe and Edwards [1, 2] induced follicular growth and oocyte maturation by injecting human menopausal gonadotropins (hMG) and human chorionic gonadotropins (hCG). In the absence of viable pregnancies, they abandoned superovulation and instead started to use the natural cycle in 1977. The follicular phase of the natural menstrual cycle was monitored to assess follicular growth and the occurrence of an endogenous luteinizing hormone (LH) surge. As a rule, in a natural cycle only one oocyte will be available.

In contrast to the previous observations, ovarian superovulation in IVF-ET was successfully reassessed by the Australian Monash University team [9]. Several drugs were administered, such as clomiphene citrate (CC) [10],

Table 1. Indications for in vitro fertilization and related procedures

Tubal disease
  Patent but abnormal tubes (recurrent ectopic pregnancies)
  Unsuccessful (micro)surgery
  Tubal disease not suitable for (micro)surgery
Unexplained infertility
Uterine disease
  Patient with myomas (after pretreatment with GnRH-a)
  Congenital malformations
Endometriosis
Cervical hostility
  After conization
  Failed intrauterine insemination
  Antibodies in cervical mucus
Sperm antibodies in female serum
Failed ovulation induction
  Risk of hyperstimulation (polycystic ovarian syndrome)
Male subfertility
  Oligospermia
  Low motility
  Abnormal morphology
  Antisperm antibodies
  Frozen sperm prior to chemotherapy/radiotherapy of males having malignancies
  Failed donor insemination
Absent or inappropriate oocytes (egg donation)
  Absent/abnormal ovaries
  Premature menopause
  Genetic diseases
  Low responders
Therapy for female cancer
  Embryo freezing prior to chemotherapy/cytotoxics/radiotherapy/surgery
  Surrogacy

hMG alone [11], or CC in association with hMG [12]. Purified follicle-stimulating hormone (pFSH) was also used [13]. The aim of follicular stimulation in an IVF-ET program is to increase the number of mature oocytes available for retrieval. The major drawbacks of using these protocols are premature LH surges, premature luteinization, or inadequate follicular growth. Ovarian stimulation fails in 15%–20% of the started cycles [14].

The number of "failed stimulations" can be reduced drastically to a few percent of the started cycles when ovarian stimulation is done in association with gonadotropin-releasing hormone analogues (GnRH-a) and hMG or pFSH [15–17]. GnRH-a are given either intranasally, subcutaneously, or intramusculary in the long, short, or ultrashort protocol.

Follicular development is monitored by serial hormonal measurements: [estradiol (E_2), LH, progesterone (P)] as well as by ultrasonography to decide accurately when to give hCG [18].

Ovarian hyperstimulation is one of the disadvantages of follicular stimulation. It is a syndrome ranging from an increase in E_2 and P secretion in the absence of clinical symptoms, through varying degrees of ovarian enlargement and associated discomfort, with increases in blood viscosity, electrolyte imbalances and hypovolemic shock in the most severe cases. It can even cause death [19]. McArdle et al. [20] gave a differentiation in mild, moderate, and serious forms of hyperstimulation. Ovarian hyperstimulation syndrome (OHSS) is the major serious form. The incidence of OHSS varies from <1% to 1.8%. It seems that with the use of GnRH-a and hMG or pFSH stimulation, OHSS occurs more frequently (5.1%) [21, 22]. Also, patients with polycystic ovarian syndrome (PCOS) are particularly susceptible to OHSS. However, in a publication of Smitz et al. [23], we could not confirm this high incidence in our center. OHSS occurred in ten out of 1673 treatment cycles (0.6%). Eight of these ten patients were hyperandrogenic and had an oligomenorrhea. At risk groups include patients with over 20 follicles at previous attempts, and PCOS patients should be monitored carefully, and follicular stimulation with a low dose of gonadotropins should be recommended [22]. No hCG should be administered for luteal support, but micronized progesterone should be administrated intravaginally.

The use of "luteal support" with hCG or progesterone has been proposed as being essential in the case of GnRH-a and gonadotropins when menopause is induced artificially. In patients with primary ovarian failure, Bourgain et al. [24] compared the morphology of the endometrium after oral, vaginal, or intramuscular administration of progesterone. After vaginal application of micronized progesterone, they found endometrial morphology closely matching that of a natural cycle. The use of luteal support has not been confirmed in the other stimulation protocols by proper studies and should therefore be considered empirical.

Oocyte Recovery Procedure

Retrieval of oocytes is performed 36–38 h after the injection of 10 000 IU hCG or 26–28 h after the detection of the endogenous LH surge. Retrieval of human oocytes was originally done by laparoscope; general anesthesia was needed, restricting the number of IVF attempts. Since 1981 ultrasound-guided ovum retrieval has been introduced. Several approaches can be used, including the transabdominal, perurethral, and transvaginal retrieval techniques [25–28]. To obtain as many oocytes as possible, the follicles might be flushed by medium. Nowadays ultrasound vaginal recovery of oocytes is the method of choice for all patients. Transuterine puncture during transvaginal oocyte retrieval does not seem to have an adverse effect on the results of human IVF treatment [29]. The complication rate of these techniques is low, but there is a risk of injuring the bowel or a blood vessel. Hence, it is recommended that the procedure is performed where arrangements for emergency treatment are available.

Table 2. Classification of the oocyte-cumulus complexes

Type 1	Cumulus cells:	Large and well dispersed
	Nuclear status:	First polar body extruded (often difficult to visualize)
	Corona radiata	
	Type 1.0	Very expanded and radiant "sunburst aspect"
	Type 1.1	Not completely expanded and the radiant sunburst aspect not completely present
	Type 1.2	Expanded and dissociated, but clumping and aggregation of cells noted in the cumulus and corona cells (postmaturity)
	Preincubation period:	2–6 h
Type 2	Cumulus cells:	Moderate size with good cell dispersion
	Nuclear status:	No polar body extruded
	Corona radiata:	A demarcated appearance
	Preincubation period:	6–12 h
Type 3	Thick cumulus and/or corona surrounding the zona pellucida	
	Nuclear status:	No polar body visible
	Preincubation period:	12–24 h
Type 4	No cumulus, corona, or first polar body visible	
	Nuclear status:	Germinal vesicle present or absent
	Preincubation period:	24–36 h

As a consequence of the use of fertility drugs to create mild ovarian hyperstimulation, it often occurs that multiple embryos become available. Multiple pregnancies occur after multiple embryo transfers in IVF [30]. In order to avoid the well-documented risks of multiple fetuses (triplets or more), most groups, including our center, transfer a maximum of three oocytes, zygotes, or embryos in the gamete or zygote intrafallopian transfer (GIFT, ZIFT, respectively) or IVF-ET programs.

Fertilization

Handling of human gametes should be done under optimal sterile conditions in the proper environment. After oocyte retrieval, the oocyte-cumulus complexes have to be examined and scored (Table 2) [31].

The culture medium used for IVF-ET does not seem to make any major difference since pregnancies have been obtained with a variety. It is necessary to prepare and test the media meticulously. Moreover, if pooled human serum is used, it must be tested for hepatitis and human immunodefiency virus (HIV). Oocyte and embryo culture can be done in Petri dishes or tubes. After removing the seminal fluid by washing the semen with culture medium and centrifugation, different procedures can be used to harvest progressive motile spermatozoa from the sperm suspension [32].

Insemination is usually performed 4–6 h after oocyte retrieval. The concentration of motile spermatozoa added to the oocyte-cumulus complexes varies between 50 000 to 100 000 spermatozoa per milliliter. It is

Table 3. Morphologic criteria of human embryos

Type 1	All blastomeres have an equal size without the presence of anucleate fragments
Type 2	Not all blastomeres have an equal size
	Type 2.0 No anucleate fragments are present
	Type 2.1 Anucleate fragments are present in less than 10% of the volume of the embryo
	Type 2.2 Anucleate fragments are present in 10%–20% of the volume of the embryo
Type 3	Not all blastomeres have an equal size
	Type 3.1 Anucleate fragments are present in 20%–50% of the volume of the embryo
	Type 3.2 Anucleate fragments are present in more than 50% of the volume of the embryo
Type 4	The embryo is totally fragmented

recommended that oocytes are examined 16–18 h after insemination in order to identify polyspermic fertilization. This can be visualized as multiple pronuclei (more than two pronuclei). This is important since polyspermic oocytes can cleave normally and, if not identified, may inadvertently be transferred into the uterus [33].

Embryo Transfer

Embryo transfer is usually performed 48–72 h after oocyte collection, when embryos are approximately in the four- to eight-cell stage. Before transfer the embryos are classified according to morphologic criteria (Table 3) [31]. A variety of catheters have been evaluated for ET [34]. However, the type of catheter does not seem to make a major difference. To reduce uterine contractility, it could be advocated to give 100 mg indomethacin suppositories 30 min before embryo replacement. The overall pregnancy rate in our center was 623 after 2309 transfers (27%); 16 pregnancies ended in an ectopic pregnancy (2.57%) (G. Verhulst, 1991, personal communication).

GIFT, ZIFT, Tubal Embryo Transfer: Indications and Results

Gamete Intrafallopian Transfer

GIFT was first performed in 1984 by Asch et al. [35] and involved transfer of both gametes into the physiologic environment of the fallopian tube(s). Ovarian stimulation and monitoring are similar to those used for IVF-ET. Semen is collected and prepared 2 h prior to oocyte aspiration, which is performed by laparoscopy or ultrasound. Subsequently the oocytes (up to three) and 10 000–100 000 motile sperm are transferred into the fimbriated end of the tube(s) by laparoscopy. Khan et al. [36] described that 10 000 was the optimal sperm count per GIFT. Transfer in one fallopian tube may give as good a result as into both tubes. Supernumerary oocytes can be inseminated in vitro and obtained embryos cryopreserved [37].

Indications for GIFT include unexplained infertility of at least 2 years' duration, cervical factor, endometriosis (not affecting the tubes), and male infertility including immunologic factors, failed donor insemination, PCOS, as well as oocyte donation. As the fertilizing potential of the gametes can only be documented for the supernumerary oocytes, in case of male infertility, it is not a first-line treatment. Though GIFT is technically simpler than IVF, a laparoscopy is required, although this can be seen as an additional diagnostic opportunity. GIFT is associated with a high clinical pregnancy rate; clinical pregnancy is defined as a rise in β-hCG > 1000 mIE/ml and ultrasound detection of a gestational sac, maintained for at least 28 days after oocyte pick-up [38]. The incidence of clinical abortions is 17%–33% [36, 37, 39]. This somewhat high abortion rate in GIFT programs might be due to the inadvertent replacement of slightly immature or defective oocytes. Such oocytes are highly susceptible to abnormal fertilization and wastage and would definitely end up as abortions. In our center we use the GIFT procedure only for unexplained infertility. The ectopic pregnancy rate is approximately the same as with IVF-ET, ranging from 2.9% to 6.0% [39, 40]. We had an ectopic pregnancy rate of 2.4%, three out of 124 pregnancies (G. Verhulst, 1991, personal communication). The incidence of multiple gestations is also similar to IVF-ET (25%–27%) [39, 40].

Zygote Intrafallopian Transfer

ZIFT was first described by Devroey et al. in 1986 [41]. Ovarian stimulation, transvaginal oocyte pick-up, semen preparation, and insemination are performed in a similar way as in IVF-ET. Eighteen hours after insemination, oocytes are examined for the presence of two pronuclei and subsequently zygotes (up to three) are transferred laparoscopically into the fimbriated end of the fallopian tube(s). ZIFT has been applied in cases of unexplained infertility, male and immunologic infertility, endometriosis, PCOS, failed donor insemination, as well as oocyte donation. The procedure is especially indicated when the fertilizing potential of the gametes must be assessed. The overall pregnancy rate after ZIFT appears to be high (27%–48%) [42–45]. In Table 4 we give a comparison between the GIFT and the ZIFT procedures.

Furthermore, in our center the pregnancy rate after replacement of two or three zygotes was similar, i.e., 42% and 45%, respectively. The multiple pregnancy rate after replacements of two zygotes was limited to 7.1%. After the replacement of three zygotes, we noticed a 30% multiple pregnancy

Table 4. Differences between GIFT and ZIFT

	GIFT	ZIFT
Fertilized oocytes are replaced	–	+
Abnormal fertilization is detected	–	+
Laparoscopy is avoided if fertilization fails	–	+
Oocytes mature in vitro	–	+
Two-step procedure	–	+

rate. We might therefore start a prospective study to define our strategy, i.e., whether the number of replaced zygotes should be limited to two.

Tubal Embryo Transfer

Tubal embryo transfer (TET) is a variation of ZIFT. The two procedures are similar, the only difference being that, instead of zygotes, cleaving four-cell embryos are transferred into the fallopian tube(s) laparoscopically [46, 47] or transcervically [48, 49]. TET has the same indications as ZIFT and a satisfactory pregnancy rate. However, the results of TET have not been compared with ZIFT in a double-blind prospective study.

Nonsurgical TET

A transcervical transfer of embryos or zygotes has been described by Jansen et al. [48], Yovich et al. [50], and Guidetti et al. [49]. The advantage of a nonsurgical method is that it reduces the costs and the risks of a laparoscopy. A selective cannulation of the fallopian tube still presents several problems. The technique is not always effective, even with the help of ultrasound. It is not yet known if the contractility of the tube can be affected. We do not know if there is a harmful effect on the tube or the endometrium. So it is too early to say whether this will be the future method of choice.

Freezing

Sperm Freezing

Human sperm freezing was first observed in 1953 by Bunge and Sherman [51]; frozen in dry ice and later thawed, they showed that normal embryonic development was possible. Human sperm cryobanking grew in the 1970s. Now, with the risk of acquired immunodefiency syndrome, it has been recommended to use only frozen semen for therapeutic insemination by donor. Semen autopreservation is advocated mainly before chemotherapy, radiotherapy, or orchiectomy in oncologic treatment. In addition, IVF or ZIFT might facilitate fertilization and normal fetal development with thawed semen of poor quality and with, often, a limited number of straws, for instance, in the case of Hodgkin's disease where sperm banking is done before chemotherapy [52].

Embryo Freezing

The techniques used to freeze human embryos were based on those used in animals, with the first human pregnancy reported after transfer of an eight-cell embryo frozen in dimethylsulfoxide (DMSO). The pregnancy aborted at 24 weeks due to an obstetric complication [53]. The first pregnancies and live births using the DMSO protocol were reported in the mid-1980s [54, 55]. Other cryoprotective agents with successful pregnancies are glycerol [56] and 1,2-propanediol (PROH) [57]. The production of numerous embryos after ovarian hyperstimulation for IVF-ET, GIFT, or ZIFT treatments and the risks of multiple fetuses (triplets or more) and hence the transfer of a maximum of three embryos make human embryo cryopreservation an essential technique with extra chances for pregnancy, but it is also an ethical obligation in the modern approach to human infertility. The stage at which human embryos should be frozen to obtain optimal results is controversial; recommendations have included blastocyst [58], eight-cell stage [59], and pronucleate egg [57] using glycerol, DMSO, and PROH, respectively. Embryo survival is strongly correlated with morphologic features [60]. Differences in the ability to implant were not found between intact and non-intact embryos. In our center, only embryos with an intact zona pellucida and having retained at least 50% of their initial blastomeres after thawing are transferred. Stimulation treatment in the assisted procreation cycle may result in embryos' resistance to cryopreservation or their ability to develop. The pregnancy rate per frozen embryo did not decrease with increased length of storage in liquid nitrogen [61].

Results after Embryo Freezing

In our center, between December 1, 1985 and March 31, 1990, 631 transfers of human frozen-thawed supernumerary embryos were performed after IVF, GIFT, and ZIFT; 2883 embryos were thawed during this period, from which 1331 were considered suitable for transfer (46%), and 973 (33%) were actually transferred. Seventy-two pregnancies out of 461 transfers (15.6%) were obtained from DMSO-frozen embryos and 23 pregnancies out of 170 transfers (13.5%) from PROH-frozen embryos. Total pregnancy, delivery rate per transfer, and 1-trimester abortion rate were 15% (95/63),

10% (66/631), and 29.5%, respectively. First-trimester pregnancy loss included 15 biochemical pregnancies and 13 clinical miscarriages. The 66 pregnancies going to term included three vanishing twins, four ongoing twins, and one ongoing triplet, making it a total of 72 born children. A mean of 1.5 embryos/cycle (973/631) were transferred; and total implantation, ongoing implantation rate, and the live-born children rate per embryo transferred were 10.9% (106/973), 7.6% (74/973), and 7.4% (72/973), respectively. An outcome of 2.5% children per thawed embryo (72/2883) and 11.4% per transfer (72/631) can be expected. The clinical outcome of the pregnancies was strongly influenced by the type of stimulation used in the collection cycle, as 82.9% (29/35) of the pregnancies were ongoing when embryos originated from CC-hMG- or hMG-stimulated cycles, compared to only 55.8% (29/52) when originated from GnRH-a/hMG cycles ($P < 0.01$) [M. Camus, 1991, personal communication].

Results after Oocyte Freezing

From a practical standpoint, oocyte freezing is clearly useful. It would also be beneficial to young women at risk of losing ovarian function due to chemotherapy, surgery, or pelvic diseases such as severe endometriosis. This would be analogous to semen banks for men who are at risk of losing their reproductive function. Oocytes could be donated by volunteers, for example, patients undergoing laparoscopic sterilization.

Two centers have reported detailed clinical results of human oocyte freezing [62, 63]. After freezing 228 oocytes, 70 were judged suitable for embryo replacement. So far, 33 transfers have resulted in four pregnancies; two aborted and two resulted in three healthy children, including one set of twins. The success rate of IVF-ET diminishes significantly with the use of frozen-thawed compared to fresh oocytes [64]. This may be due to cytoskeletal damage, especially at the spindle level, as demonstrated in several species. Also, polyploidy is substantially increased [63]. Clearly, the procedure of freezing oocytes requires extensive research before it can be applied in humans [65]. The Interim Licensing Authority for Human In Vitro Fertilization and Embryology in the United Kingdom has advised that the clinical use of frozen-thawed oocytes should not take place until there is sufficient evidence as to the safety of the procedure.

Oocyte and Embryo Donation Programs

Oocyte Donation

Though sperm and egg donation parallel each other, they differ considerably in that the latter implies donation of a gamete given by one woman to another. Consequently, three individuals participate in this procedure: the donor by giving the oocyte(s), the recipient's husband by giving his sperm, and the recipient by carrying the infant during pregnancy. In addition, whereas spermatozoa can easily be stored in liquid nitrogen at $-196°C$, freezing oocytes has so far not been very successful [62]. Furthermore, unlike donation of sperm, that of oocytes remains more invasive. While a sperm sample can be easily produced by masturbation, donated oocytes can only be retrieved surgically, either by laparoscopy or by ultrasound-guided puncture after ovarian stimulation. Obviously, the latter treatment is bound to exert considerable psychologic pressure on both donor and recipient. In our center, we favor complete anonymity. However, in harmony with the guidelines of both the Warnock Report [66] and the Ethics Committee of the American Fertility Society [67], we do not exclude known donation, provided the persons involved have been extensively screened and counseled, donors and recipients must receive full information. Commercial benefits are unacceptable.

Indications for Oocyte Donation

There are two distinct groups of patients who may require oocyte donation. The first group (without gonadal function) include women with gonadal dysgenesis, premature menopause (occurring spontaneously, after bilateral oophorectomy or induced by chemo- or radiotherapy), as well as women with resistant ovary syndrome. The second group of patients with functional ovaries comprises women in whom different regimens of ovarian stimulation did not allow the development of ovarian follicles, as well as patients whose oocytes, in the course of various IVF-ET treatment cycles, never fertilized after insemination with spermatozoa of their partner and/or a fertile donor. Furthermore, in some conditions, women with functional ovaries do not want to use their own oocytes for genetic reasons:

1. In the presence of a chromosomal translocation, an autosomal dominant disease (e.g., Huntington's chorea and Steinert's disease), or if the female partner is a definite carrier of an X-linked disease (e.g., Duchenne's muscular dystrophy).
2. When both partners are carriers of an autosomal recessive disease (e.g., mucoviscidosis, phenylketonuria, thalassemia), it is advisable that these couples do not use one or both gamete(s) for their procreation.

Stimulation of the donor is mostly done with GnRH-a to synchronize with the recipient. The donor is injected with 150 or 225 IU hMG on the 1st day of the recipient's artificial cycle. Progesterone is administered to the recipient 24h after the donor has been given hCG. Four-cell embryos are replaced on the 4th day of progesterone administration. The donor receives a further treatment of GnRH-a during the luteal phase (Fig. 1).

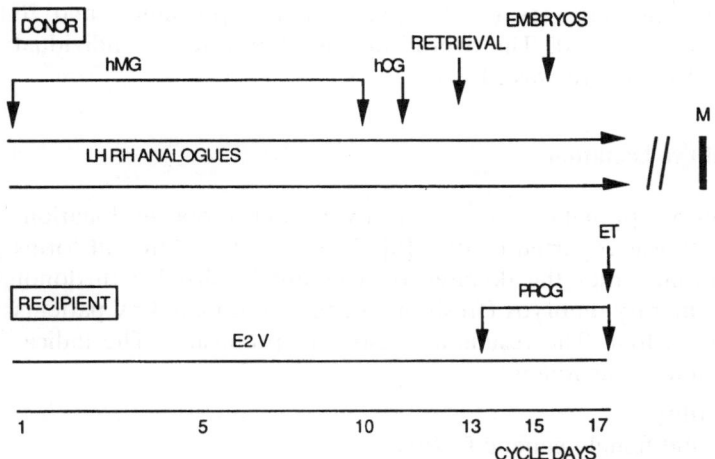

Fig. 1. Superovulation of the donor with the combination of hMG; steroid replacement therapy of the recipient with oestradiol valerate (E_2V), and progesterone (*PROG*). The stimulation in the donor started on the 1st day of the recipient's substituted cycle. The recipient is injected with progesterone intramuscularly or given micronized progesterone vaginally the day after administration of hCG to the donor. *LH RH*, luteinizing hormone-releasing hormone; *M*, menses

When a recipient has functional ovaries, we make them "agonadal" by long-term GnRH-a administration. They subsequently have to be treated with exogenous estradiol valerate and progesterone. In the absence of a corpus luteum in agonadal women, the administration of estradiol and progesterone is mandatory to provide an adequate environment for embryonic development. To maintain a pregnancy in agonadal women, 3×50 mg natural progesterone in oil should be injected daily intramusculary or 5×300 mg micronized progesterone should be administered vaginally. Estradiol valerate 5×2 mg should be given orally daily.

Results After Oocyte Donation

In our center, we have had good results in our program of 95 attempts in 28 women without ovarian function and in 21 with functional ovaries. Overall, 22 pregnancies (44%) were established, 13 after the transfer of fresh embryos and nine after the transfer of frozen-thawed embryos. Eleven of the pregnancies were established in women without ovarian function and 11 in women with functional ovaries. Nine of the pregnancies were established with donated oocytes inseminated with partner's semen and 13 with donated embryos. Eighteen healthy infants were born, including one set of twins, and five women miscarried [68]. Recently Sauer et al. [69] also described very good results in a group of 31 women with ovarian failure. Twenty-five of 47 ET resulted in pregnancy (53.2% per ET); five preclinical and 20 clinical of which 18 delivered. The overall implantation rate per individual transferred fresh pre-embryo was 21.1%.

Results after Embryo Donation

Embryo donation or "prenatal adoption" is a variation of oocyte donation. The first pregnancy was reported in 1983 [8]. There are two different forms of embryo donation, either the donated oocytes are fertilized with donor sperm or supernumerary embryos (fresh or frozen) are donated by patients of the IVF program [68]. The results are also very promising. The indications to be accepted are as follows:

- Double infertility
- Severe male and female genetic factors
- Male or female infertility combined with one severe genetic factor
- In vitro fertilization of donated sperm or oocytes has failed repeatedly

Selective Embryo Reduction

Multiple pregnancies do occur after assisted procreation. Between 1982 and 1989 the incidence of multiple pregnancies after 11 127 children were born through these methods was 23%, of which 3.6% were triplets or more [70]. At our center, we had an incidence of 8% triplets after the transfer of three oocytes, zygotes, or embryos in the case of GIFT, ZIFT, or IVF-ET [71]. In our group of 23 patients who did not undergo a reduction, there was one spontaneous reduction to two and one miscarriage at 20 weeks. Four patients lost their entire pregnancy; 13 neonates died perinatally. Several techniques have been used, including transcervical aspiration [72] or transabdominal needle-guided procedures [73]. We now use the transvaginal puncture technique. With this technique we use the same material as during a transvaginal ovum pick-up. We inject 2–5 ml potassium chloride intrathoracally. Until December 1990 we used this technique in 14 patients. There was one accidental reduction from a triplet to a singleton pregnancy. All the other pregnancies, including a quadruplet and a septuplet, were reduced to twins. Four patients have delivered. All the other ten pregnancies have a normal evolution of which four are over 32 weeks (N. Bollen, 1991, personal communication). Although these are preliminary results, there seems to be a lower chance of obstetric and neonatal complications.

Assisted Fertilization

Recently, different procedures of assisted fertilization have been described [74-78]. They could be applied in infertile couples where IVF failed due to idiopathic or severe andrologic infertility. Pregnancies and births were reported after zona drilling, partial zona dissection, or subzonal insemination of spermatozoa. After partial zona dissection, the incidence of polyspermy is increased [76]. Therefore we elected to perform a clinical trial with subzonal insemination in patients who failed to conceive by IVF because fertilization never occurred. Assisted fertilization was also applied successfully in a patient with Kartagener's syndrome [79]. If confirmed on a large group of patients, assisted fertilization might be able to alleviate infertility in couples in whom other methods of assisted procreation have failed.

References

1. Steptoe PC, Edwards RG (1976) Reimplantation of a human embryo with subsequent tubal pregnancy. Lancet 1:880-882
2. Steptoe PC, Edwards RG (1978) Birth after the reimplantation of a human embryo. Lancet 2:366
3. Edwards RG, Steptoe PC, Purdy JM (1980) Establishing full term human pregnancies using cleaving embryos grown in vitro. Br J Obstet Gynaecol 87:737-756
4. Mahadevan MM, Trounson A, Leeton JF (1983) The relationship of tubal blockage, infertility of unknown cause, suspected male infertility, and endometriosis to success of in vitro fertilization and embryo transfer. Fertil Steril 40:755-762
5. De Kretser DM, Yates C, Kovacs GT (1985) The use of IVF in the management of male infertility. Clin Obstet Gynaecol 12:767-773
6. Cohen J, Fehilly CB, Fishel SB (1984) Male infertility treated by in vitro fertilization. Lancet 1:239-240
7. Lutjen P, Trounson A, Leeton J, Findlay J, Wood C, Renou P (1984) The establisment and maintenance of pregnancy using in vitro fertilization and embryo donation in a patient with primary ovarian failure. Nature 307:174-175
8. Trounson A, Leeton J, Besanko M, Wood C, Conti A (1983) Pregnancy established in an infertile patient after transfer of a donated embryo fertilized in vitro. Br Med J 286:835-839
9. Trounson AO, Leeton F, Wood C, Webb J, Wood J (1981) Successful human pregnancies by fertilization in vitro and embryo transfer in the controlled ovulatory cycle. Science 212:681-682
10. Wood C, Trounson A, Leeton JF (1981) A clinical assessment of nine pregnancies obtained by in vitro fertilization and embryo transfer. Fertil Steril 35:502-507
11. Laufer N, DeCherney AH, Haseltine FP (1983) The use of high dose-human menopausal gonadotrophin in an in vitro fertilization program. Fertil Steril 40:731-737
12. Trounson AO (1983) In vitro fertilization at Monash University, Melbourne, Australia. In: Crosignani PG, Rubin BL (eds) In vitro fertilization and embryo transfer. Academic, New York, pp 315-322
13. Jones GS (1985) Use of purified gonadotrophins for ovarian stimulation in IVF. In: Wood EC, Trounson AO (eds) New clinical issues in in vitro fertilization. Saunders, London, pp 775-784

14. Devroey P, Naaktgeboren N, Traey E, Wisanto A, van Steirteghem AC (1985) Hormonal evaluation of failed ovarian stimulation in an IVF programme (Abstr). In: 4th World Conference on In Vitro Fertilization, Melbourne, Australia. Plenum, New York, p 6
15. Porter RW, Smith W, Craft IL, Abdulwahiad NA, Jacobs HS (1984) Induction of ovulation for in vitro fertilization using buserelin and gonadotrophins. Lancet 2:1284–1285
16. Wildt T, Diedrich K, van der Ven H, Al Hasani S, Hübner H, Klasen R (1986) Ovarian hyperstimulation for in-vitro fertilization controlled by GnRH agonist administered in combination with human menopausal gonadotrophins. Hum Reprod 1:15–17
17. Smitz J, Devroey P, Braeckmans P, Camus M, Khan I, Staessen C, van Waesberghe L, Wisanto A, van Steirteghem AC (1987) Management of failed cycles in an IVF/GIFT programme with the combination of a GnRH analogue and hMG, Hum Reprod 2:309–314
18. Diedrich K, van der Ven H, Al Hasani S, Krebs D (1988) Ovarian stimulation for in vitro fertilization. Hum Reprod 3:39–44
19. Schenker JG, Weinstein D (1978) Ovarian hyperstimulation syndrome: a current survey. Fertil Steril 30:255–268
20. McArdle C, Seibel M, Hann LE (1983) The diagnosis of ovarian hyperstimulation (OHS): the impact of ultrasound. Fertil Steril 39:464–467
21. Golan A, Ron-El R, Herman A, Weinraub Z, Soffer Y, Caspi E (1988) Ovarian hyperstimulation following D-Trp-6 luteinizing hormone-releasing hormone microcapsules and metrodin for in vitro fertilization. Fertil Steril 50:912–916
22. Forman R, Frydman R, Egan D, Ross C, Barlan DH (1990) Severe ovarian hyperstimulation syndrome using agonists of gonadotrophine-releasing hormone in vitro fertilization. European series and a proposal for prevention. Fertil Steril 53:502–509
23. Smitz J, Camus M, Devroey P, Erard P, Wisanto A, van Steirteghem AC (1990) Incidence of severe ovarian hyperstimulation syndrome after GnRH agonist/hMG superovulation for in-vitro fertilization. Hum Reprod 5:993–937
24. Bourgain C, Devroey P, van Waesberghe L, Smitz J, van Steirteghem AC (1990) Effects of natural progesterone on the morphology of the endometrium in patients with primary ovarian failure. Hum Reprod 5:537–543
25. Lenz S (1984) Ultrasonic-guided aspiration of human oocytes. Ultrasound Med Biol 10:625–628
26. Parsons J, Booker M, Goswamy R, Akkermans J, Riddle A, Sharma V, Wilson L, Whitehead M (1985) Oocyte retrieval for in vitro fertilization by ultrasonically guided needle aspiration via the urethra. Lancet 1:1076–1077
27. Feichtinger W, Kemeter P (1986) Transvaginal sector scan sonography for needle guided transvaginal follicle aspiration and other applications in gynaecologic routine and research. Fertil Steril 45:722–725
28. Hamberger L, Wikland M, Enk L, Nilsson L (1986) Laparoscopy versus ultrasound guided puncture for oocyte retrieval. Acta Eur Fertil 17:195–198
29. Wisanto A, Bollen N, Camus M, de Grauwe E, Devroey P, van Steirteghem AC (1989) Effect of transuterine puncture during transvaginal oocyte retrieval on the results of human in-vitro fertilization. Hum Reprod 4:790–793
30. Kerin JF, Quinn PJ, Irby C, Seemark RF, Warnes GN, Jeffrey N, Matteus CM, Cox LW (1983) Incidence of multiple pregnancy after in vitro fertilization and embryo transfer. Lancet 2:537–540
31. Staessen C, Camus, Khan I, Smitz J, van Waesberghe L, Wisanto A, Devroey P, van Steirteghem AC (1989) An 18-month survey of infertility treatment by in vitro fertilization, gamete and zygote intrafallopian transfer, and replacement of frozen-thawed embryos. J In Vitro Fert Embryo Transfer 6(1):22–29
32. Tarlatzis BC, de Cherney AH (1987) Semen preparation for IVF. In: Frederiks CM, Paulson J, de Cherney AH (eds) Foundation of in vitro fertilization. Hemisphere, New York

33. Laufer N, Tarlatzis BC, Naftolin F (1984) In vitro fertilization: state of the art. Semin Reprod Endocrinol 2:197–219
34. Wisanto A, Camus M, Janssens R, Devroey P, Deschacht, van Steirteghem AC (1989) Performance of different embryo transfer catheters in a human in vitro fertilization program. Fertil Steril 52:79–84
35. Asch R, Ellsworth L, Balmaceda J, Wong PC (1984) Pregnancy after translaparoscopic gamete intrafallopian transfer. Lancet 2:1034–1035
36. Khan I, Camus M, Staessen C, Wisanto A, Devroey P, van Steirteghem AC (1988) Success rate in gamete intrafallopian transfer using low and high concentrations of washed spermatozoa. Fertil Steril 50:922–927
37. Braeckmans P, Devroey P, Camus M, Khan I, Staessen C, Smitz J, van Waesberghe L, Wisanto A, van Steirteghem AC (1987) Gamete intrafallopian transfer: evaluation of 100 consecutive attempts. Hum Reprod 2:201–205
38. Page H (1989) Calculating the effectiveness of in vitro fertilization. A review. Br J Obstet Gynaecol 96:334–339
39. Asch RH, Balmaceda JP, Cittadini E, Figueroa Casa P, Gomel V, Hohl MK, Johnston I, Leeton J, Rodriguez Escudero FJ, Noss K, Wong PC (1988) Gamete intrafallopian transfer. International cooperative study of the first 800 cases. Ann NY Acad Sci 541:722–727
40. National Perinatal Statistics Unit and Fertility Society of Australia (1987) IVF and GIFT pregnancies, Australia and New Zealand 1986. National Perinatal Statistics Unit and Fertility Society of Australia, Sydney
41. Devroey P, Braeckmans P, Smitz J, van Waesberghe L, Wisanto A, van Steirteghem AC (1986) Pregnancy rate after translaparoscopic zygote intrafallopian transfer in a patient with sperm antibodies. Lancet 1:1329
42. Matson PL, Blackledge DG, Richardson PA, Turner SR, Yovich JM, Yovich JL (1987) Pregnancies after pronuclear stage transfer. Med J Aust 146:60
43. Yovich JL, Blackledge DG, Richardson PA, Matson PL, Turner SR, Draper R (1987) Pregnancies following pronuclear stage tubal transfer. Fertil Steril 48:851–857
44. Hamori M, Stuckensen JA, Pumpf D, Kniewald T, Kniewald A, Marquez MA (1988) Zygote intrafallopian transfer (ZIFT): evaluation of 42 cases. Fertil Steril 50:519–521
45. Devroey P, Staessen C, Camus M, de Grauwe E, Wisanto A, van Steirteghem AC (1989) Zygote intrafallopian transfer as a successful treatment for unexplained infertility. Fertil Steril 52:246–249
46. Balmaceda JP, Castldi C, Remohi J, Borrero C, Ord T, Asch RH (1988) Tubal embryo transfer as a treatment for infertility due to male factor. Fertil Steril 50:476–479
47. Henriksen T, Abyholm T, Tanbo T, Magnus O (1988) Pregnancies after intrafallopian transfer of embryos. J In Vitro Fert Embryo Transfer 5:296–298
48. Jansen RPS, Anderson JC, Sutherland PD (1988) Non-operative embryo transfer to the fallopian tube. N Engl J Med 319:288–291
49. Guidetti R, Balmaceda JP, Ord T, Asch RH (1990) Non-surgical tubal embryo transfer. Case report. Hum Reprod 5:221–224
50. Yovich JL, Yovich JM, Edirisinghe WR (1988) The relative chance of pregnancy following tubal or uterine transfer procedures. Fertil Steril 49:858–864
51. Bunge RG, Sherman IK (1953) Fertilizing capacity of frozen human spermatozoa. Nature 172:767
52. Tournaye H, Camus M, Bollen N, Wisanto A, van Steirteghem AC, Devroey P (1991) In vitro fertilization with frozen-thawed sperm: a method for preserving the progenitive potential of Hodgkin patients. Fertil Steril 55:443–445
53. Trounson AO, Mohr L (1983) Human pregnancy following cryopreservation, thawing and transfer of an eight-cell embryo. Nature 305:707–709
54. Zeilmaker GH, Alberda AT, van Gent I, Rijkmans CMPM, Drogendijk AC (1984) Two pregnancies following transfer of intact frozen-thawed embryos. Fertil Steril 42:293–296
55. Downing BG, Mohr LR, Trounson AO, Freeman LE, Wood C (1985) Birth after transfer of cryopreserved embryos. Med J Aust 142:409–411

56. Cohen J, Simon RF, Fehilly CB, et al. (1985) Birth after replacement of hatching blastocyst cryopreserved at expanded blastocyst stage. Lancet 1:647
57. Lasalle B, Testart J, Renard JP (1985) Human embryo features that influence the success of cryopreservation with the use of 1,2-propanediol. Fertil Steril 44:645–651
58. Cohen J, Simon FR, Edwards RG, Fehilly CB, Fishel SB (1985) Pregnancies following the frozen storage of expanding human blastocysts. J In Vitro Fert Embryo Transfer 2:59–74
59. Freeman L, Trounson AO, Kirby C (1986) Cryopreservation of human embryos progress on the clinical use of the technique in human in vitro fertilization. J In Vitro Fert Embryo Transfer 3:53–61
60. Camus M, van den Abbeel E, van Waesberghe L, Wisanto A, Devroey P, van Steirteghem AC (1989) Human embryo viability after freezing with dimethylsulfoxide as a cryoprotectant. Fertil Steril 51:460–465
61. Frydman R, Forman RG, Belaisch-Allart J, et al. (1989) An obstetric analysis of fifty consecutive pregancies after transfer of human cryopreserved embryos. Am J Obstet Gynecol 160:209–213
62. Chen C (1986) Pregnancy after human oocyte cryopreservation. Lancet 1:884–886
63. Al-Hasani S, Diedrich K, van der Ven H, et al. (1987) Cryopreservation of human oocytes. Hum Reprod 2:695–700
64. Mandelbaum J, Junca AM, Plachot M, et al. (1988) Cryopreservation of human embryos and oocytes. Hum Reprod 3:117–119
65. Van der Elst J, van den Abbeel E, Jacobs R, Wisse E, van Steirteghem AC (1988) Effect of 1,2-propanediol and dimethylsulphoxide on the meiotic spindle of the mouse oocyte. Hum Reprod 3:960–967
66. Warnock M (1984) Report of the Committee of Inquiry into Human Fertilization and Embryology. Her Majesty's Stationary Office, London
67. The Ethics Committee of the American Fertility Society (1986) Ethical considerations of the new reproductive technologies. Fertil Steril [Suppl 1] 46:515S–525S
68. Devroey P, Camus M, van den Abbeel E, van Waesberghe L, Wisanto A, van Steirteghem AC (1989) Establishment of 22 pregnancies after oocyte and embryo donation. Br J Obstet Gynaecol 96:900–906
69. Sauer MV, Paulson RJ, Macaso TM, Francis MM, Lobo RA (1991) Oocyte and pre-embryo donation to women with ovarian failure: an extended clinical trial. Fertil Steril 55:39–43
70. Testart J, et al. (1990) Le magasin des enfants. Bourin, Paris
71. Bollen N, Camus M, Staessen C, Tournaye H, Devroey P, van Steirteghem AC (1991) The incidence of multiple pregnancy after in vitro fertilization and embryo transfer, gamete or zygote intrafallopian transfer. Fertil Steril 55:314–318
72. Salat-Baroux J, Aknin J, Antoine JM, Alamowitch R (1988) The management of multiple pregnancies after induction of superovulation. Hum Reprod 3:399–401
73. Shalev J, Frenkel Y, Goldenberg M, et al. (1989) Selective reduction in multiple gestations: pregnancy outcome after transvaginal and transabdominal needle-guided procedures. Fertil Steril 52:416–420
74. Ng SC, Bongso A, Sathananthan AH, Ratnam SS (1990) Micromanipulation: its relevance to human in vitro fertilization. Fertil Steril 53:203–219
75. Malter HE, Cohen J (1989) Partial zona dissection of the human oocyte: a non traumatic method using micromanipulation to assist zona pellucida penetration. Fertil Steril 51:139–148
76. Hill DL, Adler D, Rothman C, Surrey M, Danzer H, Friedman S (1991) Micromanipulation in a center for reproduction medicine. Fertil Steril 55:36–38
77. Laws-King A, Trounson AO, Sathananthan AH, Kola I (1987) Fertilization of human oocytes by microinjection of a single spermatozoon under the zona pellucida. Fertil Steril 48:637–642
78. Fishel S, Jackson P, Antinori S, Johnson J, Grossi S, Versaci C (1990) Subzonal insemination for the alleviation of infertility. Fertil Steril 54:828–835

79. Ng SC, Sathananthan AH, Edririsinghe WR, Ho KCJ, Wong PC, Ratnam SS, Ganatra S (1987) Fertilization of a human egg with sperm from a patient with immotile cilia syndrome: case report. In: Ratnam SS, Teoh ES, Anandakumar C (eds) Advances in fertility and sterility. Parthenon, Lancaster, p 71

Epilogue

In the various chapters of this book, evidence has been given that the first 12 weeks of gestation constitute a crucial period for pregnancy well-being. There is a wealth of data which suggest that after 12 weeks the evolution of pregnancy does not follow the same path. It is thus mandatory to point out that first-trimester physiology, biochemistry, and endocrinology cannot be extrapolated from the results which are accumulated later on. The first weeks of gestation encompass a period of maximal growth and differentiation. All the concepts of embryology are described for this period only.

In the beginning, there is successful penetration of the head of a spermatozoon into the oocyte. Then the first mitotic divisions occur while the egg makes its journey through the fallopian tube. During this period (5–7 days), the developing blastocyst is able to send signals: these are the earliest which are encoded by the maternal organism as evidence of fecundation. Embryonic signals are probably multiple. They could involve, among others:

1. A physical contact established between the future embryo and epithelial (tubal) cells with a possible transfer of information.
2. The growing blastocyst has its own real metabolism although this is still limited. Its production of carbon dioxide with the appearance of bicarbonate ions is probable, while during the 1st postovulatory week a monoamine oxidase activity appears within the endometrium.
3. Preimplantation embryos can synthetize prostaglandins and thus have a direct action on endometrial vascular permeability.
4. From the eight-cell stage, β-human chorionic gonadotropin β-hCG mRNA is present in the blastocyst [1]. Indeed, production of free β-subunit is predominant in the very first postimplantation stages [2]. Later production of the free α-subunit by proliferating trophoblast results in increasing levels of dimer hCG. Free β-hCG has few biologic properties of dimer hCG, but nothing is known about its signal effect.
5. Two additional factors have been identified so far. One is platelet activating factor (PAF) which induces very early a significant maternal thrombopenia [3]. Platelet activation can be followed by local release

of different factors such as amines, and also of growth factors (e.g., platelet-derived growth factor, PDGF). The other has been called early pregnancy factor (EPF). It can be detected very early after fecundation, long before hCG can be assayed [4]. EPF could be immuno-suppressive and thus prepare immune tolerance of the embryo after it has implanted.

Our book begins at this stage. The blastocyst has reached the uterine cavity. The trophoectoderm has differentiated from the inner cell mass. Corpus luteum is functioning well, and adequate amounts of progesterone have been produced. The endometrium is secretory, with a maximal stromal edema.

Implantation can and will occur, provided that adequate local biochemical changes are present. These involve especially the glycocalyx of surface epithelium and trophoectoderm allowing adhesion to occur. Separation of epithelial cells and penetration of primary trophoblasts usually follow, with a progressive embedding of the entire egg within the mucosa. This is where the true story begins.

We have proposed to discuss all known problems pertinent to this period. After implantation has occurred, the major event implies differentiation of trophoblast. The primary trophoblast is progressively replaced by less-differentiated, more-proliferating cytotrophoblast. This forms the major part of the trophoblastic shell, from which emerge cell columns where mesenchymal core will form the first villous tree. Extravillous trophoblast (intermediate trophoblast) is also formed which will eventually intrude into the adjacent decidua where specialized sets of cells will receive and transfer the information.

In order to facilitate the interrelation between early trophoblast and maternal environment, a number of protein productions are initiated on both sides: they most probably exert a regulatory effect on each other via paracrine controls. Most probably embryonic products are also involved. These biochemical relations enable the conceptus to thrive within the uterus through the establishment of an immune tolerance and the transfer (deciduochorial) of growth factors and nutrients.

This notion that for a substantial period the relationship mother-conceptus exists exclusively by local, interstitial transfer is provocative. There are enough arguments to suggest that, during the first weeks, the intervillous space is not bathed by maternal blood. There does not seem to be an effective blood flow in the uteroplacental arteries, but since these are almost totally occluded by trophoblastic plugs, only minute amounts of acellular filtrate bathe the villi. Thus, oxygen transfer to the embryo is limited to plasma-dissolved oxygen. This implies that a relative degree of hypoxia exists within the conceptus at the time when all organogenesis takes place. We must not forget the importance of the yolk sac which provides the early embryo placental unit with blood vessels, hematopoietic cells, and primitive liver and gut functions. Embryonic hemoglobin is distinct from the fetal moiety and disappears with the extinction of yolk sac function. It has an increased oxygen-binding capacity, especially under reduced tension.

The human placenta produces quite a lot of substances, but none is as characteristic of the trimester than hCG. Dimer hCG becomes progressively prominent with the functional differentiation of trophoblast. Tissue cultures suggest that hCG is produced in pulses, and that regulatory mechanisms occur which involve placental gonadotropin-releasing hormone, opioid peptides, and decidual prolactin. It seems that paracrine control is most prominent. Simultaneously, enzymatic activities are expressed which probably reflect protective mechanisms against a noxious environment.

In summary, the first trimester of pregnancy is a very critical period. The conceptus embedded in the decidualized uterine mucosa has to thrive under equilibrium conditions which are precise and delicate. These conditions are frequently not met, and this is probably the main explanation for the high percentage of pregnancy failure (detrimental maternal and/or embryonic factors being involved). The mechanisms which lead to interruption of pregnancy are either premature separation at the trophoblastic shell level, or insufficiency of extravillous trophoblastic invasion, especially intravascular, with the initiation of an untimely blood flow in the intervillous space and subsequent villous collapse and arrest of embryonic circulation. Other pregnancies are definitely abnormal. Notwithstanding those where the embryo has an abnormal karyotype, some evolve towards hydatidiform mole by loss of the oocyte nucleus. When, rarely enough, malignant change occurs and choriocarcinoma develops, important changes occur in the biochemical properties of transformed trophoblast. This is particularly evident for hCG, which loses almost all its sialic acid moiety.

Ectopic pregnancy is usually normal as regards the conceptus, which can implant almost anywhere and probably more easily than in a nonconditioned endometrium. Monitoring of early pregnancy is of paramount importance. Serum biologic assays have been considerably refined in recent years. However, dimer hCG and β-hCG assays remain the most credible and reliable.

Ultrasound techniques have progressed considerably. Transvaginal ultrasound can define earlier stages of gestation and fundamental events such as primary beats of the heart tube. In pathologic cases, it is possible to diagnose early pregnancy demise or ovular anomalies. Moreover, Doppler techniques will provide a wealth of information on physiology and pathology of the embryonic and maternal circulation.

Embryoscopy is still an experimental technique, but one which seems safe and most promising. In order to gain access to the embryonic genome, chorion villous sampling is now a well-established technique which is both reliable and safe. Most exciting and elegant is the flushing of the uterine cavity at the preimplantation stage with blastocyst recovery and biopsy. Also encouraging are the perspectives linked with intrauterine embryonal surgery. Up to now, only reduction of multiple pregnancy has been considered, but perspectives appear for a near future which could include gene therapy.

Sterility of the couple is a frustrating condition. More and more explorations and treatments have been designed to overcome this problem. The latest to date is in vitro fertilization and embryo transfer. This technique needs correct oocyte recovery, adequate endometrial preparation, and effective in vitro fertilization. Needless to say, techniques for ovulation induction must be refined and effective. The possibility of raising human blastocysts in vitro is, of course, an extremely serious matter. All kinds of ethics are involved, religious and medical. Strict guidelines must be set for future progress. That is where we are.

There are of course some messages in this book. Gestation in the human species lasts 40 weeks. This period is divided into two unequal parts. One is the first 12 weeks during which immune tolerance of the conceptus is set, and organogenesis of the embryo is completed. This period is characterized by the conceptus as an almost closed system, efficiently protected by the trophoblastic shell. Our knowledge of the events that take place during these few weeks is still limited, but we can envision a global picture. Biologic properties are particular to this period and must never be extrapolated from what can be measured later on. Obviously, more information is needed about the diffusion transfer of gases and nutrients. This could help our in vitro technology. We live in a dangerous world where the environment becomes more harmful year by year. Astonishing adaptative properties of the early placenta exist. They must be fully explored: they could shed light on what is or could. be ominous to our species.

A major field of investigation remains the treatment of recurrent abortion. Quite a number of etiologic factors are still unknown and should be searched for. It is probable that the multiple treatment regimens proposed are effective in a number of cases. However, there are no indications whatsoever about the mode of action of the proposed schemes, except perhaps for progesterone supplementation after incidental lutectomy. It seems difficult to reconcile a possible beneficial effect of a given hormone with the notion that spontaneous abortion occurs either with hemorrhagic separation at the deciduotrophoblastic interface or as embryonic death and villous involution. All proposed regimens must be tested against the efficiency of simple bed rest. An ill-developed embryo is probably often associated with a "poor" placenta in both its extravillous and villous components. If a number of these cases are linked with an anomaly of the karyotype, others are obviously not. Polygenic malformations must be considered, but studies should be directed towards the possibilities of vitamin deficiencies, for instance, or of a noxious environment.

Whatever the results, we must never forget that human reproduction is largely inefficient as compared to the scores obtained by other species. Pregnancy rate in spontaneous cycles reaches barely 25% per cycle, and abortion rate is 15%. Has anyone ever thought of what could have happened if the pregnancy score had reached 100%! Mankind suffers from increasing birthrates in underdeveloped countries. Global world population is increasing steadily and will approach its limits sooner or later. We, the so-called specialists in reproductive problems, must be fully aware that we are working both for basic science and for that ancestral need of the human species to reach immortality through its descendance. Increasing the rate of successful pregnancy is our goal. On the other hand, fertility control is absolutely necessary throughout the world. This is the price for the survival of mankind.

Loverval, December 1991   The Editors

References

1. Bonduelle ML, Dodd R, Liebaers I, Van Steirteghem A, Williamson R, Akhurst R (1988) Chorionic gonadotrophin β-mRNA, a trophoblast marker, is expressed in human 8-cell embryos derived from tripronucleate zygotes. Hum Reprod 3:909–914
2. Hay DL (1985) Discordant and variable production of human chorionic gonadotropin and its free α- and β-subunits in early pregnancy. J Clin Endocrinol Metab 61:1195–1200
3. Weitlauf HM (1988) Biology of implantation. In: Knobil E, Neil JD (eds) The physiology of reproduction. Raven, New York, pp 231–261
4. Chen C, Jons WR, Bastin F, Forde C (1985) Monitoring embryos after in vitro fertilization using early pregnancy factor. In: In vitro fertilization and embryo transfer. Seppälä M, Edwards RG (eds) Am NY Acad Sci 442:420–448
5. Barnea ER (1991) The first three months of pregnancy. In: Teoh E-S, Ratnam SS (eds) The future of obstetrics and gynecology – A preview for the 21st century. Parthenon, New York, pp 17–36

Subject Index

The latest practice,
thinking and research in fetal diagnosis

J. O. Drife, Leeds; **D. Donnai,** Manchester (Eds.)

Antenatal Diagnosis of Fetal Abnormalities

1991. XIV, 363 pp. 44 figs. Hardcover.
ISBN 3-540-19673-0

The field of antenatal diagnosis is one of the most rapidly changing areas of research in obstetrics and gynaecology. The information in this book, published within months of the presentations given by the Study Group of the Royal College of Obstetricians and Gynaecologists, keeps pace with the fast developments in research and clinical practice.

A wide range of new techniques including maternal serum screening, chorion villus sampling, Doppler ultrasound, computerized tomography and magnetic resonance imaging have become available. There are continual improvements in ultrasound imaging. In addition, the explosion of knowledge in molecular biology has opened up the field of pre-implantation diagnosis and genetic disease. Besides describing the most recent advances in these fields, this book discusses the psychological, ethical and economic aspects of the new technology, the organization of regional genetic services and the impact on non-specialist clinicians providing routine antenatal care.

Springer-Verlag
Berlin
Heidelberg
New York
London
Paris
Tokyo
Hong Kong
Barcelona
Budapest